Isaac Newton's Scientific Method

Turning Data into Evidence about Gravity and Cosmology

Isaac Newton's Scientific Method

Turning Data into Evidence about Gravity and Cosmology

William L. Harper

OXFORD
UNIVERSITY PRESS

OXFORD
UNIVERSITY PRESS

Great Clarendon Street, Oxford ox2 6dp

Oxford University Press is a department of the University of Oxford.
It furthers the University's objective of excellence in research, scholarship, and education by publishing worldwide in

Oxford New York

Auckland Cape Town Dar es Salaam Hong Kong Karachi
Kuala Lumpur Madrid Melbourne Mexico City Nairobi
New Delhi Shanghai Taipei Toronto

With offices in

Argentina Austria Brazil Chile Czech Republic France Greece
Guatemala Hungary Italy Japan Poland Portugal Singapore
South Korea Switzerland Thailand Turkey Ukraine Vietnam

Oxford is a registered trade mark of Oxford University Press
in the UK and in certain other countries

Published in the United States
by Oxford University Press Inc., New York

First published 2011

British Library Cataloguing in Publication Data
Data available

Library of Congress Cataloging in Publication Data
Data available

Typeset by SPI Publisher Services, Pondicherry, India
Printed in Great Britain
on acid-free paper by
MPG Books Group, Bodmin and King's Lynn

ISBN 978–0–19–957040–9

10 9 8 7 6 5 4 3 2 1

This book is dedicated to the memory of my father,
LEONARD ANDREW HARPER
and
to the memory of my mother,
SOPHIA RATOWSKI HARPER

Preface

The title of this book uses the modern term "scientific method" to refer to the methodology for investigating nature argued for and applied in Newton's argument for universal gravity. I use this modern term, rather than Newton's term "experimental philosophy" for his method of doing natural philosophy, to make salient the main theme I will be arguing for. I will argue that Newton's rich method of turning data into evidence was central to the transformation of natural philosophy into natural science and continues to inform the practice of that science today.

Newton's argument for universal gravity exemplifies a method that adds features which can significantly enrich the basic hypothetico-deductive (H-D) model that informed much of philosophy of science in the last century. On this familiar H-D model, hypothesized principles are tested by experimental verification of observable consequences drawn from them. Empirical success is limited to accurate prediction of observable phenomena. Such success is counted as confirmation taken to legitimate increases in probability. We shall see that Newton's inferences from phenomena realize an ideal of empirical success that is richer than prediction. To achieve this richer sort of empirical success a theory needs, not only to accurately predict the phenomena it purports to explain, but also, to have those phenomena accurately measure the parameters which explain them. Newton's method aims to turn theoretical questions into ones which can be empirically answered by measurement from phenomena. Propositions inferred from phenomena are provisionally accepted as guides to further research. Newton employs *theory-mediated* measurements to turn data into far more informative evidence than can be achieved by hypothetico-deductive confirmation alone.

On his method, deviations from the model developed so far count as new *theory-mediated* phenomena to be exploited as carrying information to aid in developing a more accurate successor. This methodology, guided by its richer ideal of empirical success, supports a conception of scientific progress that does not require construing it as progress toward Laplace's ideal limit of a final theory of everything. This methodology of progress through successively more accurate revisions is not threatened by Larry Laudan's argument against convergent realism. We shall see that, contrary to a famous quotation from Thomas Kuhn, Newton's method endorses the radical theoretical transformation from his theory to Einstein's. We shall also see that this rich empirical method of Newton's is strikingly realized in the development and application of testing frameworks for relativistic theories of gravity. Finally, we shall see that this rich methodology of Newton's appears to be at work in cosmology today. It appears that it was realizations of Newton's ideal of empirical success as convergent agreeing measurements of parameters by diverse phenomena that turned dark energy from a wild hypothesis into an accepted background assumption that guides further empirical research into the large-scale structure and development of our universe.

This book is directed to philosophers of science and students studying it. It is also directed to physical scientists and their students. Practicing scientists may well be able to profit from this book. Almost universally, scientists describe the role of evidence in their science as though it were just an application of hypothetico-deductive confirmation. This is so even when, as I try to show in the context of General Relativity and its empirical evidence, the practice of their science exemplifies Newton's richer and more effective method of turning data into evidence. This book is also directed to historians of science and their students. I hope it can suggest how studying the role of evidence can usefully contribute toward understanding the history of radical theory change.

I have found that attention to the details of calculations and proofs of theorems offered by Newton in support of his inferences helped me understand their role in affording empirical support for the propositions inferred as outcomes of *theory-mediated* measurements. I have also found attention to historical details about data available to Newton instructive. I have found historical episodes, such as Römer's use of eclipses of a moon of Jupiter's to measure a finite speed of light and the observation enterprise of Pound and Bradley initiated by Newton to obtain more precise measurements of the orbits of Jupiter's moons, both informative and fascinating. I have, however, attempted to relegate such details of proofs, calculations, and specialized historical background to appendixes so that readers who do not share my fascination for such details can follow the main argument without getting bogged down.

I have, however, included a fairly detailed account of data cited by Newton in support of his phenomena in chapter 2 and of his argument in chapters 3 through 8. Readers interested in Newton's main lessons on scientific method can focus on chapter 1, section IV of chapter 3, sections II.2–IV of chapter 4, and chapters 7, 9, and 10. They would also profit from the specifically labeled sections on method in the other chapters, without costing them very much extra time and effort to master details. The details offered in the other sections of these chapters, and the other chapters, do strongly reinforce these lessons on method and their historical context. I hope they will be of considerable interest to the growing number of very good philosophers of science, who are now taking a great interest in the details of Newton's work on gravity and method and in how these details can illuminate scientific method today.

In my effort to show how Newton's argument can illuminate scientific method today, I have appealed to modern least-squares assessments of estimates of parameter values. I argue that Newton's moon-test inference holds up by our standards today. Student's t-95% confidence parameter estimates illustrate the basic agreement achieved in the moon-test in Newton's initial version, and in the different published editions, of his *Principia*. Gauss's least-squares method of combining estimates of differing accuracy affords insight into how the agreement of the cruder moon-test estimates of the strength of terrestrial gravity adds empirical support to the much sharper estimates from pendulums. The cruder agreeing moon-test estimates are irrelevant to small differences from the pendulum estimates, but they afford additional empirical support for resisting large differences. This increased resistance to large differences – an increased resiliency – is an important empirical advantage afforded by agreeing measurements from diverse phenomena.

Some Acknowledgements

I want to thank my wife, Susan Pepper. Without her generous support I would not have been able to finish this long project. My daughter Kathryn May Harper and my sister Vicki Lynn Harper have also provided much appreciated support and encouragement.

A great many people have contributed to assist my efforts over the more than twenty years I have worked on this project. The historian of science Curtis Wilson and the philosophers of science Howard Stein and George Smith have been my chief role models and have offered very much appreciated critical comments on early versions of several chapters. I also thank George for his permission to use his very informative phrase "Turning data into evidence" in the title of this book.

My colleague Wayne Myrvold has been responsible for a great many improvements, as his very insightful criticisms over quite a few years have led me to deeper understanding of important issues I have tried to come to grips with. This book is part of a project supported by a joint research grant awarded to Wayne and me. Gordon Fleming and Abner Shimony, both of whom are physicists as well as philosophers of science, have offered very much appreciated encouragement and guidance. I especially thank Gordon for his careful reading and critical comments on chapters 1, 9, and 10. Gordon's emphasis on the value of diagrams was reinforced by my reading of Simon Singh's book *Big Bang: The Origin of the Universe* as a role model suggested by my lawyer Anthony H. Little. I also want to thank North Davis, a physician and a fellow mountain hiker, who kindly read and sent comments on an early version of my introductory chapter.

My colleagues Robert Batterman, Chris Smeenk, John Nicholas, and Robert DiSalle have also contributed much appreciated critical comments and important guidance. The historical background section in chapter 1 benefited from very much appreciated help by the late James MacLachlan. The comments by referees and by Eric Schliesser have led to substantial improvements, for which I thank them.

This work on Newton's scientific method has benefited from students and colleagues who participated in my graduate seminars on Newton and method. These included over the years four two-term interdisciplinary seminars on gravitation in Newton and Einstein. These were jointly listed in and taught by faculty from Physics and Astronomy, Applied Mathematics and Philosophy. I want especially to thank Shree Ram Valluri, from Applied Mathematics, who convinced me to help initiate these valuable learning experiences. He has also helped me understand details of many of the calculations and is the developer of the extension of Newton's precession theorem to eccentric orbits. I also want to thank Rob Corless, another participant in the gravitation seminars from Applied Mathematics, for checking derivations.

My treatment of Newton's method at work in cosmology today owes much to Dylan Gault who, in December 2009, completed his thesis on cosmology as a case study of scientific method. Wayne Myrvold and I were co-supervisors.

This book has also benefited from questions and comments raised at the many talks I have given over the years on Newton's method. I want, particularly, to thank Kent Staley, who presented very insightful comments on my paper at the Henle conference in March 2010.

I want to thank my research assistant Soumi Ghosh. She developed diagrams, converted my WordPerfect documents to Word, acquired permissions, and put the whole thing together. Without her talent, dedication, effort, and good judgment this project might never have resulted in this book. In the last drive to format the manuscript she was ably assisted by Emerson Doyle. Their work was supported by the joint research grant awarded to Wayne Myrvold and me. I am grateful to Wayne and the Social Sciences and Humanities Research Council of Canada for this support. Soumi's work reading page proofs was supported by the Rotman Institute of Philosophy at the University of Western Ontario. I am very grateful for their funding support and their much appreciated support in work space, resources help and encouragement in my project for engaging the role of evidence in science.

I want to thank Peter Momtchiloff for encouraging me to publish with Oxford University Press, selecting the excellent initial readers, guiding me through the initial revisions, and convincing me to turn endnotes into footnotes. Daniel Bourner, the production editor, kept us on track and helped to resolve difficult problems about notation. Eleanor Collins was responsible for the excellent cover design. Sarah Cheeseman skillfully guided the transformation to page proofs, including the transformation of my long endnotes into footnotes. Soumi Ghosh helped me proofread and assemble the corrections for the page proofs. Erik Curiel was an excellent, careful proofreader for chapters 1–3, 6, and 8. Howard Emmens was the expert proofreader who used all the queries and my responses to produce the final version for printing. I am very grageful for all their efforts. This book is much better than it would have been without all their help.

Contents

Detailed Contents

Abbreviations

Corresp.	Newton, I. (1959–1977). *The Correspondence of Isaac Newton*. Turnbull, H.W. (ed., vols. I–III), Scott, J.F. (ed., vol. IV), Hall, A.R. and Tilling, L. (eds., vols. V–VII). Cambridge: Cambridge University Press.
C&W	Cohen, I.B. and Whitman, A. (trans.) (1999). *Isaac Newton, The Principia, Mathematical Principles of Natural Philosophy: A New Translation*. Los Angeles: University of California Press.
ESAA	Seidelmann, P.K. (ed.) (1992). *Explanatory Supplement to the Astronomical Almanac*. Mill Valley: University Science Books.
GHA 2A	Taton, R. and Wilson, C. (1989). *The General History of Astronomy, vol. 2, Planetary Astronomy from the Renaissance to the Rise of Astrophysics, Part A: Tycho Brahe to Newton*. Cambridge: Cambridge University Press.
GHA 2B	Taton, R. and Wilson, C. (1995). *The General History of Astronomy, vol. 2, Planetary Astronomy from the Renaissance to the Rise of Astrophysics, Part B: The Eighteenth and Nineteenth Centuries*. Cambridge: Cambridge University Press.
Huygens 1690	Huygens, C. *Discourse on the Cause of Gravity*. Bailey, K. (trans.), Bailey, K. and Smith, G.E. (ann.), manuscript.
Kepler 1992	Kepler, J. (1992). *New Astronomy*. Donahue, W.H. (trans.). Cambridge: Cambridge University Press.
Math Papers	Whiteside, D.T. (ed.) (1967–81). *The Mathematical Papers of Isaac Newton*, 8 vols. Cambridge: Cambridge University Press.

1

An Introduction to Newton's Scientific Method

> ... our present work sets forth the mathematical principles of natural philosophy. For the basic problem [*lit.* whole difficulty] of philosophy seems to be to discover the forces of nature from the phenomena of motions and then to demonstrate the other phenomena from these forces. (C&W, 382)[1]

Our epigram is from Newton's preface to his masterpiece, *Philosophiae Naturalis Principia Mathematica* (*Mathematical Principles of Natural Philosophy*). It was published in 1687 when he was 44 years old. In it he argued for universal gravity. We still count gravity as one of the four fundamental forces of nature. The publication of Newton's *Principia* was pivotal in the transformation of natural philosophy into natural science as we know it today.

Newton used some phenomena of orbital motions of planets and satellites as a basis from which to argue for his theory. We shall see that these phenomena are patterns exhibited in sets of data Newton cites in their support. We shall seek to understand the method by which he transformed these data into evidence for universal gravity. This will mostly be an effort to explicate the first part of the endeavor described in our epigram from Newton's preface,

to discover the forces of nature from the phenomena of motions.

Newton went on to apply his theory to refine our knowledge of the motions of solar system bodies and the gravitational interactions on which they depend. The second part of the endeavor described in our epigram,

and then to demonstrate the other phenomena from these forces,

is exemplified in Newton's applications of universal gravity to demonstrate the basic solar system phenomena of elliptical orbits and some of their corrections for

[1] This passage is from Newton's preface to the first edition of his *Principia.* Quotations from the *Principia* are from the translation by I.B. Cohen and Anne Whitman: the citation (C&W, 382) is to that translation, Cohen and Whitman 1999.

perturbations. These applications resulted in more accurate corrected phenomena backed up by accurate measurements of the relative masses of solar system bodies. These results contribute significant additional evidence for universal gravity, so understanding how they work is also important for understanding Newton's scientific method of turning data into evidence.

In this book we will examine the steps by which Newton argued from phenomena of orbital motion to centripetal forces and then on to universal gravity. We will see how this led to measurements of masses of the sun and planets from orbits about them and, thereby, to his decisive resolution of the problem of deciding between geocentric and heliocentric world systems. This *Two Chief World Systems* problem, the topic of Galileo's famously controversial dialogue, was the dominant question of natural philosophy in the seventeenth century. Newton's theory affords decisive evidence for his surprising conclusion that both are wrong, because the sun and the earth both move relative to the center of mass of the solar system. The sun, however, never recedes very far from this center of mass. This makes elliptical orbits about the sun a good first approximation from which to begin accounting for the complex motions of the planets by successively more accurate models, as more and more perturbation producing interactions are taken into account.

Our investigation will show that Newton's scientific method adds features which can significantly enrich the basic hypothetico-deductive model that informed much of philosophy of science in the last century. On this familiar model of scientific inference, hypotheses are verified by the conclusions to be drawn from them and empirical success is limited to accurate prediction.[2]

Newton's scientific method goes beyond this basic hypothetico-deductive model in at least three important ways. First, Newton's inferences from phenomena realize an ideal of empirical success that is richer than prediction. In addition to accurate prediction of the phenomena a theory purports to explain, this richer ideal of empirical success requires that a theory have those phenomena accurately measure the parameters which explain them. Second, Newton's scientific method aims to turn theoretical questions into ones which can be empirically answered by measurement from phenomena. Third, theoretical propositions inferred from phenomena are provisionally accepted as guides to further research.

[2] On Karl Popper's version of this hypothetico-deductive (H-D) method, legitimate scientific inferences would be limited to rejection of hypotheses that have been falsified when observable consequences have turned out false (see, e.g., Popper 1963, 56). Most philosophers and scientists accept a more liberal version of H-D method on which successful prediction can legitimate increases in the probability assigned to hypotheses.

In section III below, we shall quote an attractive articulation of such a liberal version of hypothetico-deductive method from Christiaan Huygens, who was Newton's greatest contemporary as an empirical investigator of nature (sec. I.2 below). In chapter 9, we shall see that Huygens highlights features on which Bayesian updating by conditioning on observations would afford increases in epistemic probability of hypotheses that would accurately predict those observable outcomes.

These improvements let Newton employ *theory-mediated measurements* to turn data into far more informative evidence than can be achieved by hypothetico-deductive confirmation alone. All three of them come together in a method of successive approximations in which deviations from the model developed so far count as new *theory-mediated phenomena* that aid in developing a more accurate successor model. This rich scientific method of Newton's was taken up by his successors in their extraordinary extensions of applications of his theory to solar system motions, in what developed into the *science* of physics applied to astronomy.

In our concluding chapter we shall argue that Newton's own scientific method endorses the revolutionary change from his theory to Einstein's. We shall also see that this rich empirical method of Newton's continues to be realized in the development and application of testing frameworks for relativistic theories of gravity. In addition, we shall see that an application of Newton's ideal of empirical success as agreeing measurements from diverse phenomena is appealed to in support of the radical inference to dark energy in cosmology today.

I Some historical background

1. Astronomy

Archaeological evidence from many ancient sites, ranging from pyramids to Stonehenge, demonstrates a long-standing human fascination with celestial phenomena. These phenomena include the regular configurations among the fixed stars, which maintain their positions relative to one another as they move westward across the sky each night. In contrast are the phenomena of the motions of planets. The word "planet" originally referred to *celestial wanderers* – bodies which move against the background of the fixed stars. These included the sun and the moon as well as Mercury, Venus, Mars, Jupiter, and Saturn. Salient among the phenomena exhibited by these last five are regular occasions of retrograde motion.

On these occasions they appear to come to a stop, reverse direction for some time, then stop again before resuming their normal motion against the fixed stars.

From the point of view of European history, the earliest influential model for the nature of the heavens was described by Aristotle (Greece 384–322 BC).[3] The stationary earth was at the center of numerous rotating spheres of ethereal material carrying the various planets and the stars. This was the world imagined by Western Europeans for two thousand years.

Mathematical astronomy was developed to accurately represent and predict the celestial phenomena exhibited by the motions of planets against the stars. As more

[3] Aristotle, *De Caelo* (*On the Heavens*), 398–466 in McKeon 1941. A version of this model was developed originally by Eudoxus, a mathematician who participated in Plato's academy (see Dreyer 1953, 87–107; Neugebauer, 1975, 675–89).

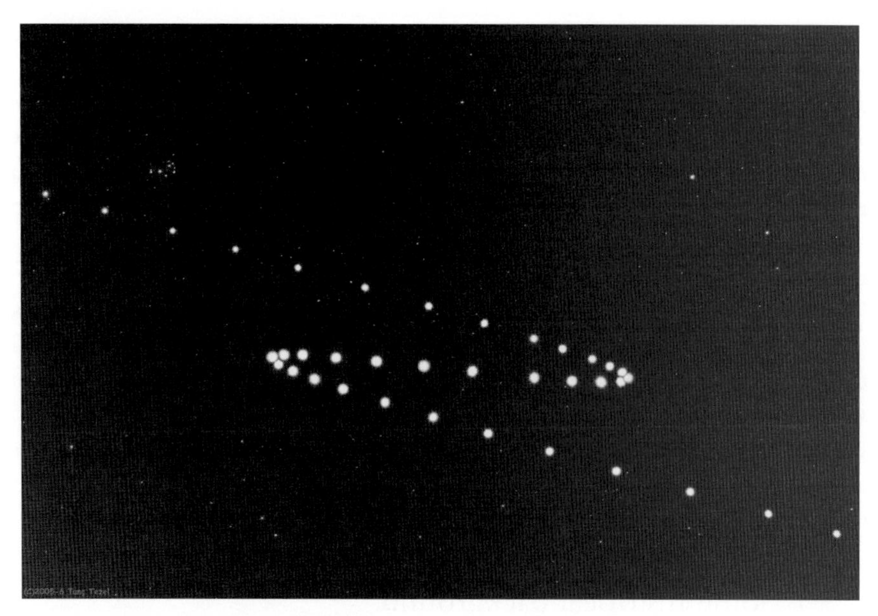

Figure 1.1 Retrograde motion of Mars
A composite image from July 2005 through February 2006 shows Mars appearing to reverse the direction of its orbit for a time.

detailed and complex mathematical models were developed for accurate representation and prediction of celestial phenomena, mathematical astronomy became a separate enterprise from the search for physical causes to explain those phenomena.

Ptolemy (Alexandria, ca AD *150)*

Ptolemy provided a comprehensive and computationally tractable account of these phenomena. His treatment of retrograde motions recovered not just the basic pattern, but also details such as regular variations in lengths of retrograde loops and in their timing. The basic pattern could be recovered from an earth-centered model by introducing a two-circle system. The observed path of each planet against the stars is represented by a motion of the planet on an epicycle, with a center that moves on another circle called the "deferent".

In order to recover the details of the variation of length and timing of retrograde loops, Ptolemy made the deferent circle eccentric with respect to the earth and introduced the equant.

The equant is a point about which the center of the epicycle describes equal angles in equal times on its motion along the deferent circle. Ptolemy located this equant point at an equal distance on the exact opposite side of the center of the deferent circle as that center is distant from the earth. This allowed his earth-centered model to accurately recover the motions of the planet against the stars.

Ptolemy's great book became the standard treatment of mathematical astronomy for 1,400 years. Ptolemaic mathematical astronomy survived the Dark Ages thanks to Arab

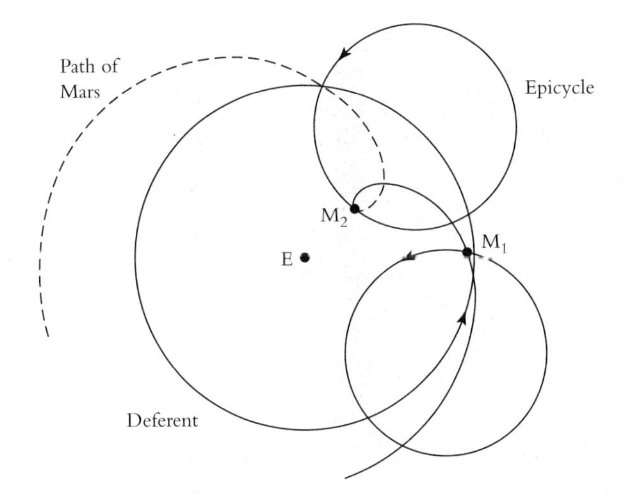

Figure 1.2 Epicycle deferent

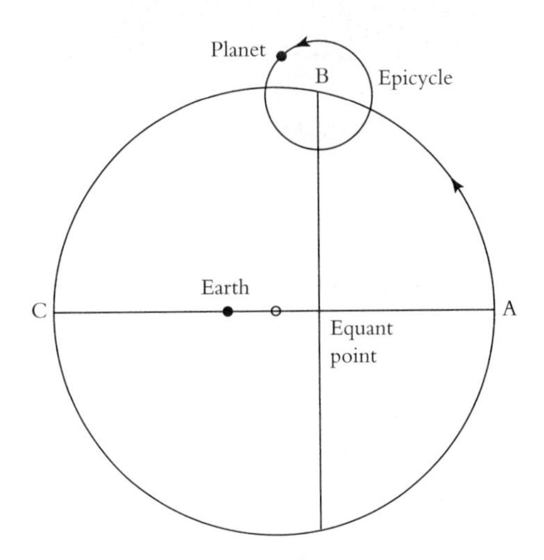

Figure 1.3 Equant

and Iranian astronomers.[4] In the 12th century Europeans obtained translations of Ptolemy's book and, perhaps somewhat later, they also obtained knowledge of various critiques of Ptolemy's system by some of these astronomers.[5]

[4] See Swerdlow and Neugebauer 1984, 41–8. Medieval Arabic astronomy was very rich. The Iranian astronomers were associated with the observatory founded at Maragha, in north-western Iran, in 1259, by Hulagu, the Mongol conqueror of Persia.

[5] Swerdlow and Neugebauer (1984, 42–8) argue that Copernicus was influenced by such critiques to which he had access in work from the Maragha School that appears to have reached Italy in the fifteenth century.

Ptolemy used data from observations to set the parameters of his model and provided tables which could be used to compute locations of planets at past and future times.[6] He specified these locations in ecliptic coordinates which are still used today. The ecliptic is the yearly path of the sun against the fixed stars.

Ecliptic latitudes specify angular distances north or south of this line. The vernal equinox, the location of the sun at sunrise on the first day of spring, marks the origin (0 degrees = 360 degrees) for ecliptic longitude.

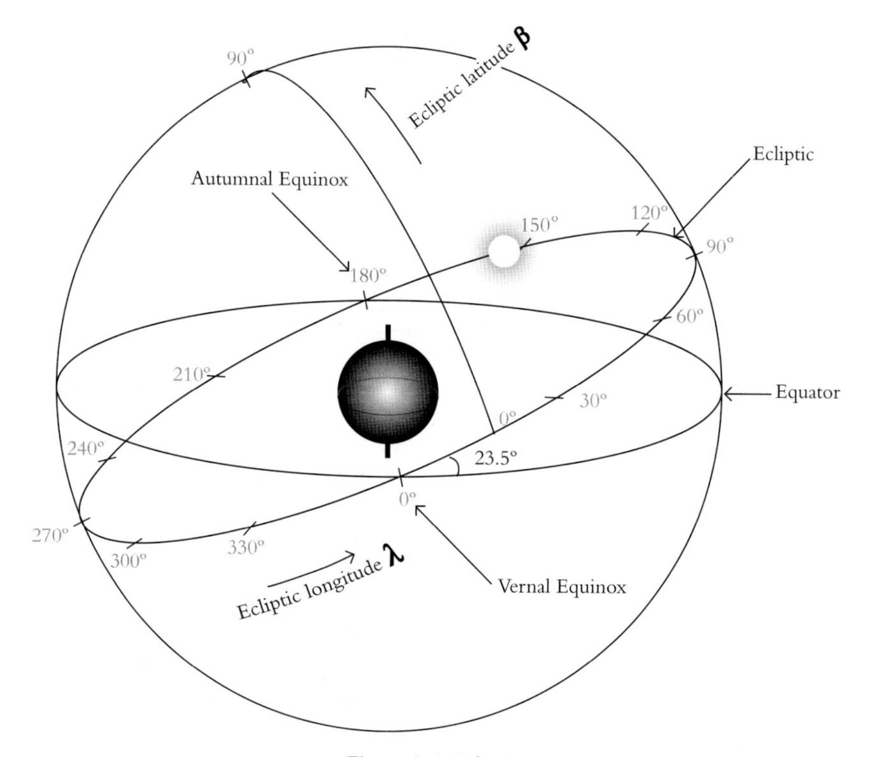

Figure 1.4 Ecliptic

Familiar constellations, the signs of the zodiac, mark off the ecliptic circle into twelve 30-degree segments. These are, in order (counterclockwise viewed from the north ecliptic pole), 1 Aries, 2 Taurus, 3 Gemini, 4 Cancer, 5 Leo, 6 Virgo, 7 Libra, 8 Scorpius, 9 Sagittarius, 10 Capricorn, 11 Aquarius, 12 Pisces.

The location of the vernal equinox on the ecliptic at the time of Hipparchus (160–127 BC) was at about the beginning of the constellation, Aries.[7] The equinoxes

[6] Tables fixing information about locations of solar system bodies at times are now called "ephemerides." They continue to be a major concern of national observatories. See Forward in Seidelmann 1992 (ESAA, xxv–xxvi).

[7] The phrase "first point of Aries" is often used as a demonstrative referring to the location on the ecliptic of the vernal equinox.

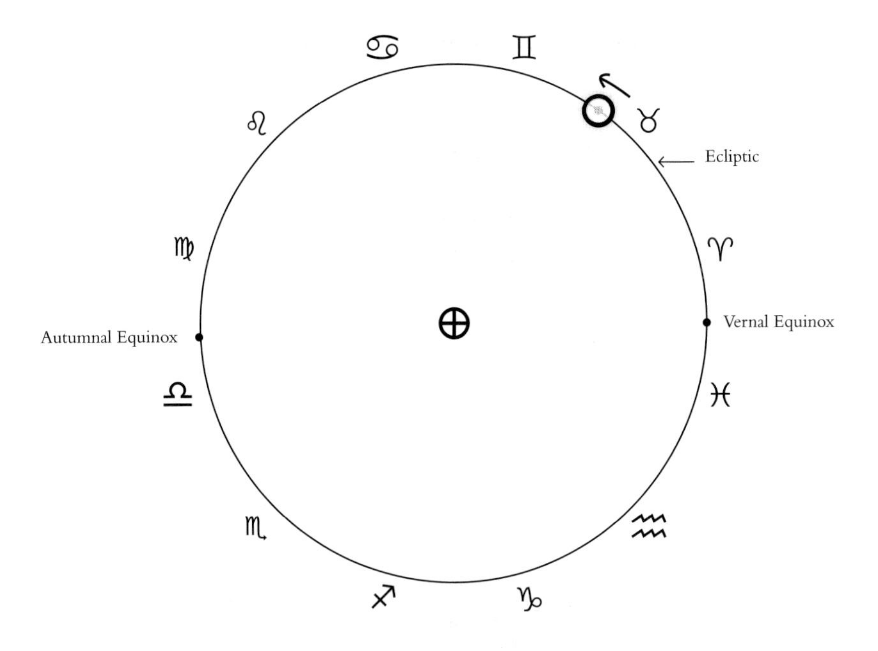

The Constellations and their Symbols

♈	Aries	The ram	♎	Libra	The scales
♉	Taurus	The bull	♏	Scorpio	The scorpion
♊	Gemini	The twins	♐	Sagittarius	The archer
♋	Cancer	The crab	♑	Capricorn	The goat
♌	Leo	The lion	♒	Aquarius	The water carrier
♍	Virgo	The virgin	♓	Pisces	The fish

Figure 1.5 The twelve constellations

move backwards (clockwise) through the zodiac at about 1 degree 23 minutes of arc every century. We call this phenomenon the "precession of the equinoxes." It corresponds to a rotation of the direction of the tilt of the earth's axis with respect to the plane of the ecliptic. The direction towards which the north pole points slowly traces a small circle against the stars in a cycle that takes about 26 thousand years to complete.

By the seventeenth century the location of the vernal equinox was well within Pisces (the 12th constellation). Early in the twenty-first century the location of the vernal equinox will be in Aquarius (the 11th constellation).

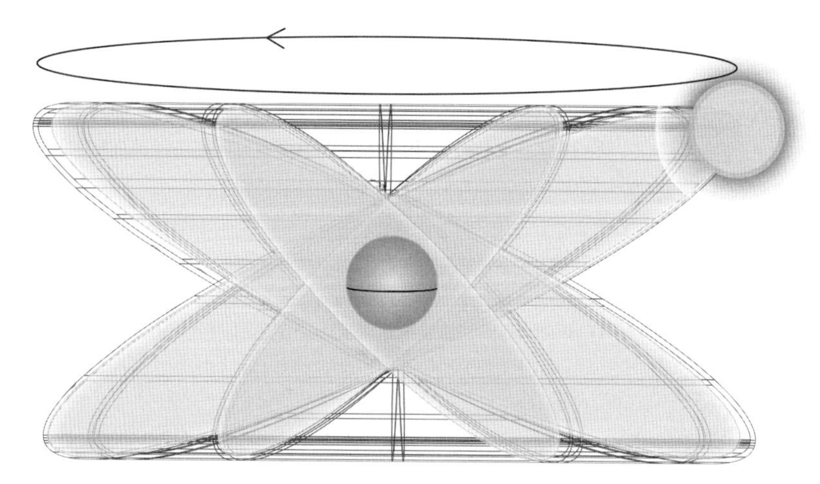

Figure 1.6 Precession of the equinoxes
Earth-centered representation. It is as though the tilted plane of the ediptic is rotating about the earth's axis at a uniform rate so that it would take 26,000 years for a complete cycle.

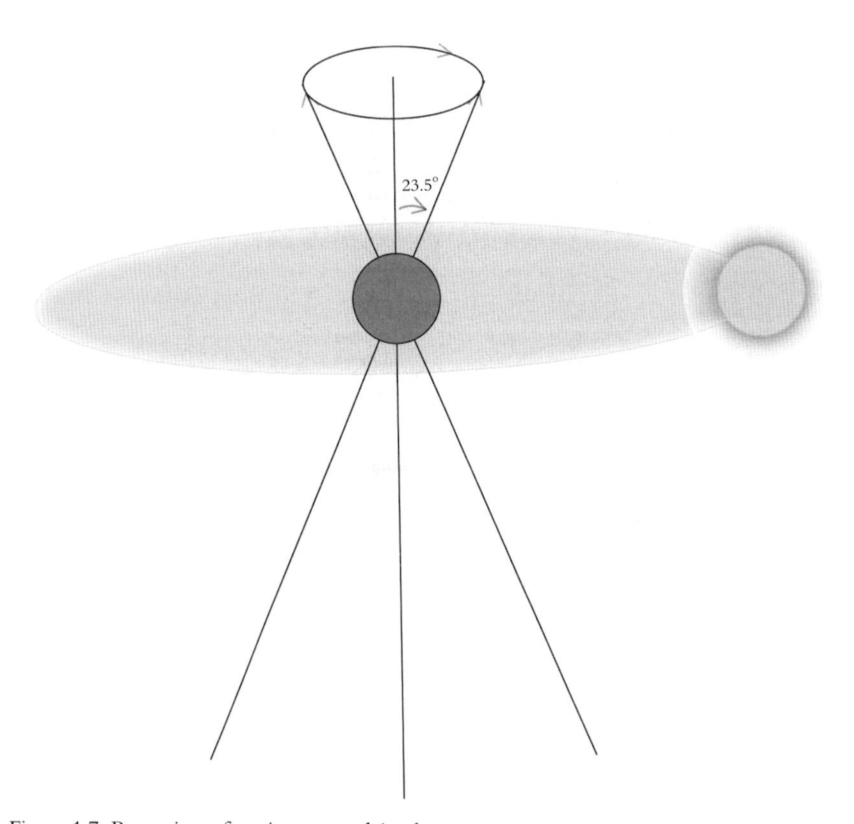

Figure 1.7 Precession of equinoxes explained
This apparent change in the Sun's orbit is actually due to a corresponding 26,000-year precession in the earth's axis of rotation.

By the twelfth century, Ptolemy's incorrect specification of the precession of the equinoxes, at only 1 degree per century (off by 23 minutes per century), was leading to rather glaring inaccuracies. The Alfonsine tables, of about 1270, were based on re-set parameters for Ptolemy's model and corrections to the precession.[8] By the early 1500s these tables were showing inaccuracies stemming, in part, from the use of the Julian year, which was slightly too short. These created problems for specifying the date of Easter that eventually led to reform of the calendar.

Nicholas Copernicus (*Poland, 1473–1543*)

Copernicus was well aware of these problematic inaccuracies. He was also bothered by models using non-uniform rotations and spheres which intersected, because they did not fit what he imagined to be the physical mechanism for moving the planets. That mechanism, like that of Aristotle, was based on uniformly revolving spheres that ought not to intersect. Copernicus was able to accommodate such a mechanism by creating models in which the spheres of all the planets encompass the sun and also *making the earth a planet*.[9] One advantage of making the earth a planet was a simpler explanation of retrograde motion. The following diagram represents the retrograde motion of Mars as an effect of the earth overtaking Mars as they both orbit the sun.[10]

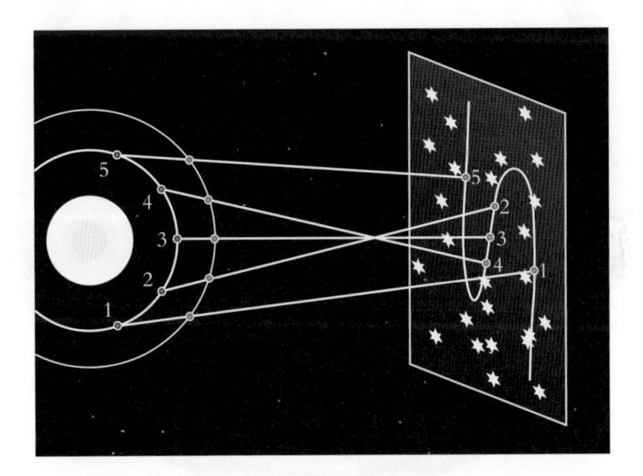

Figure 1.8 Retrograde motion of Mars explained

[8] These included corrections to offset additional small periodic variations in the precession. Swerdlow and Neugebauer (1984, 42).

[9] For a detailed account of Copernicus' system see Swerdlow and Neugebauer 1984. Gingerich and MacLachlan (2005) is a very accessible account of Copernicus's life and achievement.

[10] Figure 1.8 represents looking down at the orbits from above the plane of the ecliptic. The photographs in Figure 1.1 are a sequence of geocentric views of Mars against the background of stars. The upward slant from right to left in the sequence of photographs and the variation above and below are due to a combination of the tilt of the earth's axis with respect to the ecliptic and the inclination of Mars's orbit with respect to the ecliptic. The reversals of direction, stationary points, and the brightness corresponding to different distances are easily explained.

Even though the basic explanation of retrograde motion was simpler in Copernicus's model, his system for explaining details used about as many circles as Ptolemy's.[11] Copernicus was able to achieve a model that was as accurate as Ptolemy's would be with corresponding parameter settings. Copernicus's book, *De Revolutionibus*, was published in 1543, the year he died. The Prutenic tables, which formed a basis for the reformed calendar adopted by Pope Gregory in 1583, were generated using Copernicus's model.

Tycho Brahe (Danish, 1546–1601)

Brahe directed an observation enterprise in astronomy which produced a very extensive body of impressively accurate data.[12] His instruments for observation, partly due to their great size, afforded the most precise data until the use of the telescope. His

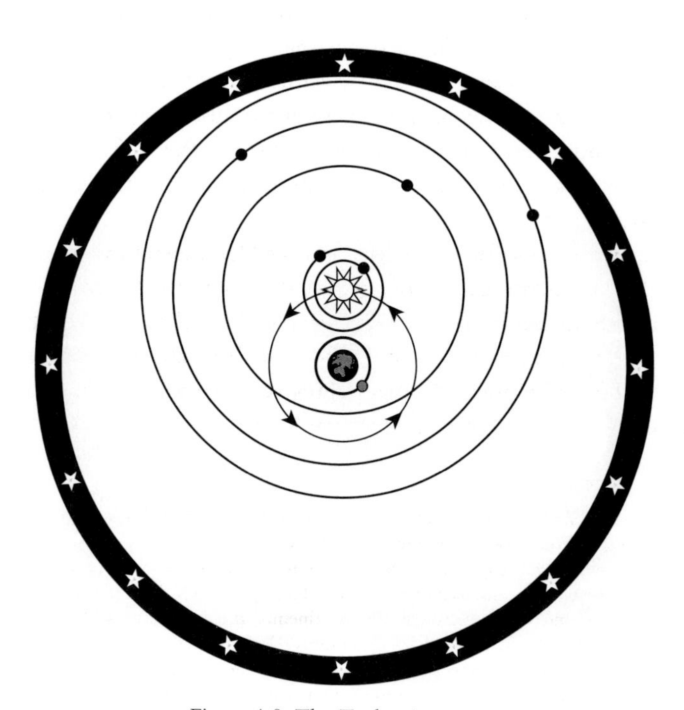

Figure 1.9 The Tychonic system

[11] Copernicus, who was committed to uniform circular motion, includes many extra epicycles so that he can do without Ptolemy's equants. These extra epicycles prevented Copernicus from taking full advantage of the simplicity made possible by using a single orbit of the earth to represent what in the Ptolemaic system had to be represented in duplicated epicycles for each planet.

Copernicus also includes extra epicycles to improve on Ptolemy's treatment of the precession of the equinoxes. Equivalent apparatus adapted to geocentric systems had to be included by Clavius when he attempted to make Ptolemaic systems adequate to data in his efforts at calendar reform (Lattis, 1994, 163–79).

[12] See *GHA 2A*, 3–21. See also Thoren 1990.

data provided substantially improved knowledge of the observed positions of planets against the fixed background of constellations of the stars.[13]

Tycho also developed a geo-heliocentric model for the solar system, in which the planets circle the sun while the sun circles the earth.[14]

Unlike Copernicus, Brahe did not count the earth as a planet.[15]

Kepler (German 1571–1630)

Near the end of his life, Brahe hired an astronomical mathematician, Johannes Kepler. During the winter of 1600/1601 Kepler wrote a work on the history and philosophy of astronomy in which he proposed that the empirical equivalence between geocentric and heliocentric world systems could be overcome by appeal to physical causes.[16] In 1609 he published his *NEW ASTRONOMY BASED UPON CAUSES or CELESTIAL PHYSICS treated by means of commentaries ON THE MOTION OF THE STAR MARS from the observations of TYCHO BRAHE, GENT.* In the first chapter he provides a diagram (see figure 1.10) illustrating the complex motion that would need to be accounted for on the assumption that the earth stands still.

This is the complexity that a physical cause of the motion of Mars would have to explain if the true motions of the planets were taken to be their motions relative to the center of the earth. Counting circles to compare the predictive models of Copernicus and Ptolemy is much less informative.

In his book, Kepler appealed to his own proposed hypotheses about physical causes as well as to his impressive analysis of Tycho's data to argue that Mars moves in an elliptical orbit with the sun at a focus.[17] He also argued that the rate at which its motion sweeps out areas by radii to that focus is constant. We call these Kepler's first and second "laws" of orbital motion.[18] Figure 1.11 is a diagram representing the motion of a planet according to these two rules found by Kepler.

[13] Many of these data are from observations made at Uraniborg, the research institute Tycho established on the island of Hven, located between Denmark and Sweden.

[14] See section III.2 below and chapter 2 section III for more on this system, which became the chief earth-centered alternative after Galileo's discovery that Venus exhibits phases which show that it circles the sun.

[15] Though Tychonic and Copernican systems are kinematically equivalent with respect to relative motions among solar system bodies, they differ in motions relative to the stars. On a Copernican system one would expect stellar parallax corresponding to the diameter of the earth's orbit. Absence of any success in observing such stellar parallax was among the grounds convincing Tycho that the earth did not move about the sun with respect to the stars. (*GHA 2A*, 8, 9; Van Helden 1989, 109).

[16] This piece, which was commissioned by Tycho Brahe to get Kepler to write a defense of him in his dispute with Ursus, was not published until 1858, long after Kepler's death. Jardine's title,

> *The Birth of History and Philosophy of Science: Kepler's A defence of Tycho against Ursus with essays on its provenance and significance,*

suggests that Kepler's discussion in this manuscript is worthy of considerable interest. See Jardine 1984.

Tycho had accused Nicholas Ursus, who had visited Uraniborg, of stealing an early version of his system. Ursus had attacked Tycho in his *Tractatus*, which included an account of planetary hypotheses as mere fictitious fabrications for calculating celestial motions (Jardine 1984, 41–2).

[17] See, e.g., *GHA 2A*, 172.

[18] It appears that it was only after Newton discovered their dynamical significance that Kepler's orbital rules came to be called laws. See below (chpt. 4, sec. III).

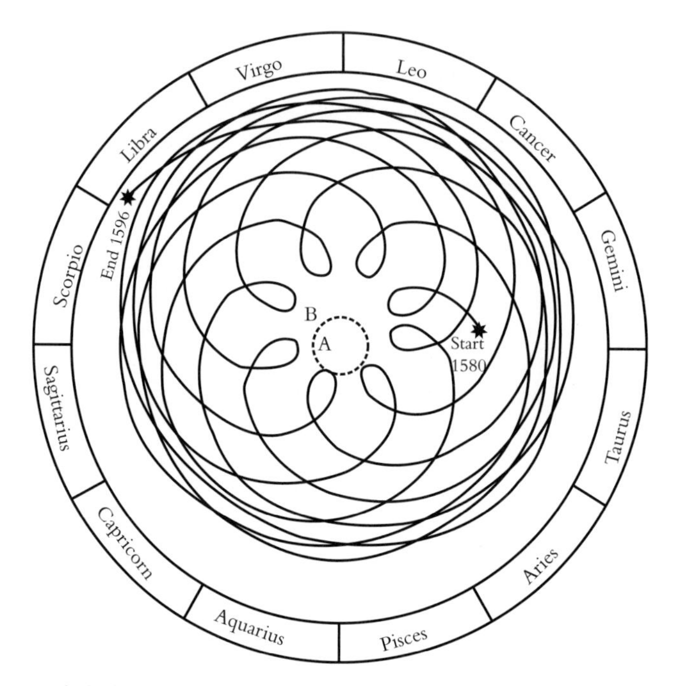

Figure 1.10 Kepler's diagram depicting the motion of Mars on a geocentric conception of the solar system

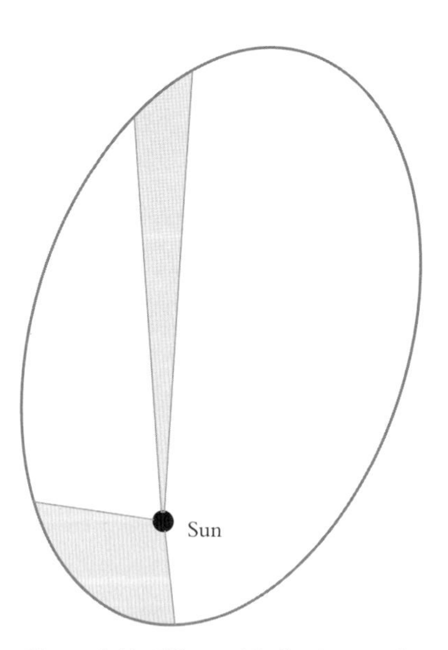

Figure 1.11 Ellipse with the Area Rule

The two equal area segments are swept out in two equal amounts of time. The planet moves faster enough in the segment with smaller distances to sweep out the same area swept out in the segment further from the sun in the same amount of time.

In 1619 Kepler published *The Harmony of the World* in which he appealed to intuitions about cosmic harmony, as well as orbital data, to arrive at the Harmonic Rule. According to Kepler's Harmonic Rule, the system of elliptical orbits of the primary planets about the sun at their common focus is such that the ratio of the cube of the mean-distance R to the square of the period t is the same for all of them.[19] That is, R^3/t^2 has the same constant value for each of these orbits. We call this Kepler's third "law" of orbital motion. We can informatively exhibit the fit of the Harmonic Rule to the data Newton cites from Kepler by plotting log periods against log distances.[20]

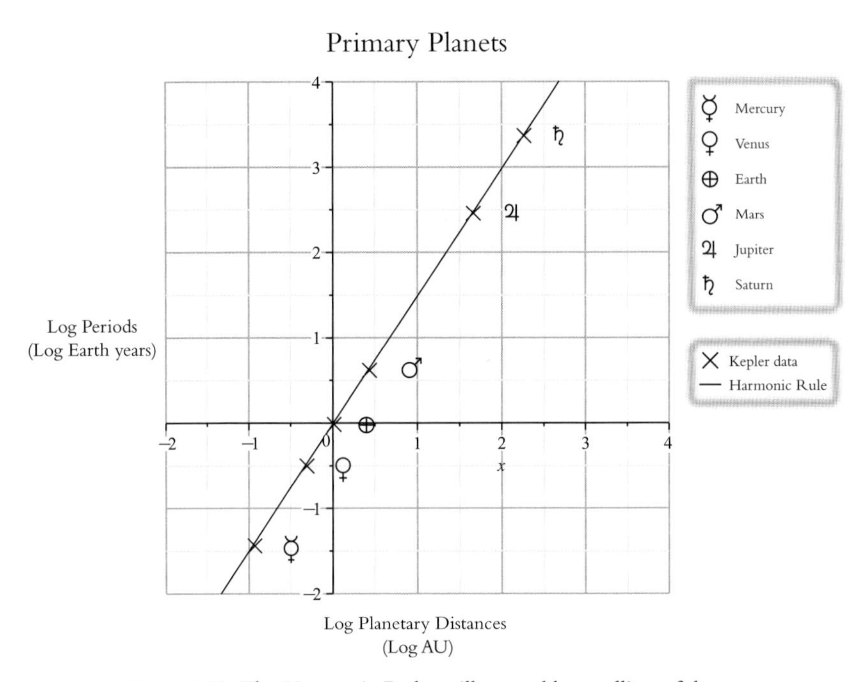

Figure 1.12 The Harmonic Rule as illustrated by satellites of the sun

[19] Kepler takes as the mean-distance R the length of the semi-major axis. In an elliptical orbit with the sun at a focus, the length of the semi-major axis is exactly halfway between the shortest distance from the sun (when the planet is at perihelion) and the longest distance (when the planet is at aphelion).

[20] In this figure we use the mean earth-sun distance represented by an astronomical unit AU as our unit of distance. Our astronomical unit was set by the mean earth–sun distance at the time of Gauss. See note 31 chapter 8 for more detail.

See chapter 2 for such diagrams representing the fit of the Harmonic Rule to Newton's cited periods and mean distances for the orbits of Jupiter's moons, Saturn's moons, and one that adds data from Boulliau for the primary planets.

That some straight line of slope n fits the result of plotting Log$t$ against Log$R$ is to have the periods be as some power n of the distances. To have the Harmonic Rule hold is to have the slope of this line be $3/2 = 1.5$.

The culmination of Kepler's life's work was the publication of his *Rudolphine Tables* in 1627. These applied his orbital rules to generate accurate predictions of the motions of all the planets. We now know that these predictions are about thirty times more accurate than those of competing tables.[21] Some indication of this became generally known when Kepler's prediction of a transit of Mercury across the face of the sun was verified in 1631. Kepler's prediction erred by about ten arc minutes while those of tables based on Ptolemy or Copernicus erred by five degrees or more.[22]

Galileo (*Italy 1564–1642*)

In 1609 Galileo manufactured telescopes that were more powerful than any previous ones. His telescopes afforded magnifications of up to 20 times or more.[23] Early in 1610 he published results of observations of the moon, stars, and planets. These revealed that our moon appeared earth-like, with mountains and valleys, and sharpened differences between the steady, round shapes of planets and the fixed stars which are seen to pulsate with bright rays. He also found many new stars not visible to the naked eye. The broad band of the Milky Way, which looks like whitish clouds to the naked eye, was seen by the telescope to be a dense crowd of stars. Some odd-shaped objects that had been called nebulous stars were revealed to be swarms of small stars placed exceedingly close together. His observations of Jupiter revealed that it was accompanied by four satellites – moons in orbit about the planet.[24] His book *Siderius Nuncius* (*Starry Messenger*) quickly made Galileo a celebrity in his age, of the sort that Einstein became in the twentieth century.

We shall say more about Galileo's discovery and later observations of Jupiter's moons in chapter 2, when we discuss Newton's acceptance of the Area Rule and Harmonic Rule for the orbits of these four Galilean moons of Jupiter as a phenomenon from which to argue. This discussion is accompanied by a diagram recording observations of Jupiter's moons by Jesuit astronomers of the Collegio Romano, when they corroborated Galileo's results after obtaining a telescope.

Galileo's later observations of phases of Venus provided direct observational evidence against Ptolemy's system, though not against Tycho's geo-heliocentric system.[25]

These phases of Venus were among the observational results of Galileo that were verified by telescope observations of Jesuit astronomers.[26] As a result, the Tychonic system became dominant among defenders of earth-centered cosmology.[27]

[21] See *GHA 2A*, 77.
[22] Ibid.
[23] See *GHA 2A*, 81; Van Helden 1989, 9.
[24] See Van Helden 1989, 64–85.
[25] See chapter 2 section III for more details.
[26] See Lattis 1994, 186–95; Van Helden 1989, 111.
[27] See Lattis, 205–16; Van Helden *GHA 2A*, 84.

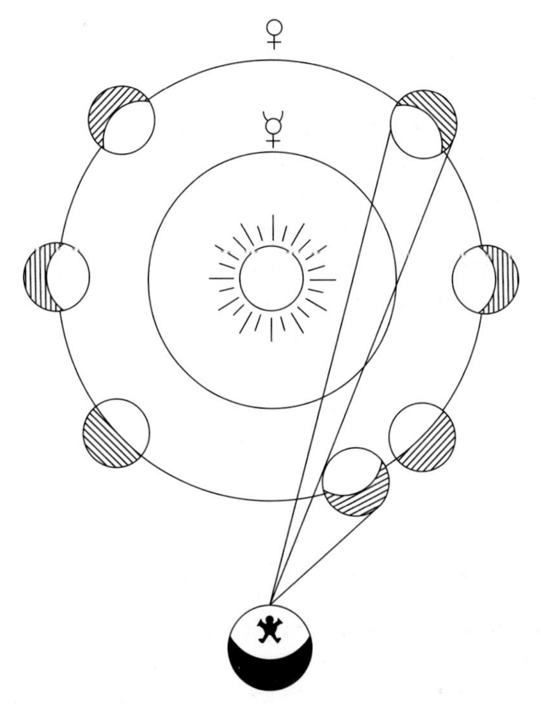

Figure 1.13 The phases of Venus
This diagram, drawn by Kepler, is from his *Epitome of Copernican Astronomy*. (Kepler 1995, 68)

2. Mechanics and physical causes

We shall see that Newton's argument for universal gravity and its application to the solar system is the successful realization of Kepler's project for using physical causes to overcome the empirical equivalence between geocentric and heliocentric world systems. We shall also see that a major aim of Newton's scientific method was to defend this use of gravity as a physical cause to explain planetary motions, even though he was unable to find a physical cause of gravity itself.

This section reviews views on physical causation by some influential earlier thinkers. It provides background that will usefully inform our account of Newton's conception of gravity as a physical cause of orbital motion.

Aristotle (Greek 384–322 BC)

Aristotle counted terrestrial bodies as being made up of combinations of four elements: earth, water, air, and fire. Earth was taken to be heavy and dry, water as heavy and wet, air as light and wet, and fire as light and dry. Heavy things tend to move toward their natural place at the center of the earth, unless prevented. Light things tend to rise toward the sphere of the moon. For a heavy terrestrial body, like a thrown spear, forced local-motion required an efficient cause to put it into non-natural motion.[28] According to Aristotle, this meant it would require an efficient cause to keep it moving once it has left the hand that threw it. On his view such a body would naturally begin to lose its motion unless some efficient cause kept it going, in analogy to the way a hand holding it up would count as an efficient cause preventing it from falling to the ground.

Aristotle sharply distinguished celestial phenomena from the terrestrial realm. Heavenly bodies were made of quintessence, an ethereal material, for which the natural state was uniform circular motion. The uniformly rotating celestial spheres did not require any appeal to efficient causes to maintain their motions.

Galileo (on mechanics)

We have already discussed Galileo's telescope observations reported in his 1610 book *Siderius Nuncius* and the important evidence afforded by his additional observations of the phases of Venus. In addition to these contributions to astronomy, Galileo also made fundamental contributions to mechanics. In his famous *Two New Sciences* of 1638, Galileo proposed uniformly accelerated fall as an exact account of idealized motion that would result in the absence of any resistant medium, even though the idealization is impossible to actually implement. He argued that the perturbing effects of resistance are too complex to be captured by any theory.[29] He also offered considerations, including inclined plane experiments that slow down the motion, to argue that his idealized uniformly accelerated motion is the principal mechanism of such terrestrial motion phenomena as free-fall and projectile motion.[30] Galileo's account of projectile motion combined uniformly accelerated fall with an appeal to horizontal inertia. Over relatively short distances, such as those explored by the projectile trajectories he considered, uniform horizontal motion approximates uniform motion on a circle of

[28] Aristotle offered an account of motion, construed widely as change in general. Local-motion, change of place, was a special case (see McKeon, 1941, 253; Solmsen, 1960, 175–6). The accounts of phenomena of local-motion we are concerned with here involve primarily the contrast between natural motions (which do not require further explanation) and forced motions construed as changes of state that require efficient causes to explain them. The sorts of efficient causes of concern to us are special cases of the general conception of efficient causation offered by Aristotle.

Aristotle proposed a doctrine of four causes. They are: 1. the *material cause*, that out of which a thing comes to be and which persists, e.g. the bronze of the statue; 2. the *formal cause*, the form or the archetype, i.e. the statement of the essence (e.g. of the octave the relation of 2:1, and generally number); 3. the *efficient cause*, the primary source of the change or coming to rest (e.g. the man who gave advice); 4. the *final cause*, in the sense of end or "that for the sake of which" a thing is done (e.g. health is the cause of walking about). See Physics book II chapter 3, McKeon, 240–1.

[29] The following passage is from Galileo's *Two New Sciences*:

Even horizontal motion which, if no impediment were offered, would be uniform and constant is altered by the resistance of the air and finally ceases; and here again the less dense [*piu leggiero*] the body the quicker the process. Of these properties [*accidenti*] of weight, of velocity, and also of form [*figura*], infinite in number, it is not possible to give any exact description; hence, in order to handle this matter in a scientific way, it is necessary to cut loose from these difficulties; and having discovered and demonstrated the theorems, in the case of no resistance, to use them and apply them with such limitations as experience will teach. (Galilei [1638] 1914, 252–3)

[30] See *Two New Sciences* days 3 and 4 (1914, 153–294).

radius equal to that of the earth. We can think of Galileo's horizontal inertia as bringing down to earth, to what Aristotle counted as forced local-motion, this feature of the natural motion Aristotle attributed to the celestial spheres.

Descartes (French 1596–1650)

René Descartes was a principal leader of the mechanical philosophy, a movement that came to dominate natural philosophy in the seventeenth century. This movement criticized Aristotelian attributions of qualities, such as heaviness to explain falling and lightness to explain rising, as pseudo-explanations. According to the mechanical philosophy, to make a motion phenomenon intelligible one must show how the motion could possibly result from contact pushes between bodies. In his *Principia Philosophiae* [*Principles of Philosophy*], Descartes offered an account of mechanics in which deviation from uniform straight-line motion (as well as deviation from rest) required explanation.[31] This introduced a need for a cause to explain orbital motion of planets that anticipated Newton's application of his first Law of Motion. Descartes offered his vortex theory as a mechanical account of how the planets could be maintained in orbits by vortex particles in analogy with the way pieces of wood are carried in curved paths by a whirlpool.[32] In his *Principia*, Descartes provided such qualitative hypothetical mechanical explanations for terrestrial gravity, the birth and death of stars, the formation of mountains, the tides, sunspots, light, and even magnetic phenomena.

Descartes was an important mathematician. His *Geometry*, which was published as an appendix to his *Discourse on Method* in 1637, is the origin of our modern discipline of analytic geometry, in which coordinates and equations allow the fruitful combination of algebra and geometry.[33] Figure 1.14 is what we call a Cartesian coordinate representation of the path of a body projected with a velocity of 10 Paris feet per second at a 45-degree angle up from atop the lower side of the leaning Tower of Pisa, at 172 Paris feet. We are here using resources provided by Descartes the mathematician to accurately

[31] See Descartes, R. (1983) *Principles of Philosophy* part II, section 37–8. Miller and Miller (trans.), 59–61.

[32] See Descartes 1983, 96–8 and 89–91. After appealing to the phases of Venus to dismiss Ptolemy, Descartes points out that the vortex motion that would be needed for Tycho's system is more complicated than for a Copernican system. Indeed he even argues that on his account of motion, Tycho's theory attributes more motion to the earth than Copernicus's.

One striking innovation, often dismissed as an attempt to avoid the sort of church condemnation directed at Galileo, is a distinction between motion in the ordinary sense (*the action by which some body travels from one place to another*) [Descartes 1983, 50] and what motion properly speaking is (*the transference of one part of matter or of one body, from the vicinity of those bodies immediately contiguous to it and considered as at rest, into the vicinity of [some] others*) [1983, 51]. This allows him to say [1983, 91], that he denies the motion of the earth more carefully than Copernicus and more truthfully than Tycho. On his definition (properly speaking) he can say that the earth does not move because it remains in contact with the contiguous vortical particles maintaining it in its orbit around the sun.

Newton's bucket experiment (see note 17 chpt. 3) in his scholium on time, space, place, and motion raises serious problems for this definition of proper motion proposed by Descartes. See DiSalle 2006, 32–3.

[33] See, e.g., Boyer 1968, 367–80. Our familiar orthogonal x,y axes are called Cartesian coordinates.

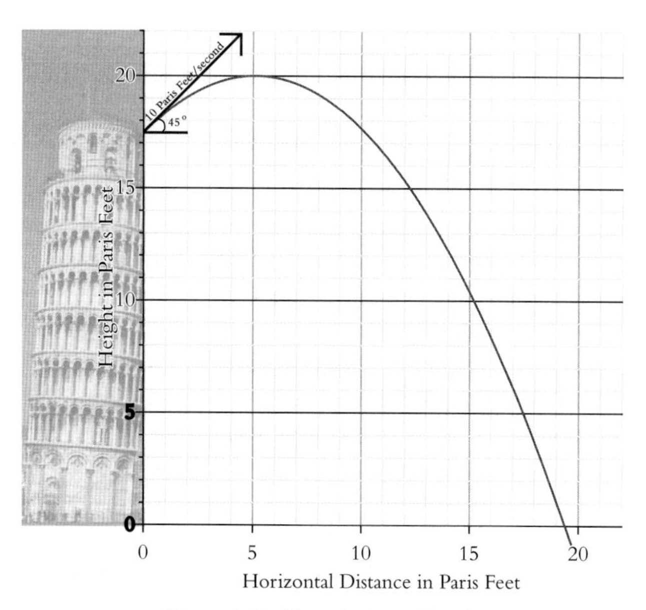

Figure 1.14 Cartesian coordinates

represent motion according to the account developed by Galileo's empirical investigation of projectile motion.

Descartes did not require his coordinates to be orthogonal. We now exploit non-orthogonal coordinates in geometrical representations of special relativity.[34]

Descartes is most widely known as a philosopher. In his philosophy, Descartes appealed to reason to defend knowledge against skepticism. He offered an ideal of infallible premises expressed in clear and distinct ideas, together with an ideal of inference limited to what follows by clear and distinct deductive steps from explicit assumptions.[35] Descartes's ideal of inference as deduction, interpreted with the help of modern logic, is still with us today. This ideal is sometimes taken as a standard for what ought to count as an acceptable inference. In the philosophy of science, we see this in the idea that an alternative hypothesis can undercut an inference unless it is shown to be logically incompatible with assumptions that are explicitly accepted as premises. This has been used to argue against the capacity of scientific practice to empirically discriminate among alternative theories.

[34] As, for example, in Mermin's wonderful new textbook (Mermin 2009).

[35] His *malin génie* argument (Descartes [1641] 1955 vol. 1, 149–57) suggests an ideal of empirical data limited to infallible subjective contents. This ideal inspired the new way of ideas that led to skeptical arguments about our knowledge of the external world in the hands of Berkeley and Hume. It is no longer especially influential in empiricist approaches to philosophy of science.

Descartes's discussion of deduction in his 2nd and 3rd "Rules for the direction of the mind" ([1628] 1955 vol. 1, 4–8) explicates an ideal of inference limited to what follows deductively from explicit assumptions.

Descartes proposed this extreme standard of inference to defend knowledge against skepticism. In the philosophy of science it has, instead, contributed what I shall argue is illegitimate support for skepticism about scientific knowledge. We shall see that Newton's fourth rule for doing natural philosophy affords a powerful counter against such overly restrictive standards as those of this Cartesian ideal.

Huygens (Dutch 1629–95)

Christiaan Huygens was the leading natural philosopher when Newton began his career. Huygens, together with his brother, invented a telescope that used long focal length lenses to obtain sharper images.

Huygens used such a telescope to discover Saturn's moon Titan and to resolve Saturn's rings. In appendix 2 of chapter 2 we will see that a Huygens telescope with a focal length of 123 feet played an important role in an observation initiative sponsored by Newton to obtain more accurate data on satellites of Jupiter and Saturn.[36]

Huygens replaced the rather puzzling Laws of Motion proposed by Descartes with principles that correspond to Newton's laws applied to collisions between elastic bodies.[37] His resulting derivation of a law of conservation of momentum appealed to a clever application of a relativity principle and, arguably, also anticipated Newton's distinction between weight and mass.[38] Huygens's summary report on his laws of collision clearly anticipates corollary 4 of Newton's Laws of Motion.[39] His treatment of what he called "centrifugal force" includes derivation of an accurate calculation of what we count as the ratios of centripetal accelerations corresponding to uniform circular motions.[40]

Huygens was able to successfully extend Galileo's treatment of free fall to account for the constrained fall exhibited by pendulum motion. His successful theory of pendulums led to his invention of the pendulum clock, the first really precise timepiece. Together with his theory of pendulum motion, his clock made possible his accurate *theory-mediated* measurement of the strength of terrestrial gravity. This measurement, together with his theory of pendulum motion and pendulum clocks, is reported in his book *The Pendulum Clock*. It was published in 1673. Huygens also developed a wave theory of light and proposed an ingenious hypothesis for a mechanical cause of gravity. After reading Newton's *Principia*, Huygens added comments on Newton's argument to his *Discourse on the Cause of Gravity*, which he published together with his *Treatise on Light* in 1690. In chapter 5 we will find these comments,

[36] These observations were carried out by James Pound, who was assisted by his nephew James Bradley, who would later succeed Halley as Astronomer Royal. We will see that their measurements of orbital parameters for Jupiter's moons were far more accurate than any of the others cited by Newton.

[37] See Stein 1990, 18–21. Blackwell 1977 is a translation of Huygens's posthumously published *The Motion of Colliding Bodies*.

[38] See Stein 1990, 25.

[39] Huygens's report was published in *Phil Trans.* in 1669. See Murray, Harper, and Wilson (2011) for a translation by Wilson together with a discussion of this important contribution by Huygens.

[40] See Blackwell (1986, 176) and Bos (1972, 606–7).

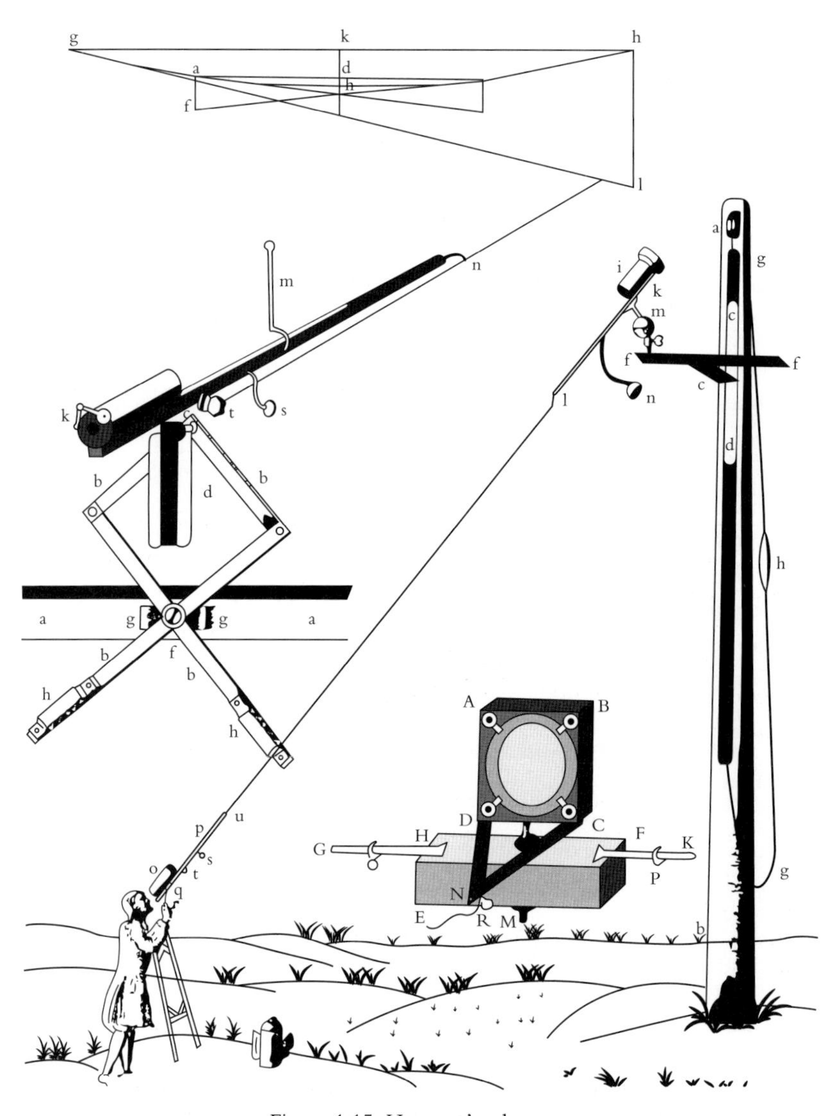

Figure 1.15 Huygens's telescope

together with Huygens's proposed cause of gravity, afford an illuminating foil for understanding Newton's scientific method.

II Newton's *Principia*: theoretical concepts

In August 1684 Edmund Halley (1656–1742), who would become the Astronomer Royal in 1720, visited Newton in Cambridge. According to a much retold story, Halley's visit convinced Newton of the importance of a calculation in which Newton had connected the ellipse with an orbit produced by an inverse-square force.[41] By November, Newton had sent Halley a small but revolutionary treatise, *De Motu*. An extraordinarily intense and productive effort by Newton over the next few years transformed this small treatise into the *Principia*.

1. *Newton's background framework*

The *Principia* opens with two sections that precede book 1. The first is entitled "Definitions." In it Newton discusses concepts basic to his fundamental theoretical framework and introduces his theoretical concept of a centripetal force. This section ends with his controversial scholium on time, space, place, and motion.[42] The basic concepts, together with his space-time framework, allow the Laws of Motion to be formulated in the second of these preliminary sections, which is entitled "Axioms or the Laws of Motion." Newton regards these Laws of Motion as empirical propositions, but he considers them sufficiently established to be treated as axioms. In chapter 3 we will see that Newton points out that their empirical support includes their confirmation from mechanics of machines, as well as phenomena of fall and projectile motion. We will also see Newton's own carefully designed trials of pendulum experiments to take into account air-resistance and to confirm the third Law of Motion for collisions with bodies of imperfect elasticity.[43] In chapter 9 we shall look in some detail at Newton's arguments to extend his third Law of Motion to attraction at a distance.

The propositions in these two preliminary sections, together with the propositions derived from them in books 1 and 2 (both of which are titled "The Motion of Bodies"), are the principles referred to in the *Principia*'s full title, *Mathematical Principles of Natural Philosophy*. They provide Newton's background framework for turning data into evidence for universal gravity and its application to empirically determine the true motions of the bodies in our solar system.

Central to Newton's application of universal gravity to this famous two chief world systems problem is corollary 4 of his Laws of Motion:

> The common center of gravity of two or more bodies does not change its state whether of motion or of rest as a result of the actions of the bodies upon one another; and therefore the common center of gravity of all bodies acting upon one another (excluding external actions and impediments) either is at rest or moves uniformly straight forward. (C&W, 421)

This allows the measurements of the relative masses of planets and the sun, afforded by his theory of gravity, to support his surprising resolution of this question which dominated natural philosophy in the seventeenth century.

[41] See Westfall's biography (1980, 403).
[42] See chapter 3 section II.1.
[43] See chapter 3 section II.3.

2. *Newton's theoretical concept of a centripetal force*

Newton's theoretical concept of a centripetal force deserves attention because of the important role it plays in his argument for universal gravity. He devotes the last four of his eight definitions to this concept and its three distinct quantities, or measures: absolute, accelerative, and motive. These definitions, as well as Newton's basic framework and Laws of Motion, are discussed in more detail in chapter 3. Here it may be enough to illustrate his concept of a centripetal force by considering the inverse-square centripetal force toward the sun that he argues for.

For each planet, its mass times its acceleration toward the sun counts as a motive measure of this inverse-square centripetal force.[44] These motive measures, which Newton says may be called separate motive forces assigned to each body being accelerated by it, are familiar to today's physics students as Newtonian forces.

Less familiar are Newton's accelerative measures, examples of which are given by assigning the centripetal acceleration of each planet to the location relative to the sun which it occupies. As Newton puts it, an accelerative measure is assigned "to the place of the body as a certain efficacy diffused from the center through each of the surrounding places in order to move the bodies that are in those places" (C&W, 407). What makes this an inverse-square centripetal force is that the magnitudes of these accelerations assigned to the places around the sun vary inversely as the squares of the distances of those places from the center of the sun. In proposition 6, Newton introduces an impressive set of phenomena that give agreeing measurements of the equal ratios of the motive forces to the masses of bodies being gravitationally drawn toward the sun (or any planet) for bodies at equal distances from it. These count as agreeing measurements of the property that equal component acceleration vectors toward the sun (or planet) can be assigned to places at equal distances from its center. Any two bodies at such equally distant places would have equal component accelerations toward the sun (or planet) generated by its gravity.

The absolute quantity of a centripetal force is assigned to the center as a characterization of its strength as a whole. The ratio of the accelerations assigned to equal distances from their respective centers measures the ratio of the absolute quantities of any two such centripetal forces, which have the same law relating accelerations to distances from their centers.

Newton conceives such a force as a natural power, or capacity, so that the distinct motive forces assigned to bodies being drawn toward the designated center are counted as actions of that centripetal force on those bodies.[45] We shall see that Newton's

[44] See chapter 3 section I.2.

[45] In an important paper, Sheldon Smith has convincingly argued that universal gravity should not be interpreted as a capacity:

> For example, fragility is manifested when something breaks when it is struck. But there is no single behavior that manifests being under the influence of gravity. *Any* behavior consistent with a differential equation derived from the Euler recipe counts as a manifestation of gravity as long as

conception of this inverse-square centripetal force toward the sun as a capacity counts it as the common cause of the distinct motive forces maintaining the planets in their separate orbits. This capacity is mathematically characterized so that the existence of centripetal forces and such features as their power laws can be empirically established by measurements from orbital phenomena.

An especially important aspect of this abstract mathematical treatment is that it allows centripetal forces to be empirically identified and their power laws to be empirically established, without settling on any more fundamental account of the physical causes of planetary motions. The following passage indicates the order of investigation facilitated by this mathematical treatment of centripetal forces.[46]

> Mathematics requires an investigation of those quantities of forces and their proportions that follow from any conditions that may be supposed. Then coming down to physics, these proportions must be compared with the phenomena, so that it may be found out which conditions [or laws] of forces apply to each kind of attracting bodies. And then, finally, it will be possible to argue more securely concerning the physical species, physical causes, and physical proportions of these forces. (C&W, 588–9)

According to this passage, the inferences from phenomena facilitated by the abstract mathematical treatment of forces are to precede, and to inform, any further investigations of physical causes of these forces. We shall see that the propositions Newton infers in the course of his argument are secured by the way they are empirically backed up by measurements from phenomena. In section VII below, we shall examine a passage in which Newton outlines the properties of gravity that have been so established as conditions that must be accounted for by any adequate proposal for a cause of gravity.

III Newton's classic inferences from phenomena

Phenomena are not just data. They are patterns exhibited in open-ended bodies of data. Newton's phenomena are patterns exhibited by the relative motions of satellites and planets with respect to the bodies about which they orbit. Newton cites estimates from astronomers of mean-distances and agreed-upon periods of orbits as data in

> one added a force of gravity between all bodies. I do not see why the content of Universal Gravitation needs to be equated with any "natural" behavior at all or even a capacity to behave in some specific way. (Smith, S. 2002, 254)

Smith is talking about the very general formulation developed by Euler; but, I think the lesson he draws also applies to Newton's earlier formulation. The theory is a framework for developing detailed models and not itself a predictive generalization.

Unlike the general theory he argues for, Newton's initial inverse-square centripetal forces of attraction toward planets should be counted as capacities. They are specifically designed to be measured by features of orbits about them. As we shall note again in chapter 3, however, they are capacities to induce component accelerations, rather than ones which would be counted as failing to act if other forces combined to produce motion that diverged from the predicted centripetal acceleration (see chpt. 3, note 58).

[46] This passage is from the scholium following proposition 69 book 1. We shall consider this important methodological scholium in more detail in chapter 9 section I.

support of these patterns of relative motion. In chapter 2 we shall analyze Newton's phenomena in fuller detail. We shall assess the fit of the patterns to his cited data and compare his data with the best estimates available today.

Let us now examine how Newton appeals to propositions inferred from his Laws of Motion as resources to make orbital motion phenomena measure centripetal forces. We shall see that his inferences to inverse-square forces toward Jupiter, Saturn, and the sun from phenomena exhibited by orbits about them are all backed up by such measurements.[47]

1. *Jupiter's moons*

In proposition 1 of book 3, Newton argues that the forces by which Jupiter's moons are maintained in their orbits are directed toward the center of Jupiter and are inversely as the squares of their distances from that center.[48] He cites the Area Rule for these moons with respect to that center and the Harmonic Rule for the system of those orbits about Jupiter as a two-part phenomenon.

Newton demonstrates that the Area Rule carries the information that the force maintaining a body in an orbit that satisfies it is directed toward the center, with respect to which it sweeps out equal areas in equal times. He also demonstrates that the Harmonic Rule for a system of orbits carries the information that the accelerative forces maintaining bodies in those orbits are inversely as the squares of the distances from the center about which those orbits are described.

1.i The Area Rule as a criterion for centripetal force Let us consider motion of Jupiter's moons under the idealized assumption that the center of Jupiter can be counted as inertial for the purpose of finding the forces that maintain those moons in their orbits about it. Consider at any given moment the total force acting to change the state of motion of a moon in its motion with respect to the center of Jupiter. Think of it as a vector acting on the center of mass of that moon. Newton proves theorems in book 1 that make having the rate at which areas are being swept out by radii to the center of Jupiter be constant equivalent to having that total force vector directed right at the center of Jupiter. He also proves theorems that make having that rate be increasing equivalent to having that total force vector be angled off-center toward the direction of the tangent toward which its orbital motion is directed, and that make having that rate be decreasing equivalent to having that force vector angled off-center backwards.[49]

[47] We give a more detailed treatment of these classic inferences from phenomena in chapter 3 section III below.

[48] See chapter 3 section III.1

[49] In chapter 3 we shall see that Newton also proves theorems that make the absence of sensible deviations from a constant area rate for the orbits of its moons about it show that the accelerations of Jupiter and its moons toward the sun are very nearly equal and parallel. This indicates that no appreciable errors result from ignoring these accelerations by treating Jupiter's center as inertial for purposes of using a constant area rate as a criterion for the centripetal direction of the forces maintaining those moons in their orbits.

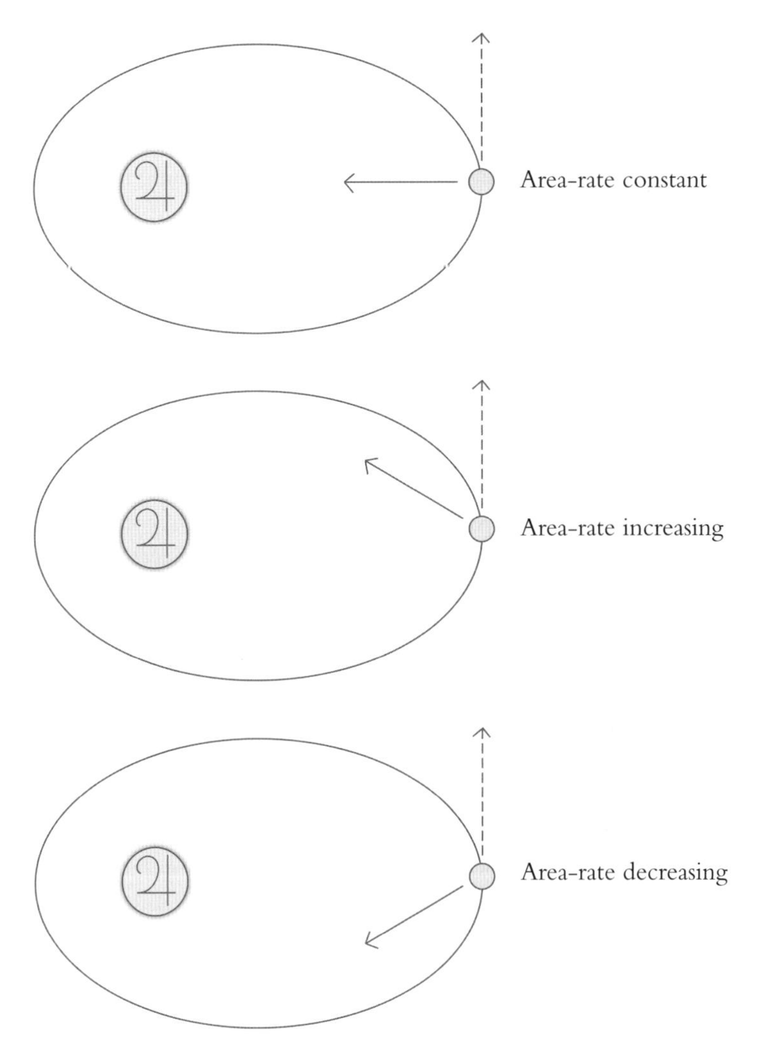

Figure 1.16 From Area Rule to centripetal force
An illustration of the systematic dependencies that make a constant rate at which areas are swept out by radii from a center carry the information that the total force acting to change the state of motion of a body with respect to that center is directed towards it.

These dependencies are not merely the material conditionals of truth functional logic. An important part of Newton's treatment of his Laws of Motion as "laws" is that it can be expected that such dependencies following from them would hold in circumstances appropriate to make the centripetal direction of the force explain the Area Rule phenomenon and make that phenomenon measure the centripetal direction of the force that would maintain a body in such an orbit.[50]

[50] John Roberts has proposed that it is this role in making quantities posited by physical theories measurable that makes a proposition count as a law (Roberts 2005; Roberts 2008).

Newton's proofs of the theorems underwriting the Area Rule as a criterion for centers toward which orbital forces are directed make no assumptions about any power law for these forces. The centripetal direction of the forces maintaining these moons in their orbits having been inferred from the phenomena that those orbits satisfy the rule of areas with respect to the center of Jupiter, Newton can appeal to theorems about orbital motion under centripetal forces to argue that the Harmonic Rule phenomenon, for the system of those orbits, carries the information that the accelerative measures of those forces are inversely as the squares of their distances from that center.

1.ii The Harmonic Rule as a criterion for inverse-square forces According to corollary 6 of proposition 4 of book 1, the Harmonic Rule for a system of concentric circular uniform motion orbits is equivalent to having the accelerative centripetal forces maintaining bodies in those orbits be inversely as the squares of the distances from the center. Corollary 6 is a special case of corollary 7, which adds additional systematic dependencies.[51] To have the periods be as some power $s > 3/2$ would be to have the centripetal forces fall off faster than the -2 power of the distances, while having the periods be as some power $s < 3/2$ would be to have the centripetal forces fall off more slowly than the -2 power of the distances. These systematic dependencies make the Harmonic Rule phenomenon ($s = 3/2$) for such a system of orbits measure the inverse-square (-2) power for the centripetal forces maintaining bodies in those orbits. This constitutes a very strong sense in which the Harmonic Rule for such a system of orbits carries the information that the forces maintaining bodies in those orbits satisfy the inverse-square power rule.

Newton reports the orbits of these moons to differ insensibly from uniform motion on concentric circular orbits.[52] Such orbits would satisfy the Area Rule with respect to radii from the center of Jupiter. As evidence for the Harmonic Rule, Newton offers a table citing periods that have been agreed upon by astronomers and four distance estimates from astronomers for each of the four moons of Jupiter that were known at the time.[53] The fit of the Harmonic Rule to these data is quite good.[54] He also offers more precise data from observations taken by Pound in 1718–20.[55]

2. Primary planets

In proposition 2, Newton infers that the primary planets are maintained in their orbits by forces that are directed to the sun and are inversely as the squares of their distances from its center.[56] The cited phenomenon for the centripetal direction is the Area Rule

[51] See chapter 3 section III.1.ii. for more detail.

[52] See chapter 2 section I.

[53] See C&W, 797. See chapter 2 section I.

[54] See chapter 2, Figure 2.1 for diagram. See chapter 2 appendix 2 for more details, least-squares assessment, and comparisons of estimates with our best estimates today.

[55] See chapter 2 appendix 2.

[56] See chapter 3 section III.2.

for the primary planets with respect to the sun, while the cited phenomenon for the inverse-square is the Harmonic Rule. An appeal to Newton's precession theorem provides an additional argument for the inverse-square.

Newton's phenomena are compatible with both sun-centered and earth-centered systems. To every sun-centered system, such as that of Copernicus or Kepler, a corresponding Tychonic system is defined by taking the center of the earth rather than the center of the sun as a reference frame. Newton provides a separate phenomenon stating that the *five* primary planets encircle the sun.[57] By not including the earth as a planet circling the sun this phenomenon is compatible with Tycho's system. Newton's statement of the Harmonic Rule is explicitly neutral between sun-centered and earth-centered systems.[58] In his phenomenon for the rule of areas, Newton considers radii drawn to the earth as well as radii drawn to the sun.[59]

Kepler's pretzel[60] diagram (Figure 1.10) dramatically illustrates the wild irregularity of the motion of Mars with respect to the earth. Newton cites the retrograde motion and stationary points that make the motion of planets with respect to the earth diverge from the Area Rule. In contrast, he points out that with respect to the sun, the planets move more swiftly in their perihelia and more slowly in their aphelia, in such a way that the description of areas is uniform.[61] As evidence for the Harmonic Rule phenomenon, Newton cites periods agreed upon by astronomers and the estimates of mean-distances made by Kepler and the French astronomer Boulliau.[62] In our discussion of Kepler we exhibited the fit of the Area Rule to his data by plotting log periods against log mean-distances from Kepler cited by Newton. In chapter 2 we do this for mean-distances cited from both Kepler and Boulliau.

2.i Inverse-square from aphelia at rest Newton claims that the inverse-square variation with distance from the sun of the forces maintaining the planets in their orbits is "proved with the greatest exactness" from the fact that the aphelia are at rest. If a planet in going from aphelion to return to it makes an angular motion against the fixed stars of $360 + p$ degrees, then the aphelion is precessing forward with p degrees per revolution.

He cites corollary 1 of proposition 45 book 1. According to this corollary, zero precession is equivalent to having the centripetal force be as the -2 power of distance; forward precession is equivalent to having the centripetal force fall off faster than the

[57] C&W, 799. See chapter 2 section III.

[58] C&W, 800. See chapter 2 section IV.

[59] C&W, 801.

[60] See chapter 2 note 34.

[61] Newton also suggests that the rule of areas for Jupiter is "especially provable by the eclipses of its satellites." The eclipses of the moons of Jupiter offer a direct test of the uniform description of areas. See chapter 2 section V.

[62] See chapter 2 section IV.

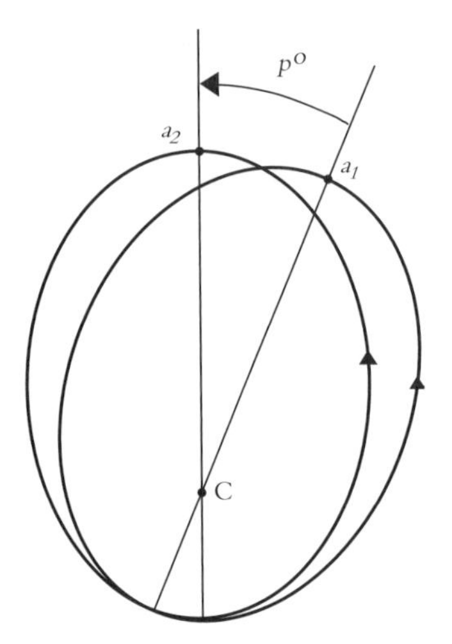

Figure 1.17 Precession of planetary orbit by p degrees per revolution

inverse-square; and backward precession is equivalent to having the centripetal force fall off less fast than the inverse-square power of distance.[63]

3. Inverse-square acceleration fields

We follow Howard Stein's interpretation of Newton's centripetal forces of gravity toward planets as acceleration fields. Stein's rich account of Newton's centripetal forces can be usefully introduced by considering a very informative question he asked about an account I once gave of Newton's argument for proposition 2.[64] I had interpreted it as an argument to the inverse-square variation of the separate centripetal motive forces maintaining the planets in their orbits about the sun. The phenomena from which these were inferred were the absence of significant orbital precession for each orbit and the Harmonic Rule relation among those six orbits. Stein asked how these phenomena provided evidence against a hypothesis that agreed with the inverse-square in the distances explored by each orbit and in the inverse-square relation among the forces at those six small distance ranges, but differed wildly from the inverse-square in the large ranges of distances not explored by the motions of those planets.

Figure 1.18 illustrates the approximate ranges of distances explored by each planet computed from the mean-distances and eccentricities assigned to their orbits today.

[63] Newton's proposition 45 book 1 and its corollaries are proved for orbits that are very nearly circular. The results, however, can be extended to orbits of arbitrarily great eccentricity. Indeed, orbital eccentricity increases the sensitivity of absence of unaccounted-for precession as a null experiment measuring inverse-square variation of a centripetal force. See chapter 3 appendix 3.

[64] This was the occasion I first met Howard Stein, when he came to a talk I gave at his department in Chicago.

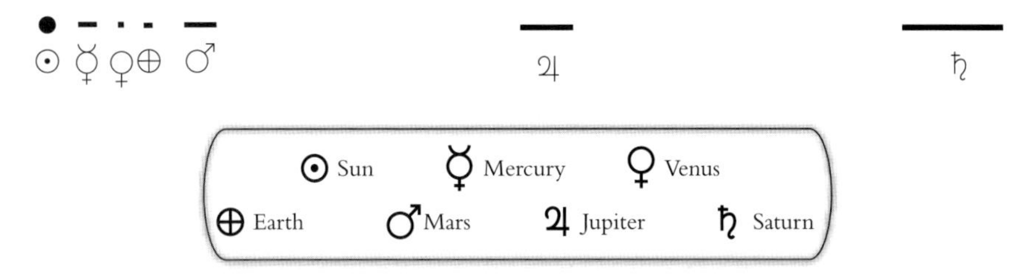

Figure 1.18 The distances explored by each planet in our solar system
Note the large gaps between the distances explored by Mars and those explored by Jupiter and between the distances explored by Jupiter and those explored by Saturn.

The following comments indicate the important role Stein sees for the concept of an acceleration field in Newton's argument for inverse-square gravitation toward the sun.

That the accelerations of the planets, severally and collectively, are inversely as the squares of their distances from the sun is not the conclusion of Newton's induction; that is his deductive inference from the laws established by Kepler. Newton's inductive conclusion is that the accelerations toward the sun are *everywhere* – i.e., even where there are no planets – determined by the position relative to the sun; namely, directed toward that body, and in magnitude inversely proportional to the square of the distance from it. And although the inductive argument is very straightforward – certainly not dependent upon tortuous constructs – that argument cannot be made, because its conclusion cannot even be sensibly formulated, without the notion of a field. From a mathematical point of view, the idea of an acceleration attached to each point in space is the idea of a function on space, hence a field; from the physical and methodological point of view, the idea of an *acceleration* characterizing *a point where there happens to be no body* makes no sense at all, unless one accepts the notion of a disposition, or tendency; subject to probing, but not necessarily probed. (Stein 1970b, 267–8)

This comment was preceded by the following comment about the extent to which Newton's induction is convincing.

The induction is very convincing. The fact that the acceleration is the field intensity is critical, for the evidence comes entirely from six bodies, each exploring the field in a fixed and severely restricted range; the inductive basis would therefore be rather weak if we were not, by good luck, able to relate directly to one another purely *kinematical* – and, thus, ascertainable – parameters of the several bodies' motions. This lucky fact is not the work of Newton's definitions, but of nature. Newton's merit was to know how to use what he was lucky enough to find. (Stein 1970b, 267)

The kinematical parameters are the centripetal direction and inverse-square relation of the accelerations of these six planets with respect to the sun. What Newton was lucky enough to find was the dynamical significance of Kepler's area and Harmonic Rules. It is this dynamical significance that transforms the exponent in Kepler's Harmonic Rule into a measure of a causally relevant parameter, namely, the exponent of the power law for a centripetal acceleration field directed toward the sun.

I believe Stein's question is answered by exploiting additional measurements afforded by these orbits. To have them fit the Harmonic Rule makes them all count as agreeing measurements of the strength of this single sun-centered inverse-square acceleration field. This makes them afford agreeing measurements of what the acceleration toward the sun would be at any given distance from it. For example, as Figure 1.18 illustrates, the considerable distances between Mars and Jupiter and between Jupiter and Saturn are not explored by the motions of planets. Consider such a distance. Using the inverse-square law to adjust the centripetal acceleration corresponding to each of the orbital data that Newton cited yields an estimate for the centripetal acceleration toward the sun that a body would be subject to if it were at that distance from the center of the sun.[65]

A central feature of Newton's method is to exploit such agreeing measurements to infer what this acceleration would be. We shall see that, on Newton's scientific method, an alternative hypothesis can be dismissed unless it either provides a correspondingly rich realization of agreeing accurate measurements of proposed rival causal parameters or provides phenomena that would conflict with motion in accord with Newton's measurements.

The Harmonic Rule measurements of inverse-square acceleration fields directed toward Jupiter and Saturn back up Newton's inference to an inverse-square acceleration field directed toward the sun. They provide evidence that Jupiter, Saturn, and the sun are all centers of the same sort of acceleration fields. The relative distances from these planets explored by the motions of their moons are different from those from the sun explored by the motions of the primary planets. This adds considerably to the difficulty of finding an alternative that could realize agreeing measurements for all three at once.

Comets explore considerably more distances from the sun than the six small ranges explored by the primary planets known to Newton. Newton's treatment of comet trajectories as being in accordance with inverse-square accelerations toward the sun adds considerable additional support to his inference to a sun-directed inverse-square acceleration field.[66]

The conception of empirical success as accurate measurement of parameters by the phenomena they explain is richer than the more restricted empiricist conception that would limit empirical success to prediction alone. The agreement among the

[65] See chapter 3 section III.3 for ten agreeing measurements (from Newton's cited data) of the acceleration toward the sun a body would be subject to if it were at a given distance between those explored by Mars and those explored by Jupiter.

These ten agreeing estimates of the centripetal acceleration a body would be subject to are computed by using the inverse-square law to adjust the centripetal accelerations corresponding to the periods and the ten distinct mean-distance estimates for orbits of planets from Kepler and Boulliau cited by Newton.

[66] See chapter 3 section III.3.

measurements of inverse-square centripetal acceleration fields afforded by orbital phenomena counts as an especially strong realization of a conception of empirical success that illuminates Newton's scientific method. Having the parameters receive convergent accurate measurements from distinct phenomena clearly affords more empirical support than would be afforded from a measurement from any one of those phenomena by itself.

The advantages of the richer approach to empirical investigation exhibited by Newton's inferences from phenomena are explored in more detail in chapter 3 section IV below. Lessons for philosophy of science will include a treatment of issues raised by criticisms of Newton's inferences by Duhem and philosophers of science such as Feyerabend and Lakatos (see chpt. 3 sec. IV.1–3). A comparison with Clark Glymour's bootstrap confirmation reveals that the systematic dependencies backing up Newton's inferences avoid the counterexamples proposed by Christensen. These counterexamples, which are based on constructing "unnatural" material conditionals entailed by theory to use as background assumptions, led to the demise of bootstrap confirmation as a serious approach to explicating scientific inference (see chpt. 3 sec. IV.4–5).

IV Unification and the moon

Newton's inference to identify terrestrial gravity as the force maintaining the moon in its orbit is a central step in his argument. This inference is supported by the conception of empirical success that informs his inferences from phenomena. On Newton's identification, two phenomena – the length of a seconds pendulum at the surface of the earth[67] and the centripetal acceleration exhibited by the lunar orbit – are found to give agreeing measurements of the strength of the same inverse-square centripetal acceleration field, the earth's gravity. This realizes the ideal of empirical success by making available the data from both phenomena to empirically back up the measured value of this common parameter.

1. *The moon's orbit*

In proposition 3, Newton argues that the force by which the moon is maintained in its orbit is directed toward the earth and is inversely as the squares of the distances of its places from the center of the earth. The centripetal direction toward the earth is inferred from the Area Rule. The inverse-square is argued for from the very slow precession of the moon's orbit. Newton suggests that the motion of the moon is

[67] The length of a seconds pendulum counts as a phenomenon (a pattern fitting an open-ended body of data) in so far as Huygens's experiment is regarded as repeatable. The stability of results over repetitions is one of the striking advantages of theory-mediated measurements using pendulums over attempts to more directly estimate the strength of gravity by comparing times and distances of fall for dropped objects. See chapter 5 below.

somewhat perturbed by the sun and that such perturbations can be neglected in his phenomena.

Newton's account of the moon's *variation*, an inequality in its motion discovered by Tycho, introduced violations of the Area Rule.[68] These violations, however, were accounted for by his successful treatment of this inequality as a perturbation due to the action of the sun.[69] This underwrites Newton's dismissal of these violations as irrelevant to his inference from the Area Rule to the earth-centered direction of the force maintaining the moon in its orbit about the earth.

The observed precession of the lunar orbit about the earth makes the argument for inverse-square variation more problematic than the corresponding argument for the planets. Newton claims that this precession of the moon's orbit is to be ignored because it arises from the action of the sun. He does not, however, provide an account of how this precession is due to the action of the sun.[70] His argument for proposition 3 ends with an appeal to the moon-test.

2. *Gravitation toward the earth*

Huygens's treatment of pendulums led to his pendulum clock, which made it possible to accurately determine the length of a seconds pendulum to considerable precision. Huygens's experiment gave the length of a seconds pendulum at Paris to be 3 Paris feet and 8 ½ lines, where a line is a twelfth of an inch.[71] His theory of the pendulum made the length of a seconds pendulum generate an accurate measurement of the strength of gravity. Both Newton and Huygens represent the strength of gravity by the distance it would make a body freely fall in the first second after release. Huygens's experimental estimate of the length of a seconds pendulum gives for this one second's fall

$d = 15.096$ Paris feet.[72]

[68] See chapter 2 note 42. See also chapter 4 section I.1 and section V.1.

[69] These results are described in section V of chapter 4. In chapter 6 we shall describe and assess a correction to the moon-test distance estimates generated by this treatment of the action of the sun to account for this *variational* inequality in the moon's motion.

[70] The argument for proposition 3 is discussed in chapter 4 section I. Most of section V of chapter 4 is devoted to the lunar precession problem. It reviews Newton's treatment as well as the very interesting treatments by Clairaut, Euler, and d'Alembert, both before and after Clairaut's successful account of the lunar precession as a perturbation due to the action of the sun in 1749.

[71] See chapter 4 section II.1 and chapter 5 section I.

[72] The acceleration of gravity, our g, is measured by this distance d of free fall in the initial second after release. In our notation this distance is given by $d = ½\, gt^2 = ½\, g(1\text{sec}^2)$. In these measurements d is given in Paris feet. So g in Paris feet/sec^2 will equal $2d/\text{sec}^2$.

The outcome of Huygens's seconds pendulum measurement of the one second's fall d gives about

$$g = 30.191 \text{ Paris feet per second squared}$$

as the acceleration of gravity for the location at Paris where Huygens carried out this experiment.

The Paris foot is somewhat larger than our foot today, so this estimate is not out of line with our modern value of g for Paris. George Smith (1997) gives $g = 980.7$ cm/sec^2 as the value in modern units of Huygens's measurement of the acceleration of gravity at Paris. He cites $g = 980.970$ cm/sec^2 as the modern measured value of g at Paris. Smith uses the value of 2.7069 cm per Paris inch given by Eric Aiton (Aiton 1972, note 65 on p. 89). See chapter 5 section I.

In chapter 5 we will further explore grounds for preferring Huygens's *theory-mediated* measurement over earlier efforts to give more direct experimental estimates of the strength of gravity by coordinating distances and times of fall for dropped objects.[73] One advantage of the pendulum estimates we can point out here is the stability exhibited in repetitions. Here are the seconds pendulum estimates of the one second's fall *d*, from experiments at Paris discussed by Huygens and Newton.

Table 1.1 Seconds Pendulum Estimates (*d* in Paris feet)

Picard	15.096
Huygens	15.096
Richer	15.099
Varin et al.	15.098

The very close clustering of these estimates shows the sharpness of the stability exhibited in repetitions of pendulum estimates of the one second's fall.

3. The moon-test

In his moon-test, Newton uses the inverse-square variation with distance to turn the centripetal acceleration exhibited by the lunar orbit into an estimate of what the corresponding centripetal acceleration would be in the vicinity of the surface of the earth. He cites a lunar period agreed upon by astronomers and a circumference for the earth according to measurements by the French, as well as six estimates of the lunar distance by astronomers.[74] In proposition 4, Newton appeals to the moon-test to argue that the force maintaining the moon in its orbit is gravity.[75] Here is the heart of his initial basic argument.

> And therefore that force by which the moon is kept in its orbit, in descending from the moon's orbit to the surface of the earth, comes out equal to the force of gravity here on earth, and so (by rules 1 and 2) is that very force which we generally call gravity. (C&W, 804)

This appeals to the equality established in the moon-test and to the first two of Newton's explicitly formulated *Regulae Philosophandi*. Here are these *Rules for Philosophizing* or *Rules for Reasoning in Natural Philosophy*:

> **Rule 1.** *No more causes of natural things should be admitted than are both true and sufficient to explain their phenomena.* (C&W, 794)

> **Rule 2.** *Therefore, the causes assigned to natural effects of the same kind must be, so far as possible, the same.* (C&W, 795)

We can read these two rules, together, as telling us to opt for common causes whenever we can succeed in finding them. This seems to be exactly their role in the

[73] See chapter 5 section I.
[74] See chapter 4 section II.
[75] Ibid.

application we are considering. We have two phenomena: the centripetal acceleration of the moon and the length of a seconds pendulum at Paris. Each measures a force producing accelerations at the surface of the earth. These accelerations are equal and equally directed toward the center of the earth. Identifying the forces makes these phenomena count as agreeing measures of the strength of the very same inverse-square force. This makes them count as effects of a single common cause.

4. Empirical success

We can calculate the one second's fall in the vicinity of the surface of the earth corresponding to using the inverse-square law to infer it from the lunar centripetal acceleration for each of the distance estimates Newton cites.[76] Table 1.2 puts the resulting moon-test estimates of the one second's fall, *d*, corresponding to each of the cited distance estimates together with the pendulum estimates of Table 1.1 above.

Table 1.2 (*d* in Paris feet)

Moon-test estimates		Seconds pendulum estimates	
Ptolemy	14.271		
Vendelin	15.009		
Huygens	15.009		
		15.096	Picard
		15.096	Huygens
		15.098	Varin et al.
		15.099	Richer
Copernicus	15.261		
Street	15.311		
Tycho	15.387		

These numbers suggest that, though the moon-test estimates may be irrelevant to small differences from Huygens's estimate, the agreement of these cruder moon-test estimates affords increased resistance to large changes from these seconds pendulum estimates. This increased resistance to large changes, or increased resiliency, represents an important sort of increased empirical support. The cruder agreeing moon-test estimates count as additional empirical support backing up the much sharper estimates available from pendulum experiments.

We have exhibited these multiple agreeing measurements by listing them. An application of modern least-squares assessment can reveal empirical advantages of multiple agreeing measurements that are not made explicit by just listing them. In chapter 4, such an application of least-squares reveals that the agreement among the

[76] See chapter 4 section IV.

estimates in Table 4.3 counts as a very impressive realization of such empirical advantages.[77]

Empiricists who limit empirical success to prediction alone would see the appeal to simplicity in rules 1 and 2 as something extraneous to empirical success. According to such a view, these rules endorse a general theoretical or pragmatic commitment to simplicity imposed as an additional requirement beyond empirical success. Such a view is suggested by Newton's comment on Rule 1:

> As the philosophers say: Nature does nothing in vain, and more causes are vain when fewer suffice. For nature is simple and does not indulge in the luxury of superfluous causes. (C&W, 794)

The application of these rules to the moon-test inference, however, does not depend on any such general commitment to simplicity. This inference is backed up by the agreement between the measurements of the strength of an inverse-square centripetal acceleration field afforded from the length of a seconds pendulum and the centripetal acceleration exhibited by the lunar orbit. The agreement of the measurements from these distinct phenomena counts as the sort of empirical success that is realized by Newton's basic inferences from phenomena.

Table 1.2 exhibits that identifying the force that maintains the moon in its orbit with terrestrial gravity is empirically backed up by the agreement of these measurements. It is surely implausible that any general commitment to simplicity as a non-empirical virtue can do justice to this sort of empirical support. We shall argue that this rich sort of empirical support afforded to the moon-test inference from these agreeing measurements is sufficient to make appeal to any general theoretical or pragmatic commitment to simplicity unnecessary.

V Generalization by induction: Newton on method

Two of Newton's most interesting *Regulae Philosophandi* (*Rules for Reasoning in Natural Philosophy*) are applied in his arguments for propositions 5 and 6 of book 3. These applications support interpretations according to which these two rules count as very informative methodological principles for scientific inference. In the scholium to proposition 5 there is an explicitly cited appeal to Rule 4 in arguing to extend inverse-square gravity to planets without satellites to measure it. This affords an interpretation that informs the provisional acceptance of propositions gathered from phenomena by induction as guides to research that should not be undercut by mere

[77] In chapter 5, such an analysis is extended to include pendulum estimates from other latitudes cited by Newton and Huygens. It illustrates the irrelevance of the moon-test estimates to the small differences between these pendulum estimates and those from Paris.

In chapter 6, such analyses are extended to include the lunar distance estimates cited in the various versions Newton gives of the moon-test. The inference holds up for each version of the moon-test. It also holds up when all these cited distances are taken together.

contrary hypotheses. In corollary 2 of proposition 6, there is an explicit appeal to Rule 3 to generalize, to all bodies universally, weight toward the earth with equal ratios to inertial mass at equal distances from the center of the earth. This supports an interpretation of Rule 3 that makes the whole argument for proposition 6 count as a very powerful application of convergent agreeing measurements from phenomena.

1. Proposition 5 and Rule 4

In proposition 5 and its corollaries, Newton identifies the inverse-square centripetal acceleration fields toward Jupiter, Saturn, and the sun with gravitation toward those bodies. All these orbital phenomena are effects of gravitation of satellites toward primaries. Given this generalization, we can understand each of these phenomena as an agreeing measurement of such general features of gravitation toward these primaries as centripetal direction and inverse-square accelerative measure.

Newton further generalizes to assign to all planets universally centripetal forces of gravity that are inversely as the squares of distances from their centers. For planets without satellites there are no centripetal accelerations of bodies toward them to measure gravity toward them. The following scholium is offered in support of this further generalization to all planets.

> Scholium. Hitherto we have called "centripetal" that force by which celestial bodies are kept in their orbits. It is now established that this force is gravity, and therefore we shall call it gravity from now on. For the cause of the centripetal force by which the moon is kept in its orbit ought to be extended to all the planets, by rules 1, 2, and 4. (C&W, 806)

This appeal to rules 1 and 2 is backed up by appeal to an additional rule.

> **Rule 4**. *In experimental philosophy, propositions gathered from phenomena by induction should be considered either exactly or very nearly true notwithstanding any contrary hypotheses, until yet other phenomena make such propositions either more exact or liable to exceptions.* (C&W, 796)

This rule instructs us to consider propositions gathered from phenomena by induction as "either exactly or very nearly true." Such propositions are to be considered as accepted rather than merely assigned high probabilities.[78] They may be accepted as exactly true or as approximations.[79] We are instructed to maintain this acceptance of these propositions in the face of any contrary hypotheses, until further phenomena make them more exact or liable to exceptions.

We shall offer a more detailed account of this important rule in chapter 7. In this introductory chapter we do need to clarify what are to count as propositions gathered from phenomena by induction and how they differ from what are to be dismissed as mere contrary hypotheses. We have seen that the classic inferences from phenomena

[78] See section VII.5 below.

[79] The provision for approximations fits with construing such propositions as established up to tolerances provided by measurements. As we shall see in chapter 10 section II.3, this makes Rule 4 very much in line with the methodology guiding testing programs in relativistic gravitation today.

which open the argument for universal gravity are measurements of the centripetal direction and the inverse-square accelerative quantity of gravities maintaining moons and planets in their orbits. To extend the attribution of centripetally directed gravity with inverse-square accelerative quantity to planets without moons is to treat such orbital phenomena as measurements of these quantified features of gravity for planets generally.

What would it take for an alternative proposal to succeed in undermining this generalization of gravity to planets without moons? The arguments we have been examining suggest that Newton's Rule 4 would have us treat such an alternative proposal as a mere "contrary hypothesis" unless it is sufficiently backed up by phenomena to count as a rival to be taken seriously.

Consider the skeptical challenge that the argument has not ruled out the claim that there is a better alternative theory in which these planets do not have gravity. Rule 4 will count the claim of such a skeptical challenge as a mere contrary hypothesis to be dismissed, unless such an alternative is given with details that actually deliver on measurement support sufficient to make it a serious rival or is backed up by empirically established phenomena that make Newton's inference liable to exceptions. On Newton's method, it is not enough for a skeptic to show that such an alternative has not been logically ruled out by the explicitly cited premises of an argument.

Newton's richer ideal of empirical success, as a criterion distinguishing propositions gathered from phenomena from mere hypotheses, also makes his Rule 4 back up his moon-test inference. Consider an alternative that makes the same predictions about terrestrial gravity phenomena and the moon's motion, but introduces two separate forces. Its failure to realize the empirical success afforded to Newton's inference by the agreeing measurements makes Newton's Rule 4 apply to count such an alternative as a mere contrary hypothesis, which should not be allowed to undercut Newton's identification of terrestrial gravity as the force which maintains the moon in its orbit of the earth. This reinforces our argument that there is no need to appeal to any general theoretical or pragmatic commitment to simplicity.

2. *Proposition 6 and Rule 3*

Proposition 6 backs up Newton's argument for assigning gravity as inverse-square centripetal acceleration fields to all planets by appealing to diverse phenomena that afford agreeing measurements of the equality of accelerations toward planets for bodies at equal distances from them. The centripetal forces that have been identified as gravity toward planets are acceleration fields. The ratio of weight toward a planet to inertial mass is the same for all bodies at any equal distances from its center.[80] In arguing for

[80] Where f_1/m_1 and f_2/m_2 are ratios of weights toward the center of a planet to inertial masses of attracted bodies while a_1 and a_2 are their respective gravitational accelerations toward it, it follows immediately from $f = ma$ that $a_1 = a_2$ if and only if $f_1/m_1 = f_2/m_2$.

proposition 6, Newton backs up his earlier arguments by providing explicit measurements of the equality of these ratios of weight to mass.

2.i Weight toward the earth Newton begins with gravity toward the earth. He describes pendulum experiments which measure the equality of the ratio of weight to inertial mass for pairs of samples of nine varied materials – gold, silver, lead, glass, sand, common salt, wood, water, and wheat. In chapter 7 we shall see that the agreement of the measurements from all these different sorts of material supports accepting that the outcome is not limited to the specific sort of material selected. This illustrates an important advantage of agreeing measurements from diverse phenomena. The equality of the periods of each such pair of pendulums counts as a phenomenon which measures the equality of these ratios for any laboratory-sized bodies near the surface of the earth.

The outcome of the moon-test affords an agreeing measurement from a considerably more diverse phenomenon, the centripetal acceleration exhibited by the orbit of our moon. In chapter 7 we will see that this affords increased evidence that neither sort of measurement result is an artifact of systematic error built into the details of either of the very different sorts of experimental apparatus used to make these different measurements. It also extends the result to moon-sized bodies. The agreement between the seconds pendulum estimates and the moon-test estimates of the strength of gravity at the surface of the earth affords a measurement of the equality of the ratio of the moon's weight toward the earth to its mass and the common ratio to their masses of the weights toward the earth that terrestrial bodies would have at the lunar distance.

2.ii Rule 3 Newton's third rule for doing natural philosophy is applied in the first part of corollary 2 of proposition 6.

Corollary 2. All bodies universally that are on or near the earth are heavy [or gravitate] toward the earth, and the weights of all bodies that are equally distant from the center of the earth are as the quantities of matter in them. This is a quality of all bodies on which experiments can be performed and therefore by rule 3 is to be affirmed of all bodies universally . . . (C&W, 809)

Rule 3. *Those qualities of bodies that cannot be intended and remitted [that is, qualities that cannot be increased and diminished] and that belong to all bodies on which experiments can be made should be taken as qualities of all bodies universally.* (C&W, 795)

Those qualities of bodies that cannot be intended or remitted are those that count as constant parameter values. This rule, therefore, endorses counting such parameter values found to be constant on all bodies within the reach of experiments as constant for all bodies universally. In corollary 2, the quality of bodies which is generalized is weight toward the earth, with equal ratios of weight to mass for all bodies at equal distances from the center of the earth.

As we shall see in chapter 7,[81] the equality of the periods of pairs of pendulums in Newton's experiments are phenomena established with precision sufficient to measure the equalities of ratios of weight to mass for terrestrial bodies to about one part in a thousand. These experiments extend to this, much greater, precision the many long-established, rougher but agreeing, observations that bodies fall at equal rates, "at least on making an adjustment for the inequality of the retardation that arises from the very slight resistance of the air."

The outcome of the moon-test counts as a rougher measurement bound that agrees with the more precise bound that would result from extending the null outcome of Newton's pendulum experiments, to include the equality of ratios to masses of the weights toward the earth any bodies would have at the lunar distance.[82] In chapter 7 we shall see that these phenomena count as agreeing measurements bounding toward zero an earth-parameter, Δ_e, representing differences between ratios of weight toward the *earth* to mass that bodies would have at any equal distances from its center.[83]

Rule 3 tells us to conclude that the ratio of mass to weight toward the earth is equal for all bodies at any equal distances, however far they may be from the center of the earth, if that equality holds for all the bodies in reach of our experiments. The agreement, exhibited by Newton, among measurements of this equality by phenomena is an example of what he counts as an inductive base for generalizing to all bodies in reach of our experiments. This makes his Rule 4 tell us to put the burden of proof on a skeptic to provide evidence for bodies within reach of our experiments that would exhibit phenomena making this equality liable to exceptions.

2.iii The argument for proposition 6 continued Newton follows up his argument for the earth with an appeal to the Harmonic Rule for Jupiter's moons as a phenomenon which measures the equality of the ratio of mass to weight toward Jupiter that bodies would have at the distances of its moons. Rule 3 would extend this equality to bodies at any distances. The data Newton cites in his table for the Harmonic Rule for Jupiter's moons measure the equality of these ratios to fair precision. They bound toward zero a Jupiter-parameter, Δ_j, representing differences between ratios of weight toward Jupiter to mass for bodies at any equal distances from Jupiter.[84] Similarly, the data Newton cites for the Harmonic Rule for the primary planets bound toward zero a sun-parameter, Δ_H ("*H*" for "*Helios*").

To show the equality of ratios of mass to weight toward the sun at equal distances Newton also appeals to three additional phenomena – absence of polarization toward or away from the sun of orbits of, respectively, Jupiter's moons, Saturn's moons, and the earth's moon. If the ratio of mass to weight toward the sun for a moon were greater

[81] See chapter 7 section II.1.i.
[82] See chapter 7 section II.1.ii.
[83] See chapter 7 section II.4.
[84] See chapter 7 section II.5.

or less than the corresponding ratio for the planet then the orbit of that moon would be shifted toward or away from the sun. Absence of such orbital polarization counts as a phenomenon measuring the equality of ratios of mass to weight toward the sun at equal distances.[85]

All these phenomena count as agreeing measurements bounding toward zero a single general parameter, Δ, representing differences between bodies of the ratios of their inertial masses to the weights toward any planet they would have at any equal distances from it.[86]

2.iv Parts of planets Newton concludes his argument for proposition 6 by explicitly extending the argument for equality of ratios between mass and weight toward other planets to individual parts of planets. We will examine this argument in chapter 7.[87]

VI Gravity as a universal force of pair-wise interaction

1. Applying Law 3

In proposition 7, Newton argues that gravity exists in all bodies universally and is proportional to the quantity of matter in each. We shall see in chapter 8 that the proof is given in two paragraphs. The first paragraph argues that the gravity toward each planet is proportional to the quantity of matter of that planet. This argument appeals to book 1 proposition 69. That proposition from book 1 applies the third Law of Motion to argue that if bodies attract each other by inverse-square accelerative forces, then the attraction toward each will be as its quantity of matter or mass.[88] As we have seen, Newton has just argued in proposition 6 that each planet attracts all bodies by inverse-square accelerative forces.

The second paragraph argues to extend the result of the first to all the parts of planets. This second argument appeals directly to Law 3.

To any action there is always an opposite and equal reaction; in other words, the actions of two bodies upon each other are always equal and always opposite in direction. (C&W, 417)

In both arguments the forces are interpreted as interactions between the bodies in such a way as to allow Law 3 to apply, by counting the attraction of each toward the other as action and equal and opposite reaction.

[85] See chapter 7 section II.6 and appendix 1.

[86] Bounds limiting this universal parameter toward zero are what count today as bounds limiting violations of the weak equivalence principle – the identification of what some physicists today (see, e.g., Damour 1987) call "passive gravitational mass" with inertial mass. The phenomena cited by Newton together with additional phenomena of far greater precision count today as agreeing measurements supporting this identification (see chpt. 7 below for discussion and references).

[87] See chapter 7 section I.1v.

[88] See appendix 1 of chapter 8 for an account of Newton's proof.

In corollary 2 of proposition 7, Newton argues that gravitation toward each of the individual equal particles of a body is inversely as the square of the distance of places from those particles. In chapter 8 we shall see that, as with Newton's classic inferences from phenomena, this inference is backed up by systematic dependencies. These dependencies make phenomena measuring inverse-square variation of attraction toward a whole uniform sphere count as measurements of inverse-square variation of the law of the attractions toward the particles which sum to make it up.

2. Universal force of interaction

In the assumption of the argument for proposition 7, gravitation of any planets *A* and *B* toward one another is treated as an *interaction*, so that the equal and opposite reaction to the weight of *B* toward *A* is the weight of *A* toward *B*. This makes the argument of proposition 69 apply, so the strengths of the centripetal attractions toward each are proportional to their masses. The extension to include interactive gravitation of bodies toward parts of planets would count, in Newton's day, as an extension to include interactive gravitation between all bodies within reach of experiments. This makes Rule 3 endorse extending interactive gravity to all bodies universally.

Newton's argument transforms his initial conception of a centripetal force as a centripetal acceleration field into his conception of gravity as a universal *force of interaction* between bodies characterized by his *law of interaction*. Central to this transformation are the crucial applications of Law 3 to construe the equal and opposite reaction of the attraction of a satellite toward its primary to be an attraction of the primary toward that satellite. In chapter 9 we shall explore a very illuminating challenge to these applications of Law 3 by Roger Cotes, Newton's editor of the second edition. Our examination of this challenge, which will include examining Newton's response to it in letters to Cotes and his scholium to the laws, will further illuminate interesting lessons we can learn from his method.

3. Resolving the two chief world systems problem

In proposition 8, Newton appeals to theorems on attraction toward spheres to extend his conclusions to gravitation toward bodies approximating globes made up of spherically homogeneous shells. Attraction between such bodies is directly as the product of their masses and inversely as the square of the distance between their centers. Proposition 7 is applied to use orbital phenomena to measure the masses of the sun and planets with moons (corollary 2 proposition 8). The resulting agreeing measurements of the masses of these bodies count as a significant realization of what I have identified as Newton's ideal of empirical success. In chapter 9, we will argue that this adds more support to his appeal to Law 3 in the argument for proposition 7 than would be available if empirical success were limited to prediction alone.

These measurements lead to his surprising center-of-mass resolution of the two chief world systems problem – the problem of deciding between geocentric and heliocentric world systems.

Proposition 12 (book 3) *The sun is engaged in continual motion but never recedes far from the common center of gravity of all the planets.* (C&W, 816)

Both the Copernican and Tychonic systems are wrong; however, the sun-centered system closely approximates true motions, while the earth-centered system is wildly inaccurate.

In this center of mass frame the separate centripetal acceleration fields toward solar system bodies are combined into a single system where each body undergoes an acceleration toward each of the others proportional to its mass and inversely proportional to the square of the distance between them. This recovers appropriate centripetal acceleration components corresponding to each of the separate inverse-square acceleration fields toward the sun and each planet.

In chapter 8 we will argue that Newton's Rule 4 supports maintaining acceptance of these inverse-square acceleration fields toward planets as approximations recovered by his theory of gravity as a universal force of pair-wise interactions between bodies. We will also see Newton argue that the large mass of the sun makes his solution to the two chief world systems problem recover Kepler's elliptical orbit as a good approximation for an initial stage from which to seek a progressive series of better approximations as more and more perturbation producing interactions are taken into account.

VII Lessons from Newton on scientific method

1. More informative than H-D method

In 1690 Huygens published his *Treatise on Light* followed by his *Discourse on the Cause of Gravity* bound together in a single volume. He includes a very nice characterization of hypothetico-deductive method in the preface to his *Treatise*.[89]

One finds in this subject a kind of demonstration which does not carry with it so high a degree of certainty as that employed in geometry; and which differs distinctly from the method employed by geometers in that they prove their propositions by well established and incontrovertible principles, while here principles are tested by the inferences which are derivable from them. The nature of the subject permits of no other treatment. It is possible, however, in this way to establish a probability which is little short of certainty. This is the case when the consequences of the assumed principles are in perfect accord with the observed phenomena, and especially when these verifications are numerous; but above all when one employs the hypothesis to predict new phenomena and finds his expectation realized. (Matthews 1989, 126–7)

On this method, hypothesized principles are tested by experimental verification of observable conclusions drawn from them. Empirical success is limited to accurate prediction of observable phenomena. What it leads to are increases in probability.

[89] In chapter 5 we shall look in more detail at Huygens contrasting methodology, including the specific comments on Newton's argument that he added to his *Discourse* after reading the *Principia*.

A great number of accurate predictions, especially of new phenomena, can lead to very high degrees of probability.[90] All this is very much in line with the basic hypothetico-deductive model for scientific method that dominated much discussion by philosophers of science in the last century.[91]

As we have argued, Newton's method differs by adding additional features that go beyond this basic H-D model. One important addition is the richer ideal of empirical success that is realized in Newton's classic inferences from phenomena. This richer ideal of empirical success requires not just accurate prediction of phenomena. It requires, in addition, accurate measurement of parameters by the predicted phenomena. Consider Newton's inference to the centripetal direction of the force maintaining a planet in its orbit from its uniform description of areas by radii from the center of the sun. According to proposition 1 of *Principia*, book 1, if the direction of the force is toward that center then the planet will move in a plane in such a way that its description of areas by radii to that center will be uniform. So, if your inference were hypothetico-deductive, that would do it. Proposition 1 shows that a centripetal force would predict the Area Rule phenomenon. Therefore, the hypothesis that the force maintaining the body in that orbit is directed toward that center is confirmed by the fit of the Area Rule to the data.

Newton cites proposition 2 of book 1, which gives the converse conditional that if the description of areas is uniform then the force is centripetal. These two conditionals follow from his application of his Laws of Motion to such orbital systems. He has proved, in addition, the corollaries that if the areal rate is increasing the force is off-center in a forward direction and if the areal rate is decreasing the force is off-center in the opposite way. Taken together, these results show that the behavior of the rate at which areas are swept out by radii to the center depends systematically on the direction of the force maintaining a body in orbital motion with respect to that center. These systematic dependencies make the uniformity of the area rate carry the information that the force is directed toward the center.

These inferences of Newton are far more compelling than the corresponding hypothetico-deductive inferences. The systematic dependencies backing them up afford a far more compelling sort of explanation of the uniform description of areas by the centripetal direction of the forces than the hypothetico-deductive model of explanation as a one-way conditional with the hypothesis as antecedent and the phenomenon to be explained as consequent.

[90] We will say more about this quote from Huygens in chapter 10 section I.1. We shall see that the sorts of cases he cites are all ones where a Bayesian model of scientific inference might well be expected to result in high posterior probability for the hypothesis after successful predictions of the sort specified. We will argue that the Bayesian model needs to add an account of acceptance informed by Newton's ideal of empirical success if it is to do justice to Newton's scientific method.

[91] See chapter 3 section IV.

2. *Newton's* hypotheses non fingo

Here is Newton's most famous and controversial passage on method.

> I have not as yet been able to deduce from phenomena the reason for these properties of gravity, and I do not feign hypotheses. For whatever is not deduced from the phenomena must be called a hypothesis; and hypotheses, whether metaphysical or physical, or based on occult qualities, or mechanical, have no place in experimental philosophy. In this experimental philosophy, propositions are deduced from the phenomena and are made general by induction. The impenetrability, mobility, and impetus of bodies, and the laws of motion and the law of gravity have been found by this method. And it is enough that gravity really exists and acts according to the laws that we have set forth and is sufficient to explain all the motions of the heavenly bodies and of our sea. (C&W, 943)

This famous *hypotheses non fingo* passage defends his inferences to properties of gravity against the objection that he has not made gravity mechanically intelligible by showing how it could be hypothetically explained by things pushing on bodies. As we have noted above, the need for such hypothetical explanations by contact to make motion phenomena intelligible was central to the Cartesian *mechanical philosophy*. Huygens and Leibniz were committed advocates of this requirement.

Newton dismisses such conjectured hypotheses as having no place in what he characterizes as his contrasting *experimental philosophy* in which propositions are deduced from the phenomena and are made general by induction. *Hypotheses* have no place in experimental philosophy and are explicitly identified as "*whatever is not deduced from the phenomena.*" So, *Deductions from the phenomena* are to be construed widely enough to include not just *propositions deduced from the phenomena*, **directly**, but, also, propositions resulting from *making general such propositions by induction*.

Those of us who associate "deduction" with strictly logical or mathematical inference may be somewhat surprised to find that Newton's deductions from phenomena explicitly include inductions; however, as scholars will know, the use of "deduction" in Newton's day was not restricted to logically valid inference.[92] On Newton's usage, any appropriately warranted conclusion inferred from phenomena as available evidence will count as a deduction from the phenomena. Newton's identification of "hypotheses" with "whatever is not deduced from the phenomena" makes counting

[92] The Latin original of Newton's characterization of his experimental philosophy, "In this philosophy, propositions are deduced from the phenomena and are made general by induction." is "In hac philosophia propositiones deducuntur ex phænomenis, & redduntur generales per inductionem" (Koyré and Cohen 1972, 764). The *Oxford Classical Dictionary* gives a relevant usage of the Latin verb "deduco" as "derive" in a sense general enough to include the origin of a word (see p. 527). The *Concise Oxford English Dictionary* gives an example for "deduction" as "The process of deducing or drawing a conclusion from a principle already known or assumed" as late as 1860, "the process of deriving facts from laws and effects from causes" (see p. 358). These usages are clearly not restricted to logically valid inferences.

something as a mere hypothesis equivalent to counting it as not appropriately warranted on the basis of available evidence.

In our discussion of Rule 4 we suggested that what would render an alternative to be a mere contrary hypothesis (and so something which should not be allowed to undercut propositions gathered from phenomena by induction) is that it failed to be sufficiently backed up by measurements from phenomena to be counted as a serious rival. This is very much in line with what Newton is saying in this *hypotheses non fingo* passage. Clear examples of what we might call *direct* deductions from the phenomena are Newton's inferences to inverse-square centripetal acceleration fields from the Area Rule and the Harmonic Rule orbital phenomena, which measure the centripetal direction and the inverse-square variation of the forces maintaining bodies in those orbits. The extension of such inverse-square centripetal acceleration fields to planets without orbiting bodies to measure them is an example of making general by induction. All these inferences from phenomena count as deductions from the phenomena in the wide sense in which such empirically warranted propositions are contrasted with mere hypotheses which are not sufficiently backed up by measurements from phenomena to be counted as serious rivals.

Newton starts with his Laws of Motion as accepted propositions. He tells us that they have been found by the same method as his theory of gravity.[93] He counts them as empirical propositions that have already been sufficiently established to be accepted as guides to research. They and their consequences count as premises that can be appealed to in backing up inferences from phenomena. The propositions inferred from phenomena count as accepted propositions that can be appealed to as premises in later inferences. Newton's argument is a piecemeal stage-by-stage construction of the theory. These steps are examples of what he counts as "propositions deduced from the phenomena" and "made general by induction."

3. *A methodology of seeking successively more accurate approximations*

One striking feature of Newton's argument, which was pointed out by Duhem and has been much remarked upon by philosophers of science,[94] is that Newton's theory of universal gravity is actually incompatible with the Keplerian orbital phenomena assumed as premises in his argument for it. The application of the third Law of Motion to the sun and planets leads to two-body corrections to the Harmonic Rule. Newton's successful treatment of the variational inequality in the moon's motion, as a perturbation due to the action of the sun, leads to violations of the Area Rule for the lunar orbit.

[93] Newton's application of pendulum experiments to extend Law 3 to collisions between bodies where there is loss of momentum due to imperfect elasticities or damage are clear examples of theory-mediated measurements made general by induction. They are inferences that would be endorsed by Newton's Rule 3 and backed up by Rule 4. See chapter 3 section II.3.

[94] See chapter 3 section IV below.

The need to correct the Keplerian orbital phenomena corresponding to the separate inverse-square centripetal acceleration fields results from the combination of those acceleration fields into a single system corresponding to pair-wise interactions among all the solar system bodies. Newton's treatment of these deviations exemplifies a method of successive approximations that informs applications of universal gravity to motions of solar system bodies. On this method, deviations from the model developed so far count as new *theory-mediated* phenomena to be exploited as carrying information to aid in developing a more accurate successor.

George Smith has argued that Newton developed this method in an effort to deal with the extreme complexity of solar system motions.[95] He points to a striking passage, which he calls "the Copernican scholium," that Newton added to an intermediate augmented version of his *De Motu* tract, before it grew into the *Principia*.[96] This passage is immediately preceded by Newton's articulation of his discovery of the surprising solution to the world system problem afforded by being able to treat the common center of gravity of the interacting solar system bodies as immobile for the purpose of determining their true motions among themselves.[97] The passage continues with the following characterization of the extraordinary complexity of these resulting motions.

> By reason of the deviation of the Sun from the center of gravity, the centripetal force does not always tend to that immobile center, and hence the planets neither move exactly in ellipses nor revolve twice in the same orbit. There are as many orbits of a planet as it has revolutions, as in the motion of the Moon, and the orbit of any one planet depends on the combined motion of all the planets, not to mention the action of all these on each other. But to consider simultaneously all these causes of motion and to define these motions by exact laws admitting of easy calculation exceeds, if I am not mistaken, the force of any human mind. (Wilson 1989b, 253)

It appears that shortly after articulating this daunting complexity problem, Newton was hard at work developing resources for responding to it with successive approximations.

[95] See Smith G.E. 2002a, 153–67.

[96] See Smith G.E. 2002a, 153–4. See also Smith G.E. 1999a, 46–51.

[97] See Smith G.E. 1999a, 46–9. Wilson 1989b, 253. Here is the first part of this paragraph:

> Moreover, the whole space of the planetary heavens either rests (as is commonly believed) or moves uniformly in a straight line, and hence the common center of gravity of the planets (by Law 4) either rests or moves along with it. In either case the motions of the planets among themselves (by Law 3) are the same, and their common center of gravity rests with respect to the whole space, and thus can be taken for the immobile center of the whole planetary system. Hence in truth the Copernican system is proved *a priori*. For if in any position of the planets their common center of gravity is computed, this either falls in the body of the Sun or will always be close to it. (Smith 1999a, 46)

The laws referred to here in this preliminary tract are:

> Law 3. The relative motions of bodies contained in a given space are the same whether that space is at rest or whether it moves perpetually and uniformly in a straight line without circular motion. (Smith 1999a, 45)
>
> Law 4. The common center of gravity does not alter its state of motion or rest through the mutual actions of bodies. (Ibid.)

They follow from corollaries 4 and 5 of the formulation of the Laws of Motion given in the *Principia*.

The development and applications of perturbation theory, from Newton through Laplace at the turn of the nineteenth century and on through much of the work of Simon Newcomb at the turn of the twentieth, led to successive, increasingly accurate corrections of Keplerian planetary orbital motions.[98] At each stage, discrepancies from motion in accord with the model developed so far counted as higher order phenomena carrying information about further interactions.[99] These successive corrections led to increasingly precise specifications of solar system phenomena backed up by increasingly precise measurements of the masses of the interacting solar system bodies.

4. *A contrast with Laplace*

Laplace followed up his successful treatment of the long-recalcitrant great inequality in Jupiter-Saturn motions as a periodic perturbation with influential arguments for the stability of the solar system.[100] After the publication of his monumental treatises on celestial mechanics there came to be very general acceptance of the solar system as what was taken to be a Newtonian metaphysics of a clockwork deterministic system of bodies interacting under forces according to laws. The successively more accurate approximations were seen as successively better approximations to an exact characterization of the forces and motions of the bodies in a stable solar system in which all perturbations were periodic.

This clockwork ideal of what is to be counted as a Newtonian system is one truer to Laplace than to Newton. We shall see that Newton, himself, suggested that on his theory of gravitation one would expect non-periodic perturbations sufficient to threaten the stability of the solar system.[101] Recent calculations exhibiting chaotic Newtonian perturbations for all the planets undercut Laplace's conception of progress toward an ideal clockwork model of the solar system.[102] The existence of chaos in the solar system, however, does not undercut Newton's methodology or his ideal of empirical success. Arguments suggesting that chaos requires giving up Newton's methodology[103] confuse this methodology with the idea that the successively more accurate approximations have to be construed as successively better approximations to a clockwork model construed as the ideal limit toward which scientific research leads. The efforts to estimate sizes of chaotic zones in work on chaos, e.g. Laskar 1990, as well

[98] See chapter 10 note 8.

[99] See Smith G.E., forthcoming, for a wonderful account of the powerful evidence these developments provided for Newton's theory.

[100] In 1785 Laplace proved the immunity to non-periodic perturbations to the order of the first powers of the perturbing masses. In 1808 his protégé Poisson was able to extend this to the second order of the perturbing masses. See Morando 1995, 140. By 1899 the eminent mathematician Henri Poincaré (1854–1912) demonstrated that solar system stability could not be shown when he proved that the required series were unavoidably divergent. See Laskar 1995, 245–7.

[101] See chapter 10 section III.

[102] See Sussman and Wisdom, 1992. For an especially accessible general account see Ivars Peterson 1993.

[103] See, e.g., Ekeland 1990.

as Sussman and Wisdom 1992, are very much in line with what I am suggesting is Newton's methodology.

Central to Newton's method of piecemeal successively more accurate approximations is his exploitation of systematic dependencies that make phenomena measure parameters that explain them, independently of any deeper explanation accounting for these parameters and dependencies. These lower level dependency-based explanations are robust not only with respect to approximations in the phenomena. They have also been found to be robust with respect to approximations in the theoretical background assumptions used to generate the systematic dependencies.[104]

The large-scale cumulative improvement in accuracy of phenomena and the systematic dependencies that explain them at this lower level have important implications for understanding scientific progress. Some of these implications will be explored in our concluding chapter. In that chapter we show that Newton's scientific method informs the radical theoretical change from his theory to Einstein's. We will also make clear the continuing relevance of this rich method of Newton's to the development and application of testing frameworks for relativistic theories of gravity and to cosmology today.

5. Security through strength: acceptance vs. assigning high probability[105]

In his debates with scientific realists, Bas van Fraassen has often appealed to the idea that strength and security are conflicting virtues which must be traded off one against the other.[106] In limiting his commitment to only the empirical adequacy of a theory, van Fraassen claims to be simply more cautious than his realist opponents. As he delighted in pointing out, a theory T cannot be more probable than its empirical consequences E. Probability is monotone with entailment, so $P(T) \leq P(E)$ if T entails E. If security were measured by any function which, like probability, is monotone with entailment, then it would seem that the trade-off between strength and security van Fraassen appeals to would be unavoidable.

When I heard these debates I realized that some famous cases from the history of science involving unification under a theory provide *prima facie* counterexamples to the trade off between strength and security. Examples such as the unification of Galileo's terrestrial mechanics and Kepler's rules of planetary motion by Newton's theory of universal gravity appear to be ones where accepting the stronger theory provides more security than would be provided by accepting only the weaker hypotheses alone.

Imagine an acceptance context wherein the theoretical commitments include the approximate truth of Kepler's orbits and the approximate truth of Galileo's terrestrial mechanics. If we choose the approximations in a reasonable way, these commitments

[104] See Smith G.E., forthcoming.

[105] This material is from section 1 of my 1989 paper "Consilience and natural kind reasoning."

[106] See, e.g., van Fraassen 1983, 165–8; 1985, 247, 280–1, 294–5.

will be entailed by Newton's theory. Now consider a rival planetary hypothesis. Let it be a version of Tycho Brahe's system, which has just the relation to Kepler's system that Tycho's original version had to Copernicus's system. Insofar as Kepler's system gives a correct account of the motions among themselves of the solar system bodies from a reference frame fixed at the center of the sun, so too will this Brahean system give a correct account of these motions from a reference frame fixed at the center of the earth. Now consider some data – the absence of stellar parallax. This data might be counted as evidence favoring the Brahean account. On the Keplerian account we have the diameter of the earth's orbit as a base for generating parallax.

Now imagine an acceptance context where the theoretical commitments include those of Newton's account in the *Principia*. This is a context in which the theoretical commitments are stronger. The original commitments are entailed by this much richer theory. Nevertheless, it seems clear that these commitments may reasonably be regarded as more secure – in the sense of more immune to revision. Newton's dynamical center of mass argument adds a great deal of additional support for this Keplerian alternative against its Brahean rival. Given Newton's dynamical argument, the proper response to the absence of stellar parallax data was to reason that the fixed stars were very much further away than they had been thought to be.

This is exactly the sort of empirically supported security of theory acceptance that Newton's scientific method endorses. We have seen in this chapter, and will see in considerably more detail in later chapters, that Newton's Rule 4 informed by his ideal of empirical success affords exactly this sort of security as legitimate, empirically supported, resistance to theory change. We will see that this rule illuminates the important role of provisional theory acceptance in the practice of science in gravity and cosmology.

2

Newton's Phenomena

This chapter reviews the phenomena Newton appealed to in his argument in book 3 of the third edition of his *Principia.* Section I is devoted to Jupiter's moons. Section I.1 introduces Römer's measurement of a finite speed of light from timing of eclipses of a moon of Jupiter. Appendix 1 gives details of Römer's calculation. Section I.2 reviews Newton's phenomenon 1 and the data he cites in support of the Area Rule and Harmonic Rule for those moons of Jupiter. Section II introduces Newton's phenomenon 2 and the data he cites in support of the Area and Harmonic Rules for the satellites of Saturn. Appendix 2 gives a more detailed assessment of the satellite data cited by Newton, including the very precise data generated by James Pound and his nephew James Bradley using a 123-foot focal length Huygens telescope.

Sections III–V are devoted to Newton's phenomenon 3, that the orbits of the planets encompass the sun; phenomenon 4, Kepler's Harmonic Rule for the planets; and phenomenon 5, Kepler's Area Rule. Appendix 3 gives details of Kepler's determinations of periods and mean-distances, as well as Kepler's equation and Area Rule motion in his elliptical orbit. Kepler's determination of the period and mean-distance of the orbit of Mars exhibits what is involved in estimating these orbital parameters from observations of geocentric angular positions recorded by Tycho.

Phenomena are not just data. They are patterns exhibited in open-ended bodies of data. In our final chapter we will see such a pattern exhibited in data from the radar time delay test for General Relativity developed by Irwin Shapiro.[1] The phenomena Newton gives at the beginning of his argument in book 3 are relative motions of orbiting bodies with respect to centers of bodies about which they orbit. He argues for these phenomena from sets of explicitly cited data. In the subsequent argument for universal gravity, motions in accord with these patterns are assumed as initial premises. The propositions inferred from them are features of centripetal forces which they are shown to measure. For Newton, theoretical considerations establish a pattern as a phenomenon worth singling out.

[1] See chapter 10 section II.3.

Let us turn now to Newton's phenomena. In the first edition he called them hypotheses.[2] In the second edition they were promoted to the status of phenomena and continued to be assigned that status in the third edition.

I The moons of Jupiter

Newton begins with the moons of Jupiter. The four known in Newton's day were discovered in 1610 by Galileo, in the famous telescope observations reported in his *Starry Messenger.*[3] Galileo's observations were corroborated by Jesuit astronomers of the Collegio Romano. Figure 2.1 is a record of their observations between November 28, 1610 and April 6, 1611.[4]

The sequence of positions of the small dots aligned with the circle representing Jupiter gave good evidence that they represented views of moons in orbit about Jupiter. A more detailed analysis of such telescope observations led to evidence that these four moons were moving in very nearly uniform concentric orbits with periods and radii that could be determined with fair accuracy.[5]

1. Römer on the speed of light

Ole Römer (1644–1710) was responsible for a striking early development involving Jupiter's moons.[6] He argued that light has a finite speed that can be measured by variation in the timing of observations of eclipses of Jupiter's inner moon Io, depending on the changing distances of the earth from Jupiter. This episode can help us appreciate some of the differences between phenomena and data and something about the sort of reasoning appropriate to getting from data to phenomena.

The orbits of Jupiter's moons have very small inclination to the orbit of Jupiter. This facilitates a large number of eclipses that can be observed by astronomers on earth. The first moon, Io, has a period of about 42½ hours and often affords long sequences with eclipses on every orbit. When Cassini[7] developed accurate tables for the eclipses of Jupiter's moons he noted small irregularities in the timings of these eclipses. These were

[2] Phenomenon 1 was hypothesis 5 in the first edition. It became phenomenon 1 in the second edition. Phenomenon 2 was first introduced in that edition. Hypotheses 6–9 of the first edition became phenomena 3–6 of the second and third editions. For the specific variations introduced in these editions see Koyré and Cohen 1972, 556–63. See also Cohen 1971 and C&W, 794 and 198–201.

[3] See Van Helden 1989 for a translation of Galileo's famous book with very useful introductory, and concluding, notes on its reception and significance.

[4] These are from Galileo's copies of their record (Lattis 1994, 187–9). The photograph is from Lattis 1994, 189.

[5] *GHA 2A*, chapter 9.

[6] See appendix 1 for more details.

[7] Cassini, Giovanni Domenico (1625–1712) was the first of four generations of Cassinis who were prominent in directing the work of the Paris observatory from 1669 until the French Revolution. He published the first tables for eclipses of Jupiter's moons in 1668 and more accurate tables in 1693. As we shall see, he discovered four of the five satellites of Saturn that Newton refers to in phenomenon 2, as well as providing the data on periods and distances cited there.

Figure 2.1 Telescope observations of Jupiter's moons
These are from Galileo's copy of a record of telescope observations by Jesuit astronomers of the Collegio Romano Corroborating his discovery of four moons of Jupiter

correlated with the distance between the earth and Jupiter. In his 1676 paper, Römer argued that increases of the distance between the earth and Jupiter, as the earth moved away after overtaking Jupiter, would lead to time delays due to the extra time it took light to cross the extra distance.[8] He similarly argued that decreases in distance, as the earth moved closer to Jupiter while overtaking it, would lead to observations earlier than expected.[9]

Römer's proposal maintains the uniformity of periods of revolution of Io about Jupiter. It does so by allowing differing times for light to cross differing distances from Jupiter to the earth to explain time variations in the data about when eclipses are observed. The main generalization that is maintained is about the motion of the moon about Jupiter, even though it is not a regularity found precisely in the data. The irregularities in the data are, themselves, a phenomenon which correlates variations from uniform time differences between eclipse observations with variation in the distance between Jupiter and the earth. On Römer's proposal, this secondary phenomenon measures the speed of light.

2. *Newton's phenomenon 1*

Let us turn now to what Newton counts as the phenomenon exhibited by the moons of Jupiter. Here is Newton's third edition statement of his first phenomenon.

Phenomenon 1

The circumjovial planets [or satellites of Jupiter], by radii drawn to the center of Jupiter, describe areas proportional to the times, and their periodic times – the fixed stars being at rest – are as the 3/2 powers of their distances from that center.

This is established from astronomical observations. The orbits of these planets do not differ sensibly from circles concentric with Jupiter, and their motions in these circles are found to be uniform. (C&W, 797)

Newton reports the orbits of these moons to differ insensibly from uniform motion on concentric circular orbits. Such orbits would satisfy the Area Rule with respect to radii from the center of Jupiter. The eccentricities of the four moons are all so small that their motions are quite close to concentric circles.[10] Römer's equation of light supported attributions of fairly precise tolerances to which the motions of these

[8] In September 1676 Römer predicted that the eclipse scheduled for the night of November 9 would be ten minutes late. His paper was delivered to the French Royal Academy of Science on November 22. A summary was published in *Journal des Scavans* of December 7, 1676. A facsimile reproduction of it is reproduced in I.B. Cohen's classic article (see Cohen 1940).

[9] See appendix 1 for a calculation Cohen uses to illustrate Römer's reasoning, and his estimate of 22 minutes as the time it would take light to cross the orbit of the earth.

[10] See *ESAA*, 708. According to *ESAA* 1992 the mean values of these eccentricities are now 1 (Io) 0.004, 2 (Europa) 0.009, 3 (Ganymede) 0.002, and 4 (Callisto) 0.007.

One might wonder whether these orbits may have been more eccentric earlier. A rough check provided in early 1996 by Myles Standish, of the Jet Propulsion Laboratory, using the J.P.L. ephemeris program, revealed that the mean eccentricities of these moons for time periods around 1719, when Pound's measurements were made, were not greater than those cited.

Galilean satellites of Jupiter could be estimated to approximate uniform motions on concentric circles.

Newton goes on to discuss evidence that these moons satisfy Kepler's Harmonic Rule – that the periods are as the 3/2 power of the semi-diameters of their orbits. These Galilean satellites of Jupiter are in order 1. Io, 2. Europa, 3. Ganymede, and 4. Callisto.

Astronomers agree that their periodic times are as the 3/2 power of the semidiameters of their orbits, and this is manifest from the following table.

[Table 2.1]

Periodic times of the satellites of Jupiter				
	$1^d18^h27'34''$	$3^d13^h13'42''$	$7^d3^h42'36''$	$16^d16^h32'9''$
Distances of the satellites from the center of Jupiter in semidiameters of Jupiter				
From the observations of	1	2	3	4
Borelli	5⅔	8⅔	14	24⅔
Towneley (by micrometer)	5.52	8.78	13.47	24.72
Cassini (by telescope)	5	8	13	23
Cassini (by eclips. satell.)	5⅔	9	14 23/60	25 3/10
From the periodic times	5.667	9.017	14.384	25.299

(C&W, 797)

Newton illustrates the fit of the Harmonic Rule to these data by offering distances calculated from the periods in accordance with the Harmonic Rule, for comparison with the astronomers' distances estimated from observations.[11] He represents the fit of the Harmonic Rule to the data in his table by simply exhibiting these computed distances along with the empirically estimated distances in his table.

As we noted in our discussion of Kepler, one quite informative way of representing the fit of the Harmonic Rule to these cited estimates of distances and periods is by plotting log periods against log distances. Unlike appeal to least-squares, this way of representing the fit of the Harmonic Rule to the data would have been accessible to Newton. Napier's publication of his system of logarithms in 1614 had led to widespread interest in and further developments of logarithmic representations of numerical relationships.[12]

The straight line in Figure 2.2 is of slope 3/2. This plot exhibits graphically the pattern in these data corresponding to the Harmonic Rule phenomenon for Jupiter's moons.

How accurate are these data? A comparison with modern values suggests that the periods are very accurate indeed, while the cited distance estimates are considerably less accurate. See appendix 2 section 1 for an assessment of this accuracy and a numerical

[11] See appendix 2. [12] See Boyer 1968, 344.

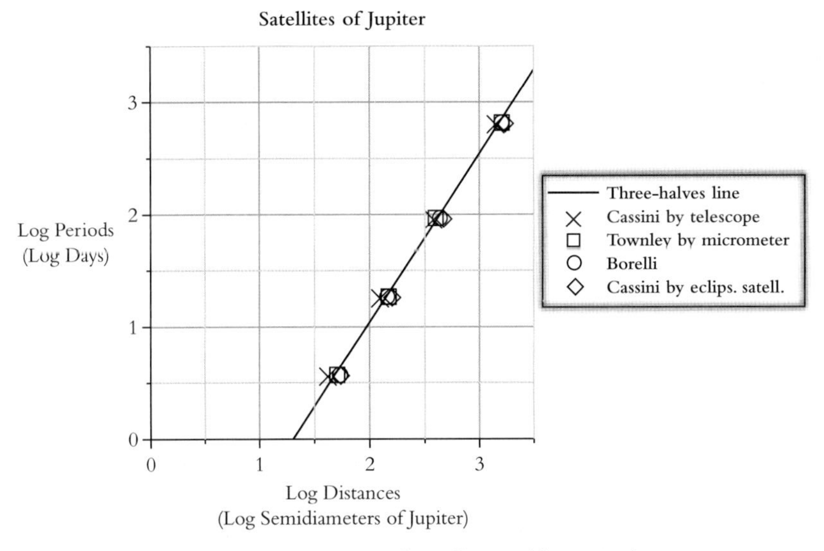

Figure 2.2 The Harmonic Rule as illustrated by Jupiter's moons

least-squares assessment of the fit of the Harmonic Rule we have exhibited graphically in the diagram.

In the third edition Newton added a discussion of data from very accurate observations made by Pound in a research initiative sponsored by Newton in 1718–1720. Comparison with estimates available today suggests that Pound's distance estimates are very much more accurate than any of the other cited data. It is very likely that Pound's observations were the most precise in the world at the time they were made. For an account of this very interesting research initiative and Newton's discussion of its results see appendix 2 section 3.

II Satellites of Saturn

In the second edition, which appeared in 1713, Newton added the Area Rule and Harmonic Rule for the satellites of Saturn as phenomenon 2. In both this second and the third edition, of 1726, Newton listed data for five satellites of Saturn.[13] Here is Newton's phenomenon 2 as it appears in the third edition.

Phenomenon 2

The circumsaturnian planets [or satellites of Saturn], by radii drawn to the center of Saturn, describe areas proportional to the times, and their periodic times – the fixed stars being at rest – are as the 3/2 powers of their distances from that center. (C&W, 798)

[13] These had been discovered by Huygens and Cassini. Huygens discovered Titan, the fourth listed by Newton, in 1655. Cassini discovered the fifth listed, Iapetus, in 1671, and the third listed, Rhea, in 1672. In 1684 he also discovered Tethys and Dione, the first and second listed by Newton (*GHA 2B*, 261).

Newton offers no specific evidence for the constant rate of areas swept out by radii from the center of Saturn. Saturn's satellites have quite small eccentricities. Their orbits approximate uniform motion on concentric circles.[14] Orbits of Saturn's moons were not known to exhibit irregularities until quite late.[15]

Newton cites Cassini as having determined the periodic times and distances for these satellites of Saturn.

Cassini, in fact, from his own observations has established their distances from the center of Saturn and their periodic times as follows.

[Table 2.2]

	The periodic times of the satellites of Saturn				
	$1^d21^h18'27''$	$2^d17^h41'22''$	$4^d12^h25'12''$	$15^d22^h41'14''$	$79^d7^h48'0''$
	The distances from the center of Saturn, in semidiameters of the ring				
From the observations	$1^{19}/_{20}$	2½	3½	8	24
From the periodic times	1.93	2.47	3.45	8	23.35

(C&W, 798–9)

Here, as in the table for Jupiter's moons, Newton exhibits the fit of the Harmonic Rule by putting distances computed from the periods in accordance with the Harmonic Rule together with the distances estimated from observations.[16]

As in the case of Jupiter's moons, we can exhibit the fit of the Harmonic Rule to these data by plotting log periods against log distances. Figure 2.3 gives the results from Cassini's data for Saturn's moons.

Here also the very close fit of a straight line with slope $3/2 = 1.5$ exhibits the excellent fit of the Harmonic Rule to these data.

The small differences between these computed distances and the distances from Cassini's observations also suggest that the Harmonic Rule fits these data quite well. See appendix 2 for discussion of this fit and of the accuracy of Newton's cited data. In the third edition Newton discussed correcting overestimating diameters from dilation of light and provided new estimates from Pound. Comparison with modern values reveals that Pound's estimates are considerably more accurate than those cited from Cassini.[17]

[14] The *ESAA*, 706 gives a small mean eccentricity of 0.0291 for Titan and a similarly small 0.028 for Iapetus. For Dione and Rhea *ESAA* gives very much smaller mean eccentricities of respectively 0.0022 and 0.0010. The inner satellite Tethys has a strikingly concentric circular orbit with mean eccentricity set at 0.00000, cited as zero to five decimal places.

[15] *GHA 2B*, 266. Even though Titan is a large bright object, it was not until work by Bessel in 1829 that perturbations for it were accurately determined (see *GHA 2B*, 266).

[16] Newton uses Titan's orbit to calculate the Harmonic Rule distances for the orbits of the other satellites. See appendix 2, section 4.

[17] See appendix 2, section 4.

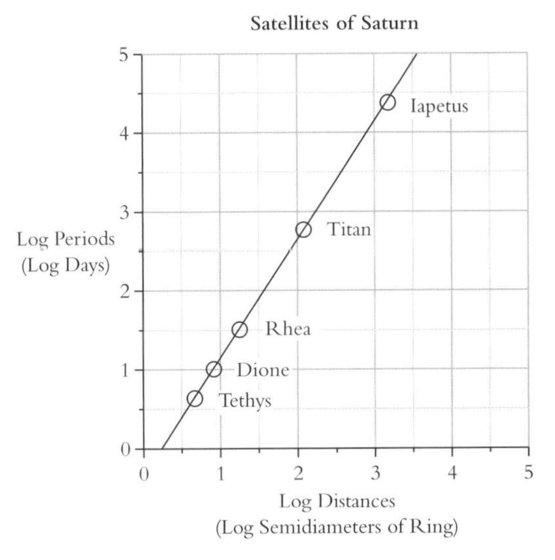

Figure 2.3 The Harmonic Rule as illustrated by Saturn's moons

III The orbits of the primary planets encompass the sun

Galileo's telescope observations of phases of Venus, announced in semi-public correspondence in late 1610 and in print in Kepler's *Dioptrice* in 1611, gave strong evidence that the orbit of Venus encompasses the sun.[18] Figure 2.4 nicely illustrates this.

In the Ptolemaic system, Venus, which remains below the sun, should not show a complete system of phases.[19] A hybrid Ptolemaic system in which Mercury and Venus orbit the sun while the other planets, as well as the sun, orbit the earth would also be compatible with Galileo's observations.[20] Nevertheless, the Ptolemaic system soon became obsolete among practicing astronomers.[21]

[18] The distinguished Jesuit mathematical astronomer Clavius and his colleagues at the Collegio Romano began observations soon after they received an adequate telescope (Lattis 1994, 186–95). Their report of their corroboration of Galileo's observations in response to Cardinal Bellarmine's request, in March 1611, for an assessment of Galileo's claims was probably more important to Galileo than Kepler's endorsement. Here is the part devoted to Venus.

> . . . it is very true that Venus wanes and waxes like the Moon. And having seen her almost full when she was an evening star, we have observed that the illuminated part which was always turned toward the Sun, decreased little by little, becoming ever more horned. And observing her then as a morning star, after conjunction with the Sun, we saw her horned with the illuminated part toward the Sun. And now the illuminated part continuously increases, according to the light, while the apparent diameter decreases. (Van Helden 1989, 111)

[19] These observations of Galileo realize a suggestion made by Copernicus that Venus ought to exhibit phases like the moon if it shines with borrowed light. (Kuhn 1957, 222–4; Van Helden 1989, 106).

[20] See Van Helden 1989, 109.

[21] See Van Helden in *GHA 2A*, 84.

In addition to his corroboration of Galileo's observations in his *Dioptrice* of 1611, Kepler had earlier provided details of mathematical astronomy that provided evidence against Ptolemy's system. In his *New Astronomy*, published in 1609, Kepler had used his discovery that solar theory required non-uniform motion on its eccentric to argue against the Ptolemaic system. Kepler's careful measurements, calculated from

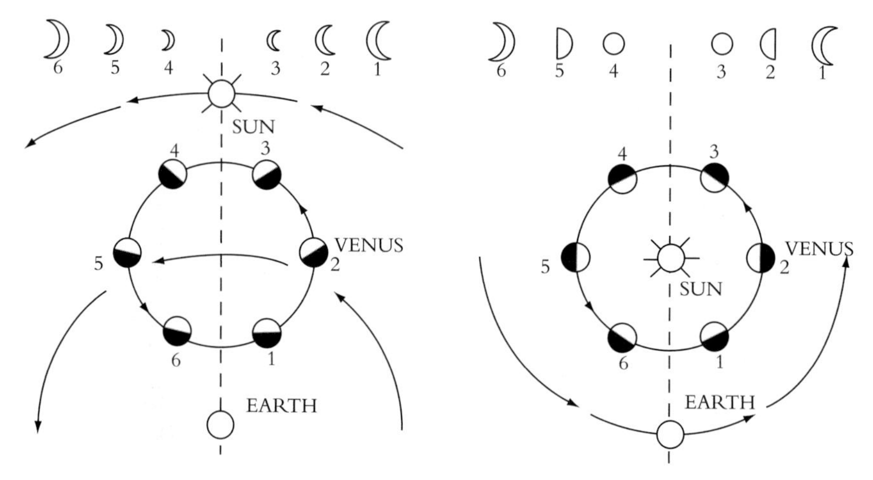

Figure 2.4 Phases of Venus as evidence against Ptolemy
The Ptolemaic system on the left does not predict a full set of phases, while the Copernican system on the right does.[22]

As Kepler made clear in a later account,[23] Galileo's observation of phases of Venus did not count against Tycho Brahe's geoheliocentric system.[24]

As we have noted in chapter 1, in Tycho's system the other five primary planets all orbit the sun while the sun orbits the earth. All relative motions among solar system bodies in any sun-centered system, such as that of Copernicus or that of Kepler, will be preserved in a corresponding Tychonic system.[25]

Tycho's data, showed that solar theory required non-uniform motion on an eccentric, of the sort Ptolemy modeled for planets with an equant (center of uniform motion) equally far beyond the center of the eccentric as the center was beyond the earth (bisected eccentricity). (See chpt. 1 sec. I)

After empirically demonstrating the necessity of non-uniform motion corresponding to a bisected eccentricity in solar theory, Kepler argued that trying to introduce duplicates of this more complex motion as separate additions to the motions for each planet would make a Ptolemaic system so very much more complex that it no longer could be regarded as a serious rival to Copernican and Tychonic systems. (Kepler 1992, 337)

[22] This diagram is from p. 108 in the concluding part of Albert Van Helden's (1989) discussion of the reception of Galileo's *Sidereal Messenger*.

[23] See Caspar [1948] 1993, p. 202.

[24] In the generation after Clavius, variations of the Tychonic system became the favored cosmology among Jesuit astronomers (Lattis, 1994, 206). Many of these combined it with a fluid heavens hypothesis, according to which the planets were carried about in spiral currents (Lattis, 211–16).

As the spheres of Mars and the sun intersect in the Tychonic system, it is incompatible with the traditional physics according to which the planets are carried around on solid crystalline spheres. The traditional physics of crystalline spheres had been in decline since the comet of 1577, which (as triangulations by Tycho and others indicated) had passed through locations where spheres should have been (*GHA 2A*, 23–32).

[25] As its inability to account for phases of Venus illustrates, the Ptolemiac system is not kinematically equivalent to a Copernican system, even though it may preserve the same geocentric angular position observations of solar system bodies.

As we pointed out in chapter 1, even though Tychonic and Copernican systems are kinematically equivalent with respect to relative motions among solar system bodies, they differ in motions relative to the stars. Absence of any success in observing stellar parallax was among the grounds convincing Tycho that the earth did not move about the sun with respect to the stars. (*GHA 2A*, 8, 9)

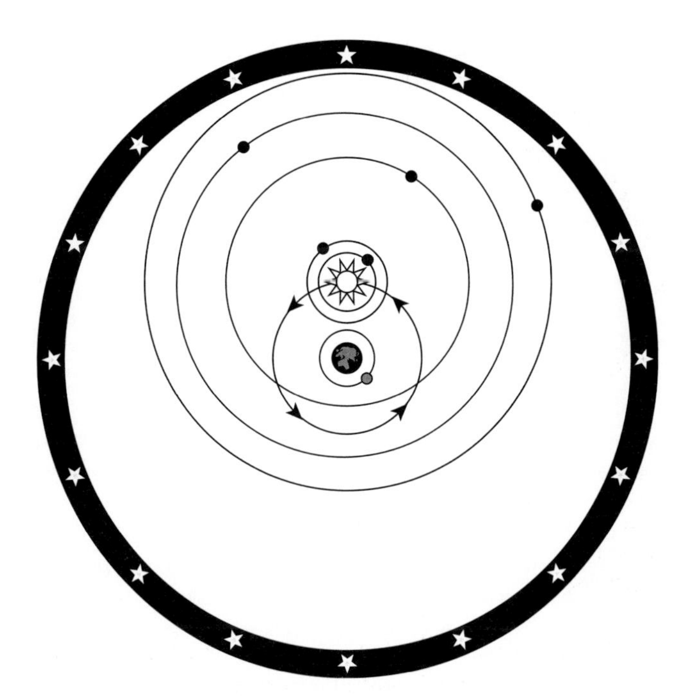

Figure 2.5 The Tychonic system
In this system, Venus moves on the second small circle around the sun; so, it does predict that Venus should show a complete set of phases.

Here is Newton's statement of phenomenon 3 and of the empirical evidence he cites for it.

Phenomenon 3

The orbits of the five primary planets – Mercury, Venus, Mars, Jupiter, and Saturn – encircle the sun.

That Mercury and Venus revolve about the sun is proved by their exhibiting phases like the moon's. When these planets are shining with a full face, they are situated beyond the sun; when half full, to one side of the sun; when horned, on this side of the sun; and they sometimes pass across the sun's disk like spots. Because Mars also shows a full face when near conjunction with the sun, and appears gibbous in the quadratures, it is certain that Mars goes around the sun. The same thing is proved also with respect to Jupiter and Saturn from their phases being always full; and in these two planets, it is manifest from the shadows that their satellites project upon them that they shine with light borrowed from the sun. (C&W, 799)

It is worth remarking that Newton does not include the earth as a planet orbiting the sun here.[26] In this phenomenon, we have appeal to Galileo's famous phases, as well as

Tycho's system also did not include a rotating earth. On it, just as for Ptolemy, the sphere of stars rotated about the fixed earth (*GHA 2A*, 8 also, 33–43).

[26] This avoids begging the question in his celebrated solution to the two chief world systems problem.

to additional data relevant to the outer planets.[27] Having all five of the other planets orbit the sun leaves the Tychonic and Copernican systems as the viable alternatives.

IV Kepler's Harmonic Rule

The next phenomenon, Kepler's Harmonic Rule for the primary planets, is explicitly formulated so as to be neutral between Tychonic and Copernican versions.

Phenomenon 4

The periodic times of the five primary planets and of either the sun about the earth or the earth about the sun – the fixed stars being at rest – are as the 3/2 power of their mean distances from the sun.

This proportion, which was found by Kepler, is accepted by everyone. In fact, the periodic times are the same, and the dimensions of the orbits are the same, whether the sun revolves about the earth, or the earth about the sun. There is universal agreement among astronomers concerning the measure of the periodic times. But of all astronomers, Kepler and Boulliau have determined the magnitudes of the orbits from observations with the most diligence; and the mean distances that correspond to the periodic times as computed from the above proportion do not differ sensibly from the distances that these two astronomers found [from observations], and for the most part lie between their respective values, as may be seen in the following table.

[Table 2.3]

Periodic times of the planets and of the earth about the sun with respect to the fixed stars, in days and decimal parts of a day						
	Saturn	Jupiter	Mars	Earth	Venus	Mercury
	10759.275	4332.514	686.9785	365.2565	224.6176	87.9692
Mean-distances of the planets and of the earth from the sun						
According to						
Kepler	951000	519650	152350	100000	72400	38806
Boulliau	954198	522520	152350	100000	72398	38585
Periodic times	954006	520096	152369	100000	72333	38710

(C&W, 800)

As we did for Jupiter's moons, we can informatively exhibit the fit of the Harmonic Rule to this data by plotting log periods against log distances. In these plots we use as our temporal unit the period of the earth's orbit, which Newton sets at 365.2565 decimal days, and as our distance unit, AU, the mean earth–sun distance, which Newton sets at 100000.

Newton's discussion of phenomenon 4 concludes with the following comment on the empirical support for the distance data in his table.

[27] See Densmore (1995, 259–68) for an excellent account of reasoning by which the data cited by Newton supports this phenomenon.

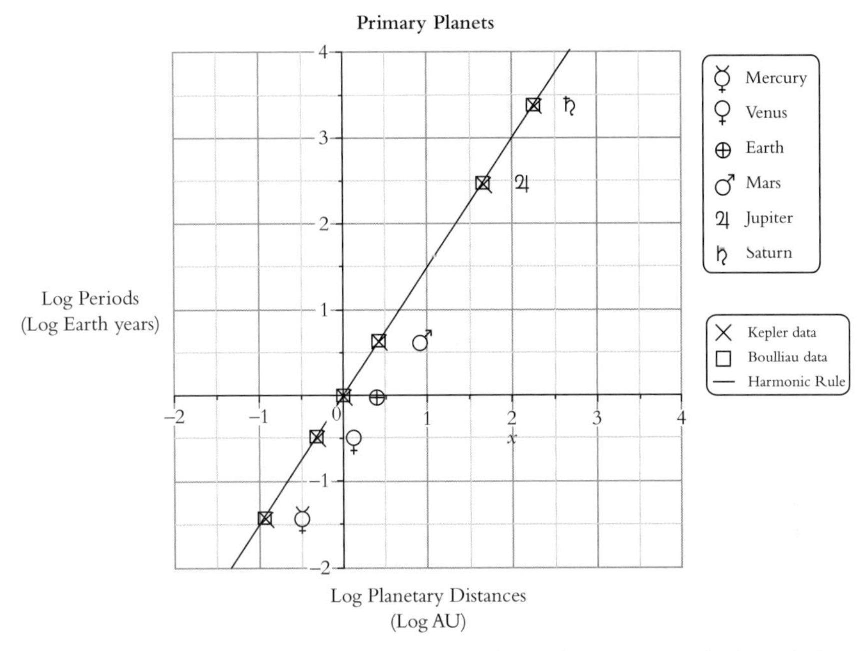

Figure 2.6 The Harmonic Rule as illustrated by satellites of the Sun using both Kepler's and Boulliau's data

There is no ground for dispute about the distances of Mercury and Venus from the sun, since these distances are determined by the elongations of the planets from the sun. Furthermore, with respect to the distances of the superior planets from the sun, any ground for dispute is eliminated by the eclipses of the satellites of Jupiter. For by these eclipses the position of the shadow that Jupiter projects is determined, and this gives the heliocentric longitude of Jupiter. And from a comparison of the heliocentric and geocentric longitudes, the distance of Jupiter is determined. (C&W, 800)

He suggests that elongations (angular separations) from the sun give especially good access to the distances for the inner planets. Given accurate tables for the sun–earth distance and direction, maximum elongations from the sun provide a resource for distance triangulations that do not depend on tables for the heliocentric longitude of an inner planet.[28]

Newton also suggest that eclipses of Jupiter's moons provide especially good resources for determining its distances. The eclipses of Jupiter's moons provide a comparably direct access to heliocentric longitudes of Jupiter as are provided for the inner planets by maximum elongations.[29] They are, however, much more frequently available since eclipses of Jupiter's moons occur so much more often than maximum elongations of inner planets. Given the heliocentric longitude of Jupiter at the time it is observed, the observed geocentric longitude and tables for the heliocentric direction

[28] See Densmore 1995, 271–4.

[29] See Densmore 1995, 275–7 for a very nice account with diagrams.

and distance of the earth allow the heliocentric distance of Jupiter to be triangulated without any more specific assumptions about its orbital motion.

V The Area Rule for the primary planets

Newton has not included Kepler's elliptical orbit among what he counts as phenomena. One reason for his not counting it as one of his assumed phenomena is that this would undercut the significance of his dynamical solution to the two chief world systems problem by building a heliocentric commitment into his starting assumptions.[30] The Area Rule, however, is included. This may seem somewhat surprising given that the ellipse was far more widely accepted among astronomers than the Area Rule. Newton's early readings in astronomy, Street's *Astronomia Carolina* (1661), and Wing's *Astronomia Britannica* followed Boulliau in using elliptical orbits with alternative procedures to Kepler's Area Rule for giving equations for determining motion on the ellipse.[31] Mercator's *Hypothesis astronomica nova* of 1664 also provided an ellipse with an alternative to the Area Rule.[32] The Area Rule was far less generally accepted than the elliptical orbit by astronomers. In 1670, however, Mercator argued that only procedures that closely approximated the Area Rule could be empirically accurate.[33]

The Area Rule is an example where theory plays an important role in making a phenomenon worth looking for. Newton's acceptance of the Laws of Motion and propositions derived from them as background assumptions allows him to argue that the uniform description of areas with respect to radii from a center carries the information that the force deflecting a moon or planet from inertial motion is directed toward that center.

The centripetal force to which Area Rule motion testifies provides additional warrant for expecting the Area Rule to continue to fit additional data as it comes in. Newton's theorems expounding upon the dynamical significance of the Area Rule add to the warrant for regarding such motion as a phenomenon continuing to fit an open-ended body of data, by showing how the data support measurements of a dynamical cause that can be expected to make the motion continue to describe approximately uniform areas about the center toward which it is directed.

Here is Newton's third edition statement of phenomenon 5 and the evidence he adduces for it.

Phenomenon 5

The primary planets, by radii drawn to the earth, describe areas in no way proportional to the times, but, by radii drawn to the sun, traverse areas proportional to the times.

For with respect to the earth they sometimes have a progressive [direct or forward] motion, they sometimes are stationary, and sometimes they even have a retrograde motion; but with respect to the sun they move always forward, and they do so with a motion that is almost uniform – but,

[30] As we shall see in chapter 3 section IV.7, George Smith has offered another reason that is informative for our understanding of the evidential implications of Newton's methodology.

[31] Wilson 1989b, *GHA 2A*, 172–9. [32] *GHA 2A*, 181–2. [33] *GHA 2A*, 182–3.

nevertheless, a little more swiftly in their perihelia and more slowly in their aphelia, in such a way that the description of areas is uniform. This is a proposition very well known to astronomers and is especially provable in the case of Jupiter by the eclipses of its satellites; by means of these eclipses we have said that the heliocentric longitudes of this planet and its distances from the sun are determined. (C&W, 801)

The first part of this evidence displays exactly the sort of investigation suggested by the scholium to proposition 3 book 1.

Scholium (prop. 3, book 1). Since the uniform description of areas indicates the center toward which that force is directed by which a body is most affected and by which it is drawn away from rectilinear motion and kept in orbit, why should we not in what follows use uniform description of areas as a criterion for a center about which all orbital motion takes place in free spaces? (C&W, 449)

We look for a center with respect to which the Area Rule holds.

The earth is not even approximately such a center, as is testified to by the well-known phenomena of stationary points and retrograde motions. As we have seen, Kepler's "Pretzel" diagram illustrates quite graphically how very far motion of Mars with respect to the earth is from satisfying the Area Rule.[34]

Kepler is quite clear that the relevant comparison in complexity of heliocentric and geocentric systems is the path or orbit they attribute to a planet. This is the complexity that must be accommodated in any account of physical forces corresponding to planetary motions.[35] Kepler's diagram illustrates that the complexity that would have to be accommodated to treat geocentric relative motions as "true" motions is overwhelming indeed.

As Newton points out, motion with respect to the sun very clearly approximates the Area Rule. The departures from uniform motion on concentric circles are of the right sort to fit the uniform description of areas – a little swifter in their perihelia and slower in their aphelia.

[34] Here is Kepler's introduction of this diagram:

If one were to put all this together, and were at the same time to believe that the sun really moves through the zodiac in the space of a year, as Ptolemy and Tycho Brahe believed, he would then have to grant that the circuits of the three superior planets through the ethereal space, composed as they are of several motions, are real spirals, not (as before) in the manner of balled up yarn, with spirals set side by side, but more like the shape of pretzels, as in the following diagram. (Kepler 1992, 118–19)

Here is a comment from Donahue on Kepler calling this shape as like that of a pretzel.

The Latin is *panis quadragesimalis*, that is, 'bread of the forty [days]', or lenten bread. Pretzels were invented by monks of southern Germany, who adopted the practice of giving them to children as treats during lent. (Kepler 1992, 119)

[35] Kepler's diagram illustrates the relevant complexity one should pay attention to in comparing earth-centered to sun-centered systems, because it gives what would count as the true motion of Mars that forces would have to account for if the earth were to count as at rest.

As we noted in chapter 1, this is considerably more to the point than counting the number of epicycles proposed by Copernicus vs. the number proposed by Ptolemy.

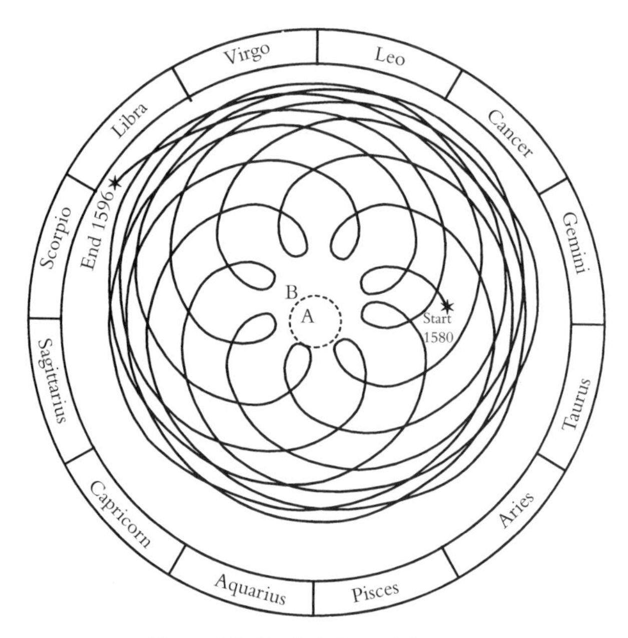

Figure 2.7 Kepler's Pretzel diagram

In addition to this sort of obvious gross approximation to Area Rule motion (and its clear contrast to the very gross divergences from the Area Rule by motion with respect to the earth), Newton suggests a more direct demonstration made available in the case of Jupiter by the eclipses of its moons. The eclipses of the moons of Jupiter offer a direct test of the Area Rule, without assuming the shape of the orbit. Each eclipse gives a heliocentric longitude that allows triangulation of its heliocentric distance from observations of its angular position with respect to the earth.[36]

VI The moon

Here is Newton's phenomenon 6. His remarks call attention to the fact that the moon's motion is perturbed by the sun.

Phenomenon 6

The moon, by a radius drawn to the center of the earth, describes areas proportional to the times.

This is evident from a comparison of the apparent motion of the moon with its apparent diameter. Actually, the motion of the moon is somewhat perturbed by the force of the sun, but in these phenomena I pay no attention to minute errors that are negligible. (C&W, 801)

[36] See Densmore 1995, 275–7.

Comparison of apparent motion with apparent diameter yields the sort of approximation to Area Rule motion that is exhibited by the primary planets. The apparent diameter measures the distance of the moon from an observer on the surface of the earth.[37] The apparent motion at apogee is slower than the apparent motion at perigee in roughly the way one would expect from the Area Rule.

Newton's remark about perturbations by the force of the sun refers to well-known inequalities in the moon's motion that had long complicated attempts by astronomers to construct lunar tables. In order to account for such inequalities, Kepler's lunar theory introduced modifications of orbital parameters over the course of its orbital motion that required some violations of the Area Rule.[38] Horrocks's initial lunar theory followed Kepler in allowing for violations of the Area Rule, but his final theory, an ellipse with varying apse and eccentricity, was explicitly designed to satisfy the uniform description of areas.[39] This lunar theory of Horrocks's was the best descriptive account of lunar motion before Newton.

Newton's own accounts of the lunar motion, which improved somewhat on the fit of Horrocks's theory to the data,[40] included small violations of the Area Rule.[41] His successful treatment of the variation, an inequality of the moon's mean motion discovered by Tycho, includes a quantitative account of these small violations of the Area Rule as perturbations due to the action of the sun.[42]

[37] As the moon moves from its greatest distance at apogee to its smallest distance at perigee, its apparent diameter varies from about 29½ minutes at apogee to about 33½ minutes at perigee. The tangent of half the apparent diameter is the ratio of the radius of the moon to its distance from the observer. The distance in moon radii is thus one over this tangent, which gives about 233.07 moon radii as the apogee distance and 205.24 moon radii as the perigee distance from an observer at the surface of the earth.

[38] See Wilson, "On the Origin of Horrocks's lunar theory" (1989a,VII, 90).

[39] The only deviations from the Area Rule are due to an approximation Horrocks introduces to simplify the calculation of the eccentric anomaly. (See appendix 3 for more on mean anomaly M and eccentric anomaly E in elliptical orbits.)

[40] See Kollerstrom 2000, 150–1. Kollerstrom has shown that Newton's corrected version of Horrocks's theory was almost twice as accurate as Flamsteed's version.

[41] See chapter 4 section V below. See also Newton's proposition 29 book 3 (C&W, 846–8) and Wilson 2001, 152.

[42] In chapter 4 section V, which is mostly devoted to the lunar precession problem, we briefly point out Newton's result for the action of the sun on the earth–moon system to make the lunar distance smaller in syzygies, when the sun and moon are lined up, and greater in the quadratures, when they are 90 degrees apart. This is an important part of his successful treatment of the variational inequality. In chapter 6 we shall see that Newton is able to apply it to correct the lunar distance estimates in the moon-test.

See Wilson 2001, 141–55 for a detailed account of Newton's treatment of solar tidal forces on the earth–moon system in proposition 26 of book 3 and for his account of the role played by small violations of the Area Rule in Newton's treatment of the variation of the moon in proposition 29 book 3.

Variation is the effect that, other things being equal, the moon maximally lags behind its mean position in the octants before syzygies and maximally exceeds its mean position in the octants after syzygies (Wilson 2001, 141). Newton derives $24'44''$ of this maximal difference from the flattened shape of the orbit and an additional $7'48''$ from violations of the areal rule resulting from the action of the sun treated in proposition 26 book 3 (Wilson 2001, 151–2). In addition to these corrections, Newton multiplies by the result (about 1.08) of dividing the synodical by the sidereal period to correct for the motion of the sun relative to the earth (Wilson 2001, 152). The resulting mean value of the maximum difference is $35'10''$. This is quite close to the modern value found by Hill of $35'6''$ (Wilson 2001, 153).

Chapter 2 Appendix 1: Römer on the speed of light: details and responses

One of the most striking developments involving Jupiter's moons was Römer's discovery that light has a finite speed. As we noted in chapter 2 above, the orbits of Jupiter's moons are very nearly uniform motion on concentric circles and have quite small inclinations to the orbit of Jupiter. This facilitates a large number of eclipses that can be observed by astronomers on earth. The first moon, Io, has a period of about 42½ hours and often affords long sequences with eclipses on every orbit.

These eclipses were used to compare longitudes on earth. An eclipse simultaneously observed at different locations – which could be anywhere on the same hemisphere exposed to Jupiter – could be used to compare local times. The differences in local time from noon would measure differences in longitude of 15 degrees per 1 hour time difference.[43] In 1671–2 Picard[44] and Cassini[45] organized an expedition of Picard to Denmark, to use comparisons between observations there with observations of the same eclipses in Paris, so that the longitude of Tycho Brahe's observatory at Uraniborg could be located with respect to that of the Paris observatory.[46] While in Denmark, Picard was assisted by Ole Römer who had been studying Tycho's observations. Römer made himself so valuable that Picard asked him to come to Paris. It was while in Paris that Römer obtained his famous result – that light has a finite speed.

I.B. Cohen uses two observations from Römer's notes in a calculation that illustrates Römer's reasoning, and his estimate of 22 minutes as the time it would take light to cross the orbit of the earth.[47] The first is an immersion of Io into the shadow of Jupiter on October 24 in 1671 at 18:15 local time. The second, a little over 79 days later as the earth has been approaching Jupiter, is an immersion of Io on January 12 1672 at 08:59:22 local time. The local times, apparent solar times, are

[43] Galileo proposed that eclipses of Jupiter's moons could be used to determine longitudes at sea. This would require very accurate eclipse tables, as well as accurate clocks and a telescope good enough to see an eclipse occur. Galileo began negotiations with the King of Spain for support to develop a method of using eclipses of Jupiter's moons to determine longitude at sea. These dragged on for two decades. Eventually, he started negotiations with the States General of Holland, which were still going on when he died in 1642 (*GHA 2A*, 144–5).

By 1657 Huygens had developed his pendulum clock (Yoder 1988, 2). This made it possible to determine the local times at which an eclipse was observed at two places sufficiently precisely to measure differences in longitude on land with enough precision to make the procedure worth carrying out (*GHA 2A*, 150–2). See chapter 5.

[44] Picard (1620–82) was a founding member of the French Royal Academy of Sciences in 1666. He became famous for the first accurate measurement of a degree of meridian. As we shall see in chapter 4 section II.1, the result of this measurement, which was published in 1671, gave Newton the data for the circumference of the earth that was used in his moon-test.

[45] See above chapter 2 note 7.

[46] A great many of Tycho Brahe's impressive body of data were collected at the excellent observing facilities he was able to construct at the research complex, Uraniborg, which he built on the island of Hven, located between Denmark and Sweden. See chapter 1 section I.1 for more on Tycho Brahe.

[47] See Cohen 1940, 350–3.

converted to mean solar times using the equation of time.[48] Cohen subtracts 15 minutes 45 seconds for the equation of time to convert the apparent solar time (18:15) of the observation on October 24, 1671 to 17:59:15 mean solar time and adds 9 minutes 23 seconds to convert the apparent solar time (08:59:22) of the observation of January 12, 1672 to 09:08:45 mean solar time. The time difference between these observations in mean solar time was 79 days 15 hours 9.5 minutes.

Römer had calculated the average observed period for Io to be 42.475 hours.[49] With this period there would be 45 revolutions of Io about Jupiter between the immersion of October 24, 1671 and that of January 12, 1672. If the transmission of light were instantaneous, as was then widely accepted,[50] the time of the immersion of January 12, 1672 would be 45 times 42.475 hours = 1911.375 hours or 79 days 15 hours 22.5 minutes later than the time of the immersion of October 24, 1671. The actual time difference was 79 days 15 hours 9.5 minutes, so the eclipse of January 12, 1672 was observed 13 minutes early.[51] On the calculation set out by Cohen, during the time between these two eclipses the distance of the earth from Jupiter decreased by 1.21 times the mean earth–sun distance.[52] According to Römer's proposal, the time decrease was due to the less time it took for light to cross this shorter distance.[53]

Cassini, who had proposed a variation in the motion of Io correlated with the distance from the earth, resisted Römer's proposal.[54] Cassini also objected that no similar time delays showed up in the eclipses of the other moons of Jupiter.[55] Cassini's followers continued to object even after the turn of the century.[56] Halley, commenting on Cassini's objection, challenged Cassini's introduction of an inequality in the motions of the third and fourth satellites and offered evidence from observations that Römer's equation of light is quite exactly fit by the third and within reasonable agreement with the fourth.[57] Halley also applied to Cassini the same sort of objection that adherents of the mechanical philosophy were applying to Newton. He suggested that it is

> ... hard to imagine how the Earth's Position in respect of *Jupiter* should any way affect the Motion of the *Satellites*. (Halley 1694, 239)

[48] Over the course of a year, the apparent solar time (given by the rotation of the earth relative to the sun) varies from local sidereal time (given by the rotation of the earth relative to the fixed stars). Mean solar time applies a correction (the equation of time) to convert apparent solar time to what it would be if solar days were uniform with respect to sidereal days over the sidereal year (*ESAA*, 74–5).

[49] Cohen (1940, 351) indicates that Römer's manuscript includes calculations giving mean periods for Io of 1d 18hr 28min 30sec for 1671–2 and of 1d 18hr 28 min 31 sec for 1672–3.

[50] See Cohen (1940, 328–38).

[51] As Römer pointed out, his ingenious calculation shows how to generate observable time differences, even when the speed of light is so great that the time separation corresponding to the difference in the earth's distance from Jupiter over a single period of Jupiter's moon would be too small to detect.

[52] Cohen 1940, 352.

[53] This would give 2(13/1.21), which is Römer's estimate of 22 minutes as the time for light to cross the orbit of the earth, which would make the time for light to reach the earth from the sun about 11 minutes. The correct value is about 8.3 minutes (Cohen, 358; *ESAA*, 696). The point of Römer's paper was that the speed of light is finite rather than infinite. He was not directly concerned with giving the most accurate achievable measurement of its speed. In his discussion of Römer's result, Halley (1694, pp. 237–56) gives a more correct figure of 8½ minutes.

[54] See *GHA 2A*, 152–4; Cohen 1940, 345–9.

[55] Cohen 1940, 349; *GHA 2A*, 154.

[56] Cohen 1940, 349.

[57] See Halley 1694, 254–6.

The difference from the case with gravity was that Halley's other remarks suggest that Römer had indeed provided a viable alternative phenomenon that fits the data, without any need for such a difficult-to-explain variation in the orbital motion of Jupiter's moons. Nevertheless, it was not until Bradley, in 1729, independently measured a finite velocity of light, in his discovery of aberration, that resistance to Römer's proposal finally died out.[58]

[58] See Cohen 1940, 357; *GHA 2A*, 156.

Chapter 2 Appendix 2: Newton's satellite data

1. The table for Jupiter's moons

According to the Harmonic Rule, $R^3/t^2 = k$, where k is a single constant giving the agreeing ratios R^3/t^2 for each moon. The distances can be computed from the periods according to $R = (k\,t^2)^{1/3}$. Newton's computed distances correspond to setting $k = 58.1378$ when periods are in decimal days. With times in decimal days, the k corresponding to Newton's computed distance estimates agrees with that calculated from the agreeing estimate of 5⅔ for Io by Cassini (from eclipses) and by Borelli. Very likely this was the estimate Newton used to set k. The distance calculated from this k agrees to within 0.001 of Jupiter's semidiameters with both Cassini's $14\frac{22}{60}$ estimate from eclipses for Ganymede and his estimate of $25\frac{3}{10}$ from eclipses for Callisto. It is also within 0.017 semidiameters of Cassini's estimate from eclipses, of 9 semidiameters, for the distance of Europa. This illustrates that the Harmonic Rule quite closely fits Cassini's eclipse data. This k requires that if the Harmonic Rule holds then the other distances from observations in Newton's table are systematically too small.[59]

Using this k as the Harmonic Rule ratio is an example of the practice, typical in Newton's day,[60] of picking what is regarded as the most accurate of multiple data estimating the same quantity. Had he used the average (51.631) of the sixteen k-values corresponding to the total of four distance estimates for each of the four moons, the overall fit of the Harmonic Rule to the data exhibited by putting the calculated distances next to the empirically estimated distances would have been much better.[61]

[59] Here are the differences of the cited distances from the distances Newton computed from the periods using the Harmonic Rule (in units of Jupiter's semidiameter).

Table 2.4

	Io	Europa	Ganymede	Callisto
Borelli	0	−0.35	−0.384	−0.632
Towneley (mic.)	−0.147	−0.222	−0.914	−0.597
Cassini (tel.)	−0.667	−1.017	−1.384	−2.299
Cassini (ecl.)	0	−0.017	−0.0007	0.001
	The mean errors are for each moon			
	−0.2	−0.4	−0.67	−0.88

[60] See the articles by Sheynin and by Schmeidler in *GHA 2B*. See, also, chapter 6 section I.6 below.

[61] The mean errors for the four moons, using the average $k = 51.631$, are respectively

$$+0.046, -0.224, -0.455, +0.417.$$

This is a far better fit than those for Newton's calculated values, using $k = 58.1378$, which were respectively

$$-0.2, -0.4, -0.67, -0.88$$

for the four moons.

Newton's selection of this k exhibits that he regards Cassini's estimates from eclipses, and also Borelli's agreeing estimate for Io, as more accurate than the others.

We see, here, an example of Newton using his judgment to select the more reliable from among these estimates, even though the result of doing so made the fit of the generalization he was defending as a phenomenon look worse than it would have if he had treated all the estimates equally by taking an average.

The fit of the Harmonic Rule can be represented by how well the ratios (R^3/t^2) from the separate orbital distance estimates and periods count as agreeing measurements of a common Harmonic Rule ratio representing the strength of a single inverse-square acceleration field centered on Jupiter. For each of the four cited sets of data we can exhibit the fit by the clustering of the corresponding estimates for what would count as the common Harmonic Rule ratio (R^3/t^2) for the four Galilean moons,

Table 2.5

	Io	Europa	Ganymede	Callisto
Borelli	58.138	51.619	53.606	53.885
Towneley (mic.)	53.739	53.671	47.746	54.236
Cassini (tel.)	39.938	40.6	42.92	43.684
Cassini (ecl.)	58.138	57.807	58.131	58.135

We can represent the fit of the Harmonic Rule by the agreement exhibited among the estimates in each row. For each row of the four estimates of the common Harmonic Rule ratio for the orbits of the four moons corresponding to the cited distance estimates from that astronomer, the fit of the Harmonic Rule is good.[62] For Cassini's estimates from eclipses this fit is very good indeed. These eclipse estimates were the basis for Cassini's new tables for Jupiter's moons in 1693 and were regarded at the time as quite accurate.[63] We shall see that they are indeed more accurate than the other estimates cited in Newton's table.

How accurate are the data? Table 2.6 is a comparison between the periodic times in Newton's table, converted to decimal days, and the rather precise mean periods cited in *ESAA*.[64]

In the worst case, that of the fourth moon Callisto, the period in Newton's table is 2.42 seconds less than that cited in the *ESAA* of 1992. This suggests that eclipses provided seventeenth-century astronomers with means for accurately determining periods to within a few

[62] One way we can show this is by the ratio of the standard deviation to the mean values for each of these four sets of data. The standard deviation (*sd*) is the square root of the average of the squares of their differences from the mean (*m*) (see Freedman, Pisani, and Purvis 1998, 66–72). For Borelli we have a mean of 54.312 and a standard deviation of 2.74 for a ratio of standard deviation over mean of 2.74/54.312 = 0.05. We have for Townly a mean of 52.348 and a standard deviation 3.08 for a ratio $sd/m = 0.059$. We have for Cassini's earlier estimate from telescope observations a mean of 41.79 and an *sd* of 1.8 for $sd/m = 0.04$. Finally, we have for Cassini's estimates from observations of eclipses a mean of 58.055 and an *sd* of 0.14 for $sd/m = 0.0026$.

In Newton's day our modern techniques for estimating quantities from multiple data were not yet available. Roger Cotes, Newton's editor of the second edition of *Principia*, did make some beginnings, for certain sorts of simple cases for determining an unknown quantity from *n* equations of condition (Schmeidler, 199 in *GHA 2B*). Modern least-squares estimates, however, stem from work by Gauss published in 1809, long after Newton (Schmeidler, *GHA 2B*, 203–5).

[63] See *GHA 2A*, 155. See also Halley 1694.

[64] The modern values are from *ESAA*, 708.

Table 2.6 (periods in decimal days)

	Io	Europa	Ganymede	Callisto
Newton	1.76914	3.55118	7.15458	16.68899
ESAA	1.769137786	3.551181041	7.15455296	16.6890184
The differences are respectively (in decimal days)[65]:				
	+0.0000022	−0.000001	+0.000027	−0.000028

seconds. It also suggests that the periods for these moons are quite stable, in that today's estimates of mean periods are so accurately fit by the estimates of seventeenth-century astronomers.

The distance estimates cited in Newton's table are not in nearly such close agreement with today's estimates. Here, for each satellite, is the *ESAA* mean-distance, converted to semidiameters of Jupiter,[66] together with the respective differences of the corresponding four estimates cited in Newton's table.

Table 2.7

	ESAA	Borelli	Towneley (mic.)	Cassini (tel.)	Cassini (ecl.)
Io	5.90276	−0.24	−0.38	−0.90	−0.24
Europa	9.38567	−0.72	−0.61	−1.37	−0.39
Ganymede	14.9667	−0.97	−1.50	−1.97	−0.60
Callisto	26.3386	−1.67	−1.62	−3.34	−1.04

These comparisons with the modern values suggest that the estimates cited in Newton's table are all too low, but that Cassini's estimates from eclipses (the fourth distance estimate for each moon) and Borelli's estimate for Io (the first estimate for that moon) are more accurate than the other estimates.

2. *Pound's measurements*

Pound's observations were made during 1718–20, between the second and third editions of the *Principia*. Newton's discussion of them is a major addition put into his discussion of phenomenon 1 in the third edition. He points out that Pound used a micrometer.[67]

> Using the best micrometers, Mr. Pound has determined the elongations of the satellites of Jupiter and the diameter of Jupiter in the following way. The greatest heliocentric elongation of the fourth satellite from the center of Jupiter was obtained with a micrometer in a telescope fifteen feet long and came out roughly 8′ 16″ at the mean distance of Jupiter from the earth. That of the third satellite was obtained with a micrometer in a telescope 123 feet long and came out 4′ 42″ at

[65] Newton's cited periods in this 3rd edition table are the same as those cited in the 2nd edition of 1713.

[66] In *ESAA*, 706, 708, the semidiameter approximate radius of Jupiter is given as 71492 km and the mean semi-major axes of the orbits of the four Galilean moons are respectively (1) 422,000 km, (2) 671,000 km, (3) 1,070,000 km, (4) 1,883,000 km.

[67] The micrometer in a telescope is an adjustable distance between edges or hairlines which allows quite precise determinations of small angular separations within the field of view. See King 1979, 93–100. See, also, Van Helden (1985) 118–28.

> the same distance of Jupiter from the earth. The greatest elongations of the other satellites, at the same distance of Jupiter from the earth, come out 2′ 56″ 47‴ and 1′ 51″ 6‴, on the basis of the periodic times.[68] (C&W, 797–8)

In order to use several elongations, obtained by telescope observations from the earth, to measure comparative distances of satellite orbits from Jupiter they have to be adjusted to the same earth–Jupiter distance. Pound's data are adjusted to the mean-distance of Jupiter from the earth, which is about the same as the mean-distance of Jupiter from the sun.[69]

Newton goes on to address in some detail Pound's efforts to determine the diameter of Jupiter. These were carried out with a 123-foot focal length telescope in 1719. According to King, the telescope Pound used is one which was presented by Constantine Huygens (Christian Huygens's brother) to the Royal Society in 1692.[70]

Newton paid to have a very long maypole made available so that Pound could erect the lens on it at Wanstead Park.[71] The eyepiece and micrometer set up could then be positioned on the ground with respect to the lens on the maypole, as in the Huygens telescope in the diagram from chapter 1.

Newton's discussion in book 3 continues with the following account of Pound's use of this long telescope to correct for overestimates of the diameter of Jupiter due to greater effect of refraction on observations with shorter telescopes.

> The diameter of Jupiter was obtained a number of times with a micrometer in a telescope 123 feet long and, when reduced to the mean distance of Jupiter from the sun or the earth, always came out smaller than 40″, never smaller than 38″, and quite often 39″. In shorter telescopes this diameter is 40″ or 41″. For the light of Jupiter is somewhat dilated by its nonuniform refrangibility, and this dilation has a smaller ratio to the diameter of Jupiter in longer and more perfect telescopes than in shorter and less perfect ones. (C&W, 798)

Newton here gives reasons why the estimates of satellite distances in semidiameters of Jupiter cited in the table are systematically too small.

He then goes on to describe Pound's use of transits of its satellites to determine Jupiter's diameter.

> The times in which two satellites, the first and the third, crossed the disc of Jupiter, from the beginning of their entrance [i.e., from the moment of their beginning to cross the disc] to the beginning of their exit and from the completion of their entrance to the completion of their exit, were observed with the aid of the same longer telescope. And from the transit of the first satellite, the diameter of Jupiter at its mean distance from the earth came out 37⅛″ and, from the transit of the third satellite, 37⅜. The time in which the shadow of the first satellite passed across the body of Jupiter was also observed, and from this observation the diameter of Jupiter at its mean distance from the earth came out roughly 37″. Let us assume that this diameter is very nearly 37¼;

[68] It is not clear why Newton includes these rather precise (the 47‴ and 6‴ are in 60ths of a second) elongations, which he says are calculated from the times here. One would have expected observed elongations, adjusted for equal distances from the earth, as he offers for the third and fourth satellites.

[69] Densmore (1995, notes 9 and 10, pp. 247–8) offers a very accessible account of how observations of elongations of Jupiter's moons at other distances of the earth from Jupiter can be corrected to what they would be at this mean distance.

[70] King 1979, 63. [71] Westfall 1980, 831.

> then the greatest elongations of the first, second, third, and fourth satellites will be equal respectively to 5.965, 9.494, 15.141, and 26.63 semidiameters of Jupiter. (C&W, 798)

Newton settles on 37¼ as the elongation of the diameter of Jupiter as observed from the mean-distance of Jupiter from the earth. Dividing the elongation estimates of the third and fourth satellites, 4′42″ and 8′16″, by half of 37¼″ gives 15.1409 and 26.63 semidiameters of Jupiter, which Newton cites as the respective distances of these satellites from Pound's observations.[72]

Though they refined the tolerances to which the motions of these satellites were known to fit the Harmonic Rule, and to approximate uniform motion on concentric circles, the observations by Pound's group also provided the first evidence for the existence of limitations on the exactitude to which such uniformities would continue to hold. Pound's nephew, James Bradley, then just past his mid twenties, assisted on the project.[73] Bradley, who became famous with his discovery of aberration in 1729, succeeded Halley as Astronomer Royal in 1742.[74] While assisting his uncle in 1719, Bradley was able to notice a very small eccentricity for the fourth satellite. He also detected some inequalities in time differences which were not removed by Römer's equation of light. Bradley's interesting remarks describing these inequalities were not published until they were included in an introduction to Halley's posthumously published tables in the 1740s.

Such inequalities were, eventually, successfully treated as perturbations by Laplace.[75] The Lieske program for Jupiter's Galilean satellites used for the Astronomical Almanac reveals variation from orbit to orbit at very fine scales.[76] It also reveals considerable stability, at scales corresponding to those cited in Newton's table, for eccentricities, distances, and periods.

Pound's distance estimates are far more accurate than those in Newton's table. Here are the differences from the mean values from *ESAA* in semidiameters of Jupiter.

0.062, 0.108, 0.174, 0.291

[72] We noted (note 10) that the elongations for the first and second satellites cited by Newton were said to be calculated from the periods rather than estimated from observations. When these computed elongations 1′51″ 6‴ and 2′ 56″ 47‴ are divided by half of the elongation Newton takes for Jupiter's diameter, the results are 5.965 and 9.4917. The 5.965 for the first satellite agrees with the distance 5.965 which Newton claims result from Pound's observations when 37¼′ is taken as the elongation of Jupiter's diameter. The 9.4917 resulting from the elongation said to be calculated from the period for the second satellite differs from the 9.494 said to result from Pound's observations. Perhaps, like the distances for the third and fourth satellites, this cited distance actually does result from Pound's observations, rather than from calculation from the period.

In Newton's manuscript collection (Add.MS.3969) there is a section titled

> 6. Transits of Satellites of Jupiter and of their shadows across the disk of the planet, observed by Pound at Wanstead.

It records observed elongations of the fourth and third satellites from May 6 and 7 of 1720 together with the following inference:

> From hence it follows that where Jupiter is in its mean Distance from the Earth, the Third satellite in its greatest elongation will be 4′41⅘″ and the Fourth 8′15⁷⁄₁₀″ from Jupiter's center.

It seems that Newton's cited maximum elongations of 4′42″ and 8′ 16″ for these satellites in the *Principia* are based on these observations.

I have been unable to find record of Pound's observed maximum elongations for the first and second satellites.

[73] See "Bradley, James" in the *Dictionary of Scientific Biography* (Gillispie 1970–80). See also the introduction to Rigaud S.P. ed. 1972.

[74] Rigaud 1972.

[75] See *GHA 2B*, 141–2.

[76] Myles Standish, of the ephemeris program for the *Astronomical Almanac* at the Jet Propulsion Laboratory in Pasadena, California, was kind enough to run data for Jupiter's moons for me.

The mean error of Pound's estimates from today's values is only + 0.135 of Jupiter's semidiameter.[77] The mean error for Borelli is −3.59, for Towneley −4.1, for Cassini's telescope observations −7.59, and for Cassini's estimates from eclipse observations −2.36. We can probably regard the measurements of these satellite distances by Pound as the most accurate available anywhere at the time they were made. They are certainly far more accurate than the data cited in Newton's table.

3. *Saturn's moons*

Newton exhibits the fit of the Harmonic Rule by putting distances computed from the periods in accordance with the Harmonic Rule together with the distances estimated from observations. Newton's computed distances are, apparently, arrived at by using Titan's distance from Cassini's observations to set the Harmonic Rule constant k.[78] The small differences between these computed distances and the distances from Cassini's observations suggest that the Harmonic Rule fits these data quite well.[79]

How accurate are the data for Saturn's moons? Here is a comparison between the periodic times in Newton's table, converted to decimal days, and the mean periods cited in the 1992 Astronomical Almanac.

Table 2.8

	Tethys	Dione	Rhea	Titan	Iapetus
Newton	1.88781	2.73706	4.5175	15.9453	79.325
ESAA	1.887802160	2.736914742	4.517500436	15.9454068	79.3301825
N-*ESAA*	0.0000078	0.000145	−0.0000003	−0.00012	−0.00518

As in the case of Jupiter's moons, the periods in Newton's table are in quite close agreement with the mean periods cited in *ESAA*.

What is now called Saturn's A-ring is the outermost observable by telescope from earth. This appears to be the ring whose semidiameter Newton uses as his distance unit. According to *ESAA*, the outer edge of this ring is 136,006 km from the center of Saturn. Let us use this value to

[77] The estimate for the first satellite, Io, is calculated from the period. The third and fourth are from observations. Their mean error is +0.23 of Jupiter's semidiameters. The estimate for the second satellite, Europa, may also be from observation. If we include it, the mean error of Pound's observed estimates is +0.19 of Jupiter's semidiameters.

[78] For periods in decimal days the R^3/t^2 value for k set from the period 15.9453 days and distance 8 (semidiameters of the specified ring) is 2.0137. This yields 1.93, 2.47, 3.45, 8, and 23.3135. The last number does not agree with Newton's 23.35.

[79] The differences between the calculated distances in the table and Cassini's distances from observations are

$$-0.02, \quad -0.03, \quad -0.05, \quad 0, \quad -0.65$$

If the correct value, 23.3135, is used for the calculated distance for the fifth moon (rather than the value 23.35 cited in the table, see previous note), then the difference from Cassini's value from observation for this moon is a little worse, −0.687 rather than −0.65.

Least-squares estimate of the fit of the five R^3/t^2 values to the mean R^3/t^2, supports the suggestion that the fit of the Harmonic Rule to the data in the table is quite good. For the data from Cassini the mean k is 2.09556 with an sd of 0.058 and an $sd^+=(5/4)^{1/2}sd$ of 0.065. This sd^+ is about 0.03 of the mean value of k.

convert the distance estimates cited by Newton into units of 1000 km for comparison with the mean-distances of the satellite orbits cited in *ESAA*.

Table 2.9

	Tethys	Dione	Rhea	Titan	Iapetus
Cassini	266.37	341.5	478.1	1,092.8	3,287 4
ESAA	294.66	377.4	527.04	1,221.83	3,561.3
Pound	286.86	367.454	512.25	1,188.42	3,462.81

To the extent that we are correct in identifying what Newton took as Saturn's ring and that the semidiameter of this ring and the mean-distances of Saturn's moons have not significantly changed since the data were obtained, comparison with *ESAA* mean-distances can represent the accuracy of the distances Newton cites. As in the case of Jupiter's moons, the estimates in the table appear to be systematically small, and the estimates from Pound – in this case all but Titan are computed from periods using Titan's distance to set k – are considerably more accurate.

Chapter 2 Appendix 3: Empirically determining periods, apsides, mean-distances, and Area Rule motion in Kepler's ellipse

1. Determining periods

The data Newton cites in support of his phenomena are estimates by astronomers of periods and mean-distances of orbits. The details of Kepler's estimates are instructive. We shall see that the determination of periods is relatively straightforward. This is as one might expect, given the agreement among astronomers remarked upon by Newton.

The determination of mean-distances is more complex. We have available the details of Kepler's reasoning from Tycho's recorded observations to his estimate of the heliocentric direction and distance of the aphelion of the orbit of Mars. This calculation applies agreeing measurements from triangulations afforded by estimates from observations. It is also neutral between an eccentric circle and the ellipse or other oval-shaped alternatives. It therefore illustrates Kepler's practice of using weak background assumptions to make data measure features of orbital parameters that constrain what can count as viable models. In this regard, Kepler's practice for determining orbits foreshadows Newton's inferences from phenomena.

We include an account of Kepler's equation; this gives the mean anomaly (the angle corresponding to the proportion of area swept out by radii from the sun at a uniform areal rate in an eccentric circular orbit) as a function of the eccentric anomaly (the corresponding angle traced out by radii from the center of that eccentric circle). These are applied to give the equations for distance and angle in an elliptical orbit as functions of the eccentricity of the ellipse and the eccentric anomaly in the corresponding circumscribed circle.

Kepler's data were geocentric angular positions giving locations of a planet against the background of fixed stars at specified times. Here is an example. According to Kepler, on October 31, 1595 at 0 hours 39 minutes Uraniborg time, Mars's position was 47.528 degrees geocentric ecliptic longitude and 8 minutes north ecliptic latitude.[80]

Our example datum is an opposition of Mars to the sun. Mars was on the exact opposite side of the earth from the sun or, equivalently, the earth was on the same heliocentric ecliptic longitude as Mars. At opposition the ecliptic longitude of Mars observable from the earth (its geocentric ecliptic longitude) is the same as its ecliptic longitude with respect to the sun. Oppositions, thus, allow relatively direct determinations of heliocentric ecliptic longitudes of superior planets.

Kepler counts time from noon, rather than from midnight.[81] The time he cites is thus 12:39 p.m., right in the middle of the day. This is hardly a time at which an actual observation might have been made. In discussing this datum he tells us:

[80] This datum is one cited by Kepler in his table of twelve oppositions (Kepler 1992, 249). As we shall see, he discusses the observations supporting it on p. 240.

[81] Kepler's practice of counting astronomical days from noon to noon was standard until 1925. As of January 1, 1925, the astronomical tabular day was brought into coincidence with the civil day and was considered to start at midnight (*ESAA*, 612).

> On 1595 October 30 at 8^h 20^m, the planet [Mars] was found at 17°47′15″ Taurus . . . The sun's position was 16°50′30″ Scorpio. The distance between the stars was 56′45″. The sun's diurnal motion was 1°0′35″; that of Mars, 22′54″, as appears by comparing the nearby observations. The sum of the diurnal motions was 1°23′29″. If the distance between the stars be divided by this, it comes out to 40′47″ of a day, or 16 hours 19 min. Therefore, the true opposition was 0^h 39^m PM on October 31. Mars's position was 17°31′40″ Taurus. (Kepler 1992, 240)

Taurus starts at 30°, so 17°47′15″ Taurus is 47°47′15″. Scorpio starts at 210°, so 16°50′30″ Scorpio is 226°50′30″. This puts the heliocentric longitude of the earth at 46°50′30″, just 56′45″ short of the geocentric longitude of Mars.

The diurnal motion of Mars against the fixed stars as observed from the earth at this time is retrograde.[82] Therefore, it can be added to the diurnal motion of the sun to compute the time, 0.6798 days, needed to eliminate the 56′45″ difference between the heliocentric longitude of the earth and the geocentric longitude of Mars. Multiplying this time by the diurnal motion of Mars and subtracting this retrograde geocentric motion from the earlier geocentric longitude of Mars at the time of the observation gives Kepler's datum for the geocentric longitude of Mars at the time of opposition. Thus, at the opposition of 1595 on October 31 at 0:39 Uraniborg time, Mars was at 47°31′40″ = 47.528° heliocentric longitude.

Kepler gives 66.476° as the heliocentric longitude of Mars at the opposition at 1:31 Uraniborg time on November 18, 1580. The opposition of 1595 was just 18.94° short of the longitude of the opposition of 1580. Mars had traversed 7.947 revolutions in the 5,459.644 days between 01:31 on November 18, 1580 and 0:39 on October 31, 1595. This gives an estimate of about 687 days for the period.

The precision of such an estimate can be increased by using oppositions separated by longer time periods. Kepler used data from Ptolemy to fine-tune his estimate of the period of Mars.[83] After some corrections introduced to transform Ptolemy's opposition from the mean sun to an opposition to the true sun, a correction to account for the precession of the equinoxes, and some other corrections as well, Kepler estimated from Ptolemy's data that the difference in heliocentric longitude between Mars and the star Cor Leonis was 128°48′30″ at 18:00 on May 26 in AD 139. From Tycho's data he found this difference to be 216°31′45″ at 18:00 on May 27, 1599. The difference between these heliocentric longitudes is 87°43′15″. The time interval is 1460 Julian years or 1460 x 365.25 = 533265 days. With the approximate value of 687 days for its period one can find that Mars completed 776 whole cycles plus the difference of 87.72° for a total of 776.24 revolutions in 533265 days. This gives a period of 533265/776.24 = 686.98 days per revolution.[84]

This calculation assumes that the period has been stable over the long-time interval. By 1625 Kepler had come to believe that the periods of Jupiter and Saturn were subject to variation.[85] It is

[82] As we saw in chapter 1, planets or "wanderers" change their positions against the fixed stars from one day to the next. Usually this diurnal motion is from west to east. About every two years Mars stops its normal eastward motion and, after a pause or stationary point, begins westward *retrograde* motion. After about 83 days this retrograde motion stops and, after another stationary point, the planet once more resumes its normal motion. Mars shines its brightest at opposition to the sun, right in the middle of its retrograde motion.

[83] Curtis Wilson helped extract the following calculation from chapter 69 of Kepler's *Astronomia Nova*.

[84] Newton cites 686.9785 days as the period of Mars.

[85] Wilson (1989a VI, 240) cites a letter of Kepler to Matthias Bernegger June 20/30, 1625 (Kepler 1937, vol. 10, p. 44, lines 21–5), in which Kepler suggests that the "mean motions are no longer mean" for these planets. According to Wilson, these remarks mistakenly include Mars as well.

instructive to compare the periods cited by Newton with the mean periods cited in the 1992 *ESAA*. Here are the comparisons with periods in Julian years.[86]

Table 2.10

	Newton 1726	*ESAA* 1992	N-*ESAA*	Difference
Mercury	0.2408465	0.24084445	0.000002	1min 5sec
Venus	0.614969	0.61518257	−0.0002	−1hr 52min
Earth	1.0000178	0.99997862	0.000039	1min 21sec
Mars	1.8808446	1.88071105	0.00013	1hr 10min
Jupiter	11.86177687	11.85652502	0.005	1day 22hr
Saturn	29.4572895	29.42351935	0.034	12days 8hr

The difference of about 12½ days for Saturn is considerable and the difference of about 2 days for Jupiter is also quite a bit larger than the differences for the others.

A good deal of this is due to the great inequality of the motion of Jupiter and Saturn, which was eventually successfully treated as a nearly 900-year period perturbation by Laplace.[87] Even though long-term variation for the period of Saturn could be quite substantial, astronomers in any given reasonably short time period could agree fairly precisely on empirical determinations of its mean period. In the case of Jupiter, the variation in period is substantially less, though still quite detectable.[88] For the other planets, variations in period range from fairly small to very small, as is suggested by the above comparison between those cited by Newton and the *ESAA*.

2. *Determining apsides and mean-distances*

Kepler takes the mean-distance in the Harmonic Rule to be the semi-major axis of an elliptical orbit. This is the radius of a circumscribed circle. The ellipse meets this circle at aphelion (the point on the orbit of furthest distance from the sun) and the perihelion (the point on the orbit of closest distance to the sun). In chapter 42, Kepler uses data to more accurately refine estimates of the location of the aphelion, the eccentricity, and the mean-distance.[89]

One especially interesting feature of this calculation is that it is neutral between an eccentric circle and the ellipse or other oval-shaped alternatives.

[86] A Julian year is 365.25 days (*ESAA*, 730).

[87] For a brief account, see chapter 9 section IV.2. For more detail, see Curtis Wilson's classic paper (Wilson 1985).

[88] The diagram of major perturbations of Jupiter and Saturn on p. 35 of Wilson's article on the Great Inequality (Wilson 1985) illustrates the comparative size of the inequalities for Jupiter and Saturn.

[89] In chapter 16, Kepler used an iterative procedure to set the parameters of what he came to call his vicarious hypothesis – an eccentric circular orbit with an equant. As we saw in chapter 1 section I.1, an *equant* is a point about which the rate at which angles are swept out is constant. In this theory the equant is on the aphelion line on the opposite side of the center of the eccentric circle of Mars's orbit from the sun. Kepler's empirical procedure set the ratio of the sun–center to the center–equant distances to be 0.11332 to 0.07232. This construction gave 148°48′55′ for the heliocentric longitude of Mars's aphelion in 1587 (Kepler 1992, 277).

Chapter 42 begins with triangulations from five sets of data for times when Mars is taken to be near aphelion and three for times when Mars is taken to be near perihelion.[90] The five sets of aphelion data are,

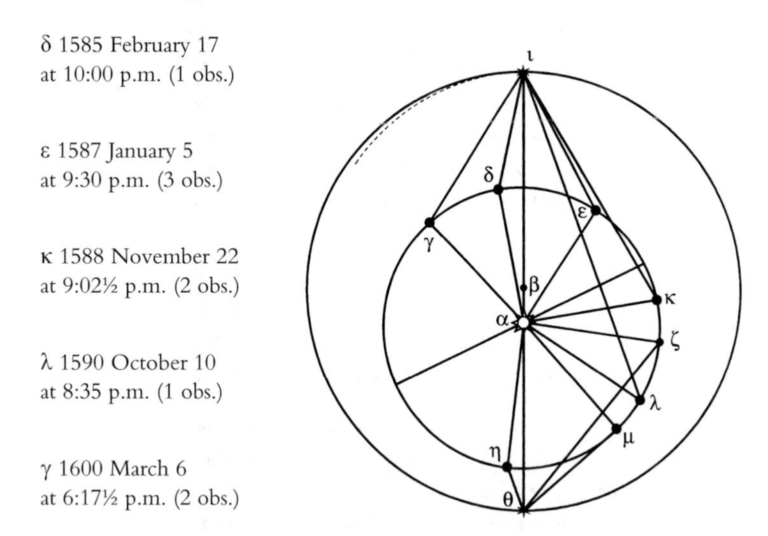

Figure 2.8 Kepler's data for Mars at aphelion and at perihelion

The time differences between δ and ε, ε and κ, and κ and λ are all 686.98 days or exactly one whole period apart. The time between λ and γ is five times 686.98 or exactly five whole periods.

As they are separated by whole multiples of periods of Mars, Kepler can use these as five triangulations of the same sun–Mars distance. Adjusting the common heliocentric longitude of Mars so that these triangulations agree, Kepler finds Mars's heliocentric longitude for 1588 November 22 at 9 hours 2.5 minutes Uraniborg time to be 149°20′12″.[91]

He gave three triangulations from data for Mars at perihelion. These are for 1589 November 1 at 6:10 p.m., 1591 on September 19 at 5 hours 42 minutes, and for 1593 on August 6 at 5 hours 14 minutes.[92] Taking all three data into account, he estimates the heliocentric longitude of Mars to be 329°54′53″ at the first of the specified times. This time, 1589 November 1 at 6 hours 10 minutes is approximately the time of perihelion passage following the aphelion passage approximated by the 1588 November 22 at 9 hours 2.5 minutes datum.

One half of Mars's period of 686.98 days is 343 days 11 hours 45.6 minutes. The interval between the two specified times is 344 days diminished by 2 hours 52.5 minutes. This is 343 days 21 hours 7.5 minutes.[93] This exceeds one half period by 9 hours 21.5 minutes. According to Kepler, the heliocentric motion of Mars at aphelion is 26′13″ or 26.217′ per day, and at

[90] The diagram is from Kepler 1992, 436. On p. 438, Kepler gives the angular positions of Mars and sun, and earth–sun distance for each of these five times.

[91] See Kepler 1992, 438–40 for the method, and 443 for this outcome.

[92] Kepler 1992, 441.

[93] Kepler gives the following calculation (1992, 443):

perihelion is 38′2″ or 38.03′ per day.[94] Dividing the excess time, 9 hours 21.5 minutes, evenly between 4.68 hours or 0.195 days before aphelion and 4.68 hours or 0.195 days after perihelion adds 26.217′ per day times 0.195 days or 5.11′ plus 38.03′ per day times 0.195 days or 7.42 minutes, for a total of 12.53′.

Kepler points out that if our motion of Mars began one day later than when it passes aphelion, it would start 26′13″ beyond aphelion and after one half period would end 38′2″ beyond perihelion, traversing 38′2″ – 26′13″ or 11′49″ more than 180°. The difference between the heliocentric longitudes of Mars between our two data points, however, after correcting for precession, is 33′53″ or 33.88′ beyond the 180° from aphelion to perihelion of an orbit. We therefore need to make up an additional 33.88′–12.53′ or 21.35′. As 11′49″, or 11.82′, is to 1 day, so is 21.35′ to the time our starting point was past aphelion. Our starting point, therefore, was 21.35/11.82 or 1.806 days beyond aphelion. The aphelion must be brought back 1.806 days times 26.21′ per day or 47.34′. The starting point was 149°20′12″, which is 149.337°. Subtracting 47.34′ or .772° gives 148.565 or 148°33′54″ as the longitude of Mars's aphelion.[95]

The triangulated distance for the starting point, corrected for inclination of the orbit, Kepler gives as 166,780 (1992, 440). For the finishing point, the inclination-corrected triangulated distance was taken as approximately 138,500 (1992, 442). He takes these as the aphelion and perihelion distances, arguing that so near to aphelion and perihelion, the distance changes corresponding to small angular differences are not perceptible (1992, 445). This gives one half the sum, 166780 + 138500, of these distances or 152,640 as the mean-distance of Mars from the sun. In chapter 54, Kepler takes into account other triangulations to adjust his estimate of the mean-distance to 152,350, the number in Newton's table (1992, 540).

According to *ESAA*, the mean earth–sun distance is 1.0000010178 AU. Let us assume that units where Newton's mean-distance for the earth is 1.00000 agree with the astronomical unit to

From 1588 November 22 at $9^{h}\ 2½^{m}$ to 1589 November 1 at $6^{h}\ 10^{m}$ are 344 days diminished by $2^{h}\ 52½^{m}$, while a whole revolution to the same fixed star has 687 days diminished by 0h. 28 min. Therefore, our interval appears to exceed half the periodic time by a few hours.

Consider:

Days	Hours	Min.	
343	11	46	Half the period
343	21	52½	Our interval
Excess	10	6½	

The correct value for the interval is 343 days 21 hours 7.5 minutes. Kepler's error was to put the 52.5 minutes as the minute term of the difference, rather than subtracting it from the result of subtracting 2 hours from 344 days.

Given this mistake, Kepler's difference of 343 days 21 hours and 52.5 minutes exceeds one half a period by 10 hours 6.5 minutes.

[94] These diurnal motions near aphelion and perihelion are empirically supported by long experience constructing and evaluating ephemerides. Kepler points out that these also follow from an equant with bisected eccentricity, given the mean diurnal motion 31′27″ (360°/686.98) together with the aphelion and perihelion distances of Mars (Kepler 1992, 443).

[95] Kepler points out that his aphelion estimate, 148°39′46″, requires bringing the aphelion back nearly 11 minutes from the 148°50′44″ calculated for November of 1588 from the vicarious theory. Our corrected version of his calculation requires bringing the vicarious theory's location of the aphelion back by nearly 17 minutes, to 148°33′54″. This is much closer to what we now estimate as the correct location of the aphelion of Mars in November of 1588. Wilson informed me that according to Simon Newcomb's formula, the aphelion of Mars in November 1588 was 148°29′45″.

five decimal places. Here is a comparison between the distances Newton cites from Kepler and Boulliau and those cited in *ESAA*.

Table 2.11

	Kepler	Boulliau	*ESAA*	Differences		
				K-B	K-*ESAA*	B-*ESAA*
Mercury	0.38806	0.38585	0.38710	0.00221	0.00096	−0.00125
Venus	0.72400	0.72398	0.72333	0.00002	0.00067	0.00065
Earth	1.00000	1.00000	1.00000	0.00000	0.00000	0.00000
Mars	1.52350	1.52350	1.52368	0.00000	−0.00018	−0.00018
Jupiter	5.19650	5.22520	5.20260	−0.02870	−0.00610	0.02260
Saturn	9.51000	9.54198	9.55491	−0.03198	−0.44910	−0.01293

These numbers exhibit more variation than the corresponding numbers with respect to the periods cited in Table 2.11. As Newton suggests, the periods were more precisely known than the mean-distances were empirically triangulated.

The great inequality generated by gravitational interaction between Jupiter and Saturn, which we remarked on when discussing periods, involves considerable variation in mean-distances.

3. Kepler's equation and Area Rule motion in Kepler's ellipse

3.i Kepler's equation

$$M = Erad + e \sin E$$

gives the mean anomaly M as a function of the eccentric anomaly E together with the eccentricity e. *Erad* is the eccentric anomaly E given the radian measure. One radian unit is the measure of the central angle subtended by an arc of a circle of length equal to its radius.[96] The *mean anomaly M* represents the *time* measured as the proportion of the period corresponding to the proportion of the area traced out in the eccentric circle at a uniform rate. Kepler measures it from aphelion C.[97]

This equation allows an easy solution for M given E. To go from M to E ("Kepler's problem") requires some method of iterating, guessing and correcting.[98]

96

$$1 \text{ radian} = \frac{180°}{\pi} \approx 57.296°$$

To convert radians ("rad"s) to degrees ("deg"s):

$$\text{rad} = \text{deg}\ (\pi/180)$$

To convert degs to rads:

$$\text{deg} = \text{rad}\ (180/\pi)$$

97 Caution: today it is more common to measure it from perihelion – this gives the more familiar alternate equation: $M = Erad - e \sin E$

98 See, e.g., Danby 1988, 149ff.

$$\frac{M}{2\pi} = \frac{Area\ CAP}{Area\ of\ Circle}$$

$$\frac{Area\ CAP}{\pi} = \frac{Area\ CBP + Area\ BPA}{\pi}$$

$$\frac{Area\ CBP}{\pi} = \frac{Erad}{2\pi}$$

$$\frac{Area\ BPA}{\pi} = \frac{e\ \sin E}{2\pi}$$

$$\frac{M}{2\pi} = \frac{Erad}{2\pi} + \frac{e\ \sin E}{2\pi}$$

$$M = Erad + e\ \sin E$$

Area Rule in Eccentric Circle
Radius = 1; BA = eccentricity e; 360° = 2π radians
Area of circle = $\pi r^2 = \pi$
Eccentric anomaly E=CBP

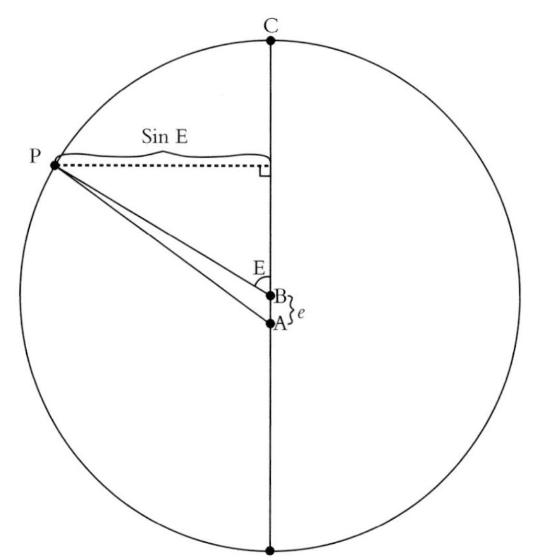

Figure 2.9 The Area Rule for an eccentric circle

3.ii The Area Rule in the ellipse Consider an elliptical orbit *ambc* and a circumscribed circle *akgc* meeting it at aphelion and perihelion. See Figure 2.10. Let *kml* be a perpendicular from circle to the diameter *abc* cutting the ellipse at *m*. Let the radius *ab* = 1 and the eccentricity *hn* = *e*.

Theorem 1:

$$\frac{ml}{kl} = \frac{bh}{gh}$$

The ratio of the part cut off by the ellipse to the whole perpendicular from the circle to the diameter *ahc* is the same for each perpendicular.

Theorem 2:

$$\frac{Area\ of\ ellipse}{Area\ of\ circle} = \frac{bh}{gh}$$

Theorem 3:

$$\frac{Area\ of\ sector\ amn}{Area\ of\ ellipse} = \frac{Area\ of\ sector\ akn}{Area\ of\ circle}$$

The areal ratio

$$\frac{(\tfrac{1}{2})[Erad + e \sin E]}{\pi}$$

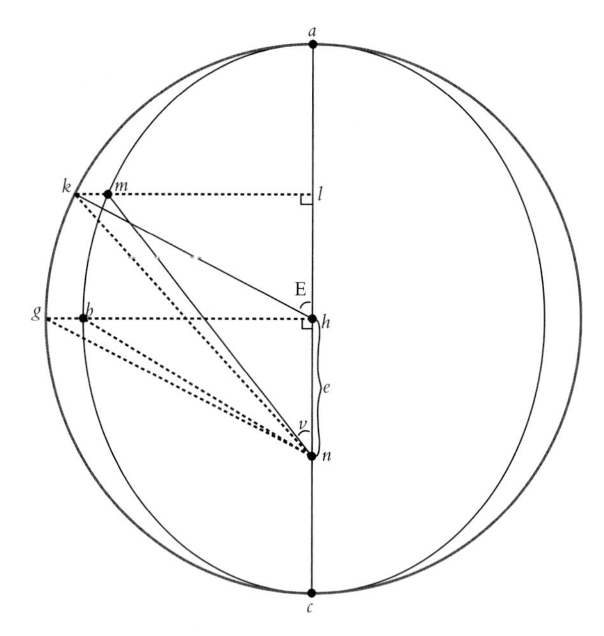

Figure 2.10 The Area Rule in Kepler's elliptical orbit

for the sector *akn* in the circle computed from the eccentric anomaly E can be used to compute the corresponding areal ratio for the sector *amn* in the ellipse.

Where *amn* is the sector swept out from aphelion *a* in the time specified by the mean anomaly M:

$$\frac{M}{2\pi} = \frac{\textit{Area of amn}}{\textit{Area of ellipse}}$$

The area swept out is proportional to the time from aphelion, so that it acts as a clock agreeing with the proportion of the circumference that would be traversed in that time by a planet revolving uniformly in a concentric circle at its mean angular rate.

3.iii The distance and true anomaly for Kepler's ellipse

Theorem 4: Distance $nm = 1 + e \cos E$

The diametrical distance computed from the eccentric anomaly E picks out the distance from the focus n to the point m that marks the sector *amn* swept out from the aphelion *a* in the time specified in the mean anomaly $M = Erad + e\sin$ according to the Area Rule.

Theorem 5:

$$\cos v = \frac{e + \cos E}{1 + e \cos E}$$

The true anomaly v (from aphelion) corresponding to the mean anomaly $M = Erad + e\sin E$ in accordance with the Area Rule can be computed from the eccentricity e and the eccentric anomaly E.

3

Inferences from Phenomena (Propositions 1 and 2 Book 3)

In section I we shall review Newton's definitions, with special emphasis on his theoretical concept of a centripetal force. In section II we shall briefly review his treatment of time, space, and motion in his scholium to the definitions and his Laws of Motion and their corollaries, before emphasizing the empirical support for these laws Newton cites in his scholium to the Laws. The Laws of Motion, together with theorems derived from them, afford resources to make orbital phenomena measure features of centripetal forces maintaining bodies in those orbits. In section III we shall give an account of Newton's classic inferences from phenomena. These are his inferences to the centripetal direction of and inverse-square variation of the forces maintaining satellites and planets in their respective orbits of Jupiter, Saturn and the sun. In section IV we shall discuss a number of philosophical lessons these inferences can teach us about scientific method.

Appendix 1 derives details of the pendulum calculations that back up Newton's appeal to pendulum experiments to afford empirical measurements that support the application of his third Law of Motion to collisions. Appendix 2 exhibits the details of Newton's proofs of propositions 1–4 of book 1. Propositions 1–3 back up his inferences from uniform description of areas to centripetal directions of forces maintaining bodies in orbits. Proposition 4 backs up his inferences from the Harmonic Rule for a system of orbits to the inverse-square variation of the centripetal accelerations at the mean-distances of those orbits. Appendix 3 extends Newton's precession theorem to include orbits of large eccentricity. This backs up his inferences to inverse-square variation of the force maintaining a body in an orbit from absence of unaccounted-for precession of that orbit.

Newton's preface to book 3 begins with an articulation of the mathematical character of the principles which he presented in books 1 and 2.

> In the preceding books I have presented principles of philosophy that are not, however, philosophical but strictly mathematical – that is, those on which the study of philosophy can be based. These principles are the laws and conditions of motions and of forces, which especially relate to philosophy . . . (C&W, 793)

These principles are his Laws of Motion, together with the propositions Newton derived from them in books 1 and 2. They are background assumptions on which the empirical investigation of forces in natural philosophy can be based. They provide impressive resources to back up his inferences from phenomena in book 3. In this chapter we will see them backing up his classic inferences to inverse-square forces of attraction toward Jupiter, Saturn, and the sun from orbits which measure them.

Newton reports that he has translated the substance of an earlier version of that book 3, which had been composed in a more accessible form, into propositions in a mathematical style.

> It still remains for us to exhibit the system of the world from these same principles. On this subject I composed an earlier version of book 3 in popular form, so that it might be more widely read. But those who have not sufficiently grasped the principles set down here will certainly not perceive the force of the conclusions, nor will they lay aside the preconceptions to which they have become accustomed over many years; and therefore, to avoid lengthy disputations, I have translated the substance of the earlier version into propositions in a mathematical style, so that they may be read only by those who have first mastered the principles. (C&W, 793)

The propositions applying these principles to exhibit the system of the world are presented in a very mathematical style indeed. Like the propositions derived from the Laws of Motion in books 1 and 2, Newton presents his inferences from phenomena in book 3 in a format modeled on Euclid's classic derivations of theorems of geometry.[1]

Newton offers advice on how to read the *Principia*, without becoming bogged down in mathematical details that would be too time-consuming even for readers who are proficient in mathematics.

> But since in books 1 and 2 a great number of propositions occur which might be too time-consuming even for readers who are proficient in mathematics, I am unwilling to advise anyone to study every one of these propositions. It will be sufficient to read with care the Definitions, the Laws of Motion, and the first three sections of book 1, and then turn to this book 3 on the system of the world, consulting at will the other propositions of books 1 and 2 which are referred to here. (C&W, 793)

We shall begin with Newton's Definitions. This will provide a more extensive discussion of his theoretical concept of a centripetal force. This basic initial conception, which is rather different from what we today think of as Newtonian forces, plays a central role in Newton's argument. We shall follow this with a discussion of his Laws of Motion and the controversial scholium on time, space, place, and motion which provides his framework for their formulation and application. Newton treats his Laws of Motion as empirical. He characterizes them as "accepted by mathematicians and confirmed by experiments of many kinds." We shall see that he cites an impressive body of empirical evidence supporting their application to motion phenomena.

[1] See Euclid's *Elements* (Heath 1956).

Our focus will be on the novel and informative features of the rich empirical method exemplified in Newton's argument in book 3. For this it may be sufficient to follow up on this discussion of Newton's Definitions and Laws of Motion by turning directly to the opening propositions of book 3, as we do in section III below. Salient propositions backing up Newton's inferences are discussed and proved in appendix 2. Readers who wish to more exactly follow Newton's instructions to his readers can read this appendix before section III.[2]

I Newton's definitions

Newton's definitions are remarks to help readers understand the concepts he introduces. They are not formal definitions of the sort a logician or technical philosopher of science would give today.

1. Basics

Definition 1. *Quantity of matter is a measure of matter that arises from its density and volume jointly.*

If the density of air is doubled in a space that is also doubled, there is four times as much air, and there is six times as much if the space is tripled. The case is the same for snow and powders condensed by compression or liquefaction, and also for all bodies that are condensed in various ways by any causes whatsoever. For the present, I am not taking into account any medium, if there should be any, freely pervading the interstices between the parts of bodies. Furthermore, I mean this quantity whenever I use the term "body" or "mass" in the following pages. It can always be known from a body's weight, for – by making very accurate experiments with

[2] There are now available a number of excellent treatments that can help today's readers, who may have difficulty following Newton's mathematics, which is formulated in a synthetic geometry with limits rather than the calculus.

Dana Densmore, *Newton's Principia: The Central Argument*, offers translations, notes, and expanded proofs to aid readers who wish to follow Newton's instructions on how to read the *Principia.*

Bruce Brackenridge, *The Key to Newton's Dynamics: The Kepler Problem and the* ***Principia***, offers an extremely useful guide to help modern readers understand and appreciate Newton's geometrical method of presenting the mathematics of centripetal forces. It also offers an English translation, by Mary Rossi, of sections 1, 2, and 3 of book 1 from Newton's first edition of 1687.

S. Chandrasekhar, a winner of the Nobel Prize in Physics, offers *Newton's* ***Principia*** *for the Common Reader.* This book developed out of a study in which he would read Newton's propositions and construct his own proofs, using modern techniques, before carefully following Newton's own demonstrations. He has been able to admirably display the physical insights and mathematical craftsmanship that illuminate Newton's proofs, once the impediments of notation and language have been removed. He has, however, ignored most of the work by historians of science. George Smith has pointed out that this has sometimes led him astray (Smith G.E. 1996). Nevertheless, I have found Chandrasekhar's book to be a very valuable resource, as we shall see on many occasions below.

I was privileged to witness Chandrasekhar present some results of his study, at a conference in 1987 celebrating the tercentenary of Newton's publication of the *Principia.* He told us that when he went to Newton's proofs he was very impressed with the fact that they were more illuminating than his own proofs. After telling us this, he then showed us with several examples. We looked up in wonder and appreciation as he displayed details of these proofs on the dome of the planetarium at the Smithsonian Institution in Washington. This presentation by Chandrasekhar was the most memorable of all at that very fine celebration conference, which was sponsored jointly by the University of Maryland and the Smithsonian Institution.

pendulums – I have found it to be proportional to the weight, as will be shown below. (C&W, 403–4)

Newton distinguishes quantity of matter or mass from weight, as is now customary. This was an innovation, as most earlier work failed to make this distinction.[3] The experiments with pendulums are described in detail in Newton's argument for proposition 6. They are precision measurements of the proportionality of weight to mass for terrestrial bodies.[4]

For Newton, quantity of motion is the measure of motion that is determined by the velocity and the quantity of matter jointly.

Definition 2. *Quantity of motion is a measure of motion that arises from the velocity and the quantity of matter jointly.*

The motion of the whole is the sum of the motions of the individual parts, and thus if a body is twice as large as another and has equal velocity there is twice as much motion, and if it has twice the velocity there is four times as much motion. (C&W, 404)

Newton takes quantity of motion to be momentum – mass times velocity.

That the basic concept of quantity of matter is inertial mass is made clear in Newton's definition and discussion of what he calls "inherent force of matter."

Definition 3. *Inherent force of matter is the power of resisting by which every body, so far as it is able, perseveres in its state either of resting or of moving uniformly straight forward.*

This force is always proportional to the body and does not differ in any way from the inertia of the mass except in the manner in which it is conceived. Because of the inertia of matter, every body is only with difficulty put out of its state either of resting or of moving. Consequently, inherent force may also be called by the very significant name of force of inertia. Moreover, a body exerts this force only during a change of its state, caused by another force impressed upon it, and this exercise of force is, depending on the viewpoint, both resistance and impetus: resistance insofar as the body, in order to maintain its state, strives against the impressed force, and impetus insofar as the same body, yielding only with difficulty to the force of a resisting obstacle, endeavors to change the state of that obstacle. Resistance is commonly attributed to resting bodies and impetus to moving bodies; but motion and rest, in the popular sense of the terms, are distinguished from each other only by point of view, and bodies commonly regarded as being at rest are not always truly at rest. (C&W, 404–5)

For Newton, a body's inherent force of matter is what we would call its inertia.

[3] Howard Stein (1990, 25) has argued that Huygens had already distinguished mass, as an intrinsic property of bodies, from weight.

[4] We have noted (chpt. 1.V.2) that these pendulum experiments are the first stage of a sequence of agreeing measurements supporting the equality of ratios of weight to mass for bodies at equal distances from planets. See chapter 7 section II for a more detailed treatment of this impressive and informative application of Newton's Rule 3 for reasoning in natural philosophy.

We, today, do not call inertia a force. The basic use of force in Newtonian mechanics today is what Newton defines as impressed force.

Definition 4. *Impressed force is the action exerted on a body to change its state either of resting or of moving uniformly straight forward.*

This force consists solely in the action and does not remain in a body after the action has ceased. For a body perseveres in any new state solely by the force of inertia. Moreover, there are various sources of impressed force, such as percussion, pressure, or centripetal force. (C&W, 405)

Newton's Laws of Motion will give considerably more content to his concept of impressed force, and to what he calls inherent force of matter or inertia, as well as to his concept of quantity of motion and his concept of quantity of matter.[5] Newton's sharp distinction between inherent force of matter and impressed forces, which he describes as actions on bodies, allows his wider use of force to recover our distinction between inertia and what we call Newtonian forces acting on bodies.

The last line of Newton's discussion of impressed force counts centripetal force as one of the various sources of impressed force.[6] This supports distinguishing between the impressed force on a body being drawn toward a center and the centripetal force which is the source of that impressed force.

2. *Newton's definitions of centripetal force*

Let us now consider Newton's definition of centripetal force and what he distinguishes as its three measures.[7]

2.i The basic definition Newton's basic definition is general enough to include whatever forces are exhibited in familiar phenomena suggestive of attraction toward a center.

Definition 5. *Centripetal force is the force by which bodies are drawn from all sides, are impelled, or in any way tend, toward some point as to a center.*

One force of this kind is gravity, by which bodies tend toward the center of the earth; another is magnetic force, by which iron seeks a lodestone; and yet another is that force, whatever it may be, by which the planets are continually drawn back from rectilinear motions and compelled to revolve in curved lines. (C&W, 405)

Newton cites gravity, the force of magnetic attraction, and that force, whatever it may be, by which the planets are maintained in their orbits. All phenomena suggesting such attractions are problematic from the perspective of the mechanical philosophy. On the

[5] See section II below.

[6] The other sources Newton cites, percussion and pressure, are interactions to which the third Law of Motion would clearly apply.

[7] This section is based on Howard Stein's interpretation of Newton's centripetal forces of gravity toward planets as acceleration fields. See Stein 1970b, 1977, 1991 and 2002.

mechanical philosophy, to make a motion phenomenon intelligible one must show how it would be possible to account for it by the action of bodies pushing on one another.[8]

Newton next, perhaps going beyond the sense of his definition, cites as an example the force by which a sling prevents a stone from flying away until it is released.

A stone whirled in a sling endeavors to leave the hand that is whirling it, and by its endeavor it stretches the sling, doing so the more strongly the more swiftly it revolves; and as soon as it is released, it flies away. The force opposed to that endeavor, that is, the force by which the sling continually draws the stone back toward the hand and keeps it in an orbit, I call centripetal, since it is directed toward the hand as toward the center of an orbit. (C&W, 405)

This example makes sense of an attraction as the tension due to stretching of the sling. This helps suggest the intelligibility of attraction. To the extent that construing tension in the sling as just pushes may not be easy to do, this example contributes to suggest that what counts as mechanical intelligibility ought to be extended. Surely, pulls via attached cords are as good as pushes as examples of mechanical intelligibility.

Newton goes on to exploit the stone whirled in a sling as an analogy to inform our understanding of bodies made to move in orbits.

And the same applies to all bodies that are made to move in orbits. They all endeavor to recede from the centers of their orbits, and unless some force opposed to that endeavor is present, restraining them and keeping them in orbits and hence called by me centripetal, they will go off in straight lines with uniform motion. (C&W, 405)

He suggests that an orbiting body would go off in a straight line unless some force played for that body the role played by the tension in the sling for the stone.

Newton then considers a thought experiment in which the role of gravity is explored by considering projectiles with increasingly great velocities.

If a projectile were deprived of the force of gravity, it would not be deflected toward the earth but would go off in a straight line into the heavens and do so with uniform motion, provided that the resistance of the air were removed. The projectile, by its gravity, is drawn back from a rectilinear course and continually deflected toward the earth, and this is so to a greater or lesser degree in proportion to its gravity and its velocity of motion. The less its gravity in proportion to its quantity of matter, or the greater the velocity with which it is projected, the less it will deviate from a rectilinear course and the farther it will go. If a lead ball were projected with a given velocity along a horizontal line from the top of some mountain by the force of gunpowder and went in a curved line for a distance of two miles before falling to the earth, then the same ball projected with twice the velocity would go about twice as far and with ten times the velocity about ten times as far, provided that the resistance of the air were removed. And by increasing the velocity, the distance to which it would be projected could be increased at will and the curvature of the line that it would describe could be decreased, in such a way that it would finally fall at a

[8] See chapter 1.I.2 for basics and chapter 5 for more detail on Huygens's version of the mechanical philosophy.

distance of 10 or 30 or 90 degrees or even go around the whole earth or, lastly, go off into the heavens and continue indefinitely in this motion. (C&W, 405–6)

If the resistance of the air were removed, projectiles projected along a horizontal line from the top of a mountain would travel distances proportional to the velocities with which they were projected before falling to the earth. Such projectiles could even be made to go around the earth or off into the heavens with sufficiently great velocities.

The following diagram is from Newton's earlier version of book 3, which he told us he had composed in popular form so that it might be more widely read.

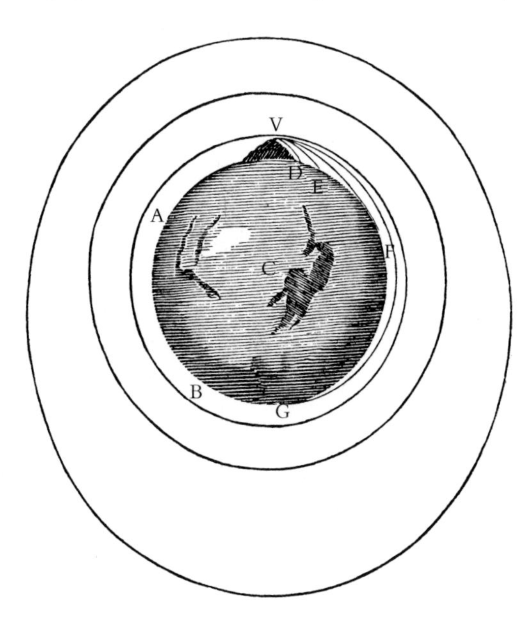

Figure 3.1 Newton's illustration of gravity

It suggests an analogy between the role of gravity in the motions of projectiles and the role of the force – whatever it may be – by which the moon is maintained in its orbit.

The passage we have been quoting from Newton's discussion of his definition of centripetal force goes on to make just such an analogy.

And in the same way that a projectile could, by the force of gravity, be deflected into an orbit and go around the whole earth, so too the moon, whether by the force of gravity – if it has gravity – or by any other force by which it may be urged toward the earth, can always be drawn back toward the earth from a rectilinear course and deflected into its orbit; and without such a force the moon cannot be kept in its orbit. If this force were too small, it would not deflect the moon sufficiently from a rectilinear course; if it were too great, it would deflect the moon excessively and draw it down from its orbit toward the earth. In fact, it must be of just the right magnitude, and mathematicians have the task of finding the force by which a body can be kept exactly in any given orbit with a given velocity and, alternatively, to find the curvilinear path into which a body leaving any given place with a given velocity is deflected by a given force. (C&W, 406)

We are offered as a task for mathematicians the task of calculating the forces from the orbits and the orbits from the forces.

Finally, Newton claims there are three kinds of quantities or measures appropriate to centripetal forces.

The quantity of centripetal force is of three kinds: absolute, accelerative, and motive. (C&W, 406)

The next three definitions are devoted to these three ways by which Newton claims centripetal forces may be quantified.

2.ii Newton's measures of centripetal force Here is Newton's definition and brief discussion of the absolute quantity of a centripetal force.

Definition 6. *The absolute quantity of centripetal force is the measure of this force that is greater or less in proportion to the efficacy of the cause propagating it from a center through the surrounding regions.*

An example is magnetic force, which is greater in one lodestone and less in another, in proportion to the bulk or potency of the lodestone. (C&W, 406)

All we are given, here, is the example of magnets differing in the strength by which they would attract, say, a given piece of iron at equal distances.

Here is Newton's definition and, only slightly less brief, discussion of the accelerative quantity of a centripetal force.

Definition 7. *The accelerative quantity of centripetal force is the measure of this force that is proportional to the velocity which it generates in a given time.*

One example is the potency of a lodestone, which, for a given lodestone, is greater at a smaller distance and less at a greater distance. Another example is the force that produces gravity, which is greater in valleys and less on the peaks of high mountains and still less (as will be made clear below) at greater distances from the body of the earth, but which is everywhere the same at equal distances, because it equally accelerates all falling bodies (heavy or light, great or small), provided the resistance of the air is removed. (C&W, 407)

Newton cites as an example the potency of a lodestone. Consider a given piece of iron being attracted toward a magnet. The attraction of that piece of iron toward that magnet at a smaller distance would be greater than it would be at a greater distance from it. His other example is the force that produces gravity. The first part of Newton's comment on gravity alludes to his discovery that, for bodies above the surface of the earth, the acceleration of gravity varies inversely with the squares of distances from the center of the earth. The second part of this comment describes the fundamental property of gravity that makes the accelerative measure at a given distance the same for any body at that distance. This feature of gravity contrasts sharply with magnetic attraction to a lodestone. Magnetic attraction could differ for different pieces of iron, even if these pieces were at the same distance from that lodestone. For such pieces of iron, the accelerative measure would have to be made relative to the body as well as to the distance. Other sorts of bodies, such as pieces of wood, would not be attracted at all by a magnet.

Here is Newton's definition and initial discussion of the motive quantity of a centripetal force.

Definition 8. *The motive quantity of centripetal force is the measure of this force that is proportional to the motion which it generates in a given time.*

An example is weight, which is greater in a larger body and less in a smaller body; and in one and the same body is greater near the earth and less out in the heavens. This quantity is the centripetency, or propensity toward a center, of the whole body, and (so to speak) its weight, and it may always be known from the force opposite and equal to it, which can prevent the body from falling. (C&W, 407)

The motive quantity is proportional to the motion which it generates in a given time. Definition 2 made clear that the quantity of motion of a body arises from its momentum, its mass times its velocity. His motive quantity of a centripetal force on a body is, therefore, the body's mass times the generated change of its velocity. It is counted as a measure of the action of that centripetal force on that body.

At any given instant, the value at that instant of the motive quantity of the action of a continuously acting centripetal force on a body is the mass of that body times its acceleration toward that center.[9] This makes the value at an instant of the motive measure of such a centripetal force acting on a body the mass times the acceleration it generates on that body – the concept of force familiar to students of elementary Newtonian physics today.

The example he gives is weight. The weight of a body at a given time and location is the mass of that body times the acceleration of gravity at that location. Weight, he tells us, can be measured by the force opposite and equal to it which can prevent the body from falling. This is just how one measures weights on a scale or with a balance. The acceleration of gravity at that location is a component acceleration corresponding to its weight. The force equal and opposite to its weight is exactly what is needed to produce an equal and opposite acceleration component on that body, so that the total acceleration of that body with respect to the center of the earth is zero.

2.iii Newton on relating his three measures Newton's comments on definition 8 continue with the following discussion relating his three measures of centripetal force.

These quantities of forces, for the sake of brevity, may be called motive, accelerative, and absolute forces, and, for the sake of differentiation, may be referred to bodies seeking a center, to the places of the bodies, and to the center of the forces: that is, motive force may be referred to

[9] Newton takes instantaneous impulses acting at discrete units of time as a basic model for forces, rather than continuously acting forces as we do. His treatment can recover continuously acting forces by taking limits, as time intervals between impulses are made smaller and smaller. (See chpt. 3, appendix 2, footnote 98 for references.)

a body as an endeavor of the whole directed toward a center and compounded of the endeavors of all the parts; accelerative force, to the place of the body as a certain efficacy diffused from the center through each of the surrounding places in order to move the bodies that are in those places; and absolute force, to the center as having some cause without which the motive forces are not propagated through the surrounding regions, whether this cause is some central body (such as a lodestone in the center of a magnetic force or the earth in the center of a force that produces gravity) or whether it is some other cause which is not apparent. This concept is purely mathematical, for I am not now considering the physical causes and sites of forces. (C&W, 407)

Newton's discussion of accelerative measure as "a certain efficacy diffused from the center through each of the surrounding places in order to move the bodies that are in those places" adds to the suggestion that a centripetal force can be counted as the cause of the various motive forces referred to the bodies on which it acts. Whereas the motive measures of a centripetal force are referred to the bodies being centripetally accelerated by it, those centripetal accelerations are assigned to the places of those bodies as the accelerative measures of that centripetal force at those places. The centripetal acceleration assigned to a place is treated as the measure of its efficacy to move any body that would be located at that place.[10] The absolute measure of a centripetal force is assigned to the center as having a cause, without which it would not generate motive forces of attraction on bodies at places around it. Newton suggests that this concept of cause is purely mathematical, as he is not now considering the physical causes and sites of forces.[11] He also points out that he is not claiming that the cause need be identified as a body at the center, such as the earth for gravitation.

Newton adds a discussion of the mathematical relation between accelerative and motive forces.

Therefore, accelerative force is to motive force as velocity to motion. For quantity of motion arises from velocity and quantity of matter jointly, and motive force from accelerative force and quantity of matter jointly. For the sum of the actions of the accelerative force on the individual particles of a body is the motive force of the whole body. As a consequence, near the surface of the earth, where the accelerative gravity, or the force that produces gravity, is the same in all bodies universally, the motive gravity, or weight, is as the body, but in an ascent to regions where the accelerative gravity becomes less, the weight will decrease proportionally and will always be as the body and the accelerative gravity jointly. Thus, in regions where the accelerative gravity is half as great, a body one half or one third as great will have a weight four or six times less. (C&W, 407–8)

[10] Eric Schliesser (2011) has proposed that the places referred to should be limited to the places actually occupied by the bodies being attracted. This would make centripetal forces more easily fit Newton's later concept of gravity as a pair-wise interaction between bodies, than Stein's interpretation on which Newton's inverse-square centripetal forces are inverse-square centripetal fields of acceleration. We shall consider Schliesser's alternative proposal in section IV.6 below.

[11] We have noted (chpt. 1 sec. II.2) that Newton's counting this as a mathematical treatment of cause will be important for his claim to empirically establish that gravity is the cause of orbital motion, even though he has not given any account of a physical cause of gravity. This important and illuminating feature of Newton's scientific method will be explored in more detail in section 2.iv below.

We can take the acceleration of gravity at a given place as the accelerative measure of gravity at that place. This makes that acceleration times the mass of a body at that place the corresponding quantity of the motive force of gravity on that body. Here we can see Newton formulating motive forces according to our familiar conception – mass times acceleration.

In this passage Newton strongly reinforces the idea that, for gravity, the accelerative measure can be referred to the places, without also being referred to the accelerated bodies as well; because, at any given distance from the center all bodies will be subject to equal centripetal accelerations. This property makes a centripetal force into a centripetal acceleration field.[12] We shall see, in section III of this chapter, that Newton's arguments from Harmonic Rule phenomena for their systems of orbiting moons or planets to inverse-square forces of attraction toward Jupiter, Saturn, and the sun are arguments to the inverse-square variation of the centripetal accelerations exhibited by those orbits.[13] Indeed, it is the accelerative measures, not the motive measures, that Harmonic Rule phenomena measure to vary as the inverse-square of the orbital mean-distances.

We have seen that the absolute quantity of a centripetal force is assigned to the center to characterize its strength as a whole. Having acceleration fields with the same law relating accelerations to distances affords a ratio measure of their absolute quantities. For any two such centripetal forces the ratio of the accelerations assigned to equal distances from their respective centers measures the ratio of their absolute quantities.

2.iv More on Newton's characterization of his method as mathematical Newton's discussion of definition 8 concludes with the following paragraph, which expands on his characterization of his conception of force as mathematical.

Further, it is in this same sense that I call attractions and impulses accelerative and motive. Moreover, I use interchangeably and indiscriminately words signifying attraction, impulse, or any sort of propensity toward a center, considering these forces not from a physical but only from a mathematical point of view. Therefore, let the reader beware of thinking that by words of this kind I am anywhere defining a species or mode of action or a physical cause or reason, or that I am attributing forces in a true or physical sense to centers (which are mathematical points) if I happen to say that centers attract or that centers have forces. (C&W, 408)

An important aspect of Newton's mathematical conception of cause is that, though he explicitly calls the centripetal force the cause of the corresponding motive forces on bodies accelerated by it, he makes no commitment about the physical cause of the centripetal force itself.[14] This made it possible for defenders of the mechanical philos-

[12] We are following Stein's interpretation of these centripetal forces as centripetal fields of acceleration. The inverse-square varying accelerations assigned to places around the center are measures of the field intensities at those places.

[13] See Stein 1970b, 265–9 as well.

[14] As we noted in chapter 1 (sec. II.2 and VII.2), this aspect of Newton's mathematical method is further expanded upon in the important scholium to proposition 69 book 1 and is also very much in line with Newton's famous *hypotheses non fingo* passage from his General Scholium at the end of book 3.

ophy, such as Huygens and Leibniz who proposed vortex theories of gravity, to consistently accept Newton's identification of the centripetal force maintaining the moon in its orbit with terrestrial gravity.[15] In chapter 4 we shall see that the centripetal accelerations of the moon and of freely falling terrestrial bodies are all identified as agreeing measures of the strength (Newton's absolute quantity) of the same inverse-square centripetal acceleration field surrounding the earth.

II Newton's scholium to the definitions and his Laws of Motion

1. The scholium on time, space, place, and motion

After Einstein's arguments for his theories of Special and General Relativity many philosophers dismissed Newton's absolute time, space, and motion as philosophically suspect appendages to his physical theories that should have been rejected on empiricist grounds. Einstein's work was taken to show that objections by Newton's contemporaries Huygens and Leibniz, as well as later objections by Berkeley and Mach, had been correct and should have been decisive all along.

In 1967 a seminal paper by Howard Stein showed that Newton's treatment of space and time is far more subtle and informed than the dismissive accounts of such philosophers had suggested. Later work by other philosophers also began to take into account the extent to which the treatment of features of space-time as objective physical facts in General Relativity fails to realize the sort of empiricist relational theory endorsed by Mach. A number of such philosophers have taken Newton's discussions in the scholium as arguments for the existence of absolute space. The bucket experiment[16] and the rotating globes thought experiment[17] were taken to be

In chapter 9, we will include a more detailed treatment of Newton's characterization of his method as mathematical in the context of his application of Law 3 to construe gravity as a force of interaction between bodies. This treatment will occasion a more detailed treatment of methodological implications of Newton's important scholium to proposition 69 book 1, as well as consideration of Andrew Janiak's (2007 and 2008) interesting development of these issues.

[15] See chapter 5.

[16] Here is Newton's basic account of this bucket experiment:

> The effects distinguishing absolute motion from relative motion are the forces of receding from the axis of circular motion. For in purely relative circular motion these forces are null, while in true and absolute circular motion they are larger or smaller in proportion to the quantity of motion. If a bucket is hanging from a very long cord and is continually turned around until the cord becomes twisted tight, and if the bucket is thereupon filled with water and is at rest along with the water and then, by some sudden force, is made to turn around in the opposite direction and, as the cord unwinds, perseveres for a while in this motion; then the surface of the water will at first be level, just as it was before the vessel began to move. But after the vessel, by the force gradually impressed upon the water, has caused the water also to begin revolving perceptibly, the water will gradually recede from the middle and rise up the sides of the vessel, assuming a concave shape (as experience has shown me), and, with an ever faster motion, will rise further and further until, when it completes its revolutions in the same times as the vessel, it is relatively at rest in the vessel. The rise of the water reveals its endeavor to recede from the axis of motion, and from such an endeavor one can find out and measure the true and absolute circular motion of the water, which here is the direct opposite of its relative motion. (C&W, 412–13)

[17] Here is Newton's basic account of this thought experiment:

better arguments for absolute space than empiricists, such as Reichenbach, had allowed.[18]

More recently, significant aspects of Stein's discussion, that had not been a focus of these philosophers, are being addressed. Robert DiSalle[19] has pointed out that Newton is not attempting to argue for absolute space as an answer to already defined questions about space, time, and motion. Rather, he is refining our intuitive common sense conceptions of space, time, and motion in a way that connects them with the laws of physics and with the empirical practice of measurement. The question he is addressing is how the concepts of space, time, and motion can be formulated in order to provide a coherent basis for his Laws of Motion and their application to the system of the world. Newton's arguments assume the distinction between true motions (those resulting from forces) and merely relative motions. Such a distinction is required for the applications of dynamics to motion phenomena shared by Newton and his critics.[20] Newton's arguments are designed to show that this fundamental distinction is not adequately captured by any merely relational account of space, time, and motion.[21]

1.i Absolute time Newton defines absolute time as time which flows uniformly. His first Law of Motion supports uniform rotation, so that equality of time intervals could be defined by the equal angular motions of a freely rotating body. Newton refers to the equation of time, which corrects solar time (given by the earth's rotation relative to the sun) to give agreement with sidereal time (given by the earth's rotation relative to fixed stars).

> It is certainly very difficult to find out the true motions of individual bodies and actually to differentiate them from apparent motions, because the parts of that immovable space in which the bodies truly move make no impression on the senses. Nevertheless, the case is not utterly hopeless. For it is possible to draw evidence partly from apparent motions, which are the differences between the true motions, and partly from the forces that are the causes and effects of the true motions. For example, if two balls, at a given distance from each other with a cord connecting them, were revolving about a common center of gravity, the endeavor of the balls to recede from the axis of motion could be known from the tension of the cord, and thus the quantity of circular motion could be computed. Then, if any equal forces were simultaneously impressed upon the alternate faces of the balls to increase or decrease their circular motion, the increase or decrease of the motion could be known from the increased or decreased tension in the cord, and thus, finally, it could be discovered which faces of the balls the forces would have to be impressed upon for a maximum increase in the motion, that is, which were the posterior faces, or the ones that are in the rear in a circular motion. Further, once the faces that follow and the opposite faces that precede were known, the direction of the motion would be known. In this way both the quantity and the direction of this circular motion could be found in any immense vacuum, where nothing external and sensible existed with which the balls could be compared. (C&W, 414)

[18] See, e.g., Friedman 1983 and, especially, Earman 1989.

[19] DiSalle's recent book (DiSalle 2006) is an extension of investigations initiated by Stein. It affords a very informative treatment of the complex methodological and evidential issues raised by Newton's conceptions of space, time, and motion and their radical transformation by Einstein.

[20] Robert Rynasiewicz (1995a) offers a detailed account of the arguments in the text of Newton's scholium that makes clear that they assume that absolute motion exists. Rynasiewicz (1995b) argues that this assumption was shared by his critics.

[21] Ori Belkind (2007) has recently focused on Newton's discussion of "place" as the place of a body to attribute to Newton an interesting argument that absolute space is needed to have a workable definition of momentum.

In astronomy, absolute time is distinguished from relative time by the equation of common time. For natural days, which are commonly considered equal for the purpose of measuring time, are actually unequal. Astronomers correct this inequality in order to measure celestial motions on the basis of a truer time. It is possible that there is no uniform motion by which time may have an exact measure. All motions can be accelerated and retarded, but the flow of absolute time cannot be changed. The duration or perseverance of the existence of things is the same, whether their motions are rapid or slow or null; accordingly, duration is rightly distinguished from its sensible measures and is gathered from them by means of an astronomical equation. (C&W, 410)

The point of Newton's distinction between absolute time and its sensible measures is to give conceptual space for always seeking improvement in the measurement of time. Just as astronomers were able to correct solar time to take into account the non-uniformity of the lengths of the solar days, so they are now able to correct sidereal time to take into account a slowing of the earth's rotation due to tidal friction.

Though he does not emphasize it in his scholium, Newton's absolute time includes absolute simultaneity for distant events. His distinction between absolute time and its sensible measures includes room for arbitrarily fast signals informing us of distant events. This allows for synchronization of distant clocks that would be invariant with respect to arbitrarily moving centers.[22]

1.ii Absolute space As we saw in chapter 1, Newton resolved the two chief world systems problem by showing that neither the center of the earth nor that of the sun could count as a center relative to which one could distinguish the true motions of the sun and planets. He showed that neither the earth nor the sun occupies the center of mass of the solar system bodies. The sun, though engaged in continual motion, never recedes far from this common center of gravity.

[22] This, of course, is what Einstein's special theory of relativity provided an alternative to in his 1905 paper "On the electrodynamics of moving bodies" (see Einstein, et al. [1923] 1952, 37–65). There was considerable initial resistance to Einstein's theory, including a rejection in 1907 by the physics faculty at Bern of Einstein's submission of his 1905 relativity paper as his *Habilitationsschrift* (Holton 1990, 58). By 1911 or so, Einstein's theory was fairly widely accepted among German physicists (Jungnickel and McCormmach 1986, vol. 2, 247–8).

Harper (1979) applies a conceptual change model to capture the transition from classical kinematics to the kinematics of Special Relativity for a Hertzian physicist who initially treated Galilean kinematics as a conceptual commitment which would be taken to be immune to empirical refutation. My agent was endowed with the capacity to carry out hypothetical reasoning to evaluate Einstein's alternative, once it was presented to him. He was then able to use empirical evidence to decide between it and his original commitments, once he had evaluated it as a serious alternative candidate. I have argued that anyone who appreciates the facility with which human beings can reason about alternative mathematical systems (once they are made aware of them) will have little doubt that a physicist, even a Hertzian who initially treated Galilean invariance as a conceptual commitment, would be capable of such a rational, even if radical, revision of accepted theory.

One of the strikingly attractive features of Newton's treatment of his space–time framework and Laws of Motion is that he explicitly counts them as empirical and open to revision. Unlike the Hertzian physicist, who would count them as *a priori* commitments that would have to be overturned, Newton's Rule 4 makes the acceptance of his Laws of Motion and space–time framework explicitly provisional.

Relative to this center of mass, the separate centripetal acceleration fields toward solar system bodies are combined into a single system. Each body undergoes an acceleration toward each of the others proportional to the mass of that other body and inversely proportional to the square of the distance between them. The very great mass of the sun makes elliptical orbits about the sun a good first approximation from which to begin accounting for the complex motions of the planets. The impressive application of universal gravity to distinguish the true motions among solar system bodies, by Newton and such distinguished successors as Euler, Laplace and others, was a sequence of increasingly more accurate approximations achieved by successively taking into account more and more gravitational interactions among solar system bodies.

Newton sets up his treatment of the two chief world systems problem by introducing what he explicitly identifies as a hypothesis.

Hypothesis 1. *The center of the system of the world is at rest.*

No one doubts this, although some argue that the earth, others that the sun, is at rest in the center of the system. Let us see what follows from this hypothesis. (C&W, 816)

Proposition 11. *The common center of gravity of the earth, the sun, and all the planets is at rest.*

For that center (by corol. 4 of the Laws) either will be at rest or will move uniformly straight forward. But if that center always moves forward, the center of the universe will also move, contrary to the hypothesis. (C&W, 816)

Proposition 12. *The sun is engaged in continual motion but never recedes far from the common center of gravity of all the planets.* (C&W, 816)

Newton's treatment of absolute space makes a distinction between absolute rest and uniform motion. This allows him to, provisionally, make sense of earth-centered and sun-centered hypotheses in order to develop his dynamical solution to the two chief world systems problem, rather than seeming to simply beg the question against the hypothesis that the earth is at rest.

Clearly, Newton is not endorsing this claim he has explicitly identified as a hypothesis. As we noted in chapter 1, a central feature of his methodology is his sharp separation between hypotheses and propositions that count as gathered from phenomena by induction.

Newton is also clearly aware that his center of mass solution to the problem of identifying forces and motions among solar system bodies is invariant with respect to uniform straight line velocities. This is explicit in corollary 5 of his Laws of Motion. Any center that is moving uniformly straight forward relative to the center of mass will recover the same relative motions among solar system bodies. This makes his distinction between absolute rest and uniform motion irrelevant to his solution. Corollary 5 shows that Newton's solution to the problem of determining what are to be counted as the true motions of the solar system bodies does not require fixing what are to be counted as the true velocities of bodies in absolute space.[23] This would limit true motions to accelerations. Corollary 6 shows that any center that is uniformly accelerating in any straight line relative to the center of mass will also recover the same relative motions and forces among solar system bodies. This makes it clear that Newton's solution also does not require fixing what are to be counted as true accelerations in absolute space.[24]

2. *The Laws of Motion and their corollaries*

Laws of Motion Law 1

Every body perseveres in its state of being at rest or of moving uniformly straight forward, except insofar as it is compelled to change its state by forces impressed.

Projectiles persevere in their motions, except insofar as they are retarded by the resistance of the air and are impelled downward by the force of gravity. A spinning hoop, which has parts that by their cohesion continually draw one another back from rectilinear motions, does not cease to rotate, except insofar as it is retarded by the air. And larger bodies – planets and comets – preserve for a longer time their progressive and their circular motions, which take place in spaces having less resistance. (C&W, 416)

Newton's discussion makes clear that he regards Law 1 as preserving rotations, as well as uniform rectilinear motions. We can think of the centripetal acceleration of a part of a uniformly rotating hoop as exactly orthogonal to its tangential motion. It, therefore, will make no change to the magnitude of the tangential velocity.

Laws of Motion Law 2

A change in motion is proportional to the motive force impressed and takes place along the straight line in which that force is impressed.

If some force generates any motion, twice the force will generate twice the motion, and three times the force will generate three times the motion, whether the force is impressed all at once or successively by degrees. And if the body was previously moving, the new motion (since motion is always in the same direction as the generative force) is added to the original motion if that motion was in the same direction or is subtracted from the original motion if it was in the opposite direction or, if it was in an oblique direction, is combined obliquely and compounded with it according to the directions of both motions. (C&W, 417–18)

Newton tells us his formulation applies whether the force is impressed all at once or successively by degrees.[25] His basic model is a force impressed all at once in an instanta-

[23] Corollary 5 explicitly makes absolute rest undetectable from motions of bodies. A corresponding Galilean relativity principle – that space–time structure only includes what is invariant with respect to choice of inertial frame – would rule out absolute rest as an objective feature of space–time.

The space–time structure Howard Stein called "Newtonian Space–time" (Stein 1967, 174–6) takes the structure of what are to be counted as true motions according to Newton's dynamics to be given by corollary 5 of his Laws of Motion. This is the structure John Earman calls "Neo-Newtonian Space–time" (Earman 1989, 33).

[24] The space–time corresponding to what are taken to be the true motions according to corollaries 5 and 6 together is given by what John Earman has called Maxwellian Space–time (Earman 1989, 31–2).

[25] See Pourciau (2006) for an interesting account of Newton's two conceptions of force and his interpretation of his second Law of Motion.

neous impact. Newton's statement of this law for this case makes force proportional to change of motion, that is, to change in mass times velocity, rather than to mass times acceleration. This application would have

$$f \propto \Delta(mv) = m(\Delta v)$$

in place of the

$$f \propto ma = m(dv/dt)$$

which would be expected by a student of Newtonian physics today.

His claim that Law 2 also holds when the force is impressed successively by degrees suggests that he intends it to cover continuously acting forces, as well as a succession of discrete instantaneous impacts. Newton's basic formulation in terms of instantaneous impacts can recover our modern formulation for continuously acting forces by taking appropriate limits as times between impacts are made increasingly small.[26]

Laws of Motion Law 3

To any action there is always an opposite and equal reaction; in other words, the actions of two bodies upon each other are always equal and always opposite in direction.

Whatever presses or draws something else is pressed or drawn just as much by it. If anyone presses a stone with a finger, the finger is also pressed by the stone. If a horse draws a stone tied to a rope, the horse will (so to speak) also be drawn back equally toward the stone, for the rope, stretched out at both ends, will urge the horse toward the stone and the stone toward the horse by one and the same endeavor to go slack and will impede the forward motion of the one as much as it promotes the forward motion of the other. If some body impinging upon another body changes the motion of that body in any way by its own force, then, by the force of the other body (because of the equality of their mutual pressure), it also will in turn undergo the same change in its own motion in the opposite direction. By means of these actions, equal changes occur in the motions, not in the velocities – that is, of course, if the bodies are not impeded by anything else. For the changes in velocities that likewise occur in opposite directions are inversely proportional to the bodies because the motions are changed equally. This law is valid also for attractions, as will be proved in the next scholium. (C&W, 417).

By including drawings as well as presses in his opening remark, Newton solicits intuitive support for his effort to widen the application of Law 3 beyond the presses to which the mechanical philosophers would restrict it. His example of a rope connecting a horse to a stone it draws, like his discussion of a stone in a sling in his discussion of centripetal forces, brings to bear familiar facts about pulling things by a rope or a string. Insofar as the stretched rope can be construed as approximating Newton's description of it as transmitting equal and oppositely directed pulls on the horse and stone, this example supports extending applications of the third law beyond bodies directly hitting or pushing on one another. Newton announces that he will

[26] See appendix 2.

prove that the third law is valid also for attractions in the next scholium. We shall consider this scholium, his scholium to the laws, after reviewing his corollaries to the laws. We shall, however, postpone a detailed treatment of his arguments for extending Law 3 to attractions until chapter 9.

Let us now turn to the corollaries.

Laws of Motion Corollary 1

A body acted on by [two] forces acting jointly describes the diagonal of a parallelogram in the same time in which it would describe the sides if the forces were acting separately.

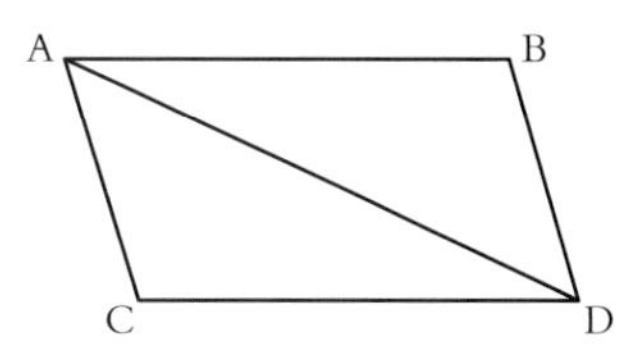

Let a body in a given time, by a force M alone impressed in A, be carried with uniform motion from A to B, and by the force N alone impressed in the same place, be carried from A to C; then complete the parallelogram ABDC, and by both forces the body will be carried in the same time along the diagonal from A to D. For, since force N acts along the line AC parallel to BD, this force, by law 2, will make no change at all in the velocity toward the line BD which is generated by the other force. Therefore, the body will reach the line BD in the same time whether the force N is impressed or not, and so at the end of that time it will be found somewhere on the line BD. By the same argument, at the end of the same time it will be found somewhere along on the line CD, and accordingly it is necessarily found at the intersection D of both lines. And, by law 1, it will go with [uniform] rectilinear motion from A to D. (C&W, 417–8)

Laws of Motion Corollary 2

And hence the composition of a direct force AD out of any oblique forces AB and BD is evident, and conversely the resolution of any direct force AD into any oblique forces AB and BD. And this kind of composition and resolution is indeed abundantly confirmed from mechanics. (C&W, 418)

Newton gives an example of resolving the forces to move a wheel of weights hung from unequal spokes going out from the center. This example illustrates pulleys, stretched strings and the law of the lever.[27] It is then further elaborated to show how to resolve forces corresponding to the actions of a wedge, a hammer, and a screw. He concludes with the following remarks:

Therefore, this corollary can be used very extensively, and the variety of its applications clearly shows its truth, since the whole of mechanics – demonstrated in different ways by those who have written on the subject – depends on what has just now been said. For from this are easily derived the forces of machines, which are generally composed of wheels, drums, pulleys, levers, stretched strings, and weights, ascending directly or obliquely, and the other mechanical powers, as well as the forces of tendons to move the bones of animals. (C&W, 419–20)

[27] Chandrasekhar (1995, 24–5) provides a very nice explication of Newton's resolution of these forces.

Newton is not just saying that the forces of machines and the forces of tendons to move the bones of animals can be derived from this corollary. He is saying that all these extensive varieties of applications of mechanics depend on it. These applications of mechanics would not be available if this corollary did not hold to sufficiently good approximations.

Laws of Motion Corollary 3

The quantity of motion, which is determined by adding the motions made in one direction and subtracting the motions made in the opposite direction, is not changed by the action of bodies on one another.

For an action and the reaction opposite to it are equal by law 3, and thus by law 2 the changes which they produce in motions are equal and in opposite directions. Therefore, if motions are in the same direction, whatever is added to the motion of the first body [*lit.* the fleeing body] will be subtracted from the motion of the second body [*lit.* the pursuing body] in such a way that the sum remains the same as before. But if the bodies meet head-on, the quantity subtracted from each of the motions will be the same, and thus the difference of the motions made in opposite directions will remain the same. (C&W, 420)

Newton offers up a number of examples illustrating these applications of Law 3 to interactions among bodies. The main role of this corollary is its application in Newton's argument for his very important next corollary.

Laws of Motion Corollary 4

The common center of gravity of two or more bodies does not change its state whether of motion or of rest as a result of the actions of the bodies upon one another; and therefore the common center of gravity of all bodies acting upon one another (excluding external actions and impediments) either is at rest or moves uniformly straight forward. (C&W, 421)

Newton proves this first for cases in which the bodies do not interact (C&W, 422). He then proves it for systems of two bodies interacting (C&W, 422), and generalizes this to systems of several bodies (C&W, 422–3).

Corollary 5 is a very important principle of what is often called Galilean relativity.

Laws of Motion Corollary 5

When bodies are enclosed in a given space, their motions in relation to one another are the same whether the space is at rest or whether it is moving uniformly straight forward without circular motion.

For in either case the differences of the motions tending in the same direction and the sums of those tending in opposite directions are the same at the beginning (by hypothesis), and from these sums or differences there arise the collisions and impulses [*lit.* impetuses] with which the bodies strike one another. Therefore, by law 2, the effects of the collisions will be equal in both cases, and thus the motions with respect to one another in the one case will remain equal to the motions with respect to one another in the other case. This is proved clearly by experience: on a ship, all the motions are the same with respect to one another whether the ship is at rest or is moving uniformly straight forward. (C&W, 423)

As we have seen, this Galilean relativity gives the space-time structure Howard Stein identified as Newtonian space-time. On this structure accelerations are counted as absolute, even though velocities are not.

Newton's relativity of motion is more extensive than this Galilean relativity of corollary 5.

Laws of Motion Corollary 6

If bodies are moving in any way whatsoever with respect to one another and are urged by equal accelerative forces along parallel lines, they will all continue to move with respect to one another in the same way as they would if they were not acted on by those forces.

For those forces, by acting equally (in proportion to the quantities of the bodies to be moved) and along parallel lines, will (by law 2) move all the bodies equally (with respect to velocity), and so will never change their positions and motions with respect to one another. (C&W, 423)

This additional invariance with respect to equal and parallel accelerations does not require that accelerations be counted as absolute. We shall see in chapter 8, that the very great distance to any given star makes the gravitational attraction of solar system bodies toward it very nearly equal and parallel.

3. Newton's scholium to the Laws (empirical support offered)

Newton opens by characterizing his Laws of Motion as accepted by mathematicians and confirmed by experiments of many kinds.

Scholium

The principles I have set forth are accepted by mathematicians and confirmed by experiments of many kinds. (C&W, 424)

He goes on to recount some of these confirmations.

By means of the first two laws and the first two corollaries Galileo found that the descent of heavy bodies is in the squared ratio of the time and that the motion of projectiles occurs in a parabola, as experiment confirms, except insofar as these motions are somewhat retarded by the resistance of the air. (C&W, 424)

These remarks refer to the confirmation afforded to the first two laws and the first two corollaries by their role accounting for the, by then familiar, idealized patterns that dominate and make intelligible the ubiquitous phenomena of free fall and projectile motion.[28] In the third edition these remarks are supplemented by a more detailed account, together with an example diagramming the parabolic trajectory for projectile

[28] Newton attributes these laws, corollaries, and their application to falling bodies and projectiles to Galileo. This exaggerates Galileo's anticipation of these laws. For example, as we noted in chapter 1, Galileo assumed horizontal inertia rather than the full content of Newton's first law.

motion (C&W, 424). As was clear from corollary 2 for basic mechanics, these well-studied and ubiquitous features of phenomena of falling bodies and projectile motions would be significantly violated unless these first two laws and first two corollaries held to sufficiently good approximations.

Newton goes on to refer to the confirmation afforded to those first two laws and corollaries from their application to pendulums and clocks.

> What has been demonstrated concerning the times of oscillating pendulums depends on the same first two laws and first two corollaries, and this is supported by daily experience with clocks. (C&W, 424)

This paragraph continues with a reference to papers in which Wren, Wallis, and Huygens reported finding the rules of the collisions and reflections of bodies.

> From the same laws and corollaries and law three, Sir Christopher Wren, Dr. John Wallis, and Mr. Christiaan Huygens, easily the foremost geometers of the previous generation, independently found the rules of the collisions and reflections of hard bodies, and communicated them to the Royal Society at nearly the same time, entirely agreeing with one another (as to these rules); and Wallis was indeed the first to publish what had been found, followed by Wren and Huygens. But Wren additionally proved the truth of these rules before the Royal Society by means of an experiment with pendulums, which the eminent Mariotte soon after thought worthy to be made the subject of a whole book. (C&W, 424–5)

Newton counts the cited pendulum experiment as confirmation of Law 3, as well as counting it as additional confirmation of the first two laws and first two corollaries.

A pendulum experiment, such as the one cited, would have appealed to the accepted theory of pendulum motion to make possible *theory-mediated* measurements of velocities before collision and after reflection. These velocities could not have been observed directly; but, as Newton points out later in this scholium (C&W, 425), they could be measured by the chords of the arcs of their pendulum swings before impact and after reflection.[29] The experiment would have afforded *theory-mediated* measurements of the equality of action and reaction in accord with the application of Newton's Law 3 to collisions, modeled as idealized instantaneous impacts. The change in motion of each body caused by such an idealized interaction would be the difference between quantity of motion before impact and its quantity of motion after reflection. We have seen (Definition 2) that Newton takes momentum – mass times velocity – as his measure of quantity of motion. So, for such an idealized instantaneous interaction,

[29] In the following quotation, Newton attributes this proposition to an idealized pendulum in a vacuum.

> For it is a proposition very well known to geometers that the velocity of a pendulum in its lowest point is as the chord of the arc that it has described in falling. (C&W, 425)

See appendix 1.1 for a proof of this for such idealized pendulum motion, in which there is no loss of energy. This velocity is also proportional to the square root of the height from which it was let fall, a fundamental property of Galileo's uniformly accelerated fall.

corresponding to a collision between body A and body B, the equality of action and reaction corresponding to Law 3 would be the conservation of momentum.[30]

Newton opens a new paragraph by pointing out that such an experiment ought to take into account air resistance and the elastic force of the colliding bodies.

> However, if this experiment is to agree precisely with the theories, account must be taken of both the resistance of the air and the elastic force of the colliding bodies. (C&W, 425)

The rest of this paragraph gives the method by which Newton carried out his own improved version to take into account air resistance. He uses an example of a pendulum experiment in which a body A is let fall to collide with a stationary body B, to show how to correct for air resistance to find the arcs of motion that represent the velocities before impact and after reflection that would have obtained if that experiment had been performed in a vacuum.[31] He then reports his results from pendulum experiments applying this method to collisions:

> On making a test in this way with ten-foot pendulums, using unequal as well as equal bodies, and making the bodies come together from very large distances apart, say of eight or twelve or sixteen feet, I always found – within an error of less than three inches in the measurements – that when the bodies met each other directly, the changes of motions made in the bodies in opposite directions were equal, and consequently that the action and reaction were always equal . . . (C&W, 426)

Examples of a collision with one body striking another at rest and of two bodies colliding head-on are followed by an example of a collision where one body overtakes another moving more slowly in the same direction. Here is Newton's summary of the agreement of these empirical measurements with the equality of action and reaction.

> As a result of the meeting and collision of bodies, the quantity of motion – determined by adding the motions in the same direction and subtracting the motions in opposite directions – was never changed. I would attribute the error of an inch or two in the measurements to the difficulty of doing everything with sufficient accuracy. (C&W, 426)

These experiments are *theory-mediated* measurements which show results that accurately agree with the equality of action and reaction in collisions, to within the precision to which the velocities were empirically determined. The theory of pendulum motion made it possible to carry out these velocity measurements.

Newton motivates extending such experiments to bodies that are not perfectly elastic by considering the possible objection that the rule they were designed to prove presupposes idealized perfectly elastic bodies, which do not occur naturally.

> Further, lest anyone object that the rule which this experiment was designed to prove presupposes that bodies are either absolutely hard or at least perfectly elastic and thus of a kind which do not

[30] $m_A v_A + m_B v_B = m_A u_A + m_B u_B$,

where v_A and v_B stand for velocities before impact, and u_A and u_B stand for velocities after reflection.

[31] See appendix 1.2 for details.

occur naturally, I add that the experiments just described work equally well with soft bodies and with hard ones, since surely they do not in any way depend on the condition of hardness. For if this rule is to be tested in bodies that are not perfectly hard, it will only be necessary to decrease the reflection in a fixed proportion to the quantity of elastic force. (C&W, 427)

He claims that the experiments can be extended to imperfectly elastic bodies by decreasing the reflections in a fixed proportion. We shall see that he offers an account of how to empirically measure the proportions corresponding to the differing degrees of elasticity for bodies of different materials.

Newton points out that the theory of Huygens and Wren was for idealized collisions in which there would be no loss of total velocity.[32]

In the theory of Wren and Huygens, absolutely hard bodies rebound from each other with the velocity with which they have collided. This will be affirmed with more certainty of perfectly elastic bodies. (C&W, 427)

Newton goes on to introduce a measurable parameter of elasticity.

In imperfectly elastic bodies the velocity of rebounding must be decreased together with the elastic force, because that force (except when the parts of the bodies are damaged as a result of collision, or experience some sort of extension such as would be caused by a hammer blow) is fixed and determinate (as far as I can tell) and makes the bodies rebound from each other with a relative velocity that is in a given ratio to the relative velocity with which they collide. (C&W, 427)

He tests this with tightly wound balls of wool, balls of steel, of cork, and of glass.

I have tested this as follows with tightly wound balls of wool strongly compressed. First, releasing the pendulums and measuring their reflection, I found the quantity of their elastic force; then from this force I determined what the reflections would be in other cases of their collision, and the experiments which were made agreed with the computations. The balls always rebounded from each other with a relative velocity that was to the relative velocity of their colliding as 5 to 9, more or less. Steel balls rebounded with nearly the same velocity and cork balls with a slightly smaller velocity, while with glass balls the proportion was roughly 15 to 16. (C&W, 427)

These reported results claim that there were no detectable losses corresponding to imperfect elasticities of the steel balls. This suggests that these experiments count as measurements of a coefficient of restitution for steel balls that was not detectably

[32] Huygens and Wren independently arrived at the same theory for computing post-reflection velocities from pre-collision velocities and the relative weights of bodies in a perfectly elastic collision where velocity is conserved. For such idealized collisions, the velocity lost by body A is equal to the velocity gained by body B. For such a collision where the v's are the velocities before impact and the u's are the velocities after impact, we have, as Huygens pointed out,

$$m_A \, v_A{}^2 + m_B \, v_B{}^2 = m_A \, u_A{}^2 + m_B \, u_B{}^2.$$

This is our formula for conservation of kinetic energy. See Huygens (1669) and Wren (1668) for the Latin originals. See Murray, Harper, and Wilson (2011) for translations by Wilson, with commentaries and expositions by Murray, Harper, and Wilson.

different from 1, for the relative velocities obtained in them. Newton's report of obtaining 15/16 for glass balls and slightly less for cork balls suggests that his experiments were accurate enough to have detected correspondingly sized deviations from perfect restitution for the steel balls had they occurred.

These experiments afford *theory-mediated* measurements of velocities that are fit by the equality of action and reaction calculated in accordance with reductions appropriate to the differing elasticities.

> And in this manner the third Law of Motion – insofar as it relates to impacts and reflections – is proved by this theory, which plainly agrees with experiments. (C&W, 427)

This suggests that Law 3 applies to collisions quite generally. What are to be counted as action and reaction are to be appropriately reduced to take into account losses due to imperfect elasticity.

This can be extended to include reductions to take into account losses due to damage of the bodies in the collision. We have seen that Newton's Rules 3 and 4 (together) endorse provisionally extending to bodies beyond the reach of experiments parameter values that have been found to have constant agreeing values for the bodies within the reach of experiments. This supports extending the equality of action and reaction to collisions that would require reductions appropriate to losses due to damage or deformation of the bodies, even though such losses would be difficult to directly measure. According to Rule 4, to undercut Newton's extension of the third Law of Motion to such collisions would require empirically establishing phenomena that would make the equality of action and reaction in such collisions liable to exceptions.[33]

After this empirical defense of its application to collisions, Newton argues for extending Law 3 to attractions. This argument includes an empirical experiment with a magnet and piece of iron floating in separate vessels (C&W, 427–8). It also includes a thought-experiment arguing for the application of the third law to the mutual gravity between the earth and its parts (C&W, 428). We shall postpone consideration of these arguments until chapter 9 below, where we will take up objections to Newton's application of Law 3 to count gravity as a mutual interaction between pairs of bodies, even when they are not touching one another.

In corollary 2 of the Laws (C&W, 418–20), Newton showed how the Laws of Motion are confirmed by the detailed compositions and resolutions of impressed forces exhibited in the actions of such fundamental machines as wheels, pulleys, levers, wedges, and stretched strings, as well as the forces of tendons to move the bones of animals. As we noted, he pointed out that all these extensive and varied applications of mechanics would not be available if this corollary did not hold to sufficiently good approximations.

[33] The experiments cited by Newton appear to make the proposition that Law 3 applies to collisions count as empirical and as having been gathered from phenomena by induction. This lends strong support for Newton's claim (C&W, 943) that he discovered the Laws of Motion by the same method as he used to discover universal gravity. See chapter 1 section VII.2.

The last three paragraphs of his scholium to the Laws are devoted to a recounting of the enormous empirical support afforded to Law 3 from these applications to machines and devices (C&W, 428–30).[34] Newton is at some pains to point out that these applications of Law 3 are not restricted to idealized cases in which resistance from sources such as friction can be ignored.

> The effectiveness and usefulness of all machines or devices consist wholly in our being able to increase the force by decreasing the velocity, and vice versa; in this way the problem is solved in the case of any working machine or device: "To move a given weight by a given force" or to overcome any other given resistance by a given force. For if machines are constructed in such a way that the velocities of the agent [or acting body] and the resistant [or resisting body] are inversely as the forces, the agent will sustain the resistance and, if there is a greater disparity of velocities, will overcome the resistance. Of course the disparity of the velocities may be so great that it can also overcome all the resistance which generally arises from the friction of contiguous bodies sliding over one another, from the cohesion of continuous bodies that are to be separated from one another, or from the weights of the bodies to be raised; and if all this resistance is overcome, the remaining force will produce an acceleration of motion proportional to itself, partly in the parts of the machine, partly in the resisting body. (C&W, 429–30)

For increasing the force by decreasing the velocity, think of a lever. For the action of the remaining force after having overcome all the resistance, think of the sliding motion over a tree branch of a rope slung over that branch to connect a very light doll to a dropped anvil, or the driving apart of the two sides of a piece of firewood split by an axe. Here we have Newton explicitly counting a force (the remaining excess force) as producing an acceleration proportional to itself.

In his previous paragraph, Newton gave an account of the applications of Law 3 to weights on a balance and how such calculations can be extended to account for weights interfered with by oblique planes or other obstacles, as well as weights raised by ropes over pulleys and combinations of pulleys (C&W, 428–9). This was followed by instructions for such calculations for engaged gears in clocks, as well as for the force of a hand turning the handle of a screw-driving machine to the force of the screw to press a body (C&W, 429). It concluded with the action-reaction calculation of the forces by which a wedge presses the two parts of the wood it splits to the force impressed upon it by the hammer (C&W, 429). His opening remark in the last quotation suggests that if the empirical phenomena corresponding to these applications were to change so as to appreciably violate these applications of Law 3, we would soon know about it.

Newton counts his Laws of Motion as empirical propositions. His scholium to the Laws describes the very strong empirical evidence supporting the appropriateness of accepting them as guides to further research. According to Newton's Rule 4, these

[34] Modern readers may be disappointed at Newton's failure to include familiar topics of Newtonian mechanics today, such as potential and kinetic energy, least action and work. These concepts were introduced in a significant mathematical reconfiguration of Newton's theory by d'Alembert, Euler and Lagrange.

Laws are to be counted as either exactly or very nearly true until yet other phenomena make them more exact or liable to exceptions. In the absence of phenomena making these Laws liable to exceptions, alternatives are to be dismissed as mere hypotheses unless they are sufficiently supported empirically to be counted as serious rivals.

III The arguments for propositions 1 and 2 book 3

We have seen that Newton's evidence for applying Law 3 to collisions is based on *theory-mediated* measurements. This part of his evidence for his Laws of Motion applies the methodology that guides his inferences to inverse-square centripetal forces directed toward Jupiter, Saturn, and the sun. We shall review, in more detail than we did in chapter 1, the propositions affording the systematic dependencies that back up these inferences from the cited phenomena.

1. Jupiter's moons and Saturn's moons

Here is proposition 1, together with Newton's terse argument for it.

Book 3 Proposition 1

The forces by which the circumjovial planets [or satellites of Jupiter] are continually drawn away from rectilinear motions and are maintained in their respective orbits are directed to the center of Jupiter and are inversely as the squares of the distances of their places from that center.

The first part of the proposition is evident from phen. 1 and from prop. 2 or prop. 3 of book 1, and the second part from phen. 1 and from corol. 6 to prop. 4 of book 1.

The same is to be understood for the planets that are Saturn's companions [or satellites] by phen. 2. (C&W, 802)

Newton cites phenomenon 1 and propositions 2 or 3 from book 1 as making it evident that Jupiter's moons are maintained in their respective orbits by forces which are directed toward the center of Jupiter, and he cites phenomenon 1 and corollary 6 of proposition 4 book 1 as making it evident that these forces are inversely as the squares of the distances of these moons from that center. He cites phenomenon 2 as evidence that the same is to be understood for Saturn's moons.

We saw in chapter 2 that, like proposition 1, phenomenon 1 consists of two parts. The first part is that the moons of Jupiter, by radii drawn to the center of Jupiter, describe areas proportional to the times. As we have noted, this constancy of these rates at which areas are described is Kepler's rule of areas for these moons with respect to that center. The second part is that the periodic times of the orbits of these moons – the fixed stars being at rest[35] – are as the $3/2$ power of their distances from the center of

[35] Newton's clause – the fixed stars being at rest – tells us that the periods are calculated with respect to those stars. This treats a reference frame at the center of Jupiter with fixed directions with respect to the stars as non-rotating. Such non-rotating frames are also used to calculate areas in the Area Rule.

Jupiter. As we have noted, this second part of phenomenon 1 is Kepler's Harmonic Rule for the orbits of Jupiter's moons.

The phenomenon Newton cites for Saturn's moons, also, has the Area Rule and the Harmonic Rule as its two parts. Given the propositions from book 1 Newton cited for Jupiter's moons, the two parts of this phenomenon make evident that Saturn's moons are maintained in their respective orbits by forces which are directed to the center of Saturn and are inversely as the squares of the distances of their places from that center.

The propositions and corollaries from book 1 cited by Newton as, together with phenomenon 1, making proposition 1 evident are an important part of his demonstration of the dynamical significance of the Area Rule and Harmonic Rule phenomena.

1.i The Area Rule as evidence for centripetal force The first part of proposition 1 is that the forces maintaining Jupiter's moons in their orbits are directed toward the center of Jupiter. Newton tells us that this part is evident from phenomenon 1 and from propositions 2 or 3 of book 1. Here is proposition 2 of book 1.

Book 1 Proposition 2 Theorem 2

Every body that moves in some curved line described in a plane and, by a radius drawn to a point, either unmoving or moving uniformly forward with a rectilinear motion, describes areas around that point proportional to the times, is urged by a centripetal force tending toward that same point. (C&W, 446)

It is the first part of phenomenon 1, that the moons of Jupiter satisfy the Area Rule with respect to the center of Jupiter, from which the centripetal direction of the forces maintaining those moons in their respective orbits is inferred. Proposition 2 (book 1) applies to the extent to which the center of Jupiter can be treated as though it were inertial.

Newton's account of the dynamical significance of the Area Rule as an indicator of centripetal forces includes more than the theorems he explicitly cited in the, above quoted, argument for proposition 1 book 3. Book 1 proposition 1, the very first proposition in the *Principia*, asserts the converse of proposition 2, for centers regarded as unmoving, and its corollary 6 extends this to any center that can be regarded as inertial.[36]

Book 1 Proposition 1 Theorem 1

The areas by which bodies made to move in orbits describe by radii drawn to an unmoving center of forces lie in unmoving planes and are proportional to the times. (C&W, 444)

Corollary 6 (prop. 1, book 1). All the same things hold, by corol. 5 of the laws, when the planes in which the bodies are moving, together with the centers of forces which are situated in those planes, are not at rest but move uniformly straight forward. (C&W, 446)

[36] That Newton begins with unmoving centers reflects that his formulation of absolute space is stated so as to not rule out representing absolute rest. See section II.1ii above.

According to proposition 1 and its corollary 6, if a body is being deflected from inertial motion into an orbit by a force directed toward an inertial center, that orbit will lie in a plane and will satisfy Kepler's rule of areas with respect to radii drawn to that center. According to proposition 2, a body will move in a plane orbit that satisfies Kepler's rule of areas with respect to an inertial center only if the force maintaining that body in that orbit is directed toward that center. These two propositions, together, yield a bi-conditional equivalence between the centripetal direction of the force maintaining a body in an orbit about an inertial center and the motion of that body being in a plane orbit satisfying Kepler's rule of areas.

Corollary 1 of proposition 2 book 1 further extends this dependency between the behavior of the rate at which areas are swept out and the direction of the force with respect to the center about which the areas are being described.

Corollary 1. In nonresisting spaces or mediums, if the areas are not proportional to the times, the forces do not tend toward the point where the radii meet but deviate forward [or in consequentia] from it, that is, in the direction toward which the motion takes place, provided that the description of the areas is accelerated; but if it is retarded, they deviate backward [or in antecedentia, i.e., in a direction contrary to that in which the motion takes place]. (C&W, 447)

In nonresisting spaces or mediums,[37] the rate at which areas are described is increasing only if the force is angled off-center toward the direction of tangential motion, while a decreasing rate obtains only if the force is angled off-center in the opposite direction.

As Newton makes clear in a scholium, these dependencies, together with those of propositions 1 and 2, make the constancy of the rate at which areas are being swept out by radii to an inertial center measure the centripetal direction of the total force acting on a body in an orbit about that center. Here is the relevant passage from Newton's scholium to proposition 2 book 1.

A body can be urged by a centripetal force compounded of several forces. In this case the meaning of the proposition is that the force which is compounded of all the forces tends toward point S . . . (C&W, 447)

An increasing rate would indicate that the net acceleration is angled forward, while a decreasing rate would indicate that the net acceleration is angled backward.[38]

Jupiter and its moons undergo a substantial centripetal acceleration toward the sun as the Jupiter system orbits it. So, Jupiter's center is not in uniform rectilinear motion with

[37] Corollary 2 of proposition 2 points out that, even in resisting mediums, an increasing areal rate requires that the forces be off-center in the direction of motion. Propositions 1 and 2 were first proved as theorems 1 and 2 of Newton's 1684 *De Motu* (Math Papers, vol 6, 34–7), the initial tract which, by 1687, had grown into the *Principia*.

Betty Jo Teeter Dobbs (1991, 130–2) has argued that it was these theorems which led Newton to see that the exactness of the empirical fit of the Area Rule counted as very strong evidence that the motions of celestial bodies were not impeded by any medium through which they moved. Theorems 1 and 2, like most of the theorems in book 1, are proved for motions devoid of resistance.

[38] Figure 1.16 from chapter 1 (III.1i) represents these systematic dependencies in diagrams.

respect to the sun.[39] In his argument for the centripetal direction of the forces maintaining Jupiter's moons in their orbits about Jupiter, Newton cites propositions 2 or 3 of book 1. Here is proposition 3 of book 1.

Book 1 Proposition 3 Theorem 3

Every body that, by a radius drawn to the center of a second body moving in any way whatever, describes about that center areas that are proportional to the times is urged by a force compounded of the centripetal force tending toward that second body and of the whole accelerative force by which that second body is urged. (C&W, 448)

Newton's proof is an application of corollary 6 of his Laws of Motion.[40] To the extent that the sun's actions on Jupiter and its moons approximate equal and parallel accelerations, the Jupiter system can be treated as unperturbed by these forces accelerating it toward the sun.[41] To the extent that this approximation holds (and the center of

[39] Jupiter is, also, not in uniform rectilinear motion with respect to the center of mass of the solar system. As we have noted, Newton will appeal to corollary 4 of the Laws of Motion as grounds to treat the center of mass of the solar system as inertial.

[40] See appendix 2.

[41] Corollary 6 of the Laws of Motion makes it clear that the sun's gravity disturbs the Jupiter system only by the tidal forces (those which produce differential accelerations on different parts of the system) it generates. What matters to produce perturbations is the ratio of this solar tidal factor to the planet's accelerative gravity at the moon's distance.

For the sun's action on the earth–moon system, this ratio is 0.0056. This equals 1/178.725, which, as we shall see in chapter 4, is the ratio Newton suggested ought to be used to represent the factor by which the action of the sun reduces the centripetal force drawing the moon toward the earth. For the sun's action on the Jupiter system, this ratio equals 0.000015 for the outermost Galilean satellite Callisto and 0.00000017 for the innermost Io. The action of the sun to disturb the Jupiter–satellites system is, therefore, very much less than its action to disturb the earth–moon system.

Here are the calculations. Tidal forces vary inversely as R^3 where R is the distance of the planet from the sun and directly as r where r is the distance of the moon from the planet. (See Ohanian and Ruffini 1994, 39–40). For the earth–moon system, R_e is about 1.496×10^{11} meters, and r_m is about 3.844×10^8 meters (*ESAA*, 700–1). The mass M_s of the sun is about 1.989×10^{30} kg (*ESAA*, 700). This gives a tidal factor $M_s(r_m/R_e^3)$ of 22836.176 kg/meter^2.

We want the ratio of this tidal factor to the earth's accelerative gravity M_e/r_m^2 at the moon's distance. The mass of the earth M_e is about 5.974×10^{24} kg (*ESAA*, 700), which makes M_e/r_m^2 about 40429508.371, making the ratio $M_s(r_m/R_e^3)/(M_e/r_m^2)$ about 0.0056. We also have the Harmonic Rule ratios measuring the masses of the central bodies. Thus, $M_s/M_e = (R_e^3 t_m^2)/(r_m^3 t_e^2)$, which makes the ratio, $M_s(r_m/R_e^3)/(M_e/r_m^2)$, of the sun's tidal factor to the accelerative gravity of the earth at the moon's distance equal $(R_e^3 t_m^2 r_m^3)/(t_e^2 r_m^3 R_e^3) = t_m^2/t_e^2$. Newton generates this number, 0.0056 = (1/178.725), directly from the periods. He takes t_m as 27.3215 decimal days and t_e as 365.2565 decimal days, which gives $(t_m/t_e)^2 = 0.00559$.

For Jupiter, R_j is about 7.783×10^{11} meters (*ESAA*, 704). For its outermost Galilean satellite Callisto, r_c is 1.883×10^9 meters (*ESAA*, 708), for a tidal factor $M_s(r_c/R_j^3)$ of 7944.087 kg/meter^2. For Jupiter's innermost Galilean satellite Io, r_i is 422×10^6 meters (*ESAA*, 708), for a tidal factor $M_s(r_i/R_j^3)$ of 1780.353 kg/meter^2. Jupiter's mass M_j is about 1898.8×10^{24} kg (*ESAA*, 706). For Callisto, the ratio $M_s(r_c/R_j^3)/(M_j/r_c^2)$ is about 0.000015. For Io, the ratio $M_s(r_i/R_j^3)/(M_j/r_i^2)$ is about 0.00000017.

The ratios 0.000015 and 0.00000017 for Jupiter's moons Callisto and Io are respectively given by $(t_j/t_c)^2$ and $(t_j/t_i)^2$, the squares of the ratios of their periods to that of Jupiter. The number of lunations of a moon's orbit is the number of periods of that moon about its planet per period of its planet about the sun. The numbers 0.000015, 0.00000017, and 0.0056 are inversely as the squares of the corresponding numbers of lunations for Callisto and Io with respect to Jupiter, and the moon with respect to the earth.

Jupiter approximates the center of mass of the Jupiter system),[42] the center of Jupiter can be treated as inertial so that the above dependencies apply.

Having the Area Rule hold very nearly for the orbits of these moons with respect to the center of Jupiter carries the information that these approximations are not appreciably inaccurate. Newton explicitly gives corollaries extending propositions 2 and 1 to cover such approximations. Let Jupiter be the central body T. In these corollaries the forces accelerating the central body T (e.g. the sun's action on Jupiter) have been subtracted from the total force accelerating a moon.

Corollary 2 (prop. 3 book 1). And if the areas are very nearly proportional to the times, the remaining force will tend toward body T very nearly. (C&W, 448)

Corollary 3 (prop. 3 book 1). And conversely, if the remaining force tends very nearly toward body T, the areas will be very nearly proportional to the times. (C&W, 449)

Newton's exposition of the dynamical significance of the Area Rule extends beyond the idealization of a body orbiting about an inertial center under the action of a single force deflecting it from uniform rectilinear motion. These extensions show that the Area Rule can be a quite general criterion for finding centers toward which forces maintaining bodies in orbits are directed.

Scholium (prop. 3, book 1). Since the uniform description of areas indicates the center toward which that force is directed by which a body is most affected and by which it is drawn away from rectilinear motion and kept in orbit, why should we not in what follows use uniform description of areas as a criterion for a center about which all orbital motion takes place in free spaces? (C&W, 449)

As we have noted in chapter 2 section V above, this scholium suggests that we ought to take up this criterion for finding forces toward centers.

In his discussion of phenomenon 1, Newton points out that the orbits of Jupiter's moons so closely approximate uniform motion on circles concentric to Jupiter that differences from such motion are not detected in observations by astronomers.

This is established from astronomical observations. The orbits of these planets do not differ sensibly from circles concentric with Jupiter, and their motions in these circles are found to be uniform. (C&W, 797)

That good observations detect no departures from uniform motion on concentric circular orbits for Jupiter's moons indicates that no appreciable errors result from treating Jupiter's center as inertial for purposes of using the Area Rule as a criterion for the centripetal direction of the forces maintaining those moons in their orbits.

The theorems underwriting the Area Rule as a criterion for centers toward which orbital forces are directed make no assumptions about any specific power law for these

[42] The relative masses of Jupiter's Galilean satellites to the mass of Jupiter are respectively, Io 4.68×10^{-5}, Europa 2.52×10^{-5}, Ganymede 7.80×10^{-5}, and Callisto 5.66×10^{-5} (*ESAA*, 710). All these satellites together are only 0.000207 of the mass of Jupiter.

forces.[43] Given that the centripetal direction of the forces maintaining these moons in their orbits is inferred from the phenomenon that those orbits satisfy the Area Rule with respect to the center of Jupiter, one can appeal to theorems about orbital motion under centripetal forces to argue that the Harmonic Rule phenomenon for those orbits carries the information that the accelerative measures of those forces are inversely as the squares of their distances from that center.

This illustrates one of the ways in which Newton's inferences are not merely hypothetico-deductive. Rather than just testing his final theory as a whole, he is able to argue for it in a succession of empirically established steps.

1.ii The Harmonic Rule as a criterion for inverse-square forces The second part of proposition 1 is that the forces maintaining Jupiter's moons in their respective orbits are inversely as the squares of the distances of their places from the center of Jupiter. Newton tells us that this part is evident from phenomenon 1 and corollary 6 of proposition 4 of book 1. Here is this corollary.

Corollary 6 (prop. 4 book 1). If the periodic times are as the $3/2$ powers of the radii, and therefore the velocities are inversely as the square roots of the radii, the centripetal forces will be inversely as the squares of the radii; and conversely. (C&W, 451)

According to this corollary, the second part of proposition 1, that the forces maintaining the moons of Jupiter in their respective orbits are inversely as the squares of their distances from the center of Jupiter, can be inferred from the second part of phenomenon 1, that those orbits satisfy the Harmonic Rule with respect to that center. The clause "and conversely" at the end of this corollary asserts that the Harmonic Rule for a system of orbits could be inferred from the inverse-square variation of the forces maintaining those bodies in their orbits. What Newton is counting as the inverse-square variation of these forces is the inverse-square variation of the centripetal accelerations of those orbits. Having these accelerative measures of the centripetal forces maintaining bodies in those orbits vary inversely as the squares of their distances is equivalent to having the periods be as the $3/2$ power of those distances.[44]

Corollary 7 of proposition 4 of book 1 extends these dependencies in a way that contributes significantly to Newton's demonstration of the dynamical significance of the Harmonic Rule.

Corollary 7 (prop. 4 book 1). And universally, if the periodic time is as any power R^n of the radius R, and therefore the velocity is inversely as the power R^{n-1} of the radius, the centripetal force will be inversely as the power R^{2n-1} of the radius; and conversely. (C&W, 451)

For the dynamical significance of the Harmonic Rule, the important connection is that between the periodic times t being as the power R^n of the radii R and the accelerative

[43] See appendix 2.

[44] Having these accelerations inversely as the squares of distances is equivalent to having the ratios of the motive forces to the masses of the orbiting bodies vary inversely as the squares of the distances.

measures of the centripetal forces being inversely as the power R^{2n-1} of R. This is equivalent to the following universal systematic dependency

$$t \propto R^n \textit{ iff } f \propto R^{1-2n},$$

where t is the periodic time, R is the distance from the center, and f is the accelerative measure of the force maintaining a body in an orbit with period t and distance R. Corollary 6 is an immediate consequence of this more general dependency. It follows when n is replaced by 3/2. We have $n = 3/2$ just in case $1-2n = -2$. The Harmonic Rule phenomenon asserts that the periods of the orbits are proportional to the 3/2 power of their radii ($t \propto R^{3/2}$). The inverse square rule for the forces is to have their accelerative measures proportional to the -2 power of those radii ($f \propto R^{-2}$).

According to corollary 6, the Harmonic Rule phenomenon ($t \propto R^{3/2}$) is equivalent to the inverse-square rule ($f \propto R^{-2}$) for the centripetal forces maintaining bodies in those orbits. The systematic dependencies exhibited in corollary 7 of proposition 4 book 1 go beyond the equivalence of corollary 6. For each of a whole range of alternative power-rule proportions of periods to orbital radii, corollary 7 establishes the equivalent power-rule proportion to radii for the accelerative measures of centripetal forces that would maintain bodies in those orbits. To have the periods be as some power $s > 3/2$ would be to have the accelerative measures of these centripetal forces fall off faster than the -2 power of the radii, while to have the periods be as some power $s < 3/2$ would be to have the accelerative measures of those centripetal forces fall off less fast than the -2 power of the radii.

Table 3.1

Harmonic Rule for a system of orbits measures inverse-square rule for centripetal force

Prop. 4 bk. 1, corols. 7&6

Corol. 7,	$t \propto R^s$	$\Leftrightarrow$	$f \propto R^{1-2s}$
Corol. 6,	$t \propto R^{3/2}$	$\Leftrightarrow$	$f \propto R^{-2}$
	Harmonic Rule Phenomenon		Inverse-square Centripetal Force
	Alternatives to Phenomenon		

$$s > 3/2 \Leftrightarrow (1-2s) < -2$$

$$s < 3/2 \Leftrightarrow (1-2s) > -2$$

Alternative values for s carry information about alternative power laws

These systematic dependencies make the Harmonic Rule phenomenon ($n = 3/2$) for a system of orbits measure the inverse square (-2) power rule for the accelerative measures of the centripetal forces maintaining bodies in those orbits. This constitutes a very strong sense in which the Harmonic Rule carries the information that the accelerative measures of the centripetal forces maintaining bodies in those orbits satisfy the inverse-square power rule.

Proposition 4 book 1 and its corollaries are proved for uniform motion on concentric circular orbits. As we have noted above, Newton points out that the orbits of Jupiter's moons do not differ sensibly from uniform motion on circles concentric with the center of Jupiter.

In chapter 2, we saw that Newton offers as evidence for the Harmonic Rule a table citing periods agreed upon by astronomers and four distance estimates from astronomers for each of the four moons of Jupiter known at the time. We noted that a particularly informative way to exhibit the good fit of the Harmonic Rule to this data was by plotting log periods against log distances. That some straight line of slope n fits the result of plotting Log t against Log R is to have the periods be as some power n of the distances. To have the Harmonic Rule hold is to have the slope of this line be $3/2 = 1.5$. The exhibited fit was quite striking.

2. *Primary planets*

In proposition 2 of book 3, Newton argues that the primary planets are maintained in their orbits by forces which are directed toward the center of the sun and are inversely as the squares of their distances from it.

Book 3 Proposition 2 Theorem 2

The forces by which the primary planets are continually drawn away from rectilinear motions and are maintained in their respective orbits are directed to the sun and are inversely as the squares of their distances from its center.

The first part of the proposition is evident from phen. 5 and from prop. 2 of book 1, and the latter part from phen. 4 and from prop. 4 of the same book. But this second part of the proposition is proved with the greatest exactness from the fact that the aphelia are at rest. For the slightest departure from the ratio of the square would (by book 1, prop. 45, corol. 1) necessarily result in a noticeable motion of the apsides in a single revolution and an immense such motion in many revolutions. (C&W, 802)

Newton cites phenomenon 5 and proposition 2 of book 1 as making evident the first part of this proposition – that the forces maintaining the primary planets in their orbits are directed toward the sun. This argument cites proposition 2 alone, where the corresponding argument for Jupiter's moons cited proposition 2 or proposition 3. Newton cites phenomenon 4 and proposition 4 of book 1, as making evident the latter part – that these forces are inversely as the squares of the distances from the center of the sun.

A major difference from the argument for Jupiter's moons is that an additional argument appealing to absence of apsidal motion and to corollary 1 proposition 45 of book 1 is said to prove this second part with the greatest exactness.

2.i From the Area Rule to the sun-centered direction Consider the argument for the first part. That Newton does not cite proposition 3 book 1 reflects the fact that, unlike the case for Jupiter and its moons, there is no explicit acceleration of the sun that needs to be taken into account when using the uniform description of areas by radii drawn to it to infer that the forces maintaining the planets in their orbits are directed toward it.

The phenomenon cited for this first part is phenomenon 5. We have discussed this phenomenon and Newton's evidence for it in chapter 2 above. Here is a brief review.

Phenomenon 5

The primary planets, by radii drawn to the earth, describe areas in no way proportional to the times but, by radii drawn to the sun, traverse areas proportional to the times. (C&W, 801)

In his discussion supporting this phenomenon Newton made clear that the motions of the planets with respect to the earth, with stationary points and retrograde loops, often describe areas which, compared with the times, are extremely unequal.[45] This shows that the earth cannot be counted as a center toward which the forces moving the planets are directed. As we have seen in our discussion of the Area Rule as a criterion for finding centers toward which forces are directed, this also suggests that the force moving a planet is directed toward some other center. With respect to the sun as center, the angular motion is almost uniform and the departures from uniform motion – a little more swiftly in their perihelia and more slowly in their aphelia – are such that the description of areas is uniform. In the case of Jupiter this is backed up with triangulations afforded by eclipses of its moons.

2.ii The Harmonic Rule for the planets Let us review Newton's treatment of the Harmonic Rule for the primary planets. His statement of this Harmonic Rule is neutral between sun-centered and earth-centered systems.

Phenomenon 4

The periodic times of the five primary planets and of either the sun about the earth or the earth about the sun – the fixed stars being at rest – are as the 3/2 powers of their mean distances from the sun. (C&W, 800)

As evidence for this Harmonic Rule phenomenon, Newton cites periods agreed upon by astronomers and the estimates of mean-distances made by Kepler and the French astronomer Boulliau.

Table 3.2 Periods and mean distances

In units equal to the period and mean-distance of the earth's orbit for

	Saturn	Jupiter	Mars	Earth	Venus	Mercury
		Periods in units t_e = 365.2565 decimal days				
	29.45677	11.86157	1.88081	1.00000	0.614959	0.240842
		Mean-distances in units AU				
Kepler	9.51	5.1965	1.5235	1.00000	0.724	0.38806
Boulliau	9.54198	5.2252	–	–	0.72398	0.38585

[45] See chapter 2 above for a more detailed account, which includes Kepler's pretzel diagram showing the extreme departures from the Area Rule motion for Mars with respect to the earth as a center.

Boulliau takes over Kepler's estimate for the mean-distance of Mars, and data from both agree in using the mean earth–sun distance to set the distance unit. We are using as our distance unit AU, the astronomical unit. This equals to five decimal places the mean earth–sun distance (which Newton set at 100000).[46] We are using as our unit of time the period t_e of the earth–sun orbit (which Newton cited as 365.2565 decimal days).

As we have seen in chapter 2, a particularly informative way to represent the fit of the Harmonic Rule to this data is by plotting log periods against log distances.[47]

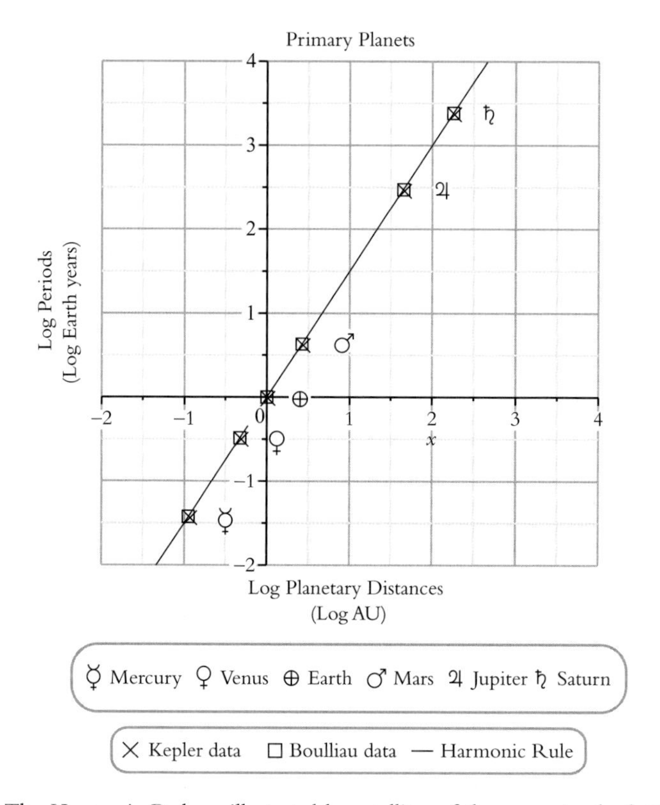

Figure 3.2 The Harmonic Rule as illustrated by satellites of the sun using both Kepler's and Boulliau's data

[46] The astronomical unit was set by the mean earth–sun distance used by Gauss. See chapter 8 note 31 for details.

[47] We can exhibit the close fit of the Harmonic Rule to these estimates of periods and mean distances by the close clustering of the corresponding estimates of the Harmonic Rule ratio R^3/t^2. These give a 95% Student's t-confidence estimate of 1.00082 ± 0.005 AU^3/t_e^2. The ratio of the 95% Student's t error bounds to the mean is just less than 0.005.

For a 95% confidence interval estimate the Student t parameter for the nine degrees of freedom corresponding to ten estimates is 2.26 (Freedman et al. 1978, A-71; 1998, A-106).

2.iii Inverse-square from the Harmonic Rule for elliptical orbits? The mean-distances cited in Newton's table are the semi-major axes of elliptical orbits, not radii of concentric circular orbits. Unlike Jupiter's moons, the orbits of the primary planets were known to have non-negligible eccentricities. Newton's proofs of proposition 4 book 1 and of its corollaries 6 and 7 are for concentric circular orbits.

Clark Glymour questions Newton's appeal to such corollaries of proposition 4 book 1, even in Newton's application to infer inverse-square forces on Jupiter's moons.

But the hypothesis of Proposition IV *requires that the bodies move in circles*. This is curious, because although it is true that Jupiter's satellites move in nearly circular orbits, the inverse square relation for the force is an immediate consequence of the hypothesis of elliptical orbits alone, and no additional assumption about the relation of periodic times and radii is needed. Newton knew that as well, of course: that is what Proposition XI of Book I of the *Principia* says. Why did Newton use Kepler's third law together with a hypothesis to the effect that the orbits are circular to obtain the inverse square relation rather than simply using a much more general orbital assumption, namely, Kepler's first law? It might seem that this makes a great puzzle of an insignificant point; for Newton knew that Jupiter's satellites have nearly circular orbits, and his readers did as well, so it may have seemed of no consequence to him that he used the less direct proof. But Newton uses the same argument to establish that the primary planets are subject to an inverse square force, even though it was well known at the time (and I presume Newton knew it) that the orbits of some of the primary planets were definitely not circular. I have no convincing explanation for this curiosity. (Glymour 1980a, 207–8)

The answer to the question why Newton did not argue to the inverse-square from Kepler's first law is quite informative. We shall see that an inference to the inverse-square variation from an ellipse with the force to a focus would satisfy the concept of bootstrap confirmation Glymour introduced in his 1980 book, but that such an inference would not be backed up by the sort of systematic dependencies that make Newton's inferences into *theory-mediated* measurements.

For now, let us concentrate on the problem of applying corollaries of proposition 4 book 1, which are proved for circular orbits, to infer the inverse-square from the Harmonic Rule for Kepler's elliptical orbits. The systematic dependency of corollary 7 of proposition 4 generalizes to any sort of orbits where the centripetal accelerations at what are counted as the mean-distances equal the centripetal accelerations of the corresponding concentric circular orbits, with radii equal to those mean-distances and periods equal to those of the corresponding orbits of that sort. Suppose, as in corollary 7 proposition 4, that the periods of several such orbits are

We have here

$$sd = 0.0067\mathrm{AU}^3/t_e{}^2$$
$$sd^+ = 0.007\mathrm{AU}^3/t_e{}^2$$
$$SE = 0.002\mathrm{AU}^3/t_e{}^2$$
$$tSE = 2.26(0.002\mathrm{AU}^3/t_e{}^2) = 0.005\mathrm{AU}^3/t_e{}^2.$$

These give $1.00082 \pm 0.005\ \mathrm{AU}^3/t_e{}^2$ as the 95% Student's *t*-confidence estimate. See chapter 4 note 41 for more detail about Student's *t*-confidence estimates.

as some power s of the mean-distances of those orbits. This is equivalent to having the periods of the corresponding concentric circular orbits as that power s of their radii. By corollary 7 proposition 4 book 1, this, in turn, is equivalent to having the accelerative measures of the centripetal forces maintaining bodies in those corresponding concentric circular orbits be as the $1-2s$ power of their radii. But, this again is equivalent to having the values at the mean-distances of accelerative measures of the forces maintaining bodies in the eccentric orbits be as the $1-2s$ power of those mean-distances. So, for any such eccentric orbits, the Harmonic Rule ($t \propto a^{3/2}$) measures the inverse-square rule ($f \propto a^{-2}$) relating the accelerative measures at those mean-distances of the centripetal forces maintaining bodies in those orbits.

Kepler's elliptical orbits are clear cases of eccentric orbits to which this extension of corollary 7 applies. When a body in an elliptical orbit with the force to a focus is at a distance from that focus equal to the semi-major axis of the ellipse, its centripetal acceleration equals that of a body in uniform circular motion about that center having the same period as that of the elliptical orbit. This follows easily from the modern formula for centripetal acceleration in an elliptical orbit with the force to a focus, and was explicitly proved by Newton.[48]

That the Harmonic Rule holds for the primary planets, therefore, carries the information that the values at their mean-distances of the separate accelerative forces maintaining them in their respective elliptical orbits agree with those of a single inverse-square centripetal acceleration field toward the sun. There was no *a priori* guarantee that a straight line would fit the result of plotting log periods against log distances and no guarantee that they would be fit by a line of slope 1.5. The fact that they are so fit makes these orbits afford agreeing measurements of what the acceleration toward the sun would be for a body if it were at any given specified distance from it.

2.iv Aphelia at rest Newton claims that the inverse-square variation with distance from the sun of the forces maintaining the planets in their orbits is proved with the greatest exactness from the fact that the aphelia are at rest. Recall from chapter 1 that in an elliptical orbit the line through the foci – the major axis – is the apsis. In a fixed elliptical orbit a planet would traverse against the fixed stars exactly 360 degrees in order to go

[48] For a modern proof, note that

$$acc_r = -(4\pi^2 a^3/t^2)(1/r^2)$$

where acc_r is the centripetal acceleration corresponding to distance (e.g. radius vector) r from the focus in an elliptical orbit of semi-major axis a and period t with the force toward that focus (see French 1971, 588). Now set $r = a$. This yields

$$acc_r = -(4\pi^2 r/t^2),$$

which equals the centripetal acceleration of a concentric circular uniform motion orbit of radius r and period t about that center.

See Brackenridge (1995, 119–23) for an exposition of Newton's proof of proposition 4 in the original tract *De Motu*.

from and return to the aphelion (the most distant point from the sun) of its orbit.[49] The orbit would count as precessing, with p degrees of precession per revolution just in case the planet would traverse against the fixed stars $360 + p$, rather than just 360 degrees, to return to its aphelion. Here again is our precession diagram.

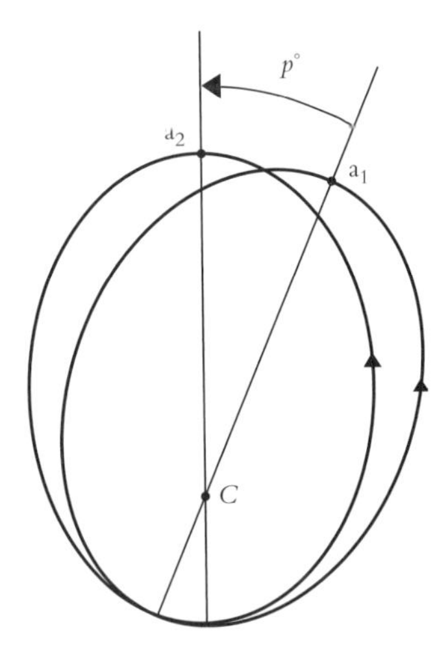

Figure 3.3 Precession per revolution

In the time it takes the planet to get back to the same point in its ellipse, the axis of that ellipse will have traversed p degrees against the stars. The apsis is rotating forward with p degrees of precession per revolution. The orbit is precessing forward if p is positive and is precessing backward (in the opposite direction from the motion of the planet in the ellipse) if p is negative. A fixed orbit has zero precession.

Newton cites corollary 1 of proposition 45 book 1. This corollary applies to the orbit of a body acted upon by a centripetal force that varies as some power of the distance from the center toward which the force is directed.[50] It is equivalent to the following systematic dependency relating the precession p of an orbit to the power n of distance by which the centripetal force varies:

$$p^\circ \text{ precession/revolution} \Leftrightarrow n = (360/360 + p)^2 - 3$$

As we noted in chapter 1, having $p = 0$ is equivalent to having $n = -2$, so that zero orbital precession is equivalent to having the centripetal force maintaining a body in that orbit vary inversely as the square of the distance from the center toward which it is directed. As in the case of the Harmonic Rule inference, this basic equivalence is backed

[49] In Newton's time it was common to refer to orbital precession as motion of the aphelion. We now refer to it as motion of the perihelion (the opposite end of the major axis) which is the closest point in the orbit to the sun.

[50] This is a special case, for a power-law force, of more general theorems relating rate of orbital precession to the dependency of a centripetal force on distance.

up by systematic dependencies. Having *p* be positive is equivalent to having *n* be less than −2, while having *p* be negative is equivalent to having *n* greater than −2. Forward precession corresponds to having the centripetal force fall off faster than the inverse-square of the distance, while backward precession corresponds to having the centripetal force fall off less fast than the inverse-square of the distance. Zero precession would exactly measure the inverse-square variation with distance of the centripetal force maintaining a body in such an orbit.

Here, from his earlier version of book 3 in popular form, is a comment by Newton on the absence of precession in the orbits of the planets about the sun.

> But now, after innumerable revolutions, hardly any such motion has been perceived in the orbits of the circumsolar planets. Some astronomers affirm that there is no such motion; others reckon it no greater than what may easily arise from causes hereafter to be assigned, which is of no moment in the present question. (Cajori 1934, 561; Newton 1728, 24–5)

Newton can apply the foregoing one-body results to systems, like the orbits of the planets about the sun, where there are forces of interaction between planets which perturb what would be motion under the inverse-square centripetal force toward the sun alone. If all the orbital precession of a planet is accounted for by perturbations, the zero leftover precession can be counted as measuring the inverse-square variation of the basic centripetal force that maintains that planet in its orbit about the sun.

As Newton points out, even a quite small departure from the inverse-square would result in a precession that would easily show up after many revolutions. In chapter 10, we consider the famous 43 seconds a century unaccounted-for precession of Mercury. This is a quite striking example, familiar to many today. According to Newton's corollary 1 of proposition 45, it would measure the −2.00000016 power of distance for gravitation toward the sun.[51] An example that was familiar in Newton's day is the lunar precession of about 3 degrees per orbit. As Newton pointed out, it would measure about the $-2\frac{4}{243}$ power of distance for gravitation toward the earth, if it were not attributed to the action of the sun to perturb the motion of the moon about the earth.[52]

Newton's proposition 45 of book 1 and its corollaries are proved for orbits that are very nearly circular. The results can be extended to orbits of arbitrarily great eccentricity. Indeed, it turns out that the larger the eccentricity the more sensitive is absence of unaccounted-for precession as a measure of inverse-square variation of forces maintaining bodies in orbit.[53]

[51] See chapter 10 note 17.

[52] In his discussion of corollary 1 proposition 45 book 1, Newton provides the example of 3 degrees forward precession per revolution. This is quite close to the 3 degrees 3 minutes forward precession he attributes to the moon in the third edition discussion of proposition 3 book 3, which we shall take up below. Both of these precession values yield a corresponding centripetal force which rounds to inversely as the −2 and $\frac{4}{243}$ power of distance. See chapter 4 below.

[53] See appendix 3.

3. Measurements supporting inverse-square centripetal acceleration fields

As we have seen in section I.2 above, Newton's three measures or quantities of a centripetal force – absolute, accelerative, and motive – correspond respectively to the strength of a centripetal acceleration field, the equal accelerations it would produce on all bodies at any given distance from the center, and the product of this acceleration with the mass of each body accelerated by it. That there should be an accelerative measure – that the field intensity at each place around the center is measured by the equal accelerations that the field would impose on bodies at that distance – is what makes a centripetal force count as an acceleration field. As Stein has argued, this makes Newton's inferences to inverse-square centripetal forces directed to Jupiter, Saturn, and the sun count as inferences to inverse-square centripetal acceleration fields.

In chapter 1, I reported the interesting question that Howard Stein had asked me at a lecture on Newton's argument I gave at Chicago. I had cited the absence of significant orbital precession of each orbit and the Harmonic Rule of those six orbits about the sun as the phenomena from which Newton inferred inverse-square variation of attraction toward the sun. Stein asked how these phenomena provided evidence against a hypothesis that agreed with the inverse-square in the distances explored by each orbit and in the inverse-square relation among the forces at those six small distance ranges, but differed wildly from the inverse-square power in the large ranges of distances not explored by the motions of those planets.[54] In chapter 1, we argued that the fit of the orbits of the planets to the Harmonic Rule makes them afford agreeing measurements of what the acceleration toward the sun would be at any given distance from it. To count these orbital phenomena as affording such measurements is to count the inverse-square centripetal forces as inverse-square acceleration fields. We saw that Newton's Rule 3 for doing natural philosophy endorses so exploiting such measurements and his Rule 4 backs this up. According to Newton's Rule 4, to avoid being dismissed as mere contrary hypotheses an alternative would have to realize agreeing measurements sufficiently to count as a serious rival or provide phenomena that make the inferences to acceleration fields liable to exceptions.

We can back this up with an actual example of the measurements afforded from these orbital data for such a specified distance. In chapter 1, we noted that the considerable distances between Mars and Jupiter and between Jupiter and Saturn are not explored by the motions of planets. Let us take as our example a distance of 7.00AU.[55]

[54] I was taking Newton's inferences to be inferences to inverse-square variation for each and the inverse-square relation among what Newton counts as the accelerative measures of the several motive forces to the individual orbiting bodies, rather than to the single inverse-square centripetal acceleration field that Newton counts as the common cause of these several centripetally directed and inverse-square related motive forces maintaining the moons or planets in their orbits.

[55] In Myrvold and Harper 2002, it is shown that the Akaike information criterion would pick a quartic curve variation that would wildly differ from the inverse-square at this distance as a better fit to Newton's cited data than the inverse-square variation. For more on Akaike see note 91 of this chapter.

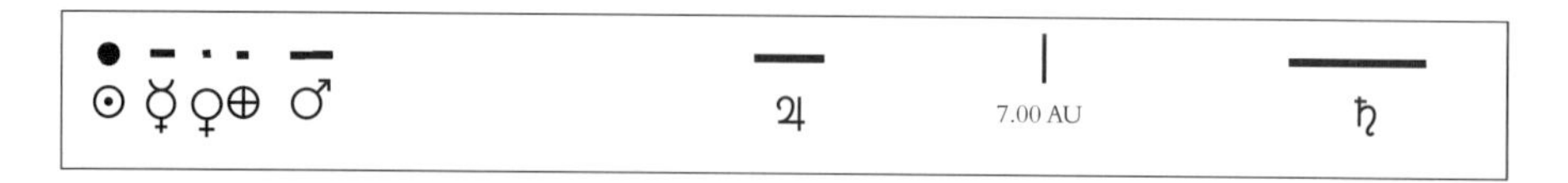

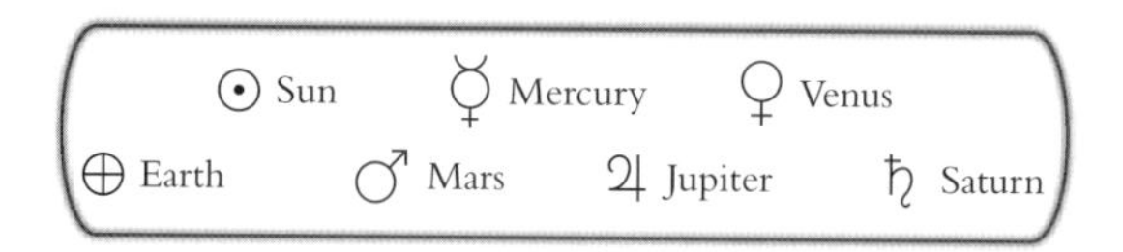

Figure 3.4 The distance 7AU is not explored by any planet

Inverse-square adjusting the centripetal acceleration corresponding to each of the orbital data that Newton cited yields a corresponding estimate for the centripetal acceleration toward the sun that a body would be subject to if it were at 7.00 AU distance from the center of the sun.[56] Table 3.3 gives these agreeing estimates in units AU/t_e^2, where t_e is the period of the earth's orbit.

Table 3.3 Acceleration toward the sun for a body at 7.00 AU distance (in units AU/t_e^2)

	Saturn	Jupiter	Mars	Earth	Venus	Mercury
Kepler	0.7986	0.8036	0.8053	0.8057	0.8085	0.8117
Boulliau	0.8066	0.8169	—	—	0.8085	0.7978

These estimates are agreeing measurements of the acceleration toward the sun that a body would be subject to if it were at a distance of 7.00 AU from its center.[57] We can think of these estimates as agreeing measurements of a sun-directed acceleration component that would apply to a body if it were at that distance from the center of the sun.[58]

[56] Where a is the cited mean distance in units AU and t is the cited orbital period in units t_e (which Newton assigns as 365.2565 decimal days), the inverse-square adjusted estimate from that orbit of the acceleration toward the sun assigned to distance 7.00 AU from its center is given by

$$(4\pi^2 a/t^2)(a/7.00)^2 \ AU/t_e^2.$$

[57] We can further illustrate the evidence provided by these agreeing measurements by computing the corresponding 95% Student's t-confidence estimate

$$0.806 \pm 0.004 \ AU/t_e^2$$

These measurements give a mean of 0.80632 AU/t_e^2, with an SE of 0.0018 AU/t_e^2. The 95% Student's t-parameter for the nine degrees of freedom corresponding to these ten estimates is 2.26 (Freedman et al. 1978, A71), which give 95% confidence bounds of $\pm$ 2.26 $SE = 0.004$ AU/t_e^2. See chapter 4 note 41 for more detail about computing of a 95% Student's t-confidence bound.

[58] This sun-directed acceleration component would combine with whatever other motive forces are acting on that body to produce what Newton's theory would count as an adequate specification of its true

The agreement among these measurements of what the centripetal acceleration toward the sun would be at this distance makes them count as agreeing measurements of the strength of the same sun-centered inverse-square centripetal acceleration field. As we have seen, the Harmonic Rule relation among the orbits of the planets and the absence of unaccounted-for precession for each orbit give agreeing measurements of the inverse-square power law for a centripetal acceleration field directed toward the sun. The Harmonic Rule measurements of inverse-square acceleration fields directed toward Jupiter and Saturn provide evidence that Jupiter and Saturn are centers of the same sort of acceleration fields. These measurements add further empirical evidence to back up Newton's inference to an inverse-square acceleration field directed toward the sun. The relative distances from these planets explored by the motions of their moons are different from those from the sun explored by the motions of the planets. This adds considerably to the difficulty of finding an alternative that could sufficiently realize agreeing measurements for all three at once to be legitimately counted as a serious rival.

In 1759 a particularly striking later vindication of extending the inverse-square law for an acceleration field directed toward the sun to distances not explored by planetary orbits was provided by Clairaut's celebrated success in predicting the return of Halley's comet.[59] As early as 1705, Halley had proposed elements for this retrograde orbit with a period of on average about 75.5 years, a perihelion distance of about 0.58 AU, corresponding to a semi-major axis of about 17.86 AU and an eccentricity of about 0.97.[60] Halley's comet actually passes through our example distance of 7.00 AU. Given Halley's orbit, his comet's centripetal acceleration toward the sun when it is at that

motion. This would make the measured centripetal acceleration apply much more generally than to those freely moving bodies that actually follow conic section trajectories with the center of the sun counted as a focus.

This makes clear that most of what Ronald Giere counts as exceptions to the combination of Newton's Laws of Motion plus the law of universal gravitation are not exceptions at all (see Giere 1999, 90–1). What do count as exceptions are limited to discrepancies like the famous 43 seconds a century of Mercury's total perihelion motion which were unable to be accommodated by Newton's theory together with other empirically acceptable assumptions.

Sheldon Smith (2002) has defended universal gravity's commitment to occurrent actual forces of interaction between bodies from objections by Giere (op. cit.) and Nancy Cartwright (1983, 1989, 1999). These objections have wrongly taken motions due to the action of other forces to interfere with motions in accord with gravity acting alone to be exceptions to Newton's theory. He has quite rightly argued (S. Smith, 249) that it is the empiricist presumption that a law must report a temporal regularity rather than universal gravity that is at fault in these examples.

[59] A.C. Clairaut (1713–65) gave preliminary results of his, still incomplete, calculation on November 14, 1758. Clairaut's academic colleague J.N. Delise (1688–1768) and his assistant C. Messier (1730–1817) spotted an object on January 21 that they later identified as the comet. The same object had also been observed on December 25, 1758 by J.G. Palitzsch (1723–88), a German amateur astronomer. On April 25, 1759 the astronomer J.J.L. Laland (1732–1807) announced that he and other Parisian astronomers had determined that the object discovered by Palitzsh and Messier had elements that agreed with Halley's proposal. (See Waff in *GHA 2B*, 78–80).

[60] See Wilson, *GHA 2B*, 83.

distance is about 0.8052 AU/t_e^2, very much in line with the agreeing estimates from Newton's data for the planets.[61]

This sun-centered inverse-square acceleration field can usefully be counted as a capacity or causal power. For different distances from the sun it is an extremely stable capacity for producing specific inverse-square related acceleration components toward the sun on bodies should they be at those different distances from it.[62] We have seen that Newton's orbital data count as agreeing measurements of this capacity. Newton's inference in proposition 2 book 3 is to this sun-centered inverse-square acceleration field from the agreeing measurements of its sun-centered direction, its inverse-square power of distance, and its strength afforded by orbital phenomena exhibited by the planets.

IV Lessons for philosophy of science

1. *Duhem on Newton's inferences*

In his influential *The Aim and Structure of Physical Theory*, Pierre Duhem (1861–1916) gives the following neat correspondence between Kepler's three laws of planetary motion and the three fundamental properties Newton inferred about gravitation toward the sun.

> This first law of Kepler's, "The radial vector from the sun to a planet sweeps out an area proportional to the time during which the planet's motion is observed," did, in fact, teach Newton that each planet is constantly subjected to a force directed toward the sun.
>
> The second law of Kepler's, "The orbit of each planet is an ellipse having the sun at one focus," taught him that the force attracting a given planet varies with the distance of this planet from the sun, and that it is in an inverse ratio to the square of this distance.

[61] We have mean-distance $a = 17.86$ AU, period $t = 75.5\ t_e$ (75.5 years). The radial acceleration at distance $r = 7.00$ AU is given by (see notes 56 and 48 above)

$$acc_r = -(4\pi^2 a/t^2)(a/r^2) = -0.8052\ \text{AU}/t_e^2$$

This is an acceleration of about 0.8052 AU / t_e^2 toward the sun.

[62] Sheldon Smith (2002, 245) correctly argues that his interpretation of occurrent forces as realized by differential equations corresponding to the Laws of Motion together with pair-wise forces of attraction between bodies according to universal gravity gets around the exception problem. Adding other forces doesn't undercut universal gravity because that theory doesn't make any commitment to gravity being the only force.

Our interpretation of the attraction as an assignment of equal component centripetal accelerations to places at equal distances, whether or not there are bodies at those places, goes beyond the limitation to actual occurrent forces of interaction between bodies. This interpretation makes the attraction into a disposition to generate component centripetal accelerations on bodies at those places (see chpt. 1 note 45). This interpretation of attraction as a disposition also removes the commitment to actual temporal behavior that supports misleadingly counting examples where other forces operate as exceptions.

The measurements cited by Newton support this inverse-square centripetal field of acceleration. This supports Stein's interpretation of the inverse-square centripetal force that, according to Newton's argument so far, maintains the planets in their orbits.

The third law of Kepler's, "The squares of the periods of revolution of the various planets are proportional to the cubes of the major axes of their orbits," showed him that different planets would, if they were brought to the same distance from the sun, undergo in relation to it attractions proportional to their respective masses. (Duhem 1962, 191)

According to Duhem, Newton inferred the centripetal direction from Kepler's law of areas (we now call this Kepler's second law), the inverse-square variation from the law that orbits are ellipses with the sun at a focus (we now call this Kepler's first law), and that if planets were brought to equal distances they would be attracted proportional to their masses from Kepler's third (or harmonic) law.

Duhem's neat correspondence between Kepler's three laws and the three basic properties Newton assigns to gravitation toward the sun is not accurate to the details of Newton's argument. Newton does infer the centripetal direction from Kepler's law of areas.[63] Newton's inferences to inverse square variation, however, are from Kepler's Harmonic Rule and from the absence of orbital precession. Newton does prove, in proposition 11 of book 1, that the force tending toward a focus of an ellipse and maintaining a body in that elliptical orbit will be inversely as the square of the distance from that focus; but, he does not use this to argue for the inverse-square variation of the force maintaining planets in their orbits. As we have seen, Newton does **not** cite Kepler's elliptical orbit as one of his assumed phenomena.

2. Objections by Duhem and some philosophers

Duhem made a very influential criticism of the role of the inferences from phenomena in Newton's argument when he pointed out that universal gravitation requires corrections to Kepler's laws due to perturbations.

The principle of universal gravity, very far from being derivable by generalization and induction from the observational laws of Kepler, formally contradicts these laws. If Newton's theory is correct, Kepler's laws are necessarily false. (Duhem 1962, 193, emphasis in original)

For Paul Feyerabend and Imre Lakatos, such incompatibilities undercut Newton's claims to use Kepler's laws as phenomena on which to base knowledge of universal gravitation. Consider the following quotations from Feyerabend:[64]

This separation of phenomenon and actual *fact* and the definition of phenomena with the help of theory completely reverses the position expressed in Newton's methodology ... For the exceptions which are observed are now accounted for by the very same theory that is derived from the phenomena as stated without the exceptions ... Used in this way phenomena are no longer a basis for knowledge ...

[63] Duhem's suggestion that Kepler's Harmonic Rule showed Newton that if the planets were brought to equal distances they would be attracted in proportion to their masses is, also, accurate to the fact that the Harmonic Rule is one of several phenomena which Newton cites in support of this acceleration field property of gravity in proposition 6. See chapter 1 above. For more detail, see below, chapter 7.

[64] See Feyerabend (1970, 164, in note 11).

and from Lakatos:

> Newton's compartmentalized mind cannot be better characterized than by contrasting Newton, the methodologist, who claimed that he *derived* his laws from Kepler's 'phenomena', with Newton, the scientist, who knew very well that his laws *directly contradicted* these phenomena. (Lakatos 1978, 210)

For such philosophers, the incompatibility pointed out by Duhem gives reason to dismiss the evidential force of Newton's appeal to Kepler's phenomena in arguing for universal gravitation.[65]

We have seen that Newton's inferences from the phenomena he cites are inferences to inverse-square acceleration fields measured by those phenomena. Newton's inferences to inverse-square variation are among those he describes as deductions from the phenomena. Newton's claims to have "deduced" inverse-square variation from orbital

[65] Here is more from Lakatos:

> The schizophrenic combination of the mad Newtonian methodology, resting on the *credo quid absurdum* of 'experimental proof' and the wonderful Newtonian *method* strikes one now as a joke. But from the rout of the Cartesians until 1905 nobody laughed. Most textbooks solemnly claimed that first Kepler '*deduced*' his laws 'from the accurate observations of planetary motion by Tycho Brahe', then Newton '*deduced*' his law from 'Kepler's laws and the law of motion' but also 'added' perturbation theory as a crowning achievement. The philosophical bric-a-brac, hurled by Newtonians at their contemporary critics to defend their 'proofs' by hook or crook, were taken as pieces of eternal wisdom instead of being recognized for the worthless rubbish that they really were. (1978, 212)

This quotation from Lakatos suggests a sharp contrast between what he dismisses as the mad Newtonian methodology and the wonderful Newtonian method. He also tells us that this wonderful Newtonian method, exemplified in the practice of Newton and his successors, may be said to have created modern science.

> Newton set off the first major scientific research programme in human history; he and his brilliant followers established, *in practice*, the basic features of scientific methodology. *In this sense one may say that Newton's method created modern science.* (1978, 220)

Lakatos takes his methodology of scientific research programs as capturing some objective standards of fallible-critical growth based on his sophisticated version of Popper's methodology of falsificationism.

> For the sophisticated falsificationist a scientific theory T is *falsified* if and only if another theory T' has been proposed with the following characteristics: (1) T' has excess empirical content over T: that is, it predicts *novel* facts, that is, facts improbable in the light of, or even forbidden, by T [Lakatos's footnote: I use 'prediction' in a wide sense that includes 'postdiction']; (2) T' explains the previous success of T, that is, all the unrefuted content of T is included (within the limits of observational error) in the content of T'; and (3) some of the excess content of T' is corroborated. (1978, 32)

He motivated this attractively objectivist alternative to Kuhn's account of scientific revolutions by interesting historical examples.

As we have seen, Lakatos's contrasting interpretation of Newton's methodological remarks is not supported by the details of Newton's actual argument for universal gravity in the *Principia*. Our project here is to use Newton's actual method, as revealed in the details of his argument for universal gravity, to inform our interpretation of his methodological remarks. On our interpretation, Newton's methodology of successive approximations anticipates Lakatos's methodology of scientific research programs.

We shall see that Newton's stronger notion of empirical success as accurate theory-mediated measurement of parameters by the phenomena which they purport to explain can informatively complement and extend Lakatos's account of the unrefuted content of the falsified theory T which is included (within the limits of observational error) in the content of the successor theory T'.

phenomena are claims that the inference from those phenomena to the inverse-square variation is appropriately warranted.[66] We have accounted for this warrant by giving details of the systematic dependencies Newton proves from his Laws of Motion that make these phenomena measure the inverse-square variation.

These measurements are not undercut by corrections to the basic phenomena to account for perturbations. Consider the action of the acceleration field of another planet to perturb the basic Keplerian orbit of a given planet. The interpretation of the corrected orbit as the result of adding the acceleration component toward that other planet to the basic acceleration component toward the sun does not undercut the inference from the basic orbit to the inverse-square acceleration field centered on the sun. One reason this account of the corrected orbit does not undercut the basic inference is that it maintains commitment to the sun-centered inverse-square acceleration component that was measured by the basic Keplerian phenomena. Commitment to an acceleration component toward the sun in accordance with this sun-centered inverse-square acceleration field is maintained when the acceleration toward the other planet is combined with it to generate corrections to the orbit. In addition, perturbation theory provides resources to carry out the basic measurements in the presence of the perturbations that correct the basic phenomena. Consider a planet for which Newtonian perturbations account for all its orbital precession. The zero leftover precession measures the inverse-square variation of its acceleration component due to the sun's gravity.

Newton took the corrections he was able to achieve as further evidence for his theory, rather than as developments which undercut his inferences to it. So too did later figures such as Laplace take their very impressive further improvements to provide further underwriting of Newton's theory. On our interpretation of Newton's inferences as *theory-mediated* measurements, it was quite appropriate for Newton, Laplace and the others to treat the success of corrections accounting for perturbations as further support for, rather than as undercutting, Newton's inferences to universal gravity.

The basic Keplerian phenomena are taken as approximations good enough to support a succession of increasingly better corrected approximations. Newton's argument can be construed as one using this first level of approximation to establish the existence of the basic inverse-square centripetal acceleration fields. The reasoning to universal gravity introduces interactions that lead to corrections that count as more

[66] The rhetoric Lakatos and Feyerabend employ suggests that they see Newton's use of "deduction" as committing him to mean following from the cited premises by logic alone:

> But even if the *Phenomena* were true statements, would Newton's theory follow logically from them? (Lakatos 1978, 211)

This use of "deduction" has become common among philosophers following the entrenchment of first order extensional logic after the revolutionary work of Frege and Russell. As we have seen in chapter 1 (sec. III), Newton's use of "deduction" is in accord with the earlier usage of his time, which was not restricted to logically valid inferences. As we noted, on Newton's usage any appropriately warranted conclusion inferred from phenomena as available evidence will count as a deduction from those phenomena.

accurate, higher-level approximations, but the Keplerian phenomena continue to hold at the first level of approximation.[67] As we noted in chapter 1, the development and applications of perturbation theory, from Newton through Laplace at the turn of the nineteenth century and on through much of the work of Simon Newcomb at the turn of the twentieth, led to successive, increasingly accurate, corrections of Keplerian planetary orbital motions.[68] At each stage, discrepancies from motion in accord with the model developed so far counted as higher order phenomena carrying information about further interactions.[69] These successive corrections led to increasingly precise specifications of solar system phenomena backed up by increasingly precise measurements of the masses of the interacting solar system bodies. In later chapters we shall consider, in more detail, the contributions of perturbation theory to the entrenchment of the acceptance of Newton's theory.

3. Duhem and H-D confirmation

Duhem's criticism of Newtonian methodology did not just rest on the inconsistency he pointed out between Keplerian phenomena and universal gravitation. According to Duhem: Kepler's laws are empirical approximations that succeed in reducing Tycho Brahe's observations to law; but, they needed to be translated into symbolic laws to be useful for constructing physical theory, and such translation presupposed a whole group of hypotheses.

> Thus the translation of Kepler's laws into symbolic laws, the only kind useful for a theory, presupposed the prior adherence of the physicist to a whole group of hypotheses. But, in addition, Kepler's laws being only approximate laws, dynamics permitted giving them an infinity of different symbolic translations. Among these various forms, infinite in number, there is one and only one which agrees with Newton's principle. The observations of Tycho Brahe, so felicitously reduced to laws by Kepler, permit the theorist to choose this form, but they do not constrain him to do so, for there is an infinity of others they permit him to choose. (Duhem 1962, 195)

Motions relative to the sun in accord with inverse-square centripetal acceleration fields, which would make Newton's Laws of Motion yield Kepler's laws exactly, count as symbolic translations of empirical laws reducing Brahe's data; but, they do not agree with Newton's principle of universal gravity. Only motions corresponding to corrections of these orbits generated by taking into account Newtonian perturbations agree with Newton's principle.[70] Like the motions in accord with inverse-square

[67] Indeed, it was only long after *Principia* was published that perturbation theory led to appreciable empirical improvements over basic Keplerian orbits. See chapter 9 section IV.2.

[68] See chapter 10 note 8.

[69] George Smith, forthcoming, provides a very informative account of the extraordinary evidence afforded by these developments.

[70] Duhem obviously exaggerates when he claims that only one such symbolic translation is in accord with Newton's principle. To the extent that masses, positions, and momentum vectors of solar system bodies are

sun-centered acceleration fields, these corrected orbits count as symbolic translations of empirical reductions to law allowed by Brahe's data. In addition to these, Duhem takes such allowable symbolic translations to include any of the infinity of formulae which could be made to generate motions compatible with that data.

Here is Duhem's positive account of how perturbation theory contributes evidence for Newton's theory.

Therefore, if the certainty of Newton's theory does not emanate from the certainty of Kepler's laws, how will this theory prove its validity? It will calculate, with all the high degree of approximation that the constantly perfected methods of algebra involve, the perturbations which at each instant remove every heavenly body from the orbit assigned to it by Kepler's laws; then it will compare the calculated perturbations with the perturbations observed by means of the most precise instruments and the most scrupulous methods. (Duhem 1962, 193–4)

Like many philosophers of science in the twentieth century, Duhem appears to be committed to a hypothetico-deductive model of confirmation. Calculating perturbations from Newton's theory with high precision and having the resulting predictions accurately fit the most precise observations is what he counts as the empirical success that supported this theory. Later data, of the greater precision available to Laplace at the end of the eighteenth century and of the considerably greater precision available to Simon Newcomb at the end of the nineteenth century, required trajectories corresponding to significant corrections resulting from Newtonian perturbations of the basic Keplerian orbits.[71]

Duhem pointed out, however, that such comparisons of Newton's theory with the most precise data available made essential appeal to additional background hypotheses.

Such a comparison will not only bear on this or that part of the Newtonian principle, but will involve all its parts at the same time; with those it will also involve all the principles of dynamics; besides, it will call in the aid of all the propositions of optics, the statics of gases, and the theory of heat, which are necessary to justify the properties of telescopes in their construction, regulation, and correction, and in the elimination of the errors caused by diurnal or annual aberration and by atmospheric refraction. It is no longer a matter of taking, one by one, laws justified by observation, and raising each of them by induction and generalization to the rank of a principle; it is a matter of comparing the corollaries of a whole group of hypotheses to a whole group of facts. (Duhem 1962, 194)

only established up to approximations, any alternatives that are within the established tolerances are compatible with Newton's principle.

[71] Laplace's *Celestial Mechanics* contains many corrections of Keplerian orbits including his famous solution of the great inequality – the approximately 900-year period mutual perturbation of Jupiter and Saturn. (See Wilson 1985, 22–36.) Wilson 1985, 227–85 gives more details. For a translation of Laplace's own text, see Bowditch vol. 3, 281–92 for Jupiter and 303–8 for Saturn. Newcomb (1895) gives his account of the basis for marked improvements made possible by more accurate estimates of masses of the interacting bodies.

According to him, a very wide range of alternative hypotheses are not ruled out, even by the most precise data available, because they could be protected by appropriately adjusting background assumptions.

This quite wide-ranging empirical under-determination of theory by evidence has become familiar to philosophers of science as the Quine–Duhem thesis. Here is Duhem's statement of it.

> The only experimental check on a physical theory which is not illogical consists in comparing the *entire system of the physical theory with the whole group of experimental laws*, and in judging whether the latter is represented by the former in a satisfactory manner. (Duhem 1962, 200)

According to this thesis, empirical support cannot differentiate among assumptions, because only the holistic body of all one's assumptions together ever gets tested. This suggests the skeptical implication that empirical support cannot differentiate among assumptions. We shall see that with Newton's richer notion of empirical success, support afforded by agreeing measurements of the same parameter from diverse phenomena provides an especially attractive answer to this sort of Quine–Duhem skepticism.[72]

4. *Glymour's bootstrap confirmation*[73]

Clark Glymour challenged the hypothetico-deductive model by proposing his bootstrap confirmation as an alternative that could better recover what seem to be compelling examples of legitimate scientific inference.[74] A major attraction was that it provided resources for differential support of theoretical hypotheses, based on the extent to which they are given agreeing bootstrap confirmations from diverse evidence.

Where Duhem focused on the need for additional background assumptions to make theoretical hypotheses yield hypothetico-deductive predictions, Glymour's central idea was that background assumptions may be appealed to in order to make empirical data entail a theoretical hypothesis. Glymour's work generated enormous interest among philosophers of science, who began to better appreciate inadequacies of the hypothetico-deductive model of scientific inference.

[72] Kuhn's (1977, 321–2) criterion of consistency – "not only internally or with itself, but also with other currently accepted theories applicable to related aspects of nature" – provides an argument against this skeptical implication suggested by the Quine–Duhem thesis.

As Jon Dorling has argued (1979), plausible Bayesian priors for real examples of scientific research also provide an answer. More recently, Wayne Myrvold (2003) has shown that Bayesians will assign more support to unified hypotheses than to un-unified ones making the same predictions.

As will be seen below (especially in chapters 9 and 10), Newton's methodology provides powerful reinforcement to supplement these arguments against Quine–Duhem skepticism.

[73] This section and the next compare Glymour's bootstrap confirmation with Newton's inferences. A version of this comparison was presented at the 20th International Wittgenstein Symposium in 1997 and is published in the proceedings of that conference. See Harper 1998.

[74] See Glymour 1980a.

Newton's inferences from phenomena are salient among examples of scientific inferences Glymour cites as bootstrap confirmations.[75] Though I will argue that bootstrap confirmation is not an adequate account of them, I will be supporting Glymour's claims that such inferences are an important part of legitimate scientific practice, that they challenge the adequacy of the hypothetico-deductive model of scientific inference, and that they do support the sort of attractive answer to Quine–Duhem skepticism that Glymour's bootstrap confirmation appeared to offer.

The basic requirement for a bootstrap confirmation is that the background assumptions be compatible with alternatives to the data that would, relative to those background assumptions, be incompatible with the hypothesis.[76] On this basic idea we can say that evidence E bootstrap confirms hypothesis H relative to background assumption A, just in case both

1) $E\&A$ entail H

and

2) There is a possible alternative evidence claim E' such that E' is compatible with A and $E'\&A$ entail not-H.

Glymour argued that each of the above inferences by Newton satisfies these conditions. Suppose we restrict Newton's inferences to the centripetal direction of, the inverse-square variation of, and the inverse-square relation among the actual accelerations produced by the several motive forces on the individual orbiting bodies. Given that the central bodies are counted as inertial, these inferences are entailed by the phenomena together with the theorems Newton cites in their support.

The initially enthusiastic response to Glymour's proposal was followed by later work raising problems for it. The most influential of these problems have been counterexamples proposed by David Christensen based on constructing "unnatural" material conditionals entailed by theory to use as background assumptions.[77] In one of these examples, Christensen constructs such "unnatural" material conditionals that would be entailed by Kepler's three laws and which would allow observations of a single planet to bootstrap-confirm Kepler's Harmonic Rule.[78] Let the k-value for a given planet be

[75] See Glymour 1980a, 203–25.

[76] Here is Glymour's statement of this basic idea:

> The central idea is that the hypotheses are confirmed with respect to a theory by a piece of evidence provided that, using the theory, we can deduce from the evidence an instance of the hypothesis, and the deduction is such that it does not guarantee that we would have gotten an instance of the hypothesis regardless of what the evidence might have been.
>
> ...
>
> The account that follows is more an illustration of the possibilities of the general idea that is central to this chapter than it is a definitive presentation of that idea itself. (1980a, 127–8)

The several different technical definitions (in *Theory and Evidence*, chapter 5; as well as several later proposals by Glymour 1983, Earman and Glymour 1988 in response to criticism) are somewhat more complicated.

[77] See Christensen 1983 and 1990.

[78] See Christensen 1983, 475.

R^3/t^2, where R is the mean-distance and t is the period of its orbit. Let the background assumption A be $C1\&C2$.

$C1$: If planet a obeys Kepler's first two laws (in a given set of observations) then the k-value for planet b will be the same as that of planet a.

$C2$: If planet a doesn't obey the first two laws, the k-value of planet b will be twice that of planet a.

E: planet a obeys the first two laws. E': not-E.

The Harmonic Rule asserts that there is a single constant k-value which would hold for any planet orbiting the sun. Without some appropriately intuitive story about background assumptions, it would appear counterintuitive to count observations of a single planet, alone, as confirmation of the Harmonic Rule.

Christensen notes that his $C1$ and $C2$ exhibit the same unnaturalness that would be found in axioms used by Glymour to illustrate the failure of hypothetico-deductive (H-D) confirmation to be able to account for the intuitive need for observations of more than one planet to confirm Kepler's Harmonic Rule.[79]

What is intuitively unnatural about such axiom sets? One answer is that while they contain only sentences entailed by the "natural" set, some of the sentences they contain as axioms are entailed by the natural set only "accidentally": they do not express any intuitive regularities of nature that would occur to us as explaining the data. (Christensen 1983, 477)

Material conditionals $C1$ and $C2$ violate Paul Grice's implicatures about intuitive assertability.[80] Material conditional $C1$ follows from the assumption of Kepler's laws only because its consequent does, while material conditional $C2$ follows only because its antecedent is incompatible with this assumption.[81]

A second difficulty was raised by Roger Rosenkrantz, who argued that the positive examples Glymour cites are examples where the H-D entailment

$H\&A$ entails E

holds as well as the bootstrap entailments.[82] According to Rosenkrantz, this makes a Bayesian, with priors suitably reflecting these entailments, sensitive to sample coverage

[79] Glymour's positive case for his alternative was backed up by criticisms of H-D proposals for countering the Quine–Duhem thesis. Among these was construction of material conditionals entailed by Kepler's laws that would allow observations of a single planet to H-D confirm Kepler's Harmonic Rule (1980a, 36–8).

[80] See Grice 1989, 24–40.

[81] Edidin (1988, 270) suggested that counterexamples such as Christensen's might be avoidable by using some stronger conditional "which, unlike its material counterpart, is not true in the 'absurd cases'." I think this is the right sort of idea.

That just using material conditionals would not be able to do the job is supported by the failures of efforts by Glymour and others to find ways of doing so. Glymour (1983) added a condition on bootstrapping designed to rule out examples of the sort Christensen pointed out. Christensen (1990) was able to show that equally unintuitive versions of his counterexamples could be constructed, which would not be ruled out by Glymour's new condition. In addition, Christensen (1990) shows that alternatives to Glymour's condition, by Zytkow (1986) and by Earman and Glymour (1988), also fail to rule out counterintuitive examples of the sort discussed.

[82] See Rosenkrantz 1983.

in exactly the right way to recover these intuitive examples of scientific inference as H-D confirmations.[83]

Finally, Duhem's basic point about Kepler's laws being only approximate was raised specifically as a problem for bootstrap confirmation by Ronald Laymon.[84] Given the background assumption *A*, knowing that the phenomenon *E* was exactly true would legitimate accepting proposition *H*. But, even given assumption *A*, if all that is known about *E* is that it holds up to some approximation the bootstrap entailment,

A&*E* entails *H*

doesn't make this inform our belief about *H*.

5. *Newton's inferences vs. Glymour's bootstrap confirmations: laws, not just material conditionals*

A central role of the Laws of Motion is to insure that the systematic dependencies following from them have appropriately subjunctive force to make the cited phenomena count as measurements of the centripetal direction and the inverse-square variation of the forces maintaining bodies in orbits. Let us begin by considering the accelerations corresponding to the separate motive forces maintaining the individual planets in their orbits about the sun. Consider the inference to the centripetal direction of the force maintaining a planet in its orbit from the phenomenon that the rate at which it sweeps out areas by radii to the sun is constant. Suppose that at this moment the rate at which Jupiter is sweeping out areas by radii from the center of the sun is constant.[85] Propositions 1 and 2 together make the constancy of this rate equivalent to having the acceleration of Jupiter relative to the sun be directed toward the center of the sun. The additional dependencies (in corollary 1 of proposition 2) would make alternatives to this Area Rule phenomenon carry information about alternative directions of this acceleration. If this rate were increasing, this acceleration of Jupiter with respect to the sun would be angled off-center in a forward direction with respect to Jupiter's present orbital velocity. If this rate were decreasing, the acceleration would be angled off-center backwards. These systematic dependencies make the Area Rule phenomenon for each planet carry the information that the total motive force maintaining that planet in its orbit about the sun is directed toward the center of the sun.

Similarly, for the system of planets orbiting the sun, corollary 7 of proposition 4 establishes systematic dependencies between values of a phenomenal magnitude, a common ratio (s) of periods to mean-distances for these orbits, and the corresponding value ($1-2s$) of a common ratio of centripetal accelerations to mean-distances. This makes the Harmonic Rule ($s = 3/2$) phenomenon for that system of orbits measure the inverse-square (-2) power rule relation to mean-distances among the motive forces maintaining those planets in their orbits. These systematic dependencies turn the

[83] See Rosenkrantz 1983, 72–6. [84] Laymon 1983.

[85] We can think of the second derivative of this area as a phenomenal magnitude, which takes the value zero when the rate is constant, is positive when the rate is increasing, and is negative when the rate is decreasing.

equivalence between the Harmonic Rule phenomenon and the inverse-square power rule into a *measurement*. Alternative values of the parameter (s) would carry information about alternative power rule relations. If the ratio s were greater than $^3/_2$ the forces would be falling off faster, and if s were smaller the forces would be falling off less fast than the inverse-square with respect to the distances.

This is also the case with Newton's inferences from absence of orbital precession to the inverse-square variation of each of the motive forces maintaining the planets in their individual orbits. The phenomenal magnitude is the amount p of precession per revolution of the orbit, and the corresponding theoretical magnitude is the power-law relating the centripetal motive force maintaining a planet in its orbit to the distances from the sun explored by that planet's orbital motion. As we have seen, zero unaccounted-for precession measures the inverse-square variation. Unless it could be otherwise accounted for, non-zero orbital precession would measure a deviation from the inverse-square. Forward precession would measure a variation of motive force falling off faster than the inverse-square of distance, while backward precession would measure a variation falling off less fast than the inverse-square.

In each of these cases, the H-D entailment of the phenomenon by the proposition together with the background assumptions holds. Unlike Glymour's weaker bootstrap confirmations, Newton's inferences include the H-D entailment appealed to by Rosenkrantz. They go beyond hypothetico-deductive confirmation by including Glymour's bootstrap conditions as well; but, they also include systematic dependencies that go beyond both hypothetico-deductive and bootstrap confirmation. These systematic dependencies make Newton's inferences answer Laymon's objection to Glymour's bootstrap confirmations. Having the value of the phenomenal parameter empirically fixed to a given approximation will carry the information that the inferred theoretical parameter holds to an appropriately corresponding approximation.

The main point of Christensen's counterexamples was to challenge Glymour's identification of first order extensional logical consequences of extensional formulations as the only content of theories that counts. He constructed material conditionals that lead to bootstrap confirmations that seem clearly not to provide any legitimate support. The systematic dependencies backing up Newton's inferences have appropriately subjunctive force. They are not merely material conditionals, but clearly express intuitive regularities of nature of the sort that would occur to us as explaining the phenomena.

We noted that Glymour took Newton's inferences to be inferences to the centripetal direction of and inverse-square variation of each, as well as the inverse-square relation among accelerations produced by what Newton counts as the several motive forces on the individual orbiting bodies. The disastrous failure of bootstrap confirmation to do justice to Newton's inferences is a failure to do justice, even, to inferences restricted to the motive forces maintaining the individual planets in their actual orbits. The positivist project of attempting to explicate scientific inference without appeal to subjunctive conditionals fails to do justice even to the single inference to the centripetal direction of the motive force maintaining Jupiter in its orbit. Material conditionals are inadequate to the systematic dependencies that make the constancy of the rate at which areas are swept out by radii from the center of the sun to Jupiter carry the information that the motive force maintaining Jupiter in its orbit is directed toward that center.

The suggestion that making do with truth functional material conditionals avoids suspect commitments of subjunctive conditionals is not driven by problems internal to the practice of science. The claim that subjunctive conditionals are suspect is, rather, a slogan for a philosophical project of revising the commitments of scientific practice to what can be accommodated by extensional logic alone. The Laws of Motion support subjunctive conditionals that express systematic dependencies that make motion phenomena carry information about forces. This is an important role that accepting them as ***laws*** plays in Newton's scientific method.

6. Inverse-square acceleration fields and gravity as a universal force of pair-wise interaction between bodies

In chapter 1, we noted that some have recently challenged Stein's field interpretation of Newton's centripetal forces. They would challenge the interpretation we have endorsed of the following passage from Newton's definitions:

> motive force may be referred to a body as an endeavor of the whole directed toward a center and compounded of the endeavors of all the parts; accelerative force, to the place of the body as a certain efficacy diffused from the center through each of the surrounding places in order to move the bodies that are in those places; . . . (C&W, 407)

We have followed Stein in assigning accelerative measures that are inversely as the square of the distance to all places beyond the surface of the sun, whether or not there is an actual body at that place. The two phrases "to the place of the body" and "in order to move the bodies that are in those places" are consistent with limiting the places to those actually occupied by bodies. There is, however, some tension between such an interpretation and the middle phrase "as a certain efficacy diffused from the center through each of the surrounding places."

Eric Schliesser has reported to me that, upon being confronted with this more limited alternative, Stein's defense of his sticking to his own richer interpretation of the above passage included the remark: "I am not a positivist!"

Before considering Schliesser's alternative, I want to say more about a commitment of the positivist program that continues to infect many philosophers of science. One of the main objectives of the positivist program for interpreting theories was an attempt to do as much as possible with material conditionals. Limiting the places to those actually occupied is very much in line with an attempt to keep the interpretation of Newton as close as possible to this positivist goal of avoiding subjunctive conditionals. The motivation to avoid commitment to laws supporting subjunctive conditionals is, however, already undermined by the systematic dependencies that make orbital phenomena measure the centripetal direction and the inverse-square variation of the accelerations exhibited by the planets in their actual orbits. It therefore does not afford any non-dubious support for restricting the places to which the accelerative measures of the inverse-square centripetal force toward the sun are assigned to the places actually occupied by planets or other bodies.

Schliesser's own argument for his interpretation is not positivistic,[86] and it does call attention to real tension between acceleration fields and Newton's later conception of gravity as a universal force of pair-wise interaction between bodies. We have followed Stein by interpreting Newton's classic inferences to inverse-square centripetal forces as inferences to inverse-square fields of acceleration surrounding the sun, Jupiter, and Saturn. As we have noted in chapter 1, and will see in more detail in chapter 8, Newton will go on to argue for gravity as a universal force of pair-wise interaction between bodies. To what extent is the claim that the sun and each of these planets is surrounded by its own inverse-square field of gravitational attraction compatible with Newton's theory of gravity?[87] Consider an isolated body with an appropriately spherically distributed distribution of total mass M and radius r.[88] Consider locations around our body at distances R from its center greater than the distance r of its surface. For any such distance R, another body of mass m can be counted as a test body for exploring gravitational attraction toward our given body of mass M so long as m is sufficiently small relative to M for the

[86] He suggests that Newton was restricting the assignment of accelerative measures to make the passage compatible with his commitment to a hypothesis about the philosophical interpretation of the action of gravity as a universal force of pair-wise interaction between bodies.

There is indeed tension between interpreting Newton's initial conception of inverse-square centripetal forces toward planets as inverse-square fields of acceleration and the conception of gravity as a universal force of pair-wise interaction in the theory he arrives at through his controversial application of Law 3 in proposition 7. I think the extent to which this conceptual transformation Newton argues for preserves the agreeing measurements supporting the inferences to inverse-square acceleration fields toward planets as approximations is very informative for how Newton's rich methodology can support progress through theory change.

Schliesser says he was led to his interpretation by considering Ori Belkind's interesting discussion of Newton's conception of place as the part of space that a body occupies that can be quantized by the amount of solid matter it contains. This suggests that treating places as point locations characterized by their distances from the center of a planet does not do justice to this characterization of them as parts occupied by bodies. Newton certainly does attribute distances to places. Moreover, his theorems about centripetal forces represent places of bodies by point masses. Consider a point location at our specified distance of 7.00 AU from the center of the sun. Suppose a body with a spherically symmetric mass distribution were to be so placed that its center of mass was located at our specified point. This, surely, would make it appropriate to count 7.00 AU as the distance that body's place would be from the center of the sun. The same centripetal acceleration toward the sun would have to be assigned to any other body whose place was that distance from the center of the sun.

To count a volume of space surrounding the sun as appropriate for specifying true motions of bodies moving from place to place within it is supported by exactly the same empirical phenomena that would make motion of the sun small with respect to the center of mass of the interacting solar system bodies. I see no problem for Stein's interpretation of inverse-square forces toward planets as acceleration fields being able to accommodate Belkind's interesting account of Newton's conception of places of bodies.

[87] Schliesser argues that Stein's acceleration field interpretation is committed to the claim that a lone point particle, however small its mass, would be surrounded by an inverse-square varying field of centripetal accelerations. I have not found any of Stein's actual assignments of inverse-square acceleration fields to interpret the inverse-square centripetal forces in Newton's argument to go beyond the centripetal forces directed to bodies like the sun and the planets.

[88] We shall see in chapter 8 that this allows the body toward which a centripetal force is directed, and the bodies between which we represent the gravitational interactions, to be accurately modeled as point masses.

interaction being modeled. Newton's theory recovers subjunctive conditionals supporting the assignment of inverse-square accelerations adjusted to test bodies at the distances of those places, whether or not those places are actually occupied by such test bodies.

For bodies too large to be counted as test bodies, the true motions among them would be fixed relative to the center of mass frame to be given by a linear combination of the inverse-square acceleration fields toward each. As we shall see in chapter 9, this will recover respective inverse-square acceleration field values as appropriately weighted component accelerations. In chapter 8, we shall see that Newton's center of mass solution to the two chief world systems problem is given by such a linear combination of the inverse-square acceleration fields of gravity toward the sun and toward each of the planets. Here, too, the inverse-square accelerations toward each body are recovered as appropriately weighted component accelerations.

This interpretation of the systematic dependencies as subjunctive conditionals supporting component accelerations is important. Newton's assignment of the acceleration of gravity to a flying bird is compatible with its maintaining altitude rather than travelling on a projectile trajectory. The actual motion of the bird at any instant is the outcome of the composition of its weight with other forces, such as those generated by the interactions between its wings and the air. The component centripetal acceleration corresponding to its weight toward the earth is composed with the component accelerations corresponding to these other interactions. Similarly, the component acceleration of gravity corresponding to the weight of a body at rest on a surface is exactly offset by an equal and opposite component acceleration generated by the total force exerted by that surface on the bottom of that body.

7. *Kepler's elliptical orbits*

We have seen that Glymour wondered why Newton did not infer the inverse-square variation of the centripetal forces maintaining the planets in their orbits about the sun from Kepler's elliptical orbit with the sun at a focus. We have noted that Newton does not cite Kepler's elliptical orbits among his phenomena. In chapter 1, we saw that assuming such heliocentric orbits for the primary planets would beg the question against the geocentric solution to the two chief world systems problem. Newton's phenomena for the primary planets are carefully formulated so as to be neutral between geocentric and heliocentric solutions. This gives quite a solid reason for Newton to avoid arguing from Kepler's elliptical orbits. Indeed, as we have noted and will explore in more detail in chapter 8, Newton argues from his theory of gravity to Kepler's orbits, not the other way around. Kepler's orbits are argued to be the orbits that would hold exactly if the only force acting on the planets were the motive forces generated by the inverse-square acceleration field surrounding the sun. This makes them appropriate initial models from which to proceed in a research program in which deviations carry information about interactions that afford successively more accurate corrected orbits.

Putting aside this two chief world systems problem, there are informative methodological reasons for not arguing to the inverse-square variation of the force maintaining a planet in its orbit from assuming as a premise that its orbit is an ellipse with the sun at a focus. Such an inference would satisfy Glymour's conditions for a bootstrap confirmation. According to proposition 11 of book 1, the law of the centripetal force directed to the focus of an elliptical orbit produced by it is inverse-square. This makes an elliptical orbit exactly satisfying the Area Rule with respect to a focus of the ellipse sufficient to require the inverse-square variation of the centripetal motive force deflecting a body into that orbit. If an elliptical orbit exactly satisfied the Area Rule with respect to the center of the ellipse, rather than a focus, then, according to proposition 10 of book 1, the centripetal force would be directly as the distance instead of inversely as the square of the distance. Adding this to proposition 11 is sufficient to make an elliptical orbit exactly satisfying the Area Rule with respect to a focus bootstrap confirm the inverse-square variation of the centripetal force deflecting a body into that orbit.

As Duhem and Laymon have pointed out, the phenomena Newton appeals to are approximations, which are not exactly true. The additional systematic dependencies backing up Newton's inferences make the conditions they satisfy more stringent than even the combination of H-D and bootstrap entailments together. One striking advantage afforded by these dependencies is that they make tolerances to which the phenomenal magnitude is known to hold carry information about corresponding tolerances to which the inferred value of the corresponding theoretical parameter is known to hold. It is these systematic dependencies that allow non-trivial inferences to approximate values of theoretical parameters from approximate values of the corresponding phenomenal magnitudes. As George Smith puts it:

> In short, wherever Newton did draw inferences from phenomena, the conclusion still holds *quam proxime* even when the premise holds only *quam proxime*; and Newton took the trouble to include mathematical results supporting this. (Smith, G. 2002a, 36)

Newton's inferences to the centripetal direction from the Area Rule, as well as his inferences to the inverse-square from the Harmonic Rule for a system of orbits and from the absence of unaccounted-for precession of each single orbit, all meet this condition of robustness with respect to approximations in the value of the phenomenal parameter.

In a quite striking argument, Smith has shown that an inference to the inverse-square variation of a centripetal force from Kepler's elliptical orbit with the Area Rule with respect to a focus definitely fails to meet this important condition of robustness with respect to approximations.[89] Note that the basic power law results for stable elliptical orbits, -2 with the force to a focus and $+1$ with the force to the center, already suggest that small departures from the direction toward the focus would not guarantee small differences from the inverse-square. Smith shows, in detail, that where

[89] Smith G.E. 2002b, 34–42.

the force is directed between the center and the focus, especially for an ellipse with small eccentricity like the orbits of the planets, even very small differences from the direction toward the focus lead to large differences from the inverse-square. He shows that the powers of distance vary quite wildly with changes in the direction of the force and that Newton was in a position to know this.

What about Newton's application of the Harmonic Rule dependencies to infer the inverse-square law for gravitation toward the sun? We have pointed out that the centripetal accelerations of Keplerian elliptical orbits, at what are counted as their mean-distances from the sun, equal those of uniform motion on concentric circular orbits of the same period with radii equal to those mean-distances. This allows the systematic dependencies of corollary 7 of proposition 4 to make having the periods be a power approximating the 3/2 power of the distances carry the information that the relation among the centripetal accelerations of the orbits at those mean-distances approximates the inverse-square variation with respect to those distances. These dependencies make the inference robust with respect to approximations to the Harmonic Rule power of distance for periods exhibited in those orbits.

The orbital data are compatible with Kepler's ellipse with the sun at a focus; but, they are not precise enough to directly rule out an ellipse with the sun slightly displaced toward the center from the focus. Does this undercut Newton's inference from the Harmonic Rule phenomenon to the inverse-square law for the centripetal force toward the sun? Consider a version of such an alternative orbit where the centripetal accelerations at what would be counted as mean-distances would not agree with those of corresponding concentric circular orbits. Suppose that some such version is consistent with the data.[90] To have the Harmonic Rule inference to the inverse-square require that such alternatives be logically ruled out by the data, together with the explicit background assumptions, is exactly the sort of illegitimate application of Descartes's extreme ideal of inference that Newton's Rule 4 is directed against. Rule 4 would have such an alternative dismissed as a mere contrary hypothesis, unless it realized Newton's standard of empirical success sufficiently to count as a serious rival.[91] In the absence of additional phenomena making

[90] I have not investigated the extent to which the centripetal accelerations, at what would be counted as mean-distances of such orbits, would agree with the centripetal accelerations of corresponding uniform motion on concentric circular orbits with periods equal to the periods of those orbits.

Having those results hold so as to support the sort of extension we have to Keplerian elliptical orbits would certainly be a sufficient condition for defending Newton's inference. I am, however, denying that such extensions are necessary to have Newton's inference count as legitimate. To require this would be an example of what I take to be an all too common error shared by many philosophers of science. This is the error of turning something shown to be a sufficient condition into a necessary condition so that a positive argument for accepting some inference can be turned into a resource for generating misleadingly skeptical arguments against other inferences.

[91] An analogy to the fit of plotting log periods against log distances may be instructive. To have the periods be some power *s* of the mean-distances is to have the result of plotting log periods against log distances be fit by a straight line of slope *s*. These results for Newton's orbital data can be fit as well or better by a higher-order curve that would not have the periods be any constant power *s* of the mean distances. On such a hypothesis, the application of corollary 7 of proposition 4 would be undercut. The orbits would, therefore,

the inverse-square liable to exceptions corresponding to it, such an alternative would be rightly dismissed as not counting to undercut the inference to the inverse-square.

8. *Newton's ideal of empirical success*

The Harmonic Rule for the system of the several planets and the absence of orbital precession for any one of those planets are distinct phenomena that count as agreeing measurements of the inverse-square variation of the same centripetal acceleration field surrounding the sun. The Harmonic Rule measurements of inverse-square acceleration fields directed toward Jupiter and Saturn provide evidence that Jupiter and Saturn are centers of the same sort of acceleration fields. These measurements add further empirical evidence to back up Newton's inference to an inverse-square acceleration field directed toward the sun. The agreement of these measurements is an example of an ideal of empirical success which informs Newton's scientific method. This ideal of empirical success motivates the methodological theme to, insofar as possible, support theoretical claims as agreeing empirical outcomes of measurements by diverse phenomena. This theme is very much in line with what Glymour took to be a major advantage of bootstrap confirmation – its capacity to represent evidential relevance.

not carry information about any simple power law relating the accelerative forces to distances from the sun. The unwarranted assumption of naive empiricism that fit to data exhausts the empirical criteria for scientific inference would have contributed to unjustified skepticism about the extent to which data can be counted as evidence.

Malcolm Forster and Elliott Sober (Forster and Sober 1994) have appealed to a concept of predicted fit to challenge this naive assumption that fit to data exhausts the empirical criteria for scientific inference. Given assumptions about errors in the data, fit of a model to that data is not always a good indication of how well the model will fit future data. A model that fits the given data too closely is likely to be tracking random errors in the data, in addition to the law-like phenomenon under investigation. They have defended the Akaike information criterion for model selection in scientific inference as an alternative to naive empiricism.

Myrvold and Harper (2002) shows that, on Newton's data for the primary planets alone, the Akaike information criterion would choose an alternative of the sort Stein challenged me to rule out, rather than the inverse-square. Newton's scientific method is more informative than one limiting scientific inference to what can be supported on the Akaike information criterion by itself. Myrvold and I argue that the increased informativeness afforded by Newton's richer criterion of empirical success is an important and legitimate part of scientific method. As we have pointed out above (see note 55 of this chapter), the Akaike information criterion would pick out an alternative higher-power variation that would assign an acceleration quite different from that assigned by the inverse-square at the distance of 7.00AU we used to illustrate the agreeing measurements afforded from Newton's data.

The Akaike information criterion is not sufficient to recover Newton's inference to inverse-square variation of gravity toward the sun. Our defense of Newton's inference, however, does not require us to dismiss the appropriateness of the lessons Akaike can teach about limitations of naive empiricism in fitting phenomena to data.

Chapter 3 Appendix 1: Pendulum calculations

1.1 Proof that the velocity of the body is proportional to the chord of the arc TA for idealized pendulum motion without loss of energy

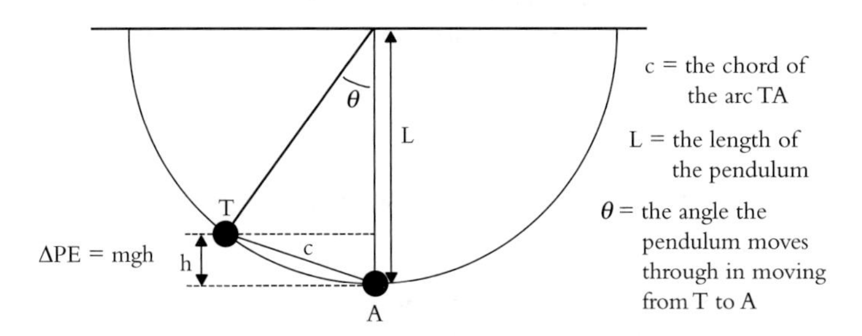

This is a modern proof using the conservation of total energy and the concepts of kinetic energy (**KE**) and potential energy (**PE**).[92] At **T** the body is stationary, **v** = 0; it therefore has no kinetic energy. As the body is released from **T** it accelerates under the force of gravity to point **A**.

ΔPE + ΔKE = 0: *Conservation of energy*

In moving from point **T** to point **A**: **ΔPE = − mgh** and **ΔKE = + mgh**

$$\Delta KE = \tfrac{1}{2}\, mv^2 = mgh \Rightarrow v = \sqrt{(2gh)}$$

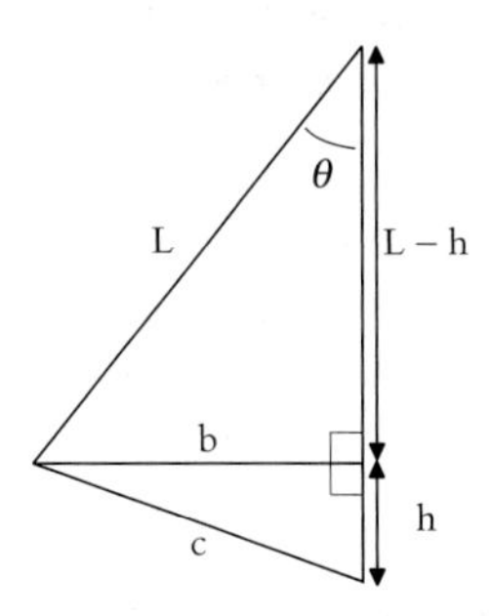

$$c^2 = h^2 + b^2$$

$$L^2 = (L - h)^2 + b^2$$

$$= L^2 + h^2 - 2Lh + b^2 \Rightarrow b^2 = 2Lh - h^2$$

$$c^2 = h^2 + 2Lh - h^2$$

$$c^2 = 2Lh \qquad \Rightarrow \qquad c = \sqrt{(2Lh)}$$

[92] This proof and diagrams are from slides Gemma Murray provided for a presentation I gave on Newton, Wren, and Huygens on Laws of Collision.

$c = \sqrt{(2Lh)}$ $\quad v = \sqrt{(2gh)}$

$c \propto \sqrt{h}$ $\quad v \propto \sqrt{h}$ $\quad$ therefore $\mathbf{c \propto v}$

1.2 Adjusting for air resistance Newton uses an example of a pendulum experiment in which a body A is let fall to collide with a stationary body B, to show how to correct for air resistance to find the arcs of motion that represent the velocities before and after collision that would have obtained if the experiment had been performed in a vacuum.

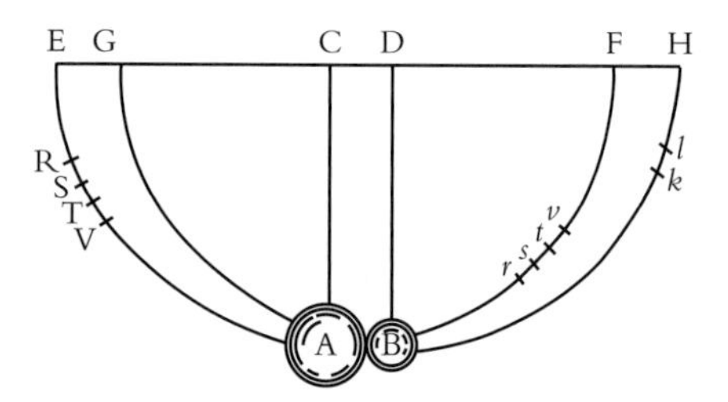

He shows how to calculate the arc TA that would, in a vacuum, have produced the velocity which body A let go from S actually has when it contacts body B. He also shows how to calculate the arcs A*t* and B*l* that would have resulted after reflection had the actual experiment, which did result in A going to *s* and body B going to *k* after reflection, been performed in a vacuum. This method is described in detail with reference to the diagram above by Newton.

> Let the spherical bodies A and B be suspended from centers C and D by parallel and equal cords AC and BD. With these centers and those distances as radii describe semicircles EAF and GBH bisected by radii CA and DB. Take away body B, and let body A be brought to any point R of the arc EAF and be let go from there, and let it return after one oscillation to point V. RV is the retardation arising from the resistance of the air. Let ST be a fourth of RV and be located in the middle so that RS and TV are equal and RS is to ST as 3 is to 2. Then ST will closely approximate the retardation in the descent from S to A. (C&W, 425)

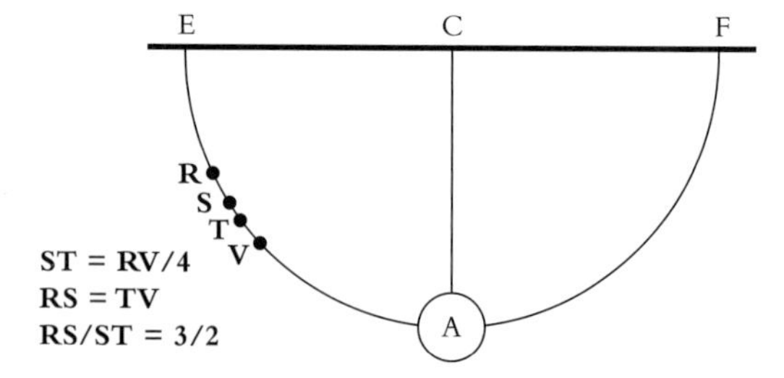

Newton determines the effect of air resistance on body A in moving through one whole oscillation by noting the distance between the point R, where body A begins its motion, and point V, where the body arrives upon completion of the oscillation. The arc RV represents the retarding effect of air resistance on the body in the completion of an oscillation. Newton then calculates the effect on the first part of this oscillation, through which the body will move before a collision at point A. He determines this by dividing the arc RV, the measured retardation for a whole oscillation, by four and averaging the length of the path during each of the four parts of the whole oscillation. This gives us the arc ST, which is one quarter of the length of RV and placed in the middle of this original arc. The arc ST represents the effect of air resistance on the body A completing the arc SA.[93]

Newton goes on to utilize such arcs to represent velocities before and after collision.

Restore body B to its original place. Let body A fall from point S, and its velocity at the place of reflection A, without sensible error, will be as great as if it had fallen in a vacuum from place T. Therefore let this velocity be represented by the chord of the arc TA. For it is a proposition very well known to geometers that the velocity of a pendulum in its lowest point is as the chord of the arc it has described in falling. (C&W, 425)

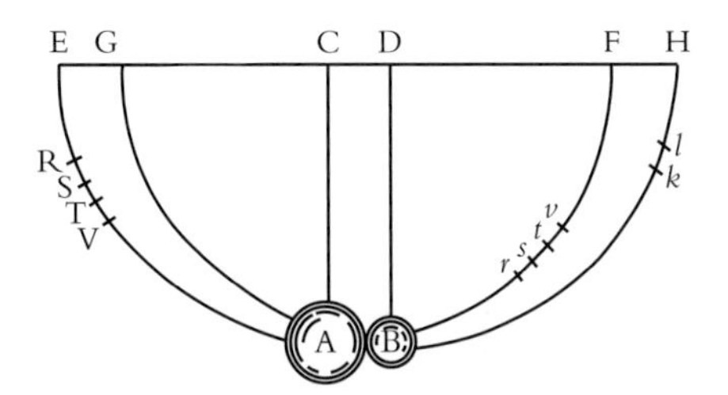

He points out that the arcs *t*A and B*l* that would have resulted after reflection had the actual experiment, which did result in A going to *s* and body B going to *k* after reflection, been performed in a vacuum can be similarly calculated (C&W, 425). He then utilizes these arcs to represent the velocities before and after reflection.

> In this manner it is possible to make all our experiments, just as if we were in a vacuum. Finally body A will have to be multiplied (so to speak) by the chord of the arc TA, which represents its velocity, in order to get its motion in place A immediately before reflection, and then by the chord of the arc *t*A in order to get its motion in place A immediately after reflection. And thus body B will have to be multiplied by the chord of the arc B*l* in order to get its motion immediately after reflection. And by a similar method, when two bodies are let go simultaneously from different places, the motions of both will have to be found before as well as after reflection, and then finally the motions will have to be compared with each other in order to determine the effects of the reflection. (C&W, 425–6)

On the, by then well-known, idealized theory of uninterrupted pendulum motion in a vacuum, the velocity of a pendulum in its lowest point is as the chord of the arc that it has described in falling and as the chord of the equal arc it traverses from that lowest point to its highest point on the other side.[94] In the application to collisions, these chords of the arcs before and after reflection represent the velocities before and after that collision.

[93] In fact, this represents the total retarding effect of all sources of resistance to idealized free motion. Newton's reduction will include taking into account the total of all losses of energy. In addition to air resistance, this will include any non-negligible losses from friction at the point of suspension, from stresses in the cord, or any other sources of resistance there may be. This fact improves the appropriateness of using Newton's correction to represent the velocity for the corresponding idealized free pendulums, so long as the resistance on a single half-swing is one-fourth of the resistance of a whole oscillation back and forth.

[94] See above, Appendix 1.1.

Chapter 3 Appendix 2: Newton's proofs of propositions 1–4 book 1

1. *Proposition 1*

Book 1 Proposition 1 Theorem 1

The areas by which bodies made to move in orbits describe by radii drawn to an unmoving center of forces lie in unmoving planes and are proportional to the times.

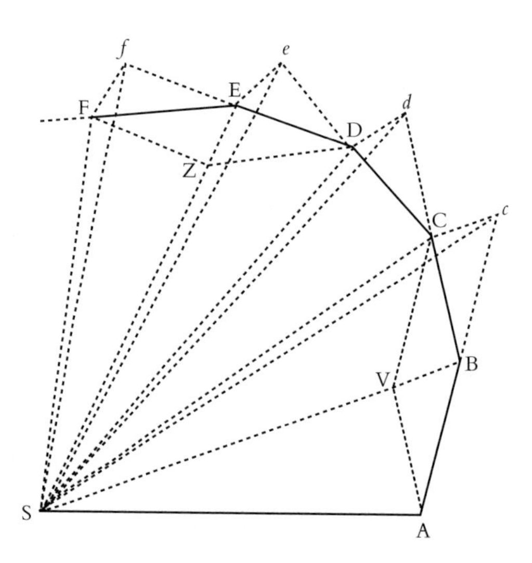

Let the time be divided into equal parts, and in the first part of the time let a body by its inherent force describe the straight line AB. In the second part of the time, if nothing hindered it, this body would (by law 1) go straight on to *c*, describing line B*c* equal to AB, so that – when radii AS, BS, and *c*S were drawn to the center – the equal areas ASB and BS*c* would be described. But when the body comes to B, let a centripetal force act with a single but great impulse and make the body deviate from the straight line B*c* and proceed in the straight line BC. Let *c*C be drawn parallel to BS and meet BC at C; then, when the second part of the time has been completed, the body (by corol. 1 of the laws) will be found at C in the same plane as triangle ASB. Join SC; and because SB and C*c* are parallel, triangle SBC will be equal to triangle SB*c* and thus also to triangle SAB. By a similar argument, if the centripetal force acts successively at C, D, E, . . . , making the body in each of the individual particles of time describe the individual straight lines CD, DE, EF, . . . , all these lines will lie in the same plane; and triangle SCD will be equal to triangle SBC, SDE to SCD, and SEF to SDE. Therefore, in equal times equal areas are described in an unmoving plane; and by composition [or componendo], any sums SADS and SAFS of the areas are to each other as the times of description. Now let the number of triangles be increased and their width decreased indefinitely, and their ultimate perimeter ADF will (by lem. 3,

> corol. 4) be a curved line; and thus the centripetal force by which the body is continually drawn back from the tangent of this curve will act uninterruptedly, while any areas described, SADS and SAFS, which are always proportional to the times of description, will be proportional to those times in this case. Q.E.D. (C&W, 444–5)

The basic result for discrete triangles corresponding to motions brought about by sharp impulses at equal time intervals follows from the first Law of Motion, corollary 1 of the Laws, and some elementary Euclidian geometry.

The first step applies the first Law of Motion to give the motion that would obtain if nothing acted on the body.

> Let the time be divided into equal parts, and in the first part of the time let a body by its inherent force describe the straight line AB. In the second part of the time, if nothing hindered it, this body would (by law 1) go straight on to *c*, describing the straight line B*c* equal to AB, so that – when radii AS, BS, and *c*S were drawn to the center – the equal areas ASB and BS*c* would be described. (C&W, 444)

Triangles ASB and BS*c* have equal areas, because they have equal bases and equal heights. These equal heights are given by the equal perpendiculars joining the base line AB*c* and a parallel to it drawn through the center S. This shows that a body in uniform rectilinear motion with respect to an inertial center will traverse equal areas in equal times by radii to that center.

Newton goes on to consider a centripetal force acting with a single impulse when the body arrives at B.

> But when the body comes to B, let a centripetal force act with a single but great impulse and make the body deviate from the straight line B*c* and proceed in the straight line BC. Let *c*C be drawn parallel to BS and meet BC at C; then, when the second part of the time has been completed, the body (by corol. 1 of the laws) will be found at C in the same plane as triangle ASB. Join SC; and because SB and C*c* are parallel, triangle SBC will be equal to triangle SB*c* and thus also to triangle SAB. (C&W, 444)

The impulse at B is directed toward center S and so is in the plane defined by line AB and center S, as is the motion from B to *c* which would have obtained if the impulse had not acted. By corollary 1 of the Laws,[95] the motion from B to C also lies in this plane and the line *c*C is parallel to the direction from B to S in which the impulse acts. This line *c*C represents the difference between the motion from B to *c* that the body would have undergone in the second unit of time had it not been deflected by the force acting at B and the motion from B to C that resulted from that impulse. Triangles SBC and SB*c* are equal, since they have the same base SB and are both between

[95] **Laws of Motion Corollary 1**

> *A body acted on by [two] forces acting jointly describes the diagonal of a parallelogram in the same time in which it would describe the sides if the forces were acting separately.* (C&W, 417)

In the present application we have the line *Bc* representing the motion in the second unit of time that the body would have undergone had the impulse at *B* not occurred. This would be the motion resulting from what Newton defines (Def. 3; C&W, 404) as the body's inherent force of matter, its inertia, alone. The line *Bv* represents the motion in the second unit of time that would have resulted from the impressed force (Def. 4; C&W, 405) of the impulse on the body had it been at rest at location *B* with respect to the center *S* at the end of the first unit of time.

that base and line *c*C which is parallel to it. Since triangle SB*c* equals triangle SAB, as was shown in the previous step, we now have triangle SBC equal to triangle SAB by transitivity of equality.

This very simple argument shows that the area swept out in the second unit of time equals the area swept out in the first unit of time before the centripetal force acted at B. Similar arguments apply if the centripetal forces act successively by a series of such impulses.

> By a similar argument, if the centripetal force acts successively at C, D, E, . . . , making the body in each of the individual particles of time describe the individual straight lines CD, DE, EF, . . . , all these lines will lie in the same plane; and triangle SCD will be equal to triangle SBC, SDE to SCD, and SEF to SDE. Therefore, in equal times equal areas are described in an unmoving plane; and by composition [or componendo], any sums SADS and SAFS of the areas are to each other as the times of description. (C&W, 445)

What matters is the centripetal directions not the relative lengths of the lines *c*C, *d*D, *e*E, *f*F representing the strengths of the impulses. The resulting triangles will be equal even if the differences among the strengths of these impulses vary by arbitrary amounts.[96]

Newton appeals to corollary 4 of lemma 3 on first and last ratios to extend these results for discrete triangles traced out by motion under a succession of centripetal impulses to corresponding results for areas traced out by motion under continuously acting centripetal forces.[97]

> Now let the number of triangles be increased and their width decreased indefinitely, and their ultimate perimeter ADF will (by lem. 3, corol. 4) be a curved line; and thus the centripetal force by which the body is continually drawn back from the tangent of this curve will act uninterruptedly, while any areas described, SADS and SAFS, which are always proportional to the times of description, will be proportional to those times in this case. Q.E.D. (C&W, 445)

There are now available several ways of expanding this brief sketch into proofs that do recover the areal rule for curves corresponding to continuously acting centripetal forces as limit results for discrete impulses.[98] There are, also, elementary proofs using calculus without Newton's appeal to impulses.

The conclusion is very general. No assumptions are made about any particular power law, or any other specific relation, of the force to distance from the center. What matters about the force is that its direction be toward the given center. In the proposition, this center is assumed to be at rest. In corollary 6, Newton appeals to corollary 5 of the Laws of Motion to extend the result to any center that counts as inertial. We shall see that, in proposition 3, he will also appeal to corollary 6 of the Laws of Motion to extend such results to centers being accelerated by

[96] Indeed, such an argument can be extended to cases where some of the centripetal impulses are of zero magnitude or, as Leibniz pointed out, are directed away from the center and so of perhaps greatly differing negative magnitudes.

[97] Whiteside (Math Papers, vol. 6, 37, in note 19 which begins on p. 35) suggests that Newton's proof only establishes the proposition for motion corresponding to infinitesimal arcs. Erlichson (1992) claims that this objection and a similar objection by Eric Aiton (1989a) are mistaken.

[98] Wayne Myrvold (1999) and Bruce Pourciau (2003) have independently provided convergence proofs showing that limit arguments of the sort appealed to by Newton can be constructed with the mathematical resources available to him. These results show that the objections by Whiteside and Aiton are, indeed, mistaken.

additional forces, so long as the accelerations produced on the center and the orbiting body by the additional forces can be counted as sufficiently equal and parallel.

2. Proposition 2

Book 1 Proposition 2 Theorem 2

Every body that moves in some curved line described in a plane and, by a radius drawn to a point, either unmoving or moving uniformly forward with a rectilinear motion, describes areas around that point proportional to the times, is urged by a centripetal force tending toward that same point.

Case 1. For every body that moves in a curved line is deflected from a rectilinear course by some force acting on it (by law 1). And that force by which the body is deflected from a rectilinear course and in equal times is made to describe, about an immobile point S, the equal minimally small triangles SAB, SBC, SCD, . . . , acts in place B along a line parallel to *c*C (by book 1, prop. 40, of the *Elements*, and law 2), that is, along the line BS; and in place C, the force acts along a line parallel to *d*D, that is, along the line SC, . . . Therefore, it always acts along lines tending toward that unmoving point S. Q.E.D.

Case 2. And, by corol. 5 of the laws, it makes no difference whether the surface on which the body describes a curvilinear figure is at rest or whether it moves uniformly straight forward, together with the body, the figure described, and the point S. (C&W, 446)

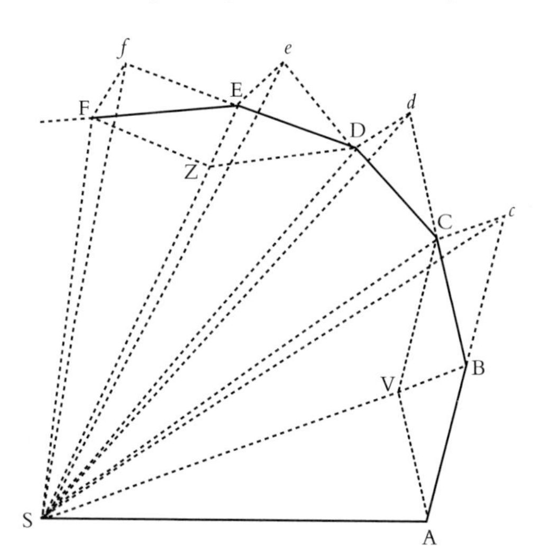

Consider triangles SAB and SBC, as given by points A, B, C on the curve, where point A is separated from point B by the same minimally small time interval by which point B is separated from point C. Assume triangle SAB is equal to triangle SBC. Construct triangle SB*c* so that line B*c* is a rectilinear parallel extension of line AB equal in length to rectilinear line AB. Triangles SAB and SB*c* are equal, since they have equal bases and the same height. By transitivity of equality, therefore, triangle SB*c* is also equal to triangle SBC.

Here is proposition 40 of Euclid's *Elements*:

> *Equal triangles which are on equal bases and on the same side are also in the same parallels.* (Heath 1956, vol. 1, 337)

The lines C*c* and SB are parallel, since both of the equal triangles SB*c* and SBC are between them. The change in motion between the motion of the body traversing AB in the first minimal time interval and the motion by which it traverses BC in the second equal minimal time interval is proportional to and in the direction represented by the line from *c* to C. Therefore, by Law 2, this line is proportional to and in the same direction as a motive force impressed at B to deflect the body into its motion along BC. But, the direction of the line from *c* to C is the same as the parallel direction at B, that is the direction from B toward the center S.

If the orbit, a time-parameterized curve, does stay in a plane and sweep out equal areas in equal times by radii to an inertial center S, then, in the limit, as intervals of equal time are made smaller and smaller, the triangles corresponding to SAB, SBC, SCD, . . . will approach equality. So, for any specified approximation to directions toward the center S, time intervals can be made small enough that, at each point B, the corresponding direction *c*C in the above construction will deviate less from the direction from B toward the center S than is required to satisfy that specified approximation.

The first corollary of proposition 2 is especially interesting to us.

> Corollary 1. In nonresisting spaces or mediums, if the areas are not proportional to the times, the forces do not tend toward the point where the radii meet but deviate forward [or in consequentia] from it, that is, in the direction toward which the motion takes place, provided that the description of the areas is accelerated; but if it is retarded, they deviate backward [or in antecedentia, i.e., in a direction contrary to that in which the motion takes place]. (C&W, 447)

It makes an increasing areal rate correspond to having the force off-center in a forward direction and a decreasing areal rate correspond to having the force off-center in a backward direction with respect to the tangential motion of the body being deflected. This makes a constant areal rate measure the centripetal direction of the force deflecting a body into an orbital motion about a center.

3. *Proposition 3*

Proposition 3 extends proposition 2 to centers moving in any way whatever.

> **Book 1 Proposition 3 Theorem 3**
>
> *Every body that, by a radius drawn to the center of a second body moving in any way whatever, describes about that center areas that are proportional to the times is urged by a force compounded of the centripetal force tending toward that second body and of the whole accelerative force by which that second body is urged.*
>
> Let the first body be L, and the second body T; and (by corol. 6 of the laws) if each of the two bodies is urged along parallel lines by a new force that is equal and opposite to the force by which body T is urged, body L will continue to describe about body T the same areas as before; but the force by which body T was urged will now be annulled by an equal and opposite force, and therefore (by law 1) body T, now left to itself, either will be at rest or will move uniformly straight forward; and body L, since the difference of the forces [i.e., the remaining force] is urging it, will continue to describe areas proportional to the times about body T. Therefore, the difference of the forces tends (by theor. 2) toward the second body as center. Q.E.D. (C&W, 448)

Chandrasekhar offers the following useful diagram.[99]

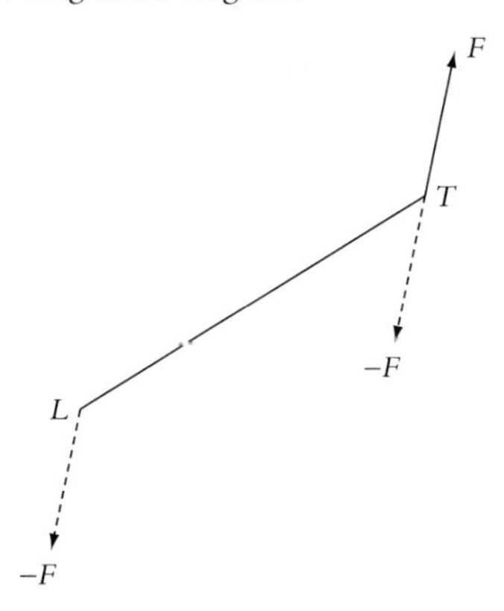

The forces added to L and to T are equal and parallel. Therefore, by corollary 6 of the Laws, the motion of L with respect to T is unchanged by the actions of these equal and parallel forces on both. But, these forces are each equal and opposite to the total force on body T. The force applied to T annuls the total force acting on T. By Law 1, this would leave T either at rest or moving uniformly straight forward. This, in turn, would make proposition 2 apply to infer the centripetal direction of the resulting force on L (that is the force resulting from subtracting from the total force on L a force equal to the total force on T) from the constant rate at which L describes areas about the center of T.

Corollary 1 follows immediately by the very same construction that proves the proposition.

> Corollary 1. Hence, if a body L, by a radius drawn to another body T, describes areas proportional to the times, and from the total force (whether simple or compounded of several forces according to corol. 2 of the laws) by which body L is urged there is subtracted (according to the same corol. 2 of the laws) the total accelerative force by which body T is urged, the whole remaining force by which body L is urged will tend toward body T as center. (C&W, 448)

So do the extensions to approximations in the next two corollaries.

> Corollary 2. And if the areas are very nearly proportional to the times, the remaining force will tend toward body T very nearly. (C&W, 448)

> Corollary 3. And conversely, if the remaining force tends very nearly toward body T, the areas will be very nearly proportional to the times. (C&W, 449)

These important corollaries make inferences to the centripetal direction of forces maintaining bodies in orbits which sweep out areas proportional to the times of description robust with respect to approximations.

A last corollary makes clear the implication that if a body describes extremely unequal areas in equal times by radii to a center, then the total force acting on it is not even approximately directed toward that center.

[99] Chandrasekhar 1995, 71.

> Corollary 4. If body L, by a radius drawn to another body T, describes areas which, compared with the times, are extremely unequal, and body T either is at rest or moves uniformly straight forward, either there is no centripetal force tending toward body T or the action of the centripetal force is mixed and compounded with very powerful actions of other forces; and the total force compounded of all the forces, if there are several, is directed toward another center (whether fixed or moving). The same thing holds when the second body moves with any motion whatever, if the centripetal force is what remains after subtracting the total force acting upon body T. (C&W, 449)

These corollaries show that the extension to moving centers afforded by proposition 3 makes the uniform description of areas into a very general criterion for finding centripetal forces.

Newton sums up this role for areal law motion in a scholium after proposition 3.

> Scholium. Since the uniform description of areas indicates the center toward which that force is directed by which a body is most affected and by which it is drawn away from rectilinear motion and kept in orbit, why should we not in what follows use uniform description of areas as a criterion for a center about which all orbital motion takes place in free spaces? (C&W, 449)

As this question suggests, the dynamical significance of the uniform description of areas makes it worth looking for such phenomena to find centripetal forces maintaining bodies in orbits.

4. *Proposition 4*

Once one has established centripetal forces from uniform description of areas in orbital motion, one can appeal to theorems about centripetal forces to compare and establish laws of centripetal forces toward the sun and planets from orbits.

> **Book 1 Proposition 4 Theorem 4**
>
> *The centripetal forces of bodies that describe different circles with uniform motion tend toward the centers of those circles and are to one another as the squares of the arcs described in the same time divided by the radii of the circles.*
>
> These forces tend toward the centers of the circles by prop. 2 and prop. 1, corol. 2, and are to one another as the versed sines of the arcs described in minimally small equal times, by prop. 1, corol. 4, that is, as the squares of those arcs divided by the diameters of the circles, by lem. 7; and therefore, since these arcs are as the arcs described in any equal times and the diameters are as their radii, the forces will be as the squares of any arcs described in the same time divided by the radii of those circles. Q.E.D. (C&W, 449–50)

The centripetal direction of the forces follows from proposition 2 and from corollary 1 proposition 1.

According to proposition 1 corollary 4, the forces are to one another as the sagittas of the arcs traced out in equal times.

Consider arc CD in the first diagram. The sagitta of this arc is equal to BD.

These forces are to one another as the versed sines of the arcs described in minimally small equal times.

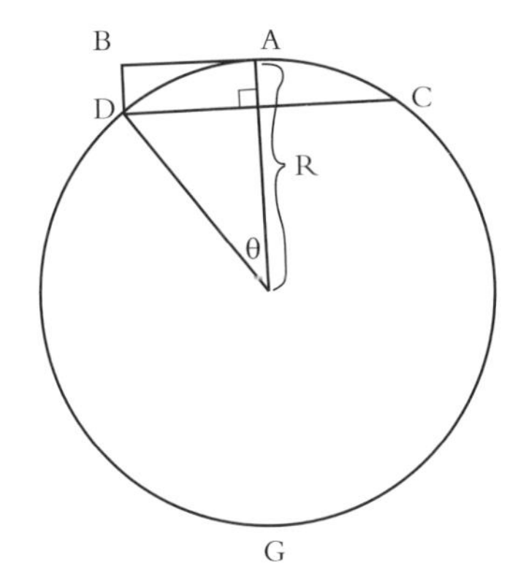

The versed sine of arc CD equals 1-cosθ

BD = R-Rcosθ

BD = R(1-cosθ)

that is, as the squares of those arcs divided by the diameters of the circles.

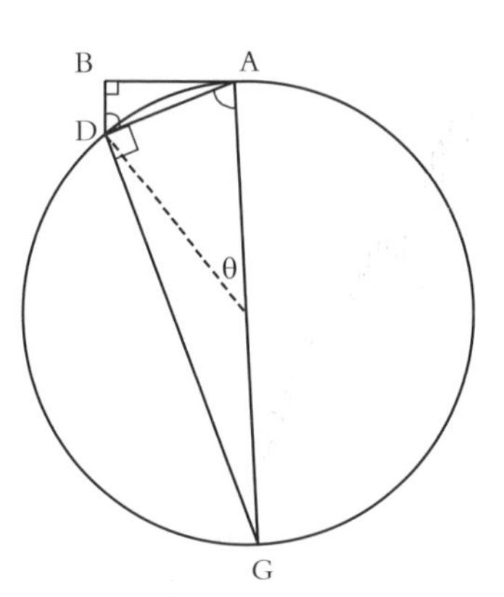

∇BDA $\approx$ ∇DAG (Euclid VI, 8)

BD/DA = DA/AG

BD = DA^2/AG = DA^2/diameter

As D $\rightarrow$ A we have arc DA $\rightarrow$ DA (By Lemma 7)

So, as D $\rightarrow$ A we have BD $\rightarrow$ (arcDA)^2/diameter

$$f_1/f_2 = \left(\text{Arc}_1{}^2/\text{diameter}_1\right)/\left(\text{Arc}_2{}^2/\text{diameter}_2\right)$$
$$= \left(\text{Arc}_1{}^2/\text{radius}_1\right)/\left(\text{Arc}_2{}^2/\text{radius}_2\right)$$

Corollary 1. Since those arcs are as the velocities of the bodies, the centripetal forces will be in a ratio compounded of the squared ratio of the velocities directly and the simple ratio of the radii inversely. (C&W, 450)

$$f_1/f_2 = \left(\text{Arc}_1{}^2/\text{radius}_1\right)/(\text{Arc}_2{}^2/\text{radius}_2)$$
$$= \left(\text{V}_1{}^2/\text{radius}_1\right)/\left(\text{V}_2{}^2/\text{radius}_2\right)$$

Corollary 2. And since the periodic times are in a ratio compounded of the ratio of the radii directly and the ratio of the velocities inversely, the centripetal forces are in a ratio compounded of the ratio of the radii directly and the squared ratio of the periodic times inversely. (C&W, 450)

$$t_1/t_2 = (2\pi R_1/V_1)/(2\pi R_2/V_2) = (R_1/V_1)/(R_2/V_2)$$
$$t_1^2/t_2^2 = (R_1^2/V_1^2)/(R_2^2/V_2^2)$$
$$(R_1/t_1^2)/(R_2/t_2^2) = (R_1(V_1^2/R_1^2))/(R_2(V_2^2/R_2^2)) = (V_1^2/R_1)/(V_2^2/R_2)$$
$$f_1/f_2 = (R_1/t_1^2)/(R_2/t_2^2)$$

The important corollary 7, which gives the systematic dependencies backing up the inference from the Harmonic Rule to the inverse-square variation of the accelerations, follows from corollary 2.

> Corollary 7. And universally, if the periodic time is as any power R^n of the radius R, and therefore the velocity is inversely as the power R^{n-1} of the radius, the centripetal force will be inversely as the power R^{2n-1} of the radius; and conversely. (C&W, 451)

$$t_1/t_2 = R_1^n/R_2^n \Leftrightarrow t_1^2/t_2^2 = R_1^{2n}/R_2^{2n} \quad (1)$$
$$\Leftrightarrow (R_1/t_1^2)/(R_2/t_2^2) = (R_1/R_1^{2n})/(R_2/R_2^{2n}) \quad (2)$$

By corollary 2 we have $f_1/f_2 = (R_1/t_1^2)/(R_2/t_2^2)$, so substituting in (2), yields

$$\Leftrightarrow f_1/f_2 = (R_1/R_1^{2n})/(R_2/R_2^{2n}) = (1/R_1^{2n-1})/(1/R_2^{2n-1}) \quad (3)$$

which gives

$$t_1/t_2 = R_1^n/R_2^n \Leftrightarrow f_1/f_2 = (1/R_1^{2n-1})/(1/R_2^{2n-1}) \quad (4)$$

As we have seen, corollary 6 is a special case of corollary 7. It follows when n is set at the value 3/2, which makes the left-hand side Kepler's Harmonic Rule.

> Corollary 6. If the periodic times are as the 3/2 powers of the radii, and therefore the velocities are inversely as the square roots of the radii, the centripetal forces will be inversely as the squares of the radii; and conversely. (C&W, 451)

$$t_1/t_2 = R_1^{3/2}/R_2^{3/2} \Leftrightarrow f_1/f_2 = (1/R_1^2)/(1/R_2^2)$$

Note: 1–2(3/2) = 1–3 = −2

> Corollary 9. From the same demonstration it follows also that the arc which a body, in revolving uniformly in a circle with a given centripetal force, describes in any time is a mean proportional between the diameter of the circle and the distance through which the body would fall under the action of the same given force and in the same time. (C&W, 451)

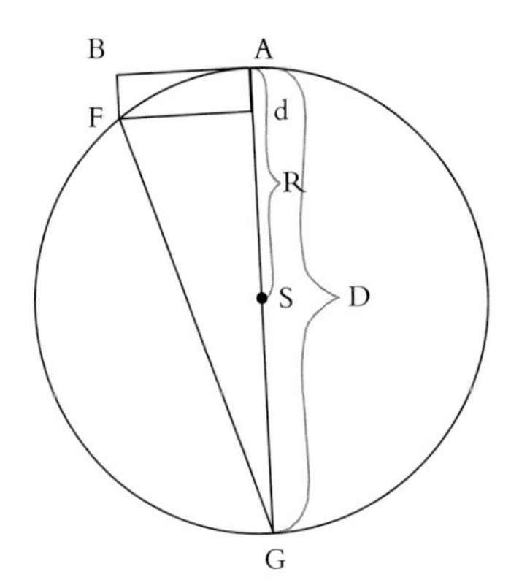

Let arc AF be traced out in time t and $a_s = V^2/\mathrm{R}$ be the acceleration toward the center s. We have

$$V = \mathrm{AF}/t$$

So,

$$a_s = (\mathrm{AF}^2/t^2)(1/\mathrm{R})$$

$$\begin{aligned} \mathrm{Ad} &= (1/2)a_0\, t^2 \\ &= (1/2)(\mathrm{AF}^2/t^2)(1/\mathrm{R})t^2 \\ &= \mathrm{AF}^2/2\mathrm{R} = \mathrm{AF}^2/\mathrm{D} \end{aligned}$$

To have the arc AF be a mean proportional between the diameter D and the distance Ad is to have

$$\mathrm{Ad}/\mathrm{AF} = \mathrm{AF}/\mathrm{D}$$

which is to have

$$\mathrm{Ad} = \mathrm{AF}^2/\mathrm{D}.$$

Chapter 3 Appendix 3: Newton's precession theorem and eccentric orbits[100]

1. *Newton's basic precession theorem*

In proposition 45 book 1, Newton derives a formula relating the power law characterizing the centripetal motive force maintaining a body in an orbit and the apsidal angle of that orbit. The apsidal angle, angle at the force center between the high and is the low apse. In a solar orbit the high apse is the aphelion and the low apse is the perihelion. In the moon's orbit of the earth these are the apogee and perigee. For a general elliptical orbit with force to a focus let us call these the "apocenter" and "pericenter." Consider a centripetal force f proportional to r^{n-3}, where r is the distance from the center to the orbiting body. Here are two examples:

$$n = 1 : f \propto 1/r^2 \quad \text{Inverse-square}$$

$$n = 4 : f \propto r \quad \text{Spring force}$$

Newton's formula relates the apsidal angle θ to n:

$$\theta = 180^\circ / n^{1/2}$$

The apse is "quiescent" if $n = 1$, so that the angle from apocenter to the next pericenter or from pericenter to the next apocenter is 180°. This gives

$$\begin{array}{lll} n = 1 \text{ iff } & \theta = 180^\circ & \text{Zero precession} \\ n < 1 \text{ iff } & \theta > 180^\circ & \text{Orbit precesses forward} \\ n > 1 \text{ iff } & \theta < 180^\circ & \text{Orbit precesses backward} \end{array}$$

Here are an inverse-square orbit and a spring force orbit.

Figure 3.5

$n=1$ **Inverse-square force**
$f \propto r^{-2}$

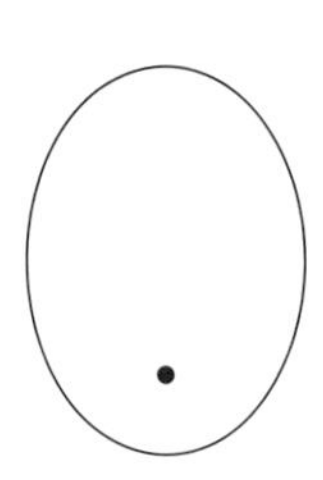

$\theta = 180^\circ$
Body traces a stable ellipse with force toward a focus

$n=4$ **Spring force**
$f \propto r$

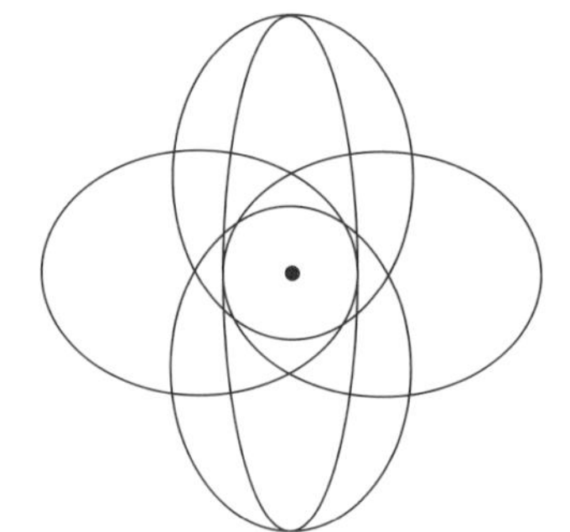

$\theta = 90^\circ$
A body moving on a backwards precessing ellipse so as to trace out another stable ellipse with the force toward its center

[100] This appendix is an exposition of a result in Valluri et al. 1997.

On Newton's formula, the apsidal angle θ measures the exponent in the force law: we have

$$n = (180/\theta)^2$$

so that

$$(180/\theta)^2 - 3$$

is the exponent of a centripetal force that maintains a body in an orbit with apsidal angle θ.[101]

Newton obtained his formula for orbits approaching very near to concentric circles, that is orbits of very small eccentricity

$$e = (r_M - r_m)/(r_M + r_m),$$

where r_M is the greatest distance from the center and r_m is the least distance from the center. He applied his formula, however, to orbits with significantly large eccentricities, such as Mercury's orbit with $e = 0.2056$ and even to Halley's comet with $e = 0.97$. This raises the question:

Does precession measure deviation from the inverse-square law even if *e* is large?

2. *A derivation extending it to include orbits of large eccentricity*

Consider the differential equation below of an orbit in polar coordinates (radius *r*, angle θ in radians), for a centripetal attractive force $f(r)$ proportional to $-r^{-N}$:

$$d^2u/d\theta^2 + u = u^\delta/h^2, \tag{1}$$

where $u = 1/r$, $\delta = N - 2$ is the deviation from $N = 2$, and $h = r^2 d\theta/dt$ is the angular momentum.

The potential energy

$$V(r) = r^{-N+1}/(-N+1) = u^{N-1}/(-N+1), \tag{2}$$

the case of $N = 1$ being excluded. The total energy is

$$E = (1/2)[(du/d\theta)^2\ h^2 + h^2u^2] + V(r), \tag{3}$$

where $h = r^2\ d\theta/dt$ is the angular momentum. We have

$$du/d\theta = (2^{1/2}/h)\ [E + u^{N-1}/(N-1) - h^2u^2/2]^{1/2} \tag{4}$$

At each apse we must have $du/d\theta = 0$, suppose u_1 is the apocenter or highest apse and u_2 is the next pericenter or lowest apse.

We can obtain the apsidal angle by integrating $d\theta/du$, the inverse of equation (4), from its value at the apocenter u_1 to its value at the pericenter u_2[102]

$$\theta = h/2^{1/2} \int_{u_1}^{u_2} [E + u^{N-1}/(N-1) - h^2u^2/2]^{-1/2} du \tag{5}$$

[101] Newton's corollary 1 of proposition 45 book 1 expresses the same result when precession is expressed in terms of the angle to return to the apocenter again, rather than as the angle to reach the pericenter.

[102] This result was developed by Valluri. It is from section 2 of Valluri et al. 1997. See section 1 of that paper for a specific treatment of the extension of Newton's precession theorem to eccentric orbits for force laws differing but slightly from the inverse-square.

This formula recovers Newton's results for the inverse-square case $N = 2$ where it gives $\theta = \pi$. It yields $\theta = \pi/2$ for the spring force case $N = -1$ also in exact agreement with Newton. It turns out that these two cases are special in that the results are independent of eccentricity. They are also special in being the two cases yielding stable elliptical orbits. The spring force ellipse with force to the center, traced out by the $\theta = \pi/2$ precessing orbit with force to its focus, is a non-precessing stable orbit.

Numerical integrations by computer of this formula support our result. Increasing the eccentricity *e* increases the sensitivity of absence of unaccounted-for precession as a measurement of the inverse-square.

The following figure illustrates the apsidal angles corresponding to $N = 2.25 = 9/4$ for a deviation $\delta = N-2$ of $+\ 0.25$ for a force of the -2.25 power of distance computed for eccentricities *e* of 0.2, 0.4, 0.6 and 0.8 respectively.[103]

Eccentricity	θ in degrees	θ–θ_0 (% deviation)
$e \ll 1$	207.846	0
0.2	208.132	+0.286 (0.138%)
0.4	209.073	+1.227 (0.590%)
0.6	210.981	+3.135 (1.508%)
0.8	214.929	+7.083 (3.408%)
1.0	239.5	+31.65 (15.23%)

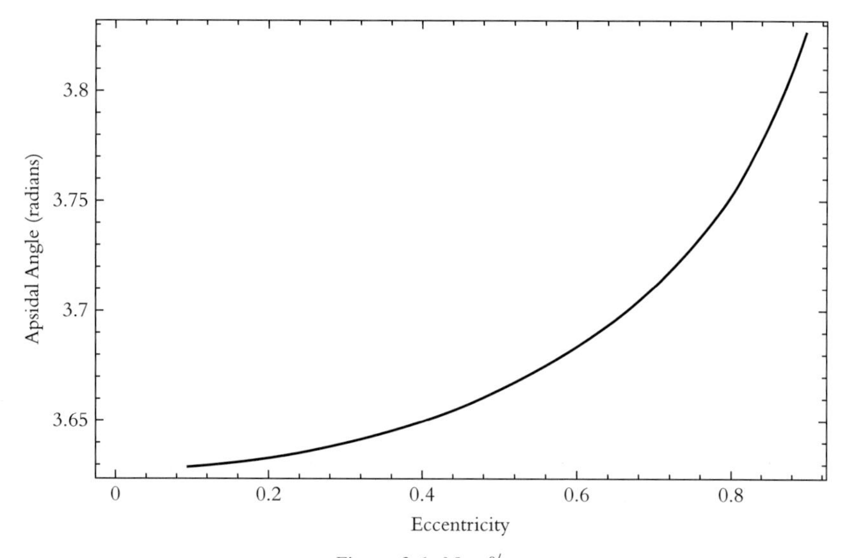

Figure 3.6 $N = 9/4$.

[103] The number π is given by a decimal expansion 3.14159.... We have π radians $\equiv$ 180°. This angular motion from one apse to the next counts as zero precession. Newton's formula gives a difference of 207. 846$-$180 $-$ +27.846 degrees ($=27.846\pi/180 = 0.486$ rads) forward precession per revolution for the above example -2.25 power of distance force law. It gives a difference of $160.997-180 = -19.003$ degrees (minus about 0.332 rads) for a backward precession for this alternative example of a -1.75 power of distance. These give what we are calling the zero eccentricity cases for nearly circular orbits.

Eccentricity	θ in degrees	$\theta-\theta_0$ (% deviation)
$e \ll 1$	160.997	0
0.2	160.808	–0.189 (0.117%)
0.4	160.194	–0.803 (0.499%)
0.6	158.957	–2.040 (1.267%)
0.8	156.438	–4.559 (2.832%)
1.0	144	–17.0 (10.6%)

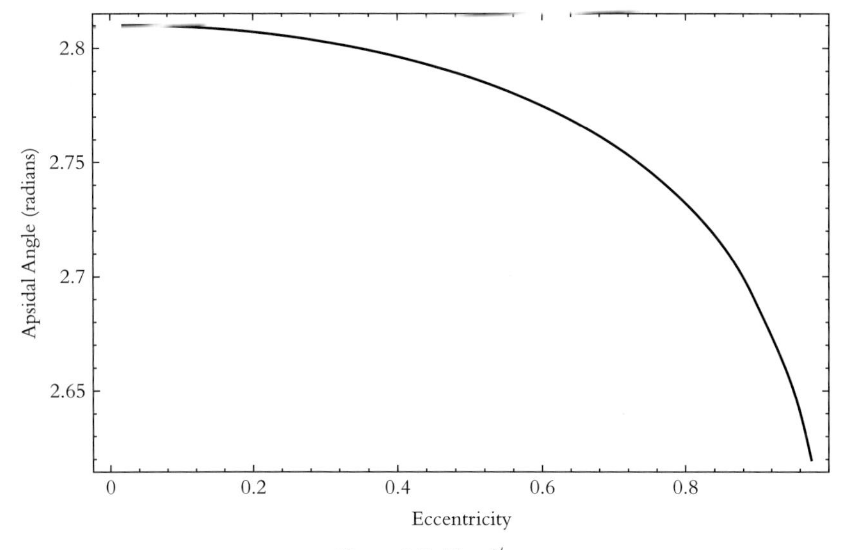

Figure 3.7 $N = 7/4$.

The next figure illustrates the apsidal angles corresponding to $N = 1.75 = 7/4$ for a deviation $\delta = -0.25$ for a force of the -1.75 power of distance computed for those same respective eccentricities.

These increases in eccentricity exhibit more than linear increases to the amount of precession that would be generated by the given departures δ from the inverse-square. Such patterns are more like a geometrical contribution of eccentricity to the precession resulting from these given δ's. This suggests a significant strengthening of the basic result that increased eccentricity increases the sensitivity of absence of unaccounted-for orbital precession as a null experiment measuring the inverse-square power law for centripetal forces maintaining bodies in orbits.

4

Unification and the Moon-Test (Propositions 3 and 4 Book 3)

We have just seen that Newton's classic deductions of inverse-square centripetal forces from Keplerian phenomena are measurements. The centripetal direction of the force maintaining a moon or planet in its orbit is measured by Kepler's Area Rule for that orbit. The inverse-square power law for such centripetal forces is measured by Kepler's Harmonic Rule for a system of orbits and from the absence of precession of any single orbit. This suggests an ideal of empirical success according to which a theory succeeds by having its theoretical parameters receive convergent accurate measurements from the phenomena it purports to explain. I shall be arguing that the application of Newton's first two rules of reasoning to argue from the moon-test to the identification of the force holding the moon in its orbit with terrestrial gravity is backed up by this ideal of empirical success. On this identification two phenomena, the length of a seconds pendulum at the surface of the earth and the centripetal acceleration exhibited by the lunar orbit, are found to give agreeing measurements of the strength of the same inverse-square centripetal acceleration field, *the earth's gravity*. This realizes the ideal of empirical success by making available the data from both phenomena to back up the measured value of this common parameter.

In section I we review Newton's argument, in proposition 3, for the claim that the moon is maintained in its orbit by an inverse-square force directed toward the earth. Section II introduces Newton's moon-test argument, in proposition 4, for identifying the force that maintains the moon in its orbit with terrestrial gravity. Newton shows that inverse-square adjusting the centripetal acceleration exhibited by the lunar orbit agrees with Huygens's measurement of the strength of terrestrial gravity at the surface of the earth.

Section II.1 gives a somewhat detailed account of Newton's moon-test calculation. It points out that Newton applies a correction factor to offset the action of the sun on the earth-moon system, which we shall see in section IV.1 is dubious. This correction factor, together with Newton's selection of 60 earth radii as the lunar distance, leads to a very precise agreement with Huygens's measurement. In chapter 6 we shall see that the precision of this agreement with Huygens's measurement is exaggerated well beyond what the data cited by Newton can reasonably support. Our modern least squares assessment will make clear that the cited data cannot support any reasonable expectation that such a precision of agreement that would be maintained by future moon-test estimates of comparable accuracy.

Section II.2 discusses Newton's appeal to his first two *Regulae Philosophandi*, or rules for doing natural philosophy. Section II.3 examines Newton's discussion of the two-body correction. In section III we examine the very interesting version of the moon-test argument presented in Newton's scholium to proposition 4.

Section IV argues that the agreement between the moon-test measurements and the pendulum measurements of the strength of terrestrial gravity is an example of empirical success. It shows that Newton's inference does not depend on the dubious correction factor he introduced in proposition 4. It also shows that this inference does not depend on Newton's selection of the single value of 60 earth radii as the lunar distance on which to base his moon-test. This section will include an exhibition of the empirical support afforded to Huygens's measurement by the rougher agreeing moon-test estimates. They add stronger evidence against very large deviations from the value inferred by Huygens than would be afforded by the pendulum estimates on their own.

Section V reviews the lunar precession problem. It discusses Newton's successful treatment of the variational inequality of the lunar orbit, and Newton's less successful treatment of the lunar precession. This is followed by brief discussions of initial unsuccessful efforts by Clairaut, d'Alembert, and Euler to account for the lunar precession, together with a brief discussion of Clairaut's successful resolution of this lunar precession problem.

I The argument for proposition 3

In proposition 3, Newton argued that the Moon is held in its orbit of the earth by an inverse-square force directed toward the earth.

Proposition 3 Theorem 3

The force by which the moon is maintained in its orbit is directed toward the earth and is inversely as the square of the distance of its places from the center of the earth.

1. Directed toward the earth

As in the corresponding argument for Jupiter's moons, the first part, the centripetal direction of the force, is inferred from the uniform description of areas and propositions 2 or 3 of book 1.

> The first part of this statement[1] is evident from phen. 6 and from prop. 2 or prop. 3 of book 1, . . . (C&W, 802)

[1] The Latin word that Cohen and Whitman have translated as "statement" is *assertionis* (Koyré and Cohen 1972, 563). According to Craig Waff, Newton's choice of *assertionis*, rather than *propositionis* as in his arguments for propositions 1 and 2, reinforces the suggestion that he has less conviction about the force of his argument for proposition 3.

> It would appear however that Newton himself had not been entirely convinced of the exactness of his proof. In the first line of the proof he referred to Proposition III as an "assertion" (assertionis, not propositionis as he had written in the first lines of the proofs of the two preceding propositions), and

Recall phenomenon 6 together with Newton's comment on it.

> *The moon, by a radius drawn to the center of the earth, describes areas proportional to the times.*
>
> This is evident from a comparison of the apparent motion of the moon with its apparent diameter. Actually, the motion of the moon is somewhat perturbed by the force of the sun, but in these phenomena I pay no attention to minute errors that are negligible. (C&W, 801)

The comparisons of apparent diameter and apparent motion mentioned by Newton are in good rough agreement with the Area Rule.[2] As we have seen in our discussion of the argument for Jupiter's moons, proposition 3 of book 1 and its corollaries and scholium extend the resources for inferring a centripetally directed force from the Area Rule to approximations appropriate to perturbations by additional forces.

Corollary 6 of the Laws of Motion limits perturbations of orbits by attractions toward third bodies to the tidal forces generated by those attractions. Tidal forces are those which produce different accelerations on the central and the orbiting body. Even though the sun's action to disturb the orbital motion of the moon about the earth is much greater than its action to disturb the orbital motions of Jupiter's satellites,[3] this did not lead to deviations from the Area Rule that raised difficulties for Newton's inference to claim that the force maintaining the moon in its orbit is directed toward the earth. As we saw in our previous discussion of phenomenon 6, the best descriptive account of lunar motion before Newton, that of Horrocks, did satisfy the Area Rule.[4]

Before Newton there was no, generally accepted, empirical evidence for violations of the Area Rule for the moon's motion. Newton's own accounts of the lunar motion, which improved somewhat on the fit of Horrocks's theory to the data,[5] did include small violations of the Area Rule.[6] His successful treatment of the variation, an inequality of the moon's mean motion discovered by Tycho, includes a quantitative account of these small violations of the Area Rule as perturbations due to the action of the sun.[7] This allows these violations to be dismissed as not undercutting Newton's inference from the Area Rule to the earth-centered direction of the force maintaining the moon in its orbit about the earth.

2. Inverse-square

The more serious problem facing Newton's attempt to establish proposition 3 is arguing for the second part – the inverse-square variation of the force maintaining the moon in its orbit about the earth. Since the earth has only one moon, the Harmonic Rule

> he terminated his proof by claiming that the inverse-square ratio "will yet more fully appear" from proof of Proposition IV . . . (Waff 1976, 19–20)

[2] See chapter 2 section VI above.

[3] See chapter 3 note 41 above.

[4] See chapter 2 section VI. See Wilson 1989a, VII, 91.

[5] See Kollerstrom 2000, 150–1. Kollerstrom has shown that Newton's corrected version of Horrocks's theory was almost twice as accurate as Flamsteed's version.

[6] See Newton's proposition 29 book 3 (C&W, 846–8). Wilson 2001, 152. See below section V.

[7] See chapter 2 note 42 above. See also sec V.1 below.

argument does not apply. The fact that the lunar orbit precesses, known from observations since ancient times, threatens to undercut his only other dynamical argument – the application of proposition 45 book 1 to infer the inverse-square variation from the absence of such orbital precession.

Newton's argument begins as follows:

> . . . , and the second part *[is evident]* from the very slow motion of the moon's apogee. For that motion, which in each revolution is only three degrees and three minutes forward [or in consequentia, i.e., in an easterly direction] can be ignored. For it is evident by (book 1, prop. 45, corol. 1) that, if the distance of the moon from the center of the earth is to the semidiameter of the earth as D to 1, then the force from which such a motion may arise is inversely as $D^{2\frac{4}{243}}$, that is, inversely as that power of D of which the index is 2⁴⁄243; that is, the proportion of the force to the distance is inversely as a little greater than the second power of the distance, but is 59¾ times closer to the square than to the cube. (C&W, 802–3)

In the first edition, Newton cited 3 degrees forward precession. Here, he cites 3 degrees and 3 minutes.[8] When we apply corollary 1 proposition 45 to this somewhat larger precession, we see that it yields a power that agrees, to within the nearest 1⁄243, with the value inversely as the 2 and 4⁄243 power cited by Newton.[9] Here he cites this as 59¾ times closer to the square than the cube.[10]

Here is the next part of the argument for the inverse-square.

> Now this motion of the apogee arises from the action of the sun (as will be pointed out below) and accordingly is to be ignored here. The action of the sun, insofar as it draws the moon away from the earth, is very nearly as the distance of the moon from the earth, and so (from what is said in book 1, prop. 45, corol. 2) is to the centripetal force of the moon as roughly 2 to 357.45, or 1 to 178²⁹⁄40. And if so small a force of the sun is ignored, the remaining force by which the moon is maintained in its orbit will be inversely as D^2. And this will be even more fully established by comparing this force with the force of gravity as is done in prop. 4 below. (C&W, 803)

We are given three reasons for neglecting the observed precession to accept the inverse-square. First, we have a promise that it will be pointed out below that motion of the moon's apogee is due to the action of the sun. Second, is a characterization of the action of the sun – insofar as it draws the moon away from the earth – as both very nearly as the distance of the moon from the earth and as roughly in the ratio of 1 over

[8] In a first edition scholium to proposition 35 book 3, Newton gives $40°41'$ as the mean annual motion of the lunar apogee cited in Flamsteed's tables (C&W, 869). Given Newton's cited period, 365.2565 decimal days, for the earth and his cited period, 27 days 7 hours 43 minutes = 27.3215 decimal days, for the moon, this yields $3°\ 2.59'$ as the lunar precession per period of the moon. According to Waff (1976, 41), Flamsteed, in his *Doctrine of the Sphere* of 1680, gave $40°39'50''$ as the mean annual motion of the lunar apogee. This yields $3°2.5'$. Both of these, to the nearest degree, round to 3° and, to the nearest minute, round to $3°3'$.

[9] The result is −2.0167 power or inversely as the 2 and 4.066/243 power.

[10] In the first edition, he tells us this is 60¾ times closer to the inverse-square than to the inverse-cube. 60¾ equals ²⁴³⁄4. In later editions Newton corrects this ratio to 59¾ closer to the square than to the cube. The difference from the inverse-square is 4 of 243 equal parts of a degree, while the difference from the inverse-cube is 243 − 4 or 239 of those parts. ²³⁹⁄4 = 59¾ is, therefore, the correct ratio.

178 and 29/40 to the centripetal force maintaining the moon in its orbit. That the ratio 1/178.725 holds, insofar as the sun draws the moon away from the earth, is said to be supported by what is said in corollary 2 of proposition 45 book 1. Moreover, this is said to support the claim that if this action is ignored, the remaining force by which the moon is maintained in its orbit will be inverse-square. Finally, we have an appeal to the moon-test, which is said to *even more fully* establish the inverse-square variation of the centripetal force maintaining the moon in its orbit.

If we look at corollary 2 proposition 45 book 1 (in the third edition) we find:

For example, if the force under the action of which the body revolves in the ellipse is as $1/A^2$ and the extraneous force which has been taken away is as cA . . . , then, . . . the angle of the revolution between apsides will be equal to an angle of $180((1-c)/(1-4c))^{1/2}$ degrees. Let us suppose the extraneous force to be 357.45 times less than the other force under the action of which the body revolves in the ellipse, that is, let us suppose c to be 100/35745, A or T being equal to 1, and then $180((1-c)/(1-4c))^{1/2}$ will come to be $180(35645/35345)^{1/2}$, or 180.7623, that is, 180° 45′ 44″. Therefore a body, setting out from the upper apsis, will reach the lower apsis by an angular motion of 180° 45′ 44″ and will return to the upper apsis if this angular motion is doubled; and thus in each revolution the upper apsis will move forward through 1° 31′ 28″. The [advance of the] apsis of the moon is about twice as swift. (C&W, 544–5)

If we double the 1° 31′ 28″, which Newton tells us is the motion corresponding to having the extraneous force be 357.45 times less than the basic inverse-square force, the result is 3° 2.9′, which is almost exactly what Newton cites as the lunar precession in proposition 3 of the third edition. What, in his argument for proposition 3 book 3, Newton cites as the 1 over 178 29/40 reduction he claims as the action of the sun is exactly twice the 1/357.45 reduction that this example shows would produce about 1/2 of the lunar precession. If we put 1 over 178 29/40 or 1/178.725 as c into the precession formula of corollary 2, we get 3° 4.6′ precession per revolution. This is a fairly good approximation to the observed 3° 3′ of the lunar precession, even though it is not as close as doubling the result for $c = 1/357.45$.[11]

What is said in corollary 2 of proposition 45 book 1 does not provide any positive account showing that the motion of the lunar apsis produced by the total action of the sun over a lunar orbit roughly equals that which would be produced by setting c equal to 1/178.725. Instead of such an account, what is said supports only the hypothetical result that having the ratio $c = 1/178.725$ in corollary 2 proposition 45 book 1 would make that corollary yield fairly close to the observed lunar precession.

Here, as in the first edition argument for proposition 3, Newton appeals to the moon-test to back up his inference to inverse-square variation of the force maintaining

[11] The number which results is 1.539° precession from upper to lower apsis or 3.077°, which is 3° 4.6′, precession per revolution.

We may be left considerably puzzled as to why Newton would have used 1/357.45 as his example in corollary 2 proposition 45. Especially, given his remark that the apsis of the moon is about twice as swift as would be produced by such a foreign force.

the moon in its orbit. In this edition the appeal to the moon-test is, further, backed up by a corollary.

Corol. If the mean centripetal force by which the moon is maintained in its orbit is increased first in the ratio of $177^{29}/_{40}$ to $178^{29}/_{40}$, then also in the squared ratio of the semidiameter of the earth to the mean distance of the center of the moon from the center of the earth, the result will be the lunar centripetal force at the surface of the earth, supposing that force, in descending to the surface of the earth, is continually increased in the ratio of the inverse square of the height. (C&W, 803)

The point of this corollary is that the correction factor (178.725/177.725), which would offset the 1/178.725 net reduction that would approximately account for the lunar precession if applied as *c* in corollary 2 of proposition 45, should be applied when carrying out the moon-test.

As we can see, the argument for proposition 3 is left somewhat open ended. Newton's reference to corollary 2 proposition 45 does not lead the reader to any clear account of how the lunar precession is due to the action of the sun on the moon's motion. Instead of such an account, we are left with an appeal to the moon-test and an instruction to use a correction to offset what would be a value of *c* ($c = 1/178.725$) that would make corollary 2 yield, approximately, the lunar precession.

II Proposition 4 and the moon-test

The moon-test is the centerpiece of Newton's argument for identifying the force keeping the moon in its orbit with terrestrial gravity. Let us now turn to this argument, as it is given in the third edition.

Proposition 4 Theorem 4

The moon gravitates toward the earth and by the force of gravity is always drawn back from rectilinear motion and kept in its orbit.

1. *The moon-test*

The mean distance of the moon from the earth in the syzygies is, according to Ptolemy and most astronomers, 59 terrestrial semidiameters, 60 according to Vendelin and Huygens, 60⅓ according to Copernicus, 60⅖ according to Street, and 56½ according to Tycho. But Tycho and all those who follow his tables of refractions, by making the refractions of the sun and moon (entirely contrary to the nature of light) be greater than those of the fixed stars – in fact greater by about four or five minutes – have increased the parallax of the moon by that many minutes, that is, by about a twelfth or fifteenth of the whole parallax. Let that error be corrected, and the distance will come to be roughly 60½ terrestrial semidiameters, close to the value that has been assigned by the others. (C&W, 803–4)

Newton begins by citing six estimates of the lunar distance by astronomers. He points out that these are estimates of the distance when the moon is in syzygies, that is, when the moon is either in conjunction with or in opposition to the sun.[12] He also argues that Tycho Brahe's estimate of 56½ should be corrected to 60½, to offset errors introduced in that estimate by Tycho's exaggerated tables of refraction.[13]

He goes on to have us assume a mean-distance of 60 semidiameters in the syzygies, which closely approximates the mean of the six estimates he cites.[14]

> Let us assume a mean distance of 60 semidiameters in the syzygies; and also let us assume that a revolution of the moon with respect to the fixed stars is completed in 27 days, 7 hours, 43 minutes, as has been established by astronomers; and that the circumference of the earth is 123,249,600 Paris feet, according to the measurements made by the French. (C&W, 804)

Newton tells us that the sidereal period of the moon he cites has been established by astronomers[15] and that the circumference he cites for the earth is according to measurements made by the French.[16]

[12] The syzygies are when the sun and moon are lined so that the angle between them is either 0 degrees (conjunction) or 180 degrees (opposition). Densmore (1995, note 4 pp. 296–7) makes the following remarks about Newton's choice of syzygies here.

> "Syzygies" means "places of being yoked together," namely, conjunctions and oppositions, when the sun, earth, and moon are in line . . .
>
> . . . Syzygy is a place popular with astronomers because that's where eclipses occur. (Among the useful opportunities offered by eclipses is the chance to observe the moon at conjunction.) This means that distances at syzygy had been measured much more often and much more carefully than at other places on the orbit. Lunar theories reflected this: some (most notably Ptolemy's) gave wildly erroneous distances at the quadratures, though they were nearly correct at the syzygies. So if Newton had not specified that distances were to be taken at syzygies, he would not have found anything approaching consensus on the mean distance . . .
>
> . . . the moon's line of apsides (containing apogee and perigee) moves slowly around the zodiac. The place of the syzygies (i.e., the place of full and new moon) also moves around the zodiac, but more quickly: since it follows the sun, its period is one year. This difference in speeds means that the distance varies from syzygy to syzygy. Thus the distances at syzygies represent fairly accurately the distances to be found anywhere on the orbit (the sun distorts the orbit somewhat, but this effect is very slight).

We can add more. One of Newton's impressive applications of gravitation to the lunar motion was his account of how the tidal action of the sun would transform a concentric circular lunar orbit with radius equal to 1 to make its distance in syzygies smaller (about 0.9982) and its distance at quadratures (when the angle between the sun and moon is either 90 degrees or 270 degrees) greater (about 1.00719) (see C&W, 846; Wilson 2001, 153). In chapter 6 we shall see that this affords an interesting correction that can be applied to reinforce Newton's moon-test inference.

[13] See chapter 6 section I.5 below for a discussion of Newton's correction to Tycho's estimate of the lunar distance.

[14] Given Newton's correction to Tycho's estimate, the mean of the six distance estimates he cites is 60.038.

[15] As is the case with Jupiter's moons, the lunar period had long been well agreed upon by astronomers. The estimate cited by Newton, 27 days 7 hours 43 minutes, equals 39,343 minutes which, also, agrees very well with the mean lunar period 27.321661 days = 39,343.19 minutes cited today (*ESAA*, 708).

What we now know of as a slow secular variation of the lunar period has not, as yet, produced any very appreciable differences from the period cited by Newton.

[16] He is here referring to Picard's (1668–70) successful efforts to compare differences in latitude corresponding to distance triangulated between points on the Paris meridian to measure the length of a degree of longitude in 1670. (See *GHA 2B*, 23)

Picard's measurement of the length of a degree was 57,060 Paris toises. A Paris toise is 6 Paris feet. Newton also cites an estimate of 57,061 Paris toises from Cassini and his son, as well as an earlier estimate of 57,300 Paris toises from Norwood in his discussion of the shape of the earth in proposition 19 book 3 (C&W, 822).

Using these numbers, he can carry out the instructions given in his corollary to proposition 3 for calculating what the inverse-square lunar centripetal force is at the surface of the earth.

If now the moon is imagined to be deprived of all its motion and to be let fall so that it will descend to the earth with all that force urging it by which (by prop. 3, corol.) it is [normally] kept in its orbit, then in the space of one minute, it will by falling describe 15 1/12 Paris feet. This is determined by a calculation carried out either by using prop. 36 of book 1, or (which comes to the same thing) by using corol. 9 to prop. 4 of book 1. For the versed sine of the arc which the moon would describe in one minute of time by its mean motion at a distance of sixty semidiameters of the earth is roughly 15 1/12 Paris feet, or more exactly 15 feet, 1 inch, and 1 4/9 lines [or twelfths of an inch]. (C&W, 804)

This calculation is for a simplified system where the moon is represented as in uniform motion on a concentric circular orbit about the center of a spherical earth.[17] Here is corollary 9 of proposition 4 book 1:[18]

From the same demonstration it follows also that the arc which a body, in revolving uniformly in a circle with a given centripetal force, describes in any time is a mean proportional between the diameter of the circle and the distance through which the body would fall under the action of the same given force and in the same time. (C&W, 451)

To have x be a mean proportional between y and z is to have

$$z/x = x/y.$$

According to this corollary, given that the motion of the moon can be appropriately represented by such a simplified system, we have

$$d/a = a/D, \quad \text{so that } d = a^2/D,$$

where d is the distance the moon in falling by its orbital force would describe in one minute's time, a is the length of the arc it would describe in its orbit in one minute of time, and D is the diameter of that circular orbit.[19] Newton's cited lunar period

Newton's moon-test is robust with respect to all these estimates. Picard's measure gives Newton's cited 123,249,600 Paris feet as the corresponding circumference of a great circle on spherical earth. For Norwood, the corresponding circumference c is 123,768,000 Paris feet. This gives a one minute's fall at the lunar orbit corresponding to 60 earth radii as the lunar distance of $c(60)(\pi/(39{,}343)^2) = 15.072$ Paris feet. This is closer to Huygens's measured one second's fall at the surface of the earth of 15.096 Paris feet than the uncorrected one minutes fall of 15.009 Paris feet resulting from Picard's measurement. The Cassini measurement of 57,061 Paris toises gives a circumference c = 123,251,760 Paris feet. This results in a corresponding one minutes fall at the lunar orbit of 15.00926 Paris feet, which is about the same as that Newton used.

[17] This system is also simplified in that motions of the earth are not taken into account. The center of the earth is treated as appropriate to count as the origin of an inertial frame. As we shall see below, when considering the two-body correction, Newton describes this calculation as one under the hypothesis that the earth is at rest.

[18] See chapter 3 appendix 2 for a proof.

[19] The distance $d = a^2/D$, where a is the length in Paris feet of angular arc a in degrees and $D = 2r$ is the diameter of a circle of radius r, equals the length $r(1\text{-cos}a)$ in Paris feet of the versed sine $(1\text{-cos}a)$ of arc a. See C&W, 305–7) for Newton's usage of "versed sine." See chapter 3 appendix 2.

together with his cited circumference of the earth and assumed lunar distance of 60 earth radii make this calculation give 15.009 Paris feet as the distance the moon would fall in one minute if it were deprived of all its motion and let fall by the force by which it is maintained in its orbit.[20]

This value, 15.009 feet, is 15 feet 0.1 inches, which is clearly not the 15 1/12 feet (or 15 feet 1 inch) cited by Newton. In his corollary to proposition 3, Newton instructs one to increase the centripetal force maintaining the moon in its orbit by a correction factor (178.725/177.725 = 1.0056). This is the correction that would be needed to offset the net reduction specified by setting $c = 1/178.725$ in corollary 2 of proposition 45 book 1. The description of the moon-test quoted above from proposition 4 refers to the corollary of proposition 3, thereby reminding the reader to apply this correction specified in it. When this correction is applied we get 15.009 x (178.725/177.725) = 15.0935 Paris feet. This is 15 feet 1.122 inches, which agrees with Newton's cited 15 1/12 Paris feet.[21] It is, also, 15 feet 1 inch and 1.464 = 1 and 4.18/9 lines. To the nearest ninth of a line, this is what Newton takes to be his more exact value of 15 feet 1 inch and 1 4/9 lines.

Newton's assumption of 60 terrestrial semidiameters as the lunar distance has the neat consequence that, under the assumption that the centripetal force maintaining the moon in its orbit varies inversely with the square of the distance from the earth's center, the fall in one minute corresponding to the strength of this force at the lunar distance exactly equals the fall in one second corresponding to the increased strength this force would have at the surface of the earth.

Accordingly, since in approaching the earth that force is increased as the inverse square of the distance, and so at the surface of the earth is 60 x 60 times greater than at the moon, it follows that a body falling with that force, in our regions, ought in the space of one minute to describe 60 x

[20] Newton's cited lunar period, 27 days 7 hours 43 minutes, is 39,343 minutes of time. Therefore, in one minute of time the moon will describe an arc of $2\pi/39{,}343$ radians of its orbit. His cited circumference of the earth gives $123{,}249{,}600/2\pi$ Paris feet as the earth's radius. His assumption of 60 earth radii as the lunar distance gives $60 \times (123{,}249{,}600/2\pi)$ as the radius of the moon's orbit. This gives $60 \times (123{,}249{,}600/39{,}343)$ which is 187,962 Paris feet for the length of the arc described by the moon in its orbit in one minute of time. This makes the diameter $D = 60 \times (123{,}249{,}600/\pi)$ of the lunar orbit equal 2,353,893,968 Paris feet. This gives $d = [(187{,}962)^2/2{,}353{,}893{,}968]$ or 15.009 Paris feet.

[21] Chandrasekhar (1995, 358) gives the following calculation for the one minute's fall = $[(60 \times \text{radius of the earth}) \times \delta\theta] \times \frac{1}{2}\,\delta\theta$, where $\delta\theta = 2\pi/39434$ radians is the angular arc the moon describes in one minute. This gives $60\pi(1.232496/(3.9343)^2) = 15.008998$, which he rounds to 15 1/120 Paris feet as the fall in one second at the surface of the earth corresponding to taking 60 semidiameters of the earth as the lunar distance. He comments as follows:

Newton gives instead 15 1/12 Paris feet. Perhaps he wrote this value from memory (?) from an earlier computation with different parameters from those listed in this proposition. (Note on p. 358)

The outcome of Chandrasekhar's calculation is just the 15.009 Paris feet that results from Newton's numbers when one does not multiply by $((178\tfrac{29}{40})/(177\tfrac{29}{40}))$, which Newton gives in his corollary to proposition 3 as the correction required to offset the action of the sun on the moon's motion. As we have seen, both Newton's 15 1/12 Paris feet and his more precise 15 feet 1 inch and 1 4/9 lines result from applying this correction to the one minute's fall corresponding to the moon's orbit.

60 x 15 1/12 Paris feet, and in the space of one second 15 1/12 feet, or more exactly 15 feet, 1 inch, and 1 4/9 lines. (C&W, 804)

He is now ready to implement the moon-test by comparing this fall in one second with the one second's fall produced by the earth's gravity on terrestrial bodies.

Newton cites Huygens's use of the experimentally established length of a seconds pendulum to measure the one second's fall produced on terrestrial bodies by the earth's gravity.

And heavy bodies do actually descend to the earth with this very force. For a pendulum beating seconds in the latitude of Paris is three Paris feet and 8½ lines in length, as Huygens observed. And the height that a heavy body describes by falling in the time of one second is to half the length of this pendulum as the square of the ratio of the circumference of a circle to its diameter (as Huygens also showed), and so is 15 Paris feet, 1 inch, 1 7/9 lines. (C&W, 804)

Huygens's experiment gives 3 Paris feet and 8½ lines, which is 3.059 Paris feet, as the length l_{sec} of a seconds pendulum at Paris.[22] As Newton points out, Huygens also showed that one-half of l_{sec} times π^2 equals the one second's fall corresponding to the acceleration of gravity.[23] This gives $\pi^2 l_{sec}/2 = 15.096$ Paris feet, which equals the 15 feet 1 inch 1 7/9 lines cited by Newton as the one second's fall of terrestrial gravity established by Huygens.[24]

Newton now makes an explicit appeal to his first two Rules for Reasoning in Natural Philosophy to infer that the force maintaining the moon in its orbit is terrestrial gravity.

And therefore that force by which the moon is kept in its orbit, in descending from the moon's orbit to the surface of the earth, comes out equal to the force of gravity here on earth, and so (by rules 1 and 2) is that very force which we generally call gravity. (C&W, 804)

The basic argument for proposition 4 is the equality established in the moon-test together with this appeal to Rules 1 and 2.

The basic argument is immediately reinforced by the following indirect argument:

For if gravity were different from this force, then bodies making for the earth by both forces acting together would descend twice as fast, and in the space of one second would by falling describe 30 1/6 Paris feet, entirely contrary to experience. (C&W, 804)

[22] Huygens cites 3 Paris feet 8½ lines as the length of a seconds pendulum in his report of this experiment in his 1673 book *The Pendulum Clock*. See Blackwell, 168. See below chapter 5 section I for a more detailed account of Huygens's measurement.

[23] See chapter 5 sec I.1.

[24] The value of 2.7069 cm per Paris inch cited by Eric Aiton (see Aiton 1972, note 65 on p. 89) gives about 980.72 cm per sec squared as Huygens's measurement of the acceleration of gravity g at Paris. It also makes a Paris foot about 1.0657 U.S. (30.48006cm) or British (30.47997cm) feet. This makes his corresponding measurement of the one second's fall $d = 15.096$ Paris feet equal about 16.1 U.S. or British feet.

This indirect argument would not count against a hypothesis that would treat the force maintaining the moon in its orbit as a different kind of force, perhaps something analogous to magnetism, which would attract the moon but not act on terrestrial bodies.[25] Therefore, we may expect that the appeal to Rules 1 and 2 ought to give reason to reject this sort of alternative.

2. Regulae Philosophandi: *Rule 1 and Rule 2*

Newton's *Regulae Philosophandi*, his *Rules for Reasoning in Natural Philosophy*, are methodological principles which he explicitly formulates as appropriate to guide reasoning about natural things. Natural philosophy – the study of natural things – is what grew into what we still distinguish as the natural sciences. Newton's rules have been the focus of considerable critical comment by philosophers and historians of science.

As we have seen, Newton's argument for proposition 4 appeals to Rules 1 and 2. Here is the third edition formulation of Rule 1 with Newton's comment on it.

Rule 1

No more causes of natural things should be admitted than are both true and sufficient to explain their phenomena.

As the philosophers say: Nature does nothing in vain, and more causes are in vain when fewer suffice. For nature is simple and does not indulge in the luxury of superfluous causes. (C&W, 794)

This rule is, explicitly, formulated as a guide for inferring causes of natural things from their phenomena. Basically, we have an appeal to simplicity – Rule 1 is stated as a sort of Occam's razor principle. We are instructed to avoid attributing more causes to natural things when fewer would suffice to explain their phenomena. We, also, are instructed to admit only causes which are true.

What does Newton mean when he says that causes which are to be admitted are such that they are able to be counted as true? This *vera causa* condition is, apparently, directed against Descartes, who introduced explicitly fictional causes, some of which may have been motivated as efforts to get around the outcome of Galileo's trial. Motion of the earth about the sun could explain the retrograde motion of Mars even though, in the philosophical sense, the earth does not move because it has no motion with respect to the contiguous parts of the ethereal vortex that carries it around the sun.[26] The vortex theory is motivated by a more general commitment to a view of

[25] Such an hypothesis would be in line with an Aristotelian view as reflected in the following remark by David Gregory, which was directed against Newton's first edition hypothesis 3 [Every body can be transformed into a body of any other kind and successively take on all the intermediate degrees of qualities].

> This the Cartesians will easily concede. But not the Peripatetics, who make a specific difference between celestial and terrestrial matter.

See Cohen 1971, 191 and C&W, 203.

[26] See Descartes (trans. 1983, 51, 75, 94). See, also, Aiton (*GHA 2A*, 210 and 1972, 42).

causal explanation as merely hypothetical. According to the mechanical philosophy, what one needs are possible causes to make phenomena intelligible by showing how they could, hypothetically, be caused by contact action. For example, the motion of a moon in orbit about a planet orbiting the sun is made intelligible by secondary eddies that may be carried along in a greater circular current.[27]

On this sort of view, a cause need only be possible. There is no requirement that it be backed up by any stronger sort of evidence than that if it were true it would explain the phenomenon. Newton's *vera causa* condition in Rule 1 is clearly a demand for something more. We shall want to explore whether the application we are investigating suggests how such a demand is to be met.

The appeal in Newton's argument for proposition 4 is to both Rule 1 and Rule 2. The statement of Rule 2 suggests that it is intended as a consequence or implication of Rule 1.

Rule 2

Therefore, the causes assigned to natural effects of the same kind must be, so far as possible, the same.

Examples are the cause of respiration in man and beast, or of the falling of stones in Europe and America, or of the light of a kitchen fire and the sun, or of the reflection of light on our earth and the planets. (C&W, 795)

We can read these two rules, together, as telling us to opt for common causes, whenever we can succeed in finding them. This seems to be exactly their role in the application we are considering.

We have two primary phenomena: the centripetal acceleration of the moon and the length of a seconds pendulum at Paris. Given that the centripetal force maintaining the moon in its orbit is inverse-square, each measures a force producing accelerations at the surface of the earth. These accelerations are equal and equally directed toward the center of the earth. This agreement in measured values is another phenomenon. Like Kepler's Harmonic Rule, it is a higher-order phenomenon relating the basic phenomena in question. Identifying the forces explains the agreement by the claim that the basic phenomena count as agreeing measures of the strength of the very same inverse-square centripetal force. This makes these phenomena count as effects of a single common cause.

We can see how each rule applies here. For Rule 1, note that if we do not identify the forces we will have to admit two causes rather than one to explain the two basic phenomena. Moreover, we will have to either leave the higher-order phenomenon of the agreement unexplained or introduce yet another cause to explain it. The identification clearly does minimize the number of causes we have to admit. For Rule 2, note that our basic phenomena are of the same kind in that each measures a force producing equal centripetal accelerations at the surface of the earth. On the identification of the forces we assign the very same cause to each. It also extends the

[27] See Descartes 1983, 97.

sense in which we understand them as phenomena of the same kind by making them count as agreeing measurements of the strength of the same inverse-square centripetal force.

The appeal to Rules 1 and 2, therefore, clearly counts against the alternative hypothesis that the centripetal force holding the moon in its orbit is a different kind of force, which acts on the moon but does not act on terrestrial bodies. This hypothesis would require a separate cause for each of the two basic phenomena and would need to be further augmented to explain their agreement. It would also treat the causes as different in kind. Newton's appeal to Rule 2 suggests that one would need positive evidence that the phenomena are sufficiently different in kind to warrant hypothesizing such a difference between their causes.

When Newton makes the identification of the lunar orbital force with terrestrial gravity he is imposing a constraint on the theory he is constructing. This helps to illuminate the *vera causa* condition in Rule 1. To count the identification as corresponding to a true cause is to be committed to make all the systematic constraints it imposes as the theory is developed. After proposition 4, Newton is committed to counting any systematic dependencies that make a phenomenon into a measure of gravity as making that phenomenon, *also*, count as a measure of the centripetal force on the moon. This requires him to be able to count the orbital distance calculated from inverse-square, adjusting Huygens's very precise measurement of the acceleration of gravity, together with the well-known lunar period, as an accurate estimate of the moon's mean-distance. Similarly, any measurement of the centripetal force on the moon measures the earth's gravity. The various parallax observations yielding estimates of the lunar distance must yield corresponding estimates of surface gravity which can be counted as agreeing, though perhaps much rougher, estimates of the strength of gravity near the surface of the earth.

Newton's identification, also, transformed the common notion of gravity. On it, weight for bodies, above the surface of the earth, counts as varying inversely with the square of distance from the center of the earth. Here is Howard Stein's marvelous description of the context and reception of this implication.

> ...; an implication that was hardly dreamt of before Newton published the Principia, and that was received with astonishment – and with assent – by the scientific community. For instance, it was regarded as a great and wholly unanticipated discovery by such distinguished opponents of the more general theory of Book III as Huygens and Leibniz.
>
> The discovery I mean is that gravity – weight – varies inversely with the square of the distance from the center of the body towards which it tends. For note that the inverse square law for the acceleration *of the moon* is asserted in Proposition III on the basis of an argument from the phenomena, and this result plays a crucial role in the argument for Proposition IV; but before Proposition IV has been established, *no grounds whatever* are apparent for asserting a like law of variation for the weight of a terrestrial body. That the two "natural effects," the acceleration of the moon and that of falling bodies, are to be regarded as "the same" in the sense of Rule II, and thus as having "the same cause," has this as at least an important part of its effective meaning: the

law governing the "diffusion from the center to the several places around it" of "an efficacy for moving the bodies" that are in those places is the same for terrestrial bodies and for the moon. (Stein 1991, 213)

We shall see Huygens express just the sort of reaction so vividly described here, in our chapter devoted to Huygens.[28] We saw in chapter 3 section I that Stein's quoted phrases

"diffusion from the center to the several places around it" of "an efficacy for moving the bodies" that are in those places

are from Newton's discussion of what he defines as the accelerative measure of a centripetal force. The common cause of the acceleration of falling bodies and the acceleration of the moon is gravity conceived as an inverse-square field of acceleration directed toward the center of the earth and extending outward from the vicinity of its surface.

We have seen that to treat a cause as true is to make the commitments that taking it as true would imply. What we will want to look for is whether this *vera causa* condition, also, signals appeal to any particular sort of mark or evidence that would reasonably be taken to justify believing that a proposed cause should be counted as true. The higher-order phenomenon exhibited by the agreement in measurements of acceleration at the surface of the earth counts as an example of what William Whewell (1794–1866) would call consilience of inductions.[29] Both Whewell and Malcolm Forster identify such consilience as a mark of the sort of evidence that justifies the inference to a common cause.[30] Forster would regard an appropriate reading of the *vera causa* condition as a demand for causes that are given agreeing measurements by distinct phenomena.[31] The moon-test inference is, clearly, an example where such a demand is met.

[28] See chapter 5 section I.5.

[29] See Harper (1989, 115–52); Forster, M. (1988, 55–101), Butts (1988, 159, 163–4).

[30] Forster (1988, 86–7) suggests that Newton should have kept his evidence for the inverse-square completely independent of the resulting agreement with Huygens and that he should have made sure the simplifying assumptions were not giving an accidental agreement, before regarding the consilience as independently established. Newton's method clearly allows for treating the consilience exhibited in the outcome of the moon-test as additional evidence supporting the inference to the inverse-square variation of the force maintaining the moon in its orbit. We shall be defending Newton's more free-wheeling appeal to agreeing measurements. See Harper (1989, 137–40) for more discussion of Forster on Newton.

Whewell (1860, chpt. 18) offers a critical discussion of Newton's rules that fails to do justice to implications of taking Newton's "deductions" from phenomena as measurements. This, I think, prevents him from seeing the extent to which Newton's application of Rules 1 and 2 here is very much in line with Whewell's own discussions of how inferences to simplicity are supported by consilience. See chapter 7 section I.1.

[31] See especially 1988, 98. Though Forster offers an excellent, and very highly informative, application of inferences to common causes supported by agreeing measurements to defend Newton's inferences to component forces from objections by Cartwright (1983), he, also, succumbs to what I take to be an all too common error shared by many philosophers of science. This is the error of turning something shown to be a sufficient condition into a necessary condition so that a positive argument for accepting some inference can be

Many empiricists would see the simplicity appealed to in Rules 1 and 2 as something extraneous to empirical success. According to such a view, these rules endorse a merely pragmatic commitment to simplicity imposed as an additional requirement beyond empirical success. The agreement of the measurements of the strength of the earth's gravity by the two phenomena, the length of a terrestrial seconds pendulum and the centripetal acceleration of the moon, is an example of what can reasonably be regarded as a kind of empirical success. This suggests that such a simplistic empiricist view is unable to do justice to the empirical support the agreement of these measurements provides for Newton's inference to identify the centripetal force maintaining the moon in its orbit with terrestrial gravity.

3. *The two-body correction*

Newton's main text for proposition 4 concludes with the following paragraph commenting on the moon-test calculation.

> This calculation is founded on the hypothesis that the earth is at rest. For if the earth and the moon move around the sun and in the meanwhile also revolve around their common center of gravity, then, the law of gravity remaining the same, the distance of the centers of the moon and earth from each other will be roughly 60½ terrestrial semidiameters, as will be evident to anyone who computes it. And the computation can be undertaken by book 1, prop. 60. (C&W, 804–5)

Newton claims that this moon-test calculation is founded on the hypothesis that the earth is at rest. The claim that this important calculation, one that is so central to his argument for universal gravity, is founded on a mere *hypothesis* may seem remarkable to readers familiar with Newton's avowed rejection of hypotheses in the general scholium.

> For whatever is not deduced from the phenomena must be called a hypothesis; and hypotheses, whether metaphysical or physical, or based on occult qualities, or mechanical, have no place in experimental philosophy. In this experimental philosophy, propositions are deduced from the phenomena and are made general by induction. (C&W, 943)

The provisional assumption that the earth is at rest is not only one that is not deduced from the phenomena, it is one that will turn out to be actually incompatible with what Newton will be arguing for.[32] One very important role played by this application of

turned into a resource for, what may well turn out to be illegitimate, skeptical arguments against other inferences.

See chapter 9 below for an argument directing Newton's method against this error of many philosophers of science.

[32] Not only is it incompatible with universal gravitation; it also goes beyond what, according to Newton's Laws of Motion, could, in principle, be empirically distinguishable by motions among bodies. See chapter 3 section II.

the two-body correction is to show that the outcome of the moon-test argument does not depend on this hypothesis.

Here is proposition 60 of book 1.

Proposition 60 Theorem 23

If two bodies S and P, attracting each other with forces inversely proportional to the square of the distance, revolve about a common center of gravity, I say that the principal axis of the ellipse which one of the bodies P describes by this motion about the other body S will be to the principal axis of the ellipse which the same body P would be able to describe in the same periodic time about the other body S at rest as the sum of the masses of the two bodies S+P is to the first of two mean proportionals between this sum and the mass of the other body S. (C&W, 564)

This makes the ratio

$$R'/R = (S+P)/[(S+P)^2 S]^{1/3},$$

where R' is the principal axis of P's ellipse about S when S moves and R is the principal axis of P's ellipse about S when S is assumed to be stationary.[33]

In this third edition, Newton gives 1 to 39.788 as the ratio of the mass of the moon to the mass of the earth.[34] This gives 60.4985 as the two-body corrected distance when 60 is taken as the semi-major axis of an orbit with the same period under the assumption that the moon orbits a stationary earth. So long as the two-body corrected

[33] As Cajori points out (1934, note 24 p. 651), x and y are the first and second of two mean proportionals between a and b just in case $a/x = x/y = y/b$. This makes $x^2 = ay$ and $y^2 = bx$, so that, eliminating y, $x = (a^2b)^{1/3}$. If $a = S+P$ and $b = S$, then the principal axis R' (of P's ellipse about S when S moves) is to the principal axis R (of P's ellipse about S stationary) as $S+P$ is to $[(S+P)^2S]^{1/3}$.

[34] In his earlier draft (*System of the World*, sec. 11, Cajori, 561) for what became book 3, Newton estimated the lunar mass, from its mean apparent diameter of 31½ minutes, to be about 1/42 of the mass of the earth. This gives $60\ (43/(43^2 42)^{1/3}) = 60.47$ or about 60½ as the two-body corrected distance.

In the *Principia*, in proposition 37, Newton used ratios of heights of neap tides (high tides when sun and moon are at quadratures – 90° apart – so that solar and lunar tidal forces maximally interfere with one another to produce lowest high tides) and spring tides (high tides when moon and sun are syzygies – 0° or 180° apart – so that solar and lunar tidal forces maximally reinforce each other to produce highest high tides) to calculate an estimate of the mass of the moon. In the first edition he gave (corollary 4) what he called a *quam proxime* (or a rough) estimate of 1 to 26 as the ratio of the earth's mass to that of the moon. If we set the mass of the earth at 1 then, according to this estimate, the moon's mass is $1/26 = 0.038$, which gives 60.76 as the two-body corrected lunar distance. Though this differs from 60.5, it is much closer to this value than the 60 which Newton used in his one-body moon-test calculation. In this case, the two-body correction can help defend using $60r$ in the moon-test, even though the mean of the cited distance estimates is 60.57.

By the third edition, the estimate of the ratio of the mass of the earth to that of the moon was taken as 1 to 39.788. This yields 60.4985 or almost exactly 60.5 as the two-body corrected distance corresponding to using 60 as the distance in the one-body calculation. The second edition estimate of the lunar mass was 1/39.371, which would give 60.50, exactly.

It turns out to be quite difficult to accurately measure the mass of the moon from the tides. Even with a relatively accurate estimate of solar parallax and far more detailed treatment of the tides than Newton's, Laplace in vol. 3 of his *Celestial Mechanics*, which was published in 1802, was still estimating 1/58.6 from the tides at Brest. His other estimates of the lunar mass were 1/69.2 from the lunar equation in tables of the sun's motion, 1/71 from the nutation of the earth's axis, 1/74.2 from an equation for the moon's parallax. He settled upon 1/68.5 as his most probable value for the lunar mass. See Bowditch, *Laplace's Celestial Mechanics* vol. 3, 336–40.

distance is at least compatible with the available distance data, the moon-test inference will not depend upon the hypothesis that the earth is at rest. If the two-body corrected distance were to be not just compatible, but were to actually fit the cited distance estimates better than 60, the two-body correction would defend using 60 rather than the mean of the cited distances in the moon-test.

Newton's estimates of the ratio of the mass of the moon to that of the earth are wildly inaccurate.[35] The modern value was established at about 1 to 81 by the time Simon Newcomb wrote up his account of the foundations of the predictive tables established by the U.S. Naval Observatory.[36] This would give a two-body correction of about 60.25 for Newton's assumed value of 60 terrestrial semidiameters as the lunar distance used in the moon-test.

III The scholium to proposition 4

Newton offers a more extended treatment of his argument for identifying gravity as the force maintaining the moon in its orbit about the earth in the following scholium.

1. *Treating the Harmonic Rule as a* **law**

Scholium

The proof of the proposition can be treated more fully as follows. If several moons were to revolve around the earth, as happens in the system of Saturn or of Jupiter, their periodic times (by the argument of induction) would observe the law which Kepler discovered for the planets, and therefore their centripetal forces would be inversely as the squares of the distances from the center of the earth, by prop. 1 of this book 3. (C&W, 805)

Newton begins his thought experiment by appealing to induction to extend Kepler's Harmonic Rule to a hypothetical system of several moons revolving around the earth. He explicitly calls it a "law."

Curtis Wilson's search for the first to write of Kepler's rules as "laws" led him to conclude that it was Leibniz in his "*Tentamen de motuum coelestium causis*" of 1689.

[35] In addition to the substantial general difficulties in estimating the lunar mass from the tides, Newton's estimates were heavily biased by his incorrect estimates of the solar parallax. The modern value is about 8.79 seconds (*ESAA*, 700), for a mean solar distance 1/tan 8.79″ in earth radii of about 23,465.85. In the first edition Newton used 20" for a solar distance in earth semidiameters of 1/tan 20″ or 10,313, in the second 10" for a solar distance of 20,626 and in the third 10½″ for 19,644. These incorrect estimates of the solar distance helped exaggerate his estimates of the mass of the moon.

[36] Using the lunar inequality in the earth's motion (an inequality due to the orbit of the earth about the center of mass of the earth and moon) Newcomb gave 1/81.32 = .01230012 as an estimate of the ratio M_m/M_e of the mass of the moon to that of the earth (Newcomb 1895, 189). The current estimate of M_m/M_e is .01230002 (*ESAA*, 710). Newcomb also gave a somewhat less accurate estimate of 1/81.58 = .012258 for M_m/M_e from the constant of nutation (Newcomb 1895, 189).

> Leibniz attempted to account for Kepler's three rules using vortices and differential equations. As Bertoloni Meli has shown,[37] Leibniz, though explicitly denying it, wrote his essay only after making a close study of the first 40 pages of Newton's *Principia*. In his rivalry with Newton, he co-opted Kepler as an ally. By calling the rules "laws" he sought to enhance their status as results achieved independently of Newton. (Wilson 2000, 225)

In his published scholium to proposition 4, which was first printed in the third edition but appeared on a sheet in Newton's interleaved and annotated copy of the second edition,[38] Newton takes advantage of the opportunity afforded by being able to call Kepler's Harmonic Rule a "law."

Newton's extension of Kepler's Harmonic Rule to such a hypothetical system of moons is a clear case of treating it as a *law* and not, merely, as an empirical generalization that we expect will continue to hold for the actual systems of orbits which have been observed. The generalization is not only extended to epistemically possible systems which had not yet been observed, such as moons of Mars. It is extended, even, to hypothetical moons which were known at the time to be counterfactual suppositions.

The Harmonic Law relation that such a system of moons would exhibit would be a phenomenon which would measure the inverse-square variation of the centripetal forces that would maintain such moons in their orbits. This thought experiment might, thus, be taken to provide some slight additional support to the argument for the inverse-square variation of the force maintaining our moon in its orbit provided in proposition 3.

Newton applies the inverse-square variation obtained by his appeal to Kepler's Harmonic Law to carry out his thought experiment version of the moon-test for a system of moons orbiting the earth.

> And if the lowest of them were small and nearly touched the tops of the highest mountains, its centripetal force, by which it would be kept in its orbit, would (by the preceding computation) be very nearly equal to the gravities of bodies on the tops of those mountains. And this centripetal force would cause this little moon, if it were deprived of all the motion with which it proceeds in its orbit, to descend to the earth – as a result of the absence of the centrifugal force with which it had remained in its orbit – and to do so with the same velocity with which heavy bodies fall on the tops of those mountains, because the forces by which they descend are equal. (C&W, 805)

The moon-test established the agreement between the inverse-square adjusted centripetal acceleration of our moon in its orbit and the acceleration of gravity at the surface of the earth. Given the inverse-square variation of the centripetal forces that would maintain moons in orbits, the outcome of the moon-test shows that there would be close agreement between the centripetal acceleration of the orbit of the little

[37] Wilson (2000, 225 note 14) refers to Bertoloni Meli 1991 and 1993, 95–142, appendix 1.

[38] See Koyré and Cohen 1972, appendix III J, 800–1.

moon nearly touching the tops of the highest mountains and the acceleration of gravity by which heavy bodies fall on the tops of those mountains.

2. The scholium moon-test argument

Newton argues from this agreement to the identification of the inverse-square force which would maintain such a little moon in its orbit with terrestrial gravity.

> And if the force by which the lowest little moon descends were different from gravity and that little moon also were heavy toward the earth in the manner of bodies on the tops of mountains, this little moon would descend twice as fast by both forces acting together. Therefore, since both forces – namely, those of heavy bodies and those of the moons – are directed toward the center of the earth and are similar to each other and equal, they will (by rules 1 and 2) have the same cause. And therefore that force by which the moon is kept in its orbit is the very one that we generally call gravity. For if this were not so, the little moon at the top of a mountain must either be lacking in gravity or else fall twice as fast as heavy bodies generally do. (C&W, 805)

The argument in the main text concluded with the reductio that terrestrial bodies would fall twice as fast if the force maintaining the moon in its orbit were another force than gravity. As we noted, an alternative hypothesis according to which the force on the moon did not affect terrestrial bodies could avoid this reductio. Here the reductio is that the little moon would descend twice as fast or be unaffected by terrestrial gravity. This version makes an alternative hypothesis according to which the little moon would not be affected by gravity one of the disjuncts of the reductio.

As in the main text, an application of Rules 1 and 2 is the central argument in the scholium. In the scholium, it is both preceded by and followed by appeal to the scholium version of the reductio. Let us examine, more closely, this application of Rules 1 and 2.

> Therefore, since both forces – namely, those of heavy bodies and those of the moons – are directed toward the center of the earth and are similar to each other and equal, they will (by rules 1 and 2) have the same cause. And therefore that force by which the moon is kept in its orbit is the very one that we generally call gravity. (C&W, 805)

The phrase "both forces" suggests that we are talking about two forces. On the other hand, the phrases "those of heavy bodies" and "those of the moons" suggest that each of the basic two, also, counts as a multiplicity. Apparently, there is a sense in which there are distinct forces for each heavy body and for each of the moons. The direct application of Rules 1 and 2, here, argues that these forces, both sorts, have the same cause because they "are directed toward the center of the earth and are similar to one another and equal." Apparently, part of the ground for identifying their cause is some "similarity" that goes beyond their centripetal direction and equality. Finally, from this identification of a common cause the conclusion "And therefore that force by which the moon is kept in its orbit is the very one which we call gravity" is taken to be immediate.

3. *Newton's definitions of centripetal force*

Recall the following discussion relating the three measures of centripetal force from Newton's discussion of his definition of the motive measure of centripetal forces.

> These quantities of forces, for the sake of brevity, may be called motive, accelerative, and absolute forces, and, for the sake of differentiation, may be referred to bodies seeking a center, to the places of the bodies, and to the center of the forces: that is, motive force may be referred to a body as an endeavor of the whole directed toward a center and compounded of the endeavors of all the parts; accelerative force, to the place of the body as a certain efficacy diffused from the center through each of the surrounding places in order to move the bodies that are in those places; and absolute force, to the center as having some cause without which the motive forces are not propagated through the surrounding regions, whether this cause is some central body (such as a loadstone in the center of a magnetic force or the earth in the center of a force that produces gravity) or whether it is some other cause which is not apparent. This concept is purely mathematical, for I am not now considering the physical causes and sites of forces. (C&W, 407)

This passage sheds light on our puzzle about how what appear to be the two centripetal forces to be identified in the application of Rule 1 and Rule 2 in the scholium to proposition 4 are each said to be a multiplicity of forces corresponding to the multiplicity of bodies being accelerated. According to Newton's elucidation, the motive quantities referred to the moons being maintained in their orbits by the first centripetal force may each be called motive forces – one for each moon. Similarly, the weights of terrestrial bodies may be called separate forces even though they are all motive measures of the same centripetal force of gravity.

As we have noted, Newton's discussion of accelerative measure as "a certain efficacy diffused from the center through each of the surrounding places in order to move the bodies that are in those places" suggests that his basic concept of centripetal force is that of a central force field. It also suggests that a centripetal force can be said to "cause" the various motive forces referred to the bodies on which it acts.[39]

Newton's accelerative measure is referred to the places. This suggests that, in the applications he is centrally concerned with, different bodies would be equally accelerated if they were at equal distances from the center. This suggestion is reinforced by the following passage:

[39] This suggestion is further supported by Newton's remarks elucidating his definition of impressed force. As we noted in chapter 3 section I, the motive force referred to a body acted upon is what Newton distinguishes as an impressed force or action exerted on that body.

> **Definition 4**
>
> *Impressed force is the action exerted on a body to change its state either of resting or of moving uniformly straight forward.*
>
> This force consists solely in the action and does not remain in a body after the action has ceased. For a body perseveres in any new state solely by the force of inertia. Moreover, there are various sources of impressed force, such as percussion, pressure, or centripetal force. (C&W, 405)

Here we have centripetal force as a source or cause of an impressed force. The motive force referred to a body on which it acts is caused by the centripetal force of gravity which acts on that body.

For the sum of the actions of the accelerative force on the individual particles of a body is the motive force of the whole body. As a consequence, near the surface of the earth, where the accelerative gravity, or the force that produces gravity, is the same in all bodies universally, the motive gravity, or weight, is as the body, but in an ascent to regions where the accelerative gravity becomes less, the weight will decrease proportionally and will always be as the body and the accelerative gravity jointly. (C&W, 407–8)

This passage strongly reinforces the idea that, for gravity, the accelerative measure can be referred to the places, without also being referred to the accelerated bodies as well; because, at any given distance from the center, all bodies will be equally accelerated.

We can now explicate what can count as the additional similarity going beyond the direction toward the earth's center and the equality of the acceleration which would be produced at the top of the mountain. Kepler's Harmonic Rule makes the centripetal accelerations of the moons proportional to the inverse-square of their distances from the center of the earth. This makes the inverse-square adjusted accelerations independent of the different bodies of any such possible moons. The extra similarity referred to is that each centripetal force produces equal accelerations on all bodies at any equal distances. A centripetal force field that would produce equal accelerations on bodies at equal distances counts as a centripetal acceleration field.

For acceleration fields with the same law relating accelerative measures to distances, the ratio of their absolute measures is fixed by the common ratio of their accelerative measures at any given distance. This enables the agreeing measurements of the acceleration of gravity at the surface of the earth from seconds pendulums and from inverse-square adjusting the centripetal acceleration exhibited by the lunar orbit to count as agreeing measurements of the absolute quantity (the strength) of a single inverse-square acceleration field extending out from the surface of the earth.

We can now expand on the account of how Rules 1 and 2 apply. The motive forces maintaining the moons in their orbits and the motive forces of gravity on heavy bodies have the same cause – a single inverse-square acceleration field directed to the center of the earth. What Newton takes to be the immediate consequence,

And therefore that force by which the moon is kept in its orbit is the very one that we generally call gravity. (C&W, 805)

is immediate. The inverse-square centripetal acceleration field maintaining the moon in its orbit is the very same centripetal force that we call gravity.

Consider the concluding reductio according to which an alternative hypothesis where the moon would not be affected by gravity is an explicit alternative known to be false.

For if this were not so, the little moon at the top of a mountain must either be lacking in gravity or else fall twice as fast as heavy bodies generally do. (C&W, 805)

The fact that gravity is known to equally accelerate all bodies near the surface of the earth, therefore, adds additional weight to his appeal to Rules 1 and 2.

IV Empirical success and the moon-test

1. The moon-test as an agreeing measurement

We are now ready to argue that the application of Rules 1 and 2 in Newton's inference to identify the force maintaining the moon in its orbit with terrestrial gravity is backed by the ideal of empirical success exhibited in his basic inferences from phenomena. According to this ideal, a theory succeeds by having its parameters receive convergent accurate measurements from the phenomena it purports to explain. On the identification Newton argues for, we have the strength of a single inverse-square centripetal acceleration field that is given agreeing measurements by the length of a seconds pendulum near the surface of the earth and by the centripetal acceleration exhibited by the orbit of the moon.

The length of a seconds pendulum could provide very accurate measurements of gravity in Huygens's day. Here are the seconds pendulum estimates of the one second fall *d*, from experiments at Paris discussed by Huygens and Newton.

Table 4.1 Seconds pendulum estimates

	d in Paris feet
Picard	15.096
Huygens	15.096
Richer	15.099
Varin et al.	15.098

Here are the lunar distances x in earth radii and the corresponding one second falls $d(x)$ in Paris feet, for each of the six distance estimates cited by Newton in the third edition.[40]

[40] Let c be the circumference, in Paris feet, of the earth and t be the period, in minutes, of the lunar orbit. For each lunar distance estimate x, the arc traversed by the moon in one minute is $a = xc/t$, while the diameter of the lunar orbit is $D = xc/\pi$. The corresponding one minut's fall, a^2/D, is given by

$$(x^2c^2/t^2)/(xc/\pi) = (x^2c^2/t^2)(\pi/xc) = xc(\pi/t^2).$$

Newton's cited period for the moon's orbit, 27 days 7 hours 43 minutes, gives $t = 39{,}343$ minutes. He also gives $c = 123{,}249{,}600$ Paris feet for the circumference of the earth. These yield

$$x(123{,}249{,}600)(\pi/(39{,}343)^2) = 0.25015x,$$

for the one minute's fall corresponding to the centripetal acceleration of the lunar orbit in Paris feet. For each x, therefore,

$$d(x) = 0.25015x(x^2/60^2),$$

gives the one second's fall at the surface of the earth corresponding to inverse-square adjusting the centripetal acceleration of the moon.

Table 4.2 Moon test estimates

	$x = R/r$	$d(x) = 0.25015x(x^2/60^2)$
Ptolemy	59	14.271
Vendelin	60	15.009
Huygens	60	15.009
Copernicus	60⅓	15.261
Street	60.4	15.311
Tycho	60.5	15.387

In chapter 1 we exhibited the agreement between these moon-test estimates and the above pendulum estimates by combining them into a single table.

Table 4.3 Moon-test estimates and seconds pendulum estimates

Moon-test estimates			Seconds-pendulum estimates
	d in Paris feet		
Ptolemy	14.271		
Vendelin	15.009		
Huygens	15.009		
		15.096	Picard
		15.096	Huygens
		15.098	Varin et al.
		15.099	Richer
Copernicus	15.261		
Street	15.311		
Tycho	15.387		

The agreement of these measurements is clearly exhibited in this table. The numbers suggest that, though the moon-test estimates may be irrelevant to small differences, the agreement of the cruder moon-test estimates affords increased resistance to large changes from the pendulum estimates.

We now have available least-squares techniques that improve on this sort of exhibition and assessment of agreement among measurements. The extremely close clustering about their mean value of the estimates in Table 4.1 shows the sharpness of the stability exhibited in repetitions of pendulum estimates of the one second's fall. Their mean is about 15.097 Paris feet. A standard way of representing their clustering about it is by their standard deviation (*sd*), the square root of the average of the squares of their differences from the mean. The *sd* of these pendulum estimates is 0.0013 Paris feet. The moon-test estimates are far less closely clustered. Their mean is about 15.041 Paris feet and their standard deviation is about 0.37 Paris feet.

We can use a Student's *t*-confidence estimate to represent expectations about future estimates of similar reliability afforded by small numbers of estimates.[41] The 95% Student's *t*-confidence estimate of the one second's fall *d* afforded by the four pendulum measurements is[42]

$$d = 15.097 \pm 0.0024 \text{ Paris feet.}$$

This represents the claim that one could expect with a probability at least 0.95 that a randomly selected pendulum estimate of the one second's fall at Paris (of reliability similar to that of these four estimates) would fall within the interval [15.0946, 15.0994] Paris feet. The 95% Student's *t*-confidence estimate afforded by the six moon-test estimates is[43]

$$d = 15.041 \pm 0.43 \text{ Paris feet}$$

for a 95% confidence interval [14.61, 15.47] Paris feet. The narrow 95% confidence interval [15.0946, 15.0994] Paris feet for the pendulum measurement of *d* certainly is compatible with the much wider 95% Student's *t*-confidence interval [14.61, 15.47] Paris feet for the moon-test estimate of *d* afforded by the six estimates of the lunar distance Newton cites. This simple least-squares assessment shows that accepting Newton's moon-test inference would make the $d(x)$'s, computed from the distance estimates $x = (R/r)$ he cites, count as rougher estimates that agree with Huygens's considerably more precise measurement of the one second's fall.

[41] See Freedman et al. 1978, 401–15 for a good account of Student's *t*-confidence estimates and those based on the normal approximation. The 95% Student's *t*-confidence estimate corresponding to a set of n measurements of a parameter is given by

$$\text{mean} \pm tSE,$$

where tSE is the result of multiplying the standard error of the mean, SE, for those n measurements by t and t is the 95% Student's t parameter for the $n-1$ degrees of freedom corresponding to n measurements.

The SE or standard error of the mean for a set of n measurements of a parameter is

$$SE = (n^{1/2}/n)sd^{+},$$

the result of multiplying the sd^{+} by the result of dividing the square root of n by n.

The sd^{+} is given by

$$sd^{+} = (n/n-1)^{1/2}sd,$$

the result of multiplying the sd by the square root of the result of dividing n by $n-1$.

[42] For a 95% confidence interval estimate, the Student t parameter for the three degrees of freedom corresponding to four estimates is 3.18 (Freedman et al. 1978, A-71; 1998, A-106).

We have here

$$sd = 0.0013 \text{ Paris feet}$$
$$sd^{+} = (4/3)^{1/2}(0.0013) = 0.0015 \text{ Paris feet}$$
$$SE = (2/4)(0.0015) = 0.00075 \text{ Paris feet}$$
$$tSE = 3.18(0.00075) = 0.002385 \text{ Paris feet.}$$

[43] We have for the six cited estimates of the one second's fall $d(x)$ from the six moon-test estimates of the lunar distance $x = R/r$ a mean d(x) of 15.041 Paris feet, with an sd of 0.37 Paris feet and an sd^{+} of 0.409 Paris feet, an SE of 0.167 Paris feet with 95% Student's *t*-confidence bounds $tSE = 2.57SE = 0.429$ Paris feet.

Our least-squares assessment gives the 95% Student's *t*-confidence estimate of the lunar distance in earth radii as[44]

$$R = 60.04 \pm 0.58r.$$

The 95% confidence bounds [59.46, 60.62]r show that the two-body corrected distance

$$R = 60.5r$$

is not in conflict with the estimate of the lunar distance supported by the six cited distances. So, the inference from the moon-test to identify the force maintaining the moon in its orbit with inverse-square gravity toward the earth does not depend on what Newton calls the hypothesis that the earth is at rest.

Our least-squares assessment did not take into account the correction factor that Newton introduced to offset what he suggested would be the action of the sun to produce apsidal motion of the moon's orbit. This shows that the rather decisive agreement between seconds pendulums and inverse-square adjustments of the centripetal acceleration exhibited in the lunar orbit did not depend on applying Newton's correction factor.

That the agreement resulted from taking the mean for all six cited distance estimates shows that Newton's selection of the specific distance estimate of $(R/r) = 60$, rather than the mean, was also not needed for his argument to identify the force maintaining the moon in its orbit with inverse-square gravity toward the earth.

2. *Resiliency*

The agreement exhibited in Table 4.3 suggested that the agreeing cruder moon-test estimates back up the sharper pendulum estimates by affording more evidence against widely diverging estimates than is provided by the pendulum estimates alone. Our simple least-squares assessment adds some details to this suggestion, but to assess it more completely we need to be able to combine these two sorts of estimates. Already in his pioneering work of 1809 on least-squares assessment, Gauss provided an appropriately weighted mean to extend his treatment to combine estimates of differing accuracy.[45] Let us use the standard deviations exhibited by the two given sets of data as estimates of the variances of the two sources to combine these moon-test estimates with the pendulum estimates.[46]

[44] For the six cited lunar distances R in earth radii r, $x = R/r$, we have a mean of 60.04 earth radii r, with an *SE* of $0.225r$, for 95% Student's *t*-confidence bounds of $0.58r$.

[45] See Gauss 1963, pp. 250–1.

[46] The appropriately weighted mean is given by

$$(1/((n/\varsigma^2) + (m/\tau^2))) \times [(n \times y/\varsigma^2) + (m \times z/\tau^2)],$$

where n and m are respectively the number of pendulum and moon-test estimates, ς and τ are their respective variances, and y and z are their respective means. We are using the standard deviations in the sets of data to represent the variances expected with repetitions of measurements from our two sources.

In the following diagram, the x axis represents values of a hypothetical new pendulum estimate of the one second's fall. The straight line of constant slope represents the new value of the mean resulting from adding this to the pendulum estimates. All the data are considered to be equally reliable and only the pendulum estimates are taken into account. The horizontal straight line $f(x) = 15.097$ represents the weighted-mean resulting from combining the cited pendulum estimates with the cited moon-test estimates. The curved line represents the new weighted-mean resulting from adding a new pendulum estimate of value x. We are using the standard deviations of the bodies of data considered as estimates of the expected variances used in the weighted-mean calculation for combining them. In addition to moving the mean of the pendulum estimates, a new divergent pendulum estimate will also contribute to increase the standard deviation of the resulting set of pendulum estimates. Taking this new increased standard deviation of the pendulum estimates as the new estimate of the variance to be expected from such estimates contributes to decrease the weight assigned to them in computing the appropriate weighted-mean resulting from combining them with the moon-test estimates.

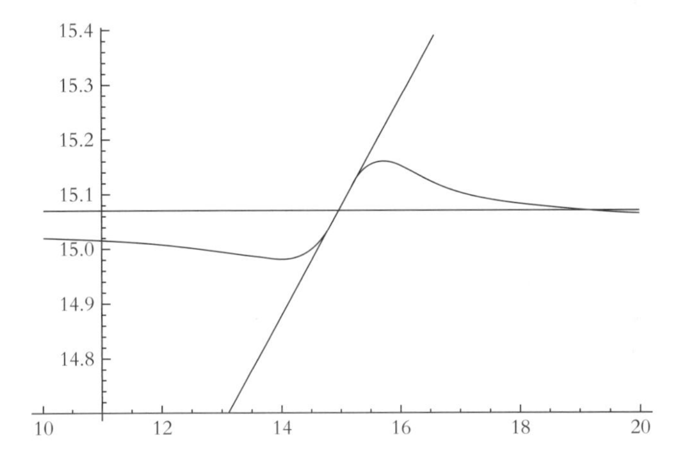

Figure 4.1 Empirical support added to Paris pendulum estimates from the agreeing moon-test estimates

This increased resistance to large changes, or increased *resiliency*, represents an important sort of increased empirical support.[47]

We have suggested that the simplicity achieved by this application of Newton's Rule 1 and Rule 2,

I am grateful to Wayne Myrvold, who independently derived this weighted-mean formula from a Bayesian maximum likelihood calculation. We later discovered that Gauss had provided an equivalent weighting formula for combining data of differing accuracies.

[47] These informative graphs were first generated for me by Myrvold. He has argued that a Bayesian account that takes into account how priors would be modified by shifting to the conditional epistemic conditional probability on hypothetical new estimates would be even more informative.

No more causes of natural things should be admitted than are both true and sufficient to explain their phenomena. (C&W, 794)

Therefore, the causes assigned to natural effects of the same kind must be, so far as possible, the same. (C&W, 795)

is that special simplicity that comes from having the unified phenomena count as agreeing measurements of the common cause assigned to explain them. These rules endorse accepting the inference to count the length of a seconds pendulum and the centripetal acceleration exhibited by the moon's orbit as agreeing measurements of their common cause, so far as possible. The data cited by Newton exhibits their agreement as measurements of d. To accept the moon-test inference is to count these phenomena as agreeing measurements of the strength of the earth's gravity, identified with an inverse-square acceleration field extending out from the surface of the earth.

V The lunar precession problem

1. *Newton's treatment of the motion of the moon*

Let us briefly review Newton's treatment of tidal action of the sun to perturb the earth-moon system. In proposition 25 of book 3, Newton resolves the action of the sun on the moon into non-orthogonal components, one along the radial earth-moon axis and the other parallel to the earth-sun axis (C&W, 839–40). He shows that the ratio of the mean quantity of this radial component of the sun's action to perturb the motion of the moon to the centripetal force by which the moon would revolve in its orbit about an earth at rest is as the square of the ratio of the periodic time of the moon about the earth to that of the earth about the sun (C&W, 840). This is the square of the ratio of $27^d7^h43^m$ to $365^d6^h9^m$, and equals the ratio 1/178.725 he assumes as the mean reduction of the centripetal force on the moon corresponding to the action of the sun to perturb the motion of the moon.[48] This radial component, however, is directed toward the earth rather than away from it. Moreover, because these components are non-orthogonal, the non-radial component also contributes to the total radially directed force.

In proposition 26, Newton resolves the action of the sun to perturb the moon's motion into orthogonal radial and transverse components (C&W, 841–3). In proposition 28, he shows that at syzygies the radial component of the tidal action of the sun draws the moon away from the earth and has the ratio 2/178.725 to the basic accelerative force maintaining the moon in its orbit, and that at quadratures this radial component is directed toward the earth and has the ratio 1/178.725 to that force (C&W, 844). Let us take positive as the radial direction away from the earth. The

[48] See chapter 3 note 41. The lunar period cited by Newton, 27 days 7 hours 43 minutes, is 27.3215 decimal days. His cited period of the earth about the sun is 365.2565 decimal days. This gives $(27.3215/365.2565)^2 = (.0748)^2 = .0056 = 1/178.725$.

mean between +2(1/178.725) and −1(1/178.725) equals +(½)(1/178.725), which exactly corresponds to setting $c = 1/357.45$ in corollary 2 of proposition 45 book 1.

George Smith (C&W, 258–9) and Curtis Wilson (2001, 143–8) point out that it follows from Newton's construction in proposition 26 book 3 that the radial component of the sun's tidal action varies sinusoidally over a synodic lunar period.[49] Such variation makes it correct to take the mean between the extremes as the average over a synodic orbit of the ratio of the radial component of the sun's tidal action to the basic force maintaining the moon in its orbit.[50]

In propositions 26–29 book 3, Newton developed his account of the perturbations which would result from the tidal action of the sun on a concentric circular lunar orbit in the plane of the ecliptic – the earth's orbit of the sun. The resulting variational orbit makes the moon's distance least in the syzygies and most in the quadratures in a ratio of about 69 to 70.[51] In proposition 30, Newton begins investigating the motion of the nodes due to the tidal action of the sun on a concentric circular lunar orbit, where the plane of this lunar orbit is inclined to the ecliptic. The nodes are the two locations where the moon will be in line with the ecliptic. In proposition 31, this investigation is extended to elliptical orbits approximating the variational orbit. In these elliptical orbits the earth is at the center, with the major axis through the quadratures and minor axis through the syzygies. In propositions 32 and 33, Newton extends this account to find the mean motion (proposition 32) and the true motion (proposition 33) of the nodes of the moon.[52] In propositions 34 and 35, this account is further extended to find the inclination of the lunar orbit to the plane of the ecliptic at any given time. Notably absent is any comparably detailed treatment of the eccentricity and apsidal motion of the moon's orbit.

[49] Smith, taking the angle θ between the moon and sun (as we did above), gives what comes to

$$(1/2)(1/178.725)(1 + 3\cos 2\theta)$$

as the variation with θ of the ratio of the radial component of the sun's tidal action to the accelerative gravity maintaining the moon in its orbit (C&W, 259). Wilson, taking the angle α of the moon from the quadrature before conjunction with the sun (as does Newton in prop. 26 bk. 3), gives

$$(1/2)(1/178.725)(1 - 3\cos 2\alpha)$$

for the variation of this ratio with α. These are, of course, equivalent. They agree in giving +2(1/178.725) at syzygies ($\theta = 0$ or 180 degrees and $\alpha = 90$ or 270 degrees) and in giving −1(1/178.725) at quadratures ($\theta = 90$ or 270 degrees and $\alpha = 0$ or 180 degrees).

[50] Suppose we integrate the θ-formula for the average of this ratio as θ goes from 0 to 2π radians = 360 degrees. We get

$$(1/2)(1/178.725)(1/2\pi)\int_0^{2\pi}(1 + 3\cos 2\theta)d\theta = (1/2)(1/178.725).$$

This is exactly the 1/357.45 Newton cited as c in his example in corollary 2 of proposition 45 book 1.

[51] C&W, 846.

[52] In the third edition a scholium was added to proposition 33, which gives an agreeing alternative calculation of the mean and true motions of the nodes of the moon by the brilliant J. Machin, Gresham Professor of Astronomy. (C&W, 860–4)

In the first edition, Newton included the following as a scholium to proposition 35 book 3.

Scholium

Up to now no consideration has been taken of the motions of the moon insofar as the eccentricity of the orbit is concerned. By similar computations, I found that the apogee, when it is in conjunction with or in opposition to the sun, moves forward 23′ each day with respect to the fixed stars but, when it is in the quadratures, regresses about 16⅓′ each day and that its mean annual motion is about 40°. By the astronomical tables which the distinguished Flamsteed adapted to the hypothesis of Horrocks, the apogee moves forward in its syzygies with a daily motion of 24′28″ but regresses in the quadratures with a daily motion of 20′12″ and is carried forward with a mean annual motion of 40°41′. The difference between the daily forward motion of the apogee in its syzygies and the daily regressive motion in its quadratures is 4′16″ by the tables but 6⅔′ by our computation, which we suspect ought to be attributed to a fault in the tables. But we do not think our computation is exact enough either. For by means of a certain calculation the daily forward motion of the apogee in its syzygies and the daily regressive motion in its quadratures came out a little greater. But it seems preferable not to give the computations, since they are too complicated and encumbered by approximations and not exact enough. (C&W, 869; Newton 1687, 462–3)

The computation referred to appears to be one from Newton's mathematical papers tentatively dated by Whiteside as of late 1686.[53] Whiteside describes this computation as an "adeptly fudged" attempt to refine the calculation based on corollary 2 proposition 45, by taking into account both the radial and transverse components of the sun's disturbing force.[54] The resulting account of the lunar precession is a version of Horrocks's geometrical model, with some modifications motivated by Newton's theory of gravity, rather than a detailed dynamical account of how the precession is due to the tidal action of the sun on the earth-moon system.[55] A later version of this Newtonian modification of a "Horrocksian" theory was published by David Gregory in 1702.[56]

Corollaries 7–9 of proposition 66 book 1 provide a dynamical account for such qualitative features of the above specified apsidal motion as apsidal advance near syzygies and regress near quadratures, with the advance greater for an average precession forward. A reader familiar with these impressive corollaries of proposition 66 and Newton's very impressive detailed quantitative dynamical account of the variation and the motion of the nodes of the moon might have expected proposition 35 to be followed up by a similarly detailed treatment of the apsidal motion. Instead, a reader of

[53] Math Papers vol. 6, 508–37.

[54] Math Papers vol. 6, 509, 518.

[55] Wilson 2001 and Nauenberg 2001 give recent assessments of what Newton achieved dynamically in such calculations.

Kollerstrom 2000 argues that the predictive model corresponding to Newton's modified version of Horrocks's geometrical lunar theory was, contrary to received opinion, quite accurate. (See note 63 below)

[56] A separate publication in a pamphlet in English titled *Sir Isaac Newton's Theory of the Moon's Motion* was published in English shortly after a Latin version had appeared in David Gregory's *Astronomical Elements* in 1702 (Kollerstrom 2000, 113).

the first edition would find only the above scholium describing the outcome of calculations declared to be too complex and encumbered with approximations to be included. In the second and third editions Newton provided a quite different scholium, which is even vaguer on details of how the apsidal motion is to be accounted for.

Nicholas Kollerstrom (2000, 165–82) has argued that this later scholium, together with the account of the variational orbit given in proposition 29, provides a summary of the steps needed to implement the sort of "Horrocksian" model used in the above calculation. As Kollerstrom points out, Newton's remarks introducing the paragraphs in which these steps are developed suggest that they can be supported by the action of the sun on the earth-moon system. The fifth paragraph begins as follows:

> By the same theory of gravity the apogee of the moon advances as much as possible when it is either in conjunction with the sun or in opposition, and regresses when it is in quadrature with the sun. And the eccentricity becomes greatest in the first case and least in the second, by book 1, prop. 66, corols. 7, 8, and 9. And these inequalities, by the same corollaries, are very great and generate the principal equation of the apogee, which I shall call the semiannual. And the greatest semiannual equation is roughly 12°18′, as far as I could gather from observations. Our fellow countryman Horrocks was the first to propose that the moon revolves in an ellipse around the earth, which is set in its lower focus. Halley placed the center of the ellipse in an epicycle, whose center revolves uniformly around the earth. And from the motion of this epicycle there arise the inequalities (mentioned above) in the advance and retrogression of the apogee and in the magnitude of the eccentricity. (C&W, 871)

Kollerstrom argues that the revisions introduced by Halley improve on the above-mentioned 1702 version of this model of the moon's motion and that they were both more accurate than the earlier version developed by Flamsteed.[57]

In the third edition Newton added his, apparently grudging, admission, "The apsis of the moon is about twice as swift", to the end of his example where *c* was set at 1/357.45 in corollary 2 proposition 45 book 1.[58] This may suggest that he thought that the correct account of the sun's action was to apply this precession theorem with that value of *c*. Newton's remarks in the scholium to proposition 35, however, indicate that he was aware that the motion of the lunar apogee is quite non-uniform. It progresses when the moon is in syzygies and regresses when in quadratures and does so in good approximation to the foregoing Horrocksian theories. This suggests that neither his simplified model corresponding to

[57] Based on comparison of the predictions of computer reconstructions of such models with predictions based on modern ephemeris calculations, Kollerstrom finds accuracies in arcminutes of ± 6.5 for Flamsteed, ± 3.8 for Newton's of 1702, and ± 2.2 for Halley's version of 1722 (Kollerstrom 2000, 227).

[58] This remark appears in Newton's annotated copy and in his annotated and interleaved copy of the second edition (Koyré and Cohen 1972, 242; C&W, 545). The latter also contains

> Query: Can this motion arise with twice the external force?

This query would invite a reader to put 2/357.45 = 1/178.725 as *c* in corollary 2 proposition 45. As we have seen, if this were correct it would account to fair approximation for all of the lunar precession. Newton did not choose to include this query in the published third edition.

a reduction of 0.0056, nor even his more realistic model corresponding to 0.0028, would be adequate to represent the action of the sun to generate motion of the lunar apogee.

Newton may have been impressed with the fact that the larger correction, 1.0056, would lead to such an extraordinarily precise agreement with Huygens's measurement in the moon-test. Here we have two independently established quantities: a reduction that would make corollary 2 of proposition 45 book 1 yield approximately the net precession of the lunar apogee, and a reduction such that the corresponding augmentation makes the moon-test agree with Huygens's measurement when the lunar distance is taken to be the 60 terrestrial radii assumed in the moon-test. The smaller correction factor (1.0028) corresponding to $c = 1/357.45$ yields a result that differs from Huygens's measurement by almost fifteen times the difference resulting from the correction factor (1.0056) recommended by Newton for the moon-test.[59]

Such an appeal to the more precise agreement with Huygens as grounds for preferring Newton's correction factor, 1.0056, over the alternative 1.0028 is, however, undercut by the large uncertainty resulting from the different lunar distance estimates. The precision of this agreement resulting from the correction factor 1.0056 is an artifact of Newton's selection of 60 terrestrial radii as the lunar distance. A modern least-squares assessment will show that it does not afford support for applying this larger correction factor.[60]

Independently of whether or not Newton was impressed with this correlation, it is clear that he did not regard his failure to produce a detailed account as reason to doubt that the lunar precession is due to the action of the sun to perturb the earth-moon system. According to Waff,

> In any case, it seems certain that Newton's sole reaction to the apogee motion discrepancy was to question the accuracy of his Corollary II calculation. Nowhere, it would appear, did he question any of the physical conditions which he had assumed in his calculation. Newton's conservatism in this regard may have been due to his perception that he was only a pioneer – indeed, the pioneer – in the theoretical calculation of perturbation effects; he may have thought that his successors would surely find new and more accurate means of mathematically studying such phenomena, and would consequently be able to account theoretically for the entire observed mean motion of the moon's apogee. (Waff 1976, 48–9)

Newton seems to have put the difficulty in accounting for the lunar precession as a problem in working out the mathematical details of the sun's action on the earth-moon system.

We have noted that Newton had good reason to believe that the calculation of corollary 2 was not adequate to accurately represent the action of the sun to produce

[59] The smaller correction factor 1.0028, yields 1.0028(15.009) = 15.051. This differs from Huygens's by 15.096−15.051 = .045 Paris feet. This is 0.045/0.003 = 14.99 times greater than the difference resulting from the larger correction factor.

[60] See chapter 6 section I.1 below.

lunar precession. The tremendous success in quantitatively accounting for the moon's variation and motion of the lunar nodes as due to the action of the sun, together with the striking qualitative account of how the action of the sun accords with the inequalities used in the foregoing "Horrocksian" models, which do fairly accurately represent such motion, may well strongly suggest that a more adequate account of the sun's action would, indeed, be able to generate the motion of the moon's apogee.

2. *Clairaut, d'Alembert, and Euler*

In the 1740s Euler, Clairaut, and d'Alembert applied differential equations to calculate the sun's action to produce precession of the lunar orbit. In their initial calculations all three obtained only about half of the lunar precession.[61] Curtis Wilson has pointed out that techniques for using differential equations to algebraically manipulate expressions for gravitational actions provided these attempts with much richer mathematical resources than those available to Newton.[62] In contrast to Newton, these later mathematicians had good grounds for believing that results from applying equations of the sort they had developed would represent the sun's action to produce precession of the lunar orbit with considerable accuracy.

Wilson offers a very clear account of the results attained by Clairaut.[63] These indicated that the resulting precession would be given by a basic formula for a rotating ellipse together with three terms proportional to three constants which were known to be quite small. In his initial calculation, where the contributions of these terms were ignored, Clairaut obtained only about half of the observed lunar precession. Euler and d'Alembert, independently making similar assumptions about contributions of such small terms, also obtained results accounting for only about half of the observed lunar precession. Euler gave the following account of his result:

> Because at first having supposed, that the forces as much from the Earth as from the Sun, which act on the Moon, are perfectly proportional reciprocally to the squares of the distances, I have always found the motion of the apogee almost two times slower than the observations mark it; and although several small terms, which I have been obliged to neglect in the calculation, may be able to accelerate the motion of the apogee, I have ascertained after several researches, that they would be unable by far to make up for this lack, and that it is absolutely necessary, that the forces by which the Moon is at the present time solicited, are a little different from the ones, which I have supposed; because the least difference in the soliciting forces produces a very considerable one in the motion of the apogee. (Waff 1976, 56–7)

This quote illustrates the confidence shared by all three in the accuracy of their calculations.

[61] See Wilson 2001, 172–82 and 2002, 213–15; Waff 1976, 50–93; Smith 1999b, in C&W, 257–64.
[62] Wilson 2001, 168–72.
[63] Wilson 2002, 214–15.

This confidence in their calculations led to significant doubts about the adequacy of Newton's inverse-square law to account for the lunar precession. This strongly reinforced other worries Euler had about inverse-square gravitation.[64] Clairaut was led to propose an alternative force law in which an inverse fourth power term was added to the basic inverse-square law. This would give the missing half of the apogee motion of the moon, while maintaining quite good agreement with the inverse-square at the greater distances of the planets from the sun.[65] D'Alembert considered an alternative hypothesis about the shape of the moon,[66] as well as one about the existence of another force, perhaps magnetic, operating between the earth and moon.[67]

Clairaut eventually undertook the rather arduous higher-order calculation required to give a detailed account of the contributions to the lunar precession of the small terms he had ignored in his earlier result. He may well have started out expecting that this more refined calculation would add to his case against the adequacy of inverse-square gravity to account for the lunar precession, by explicitly showing that the terms he had ignored would not account for the rest of the observed lunar precession.[68] By 1749, Clairaut's more detailed calculation had led him to discover that properly taking into account the contributions of these small terms would make Newton's inverse-square gravity account for the lunar precession.[69]

Euler's reaction is expressed in the following quotation from a letter to Clairaut of June 29, 1751:

> ... the more I consider this happy discovery, the more important it seems to me ... For it is very certain that it is only since this discovery that one can regard the law of attraction reciprocally proportional to the squares of the distances as solidly established; and on this depends the entire theory of astronomy. (Waff 1995, 46)

The strength of evidence Euler took this result to give for the inverse-square is even more vividly expressed in the following quotation from his prize-winning memoir of 1752:

> ... because M. Clairaut has made the important discovery that the movement of the apogee of the Moon is perfectly in accord with the Newtonian hypothesis ..., there no longer remains the least doubt about this proportion ... If the calculations that one claims to have drawn from this theory are not found to be in good agreement with the observations, one will always be justified in doubting the correctness of the calculations, rather than the truth of the theory. (Waff 1995, 46)

[64] See Waff, 50–64. We shall consider another of Euler's worries below in chapter 6.

[65] Waff 1976, 74.

[66] We see only the one face directed toward the earth, so it might be a quite elongated body we see end-on. See Waff, 87–9.

[67] Waff 1976, 90–1.

[68] Waff 1976, 214.

[69] Waff 1976, 214–15. Boulos (1999) contains an English translation, as well as an account of details, of this important paper of Clairaut's.

It appears that by the time of these letters the evidence for inverse-square gravitation toward the earth afforded by accounting for all the lunar precession, as a perturbation due to the action of the sun, was taken by Euler as sufficient to override any remaining other worries he may have had about the inverse-square.[70]

[70] As we shall see in chapter 9, Euler continued to have doubts about Newton's application of his third Law of Motion to count the gravitation of the sun toward a planet as the equal and opposite reaction to that planet's inverse-square gravitation toward the sun.

5

Christiaan Huygens: A Great Natural Philosopher Who Measured Gravity and an Illuminating Foil for Newton on Method

Part I gives background on Huygens that will be exploited to illuminate Newton's method. Section I.1 gives background on Huygens's *theory-mediated* measurement. It points out the impressive stability exhibited in such measurements of the strength of gravity at Paris. It also includes an account of Huygens's argument for trusting these *theory-mediated* measurements over more direct estimates from falling body experiments. Section I.2 reviews Huygens's reaction to pendulum measurements exhibiting different outcomes for the strengths of gravity at different latitudes. Huygens concluded that the rotation of the earth does make accurate pendulum measurements vary with latitude.

Section I.3 begins by pointing out Huygens's application of rotation to give an empirical argument that the earth is oblate rather than spherical. It then points out the specific oblateness Newton arrived at from his theory of universal gravity and the smaller specific oblateness Huygens argued for. We quote Huygens's rejection of universal gravity for violating the mechanical philosophy. We then call attention to work by George Smith and Eric Schliesser which shows that Huygens offered an interesting empirical challenge to support his theory over Newton's, based on differences over the amount of oblateness together with differences for the variation of gravity with latitude. This section closes with a brief description of the more famous challenge to Newton's theory based on measurements of the lengths of degrees of latitude that suggested the earth is prolate rather than oblate.

Section I.4 reviews Huygens's statement of his mechanical philosophy and gives an account of his proposal for a contact action cause of gravity. Attention to details of Huygens's mechanical account shows that it successfully employs *theory-mediated* measurements and is far more sophisticated than Descartes's vortex theory. Section I.5 reviews Huygens's comments on Newton's moon-test inference and his inferences to inverse-square gravity toward the sun and planets. We shall see that Huygens accepts these inferences and proposes that an inverse-square version of his proposed cause of gravity would account for such an attraction toward the earth at least out to the orbit of the moon and toward the sun at least out to the orbit of Saturn. Such a version of Huygens's theory would generate inverse-square acceleration fields toward the earth and toward the sun out to those respective distances.

Part II explores some of the lessons on scientific method from using Huygens as a foil for Newton. Section II.1 extends our exploration of the moon-test as a realization of Newton's ideal of empirical success as agreeing measurements of the same parameter from diverse phenomena. We shall see that the agreeing moon-test estimates afford increased resiliency to the pendulum estimates, even when the pendulum estimates from different latitudes are included. The combined list of the moon-test estimates, Paris pendulum estimates, and the pendulum estimates from different latitudes will make it clear that the moon-test estimates are irrelevant to the relatively small differences between the pendulum estimates from different locations. It will also make clear that the moon-test does not depend on limiting the pendulum estimates to the very precise measurements from seconds pendulums at Paris. These lessons are vividly reinforced and further illuminated by the weighted-mean calculation for the result of combining the moon-test estimates with all of these cited pendulum estimates from various locations.

Section II.2 explores the contrast between Newton's extension of inverse-square gravity toward the earth to arbitrarily great distances and Huygens's appeal to his hypothesis as to the cause of gravity to limit such extensions. Newton and Huygens both reject extending the inverse-square variation of gravity to distances below the surface of the earth. Grounds for their agreement on this suggest what, according to Newton's method, Huygens would need to back up his rejection of extending inverse-square terrestrial gravity very far beyond the orbit of the moon. This use of Huygens as a foil will help illuminate our detailed discussion in chapter 7 of Newton's third and fourth rules for reasoning in natural philosophy.

I Huygens on gravity and Newton's inferences

1. *Huygens's measurement*

Huygens's determination of the distance a body would fall in one second was a major achievement for *theory-mediated* measurement. He was able to use his theory of pendulum motion to achieve a remarkably precise measurement of the strength of surface gravity at Paris.

Both Newton and Huygens use the distance of vertical fall in the first second in the absence of air resistance to represent the strength of terrestrial gravity. Let d_{sec} represent this distance of fall in one second. Their measurements of d_{sec} in Paris feet are equivalent to measurements in Paris feet/sec^2 of $(1/2)g$, where g is our modern parameter representing the acceleration of terrestrial gravity.[1]

[1] $d_{sec} = (1/2)g$ (one second)^2

Huygens published his classic book *Horologium Oscillatorium* (*The Pendulum Clock*) in 1673.[2] Here is the opening passage of proposition 26 of part IV:

Proposition XXVI

How to define the space which is crossed by a heavy body falling perpendicularly in a given time.

Those who have previously studied this measure agree that it is necessary to consult experiments. But the experiments which have been conducted so far do not easily give an exact determination of this measure, because of the speed which the falling body has acquired at the end of its motion. (Blackwell 1986, 170–1)

Earlier attempts to measure the distance a body would fall in one second were plagued by the great difficulty of coordinating estimates of times and distances. For example, in 1636 Mersenne criticized Galileo for figures from which it would follow that a body would fall about 6½ Paris feet in one second[3] and reported that his own experiments gave 12 Paris feet as the distance a body would fall in one second.[4] In 1645 Riccioli, in Bologna, obtained about 13⅔ Paris feet.[5]

Huygens's theory of pendulum motion[6] made it possible to carry out an indirect *theory-mediated* measurement of this one second's fall, d_{sec}, from the more accurately determinable length, l_{sec}, of a seconds pendulum.[7] Huygens's value for l_{sec} at Paris was 440.5 Paris lines. A line is 1/12 of an inch. George Smith has pointed out that the largest value put forward over the next 140 years was 440.67 lines.[8] This value, which included corrections, differs from Huygens's by 0.17 Paris lines. This is less than 0.4 millimeters.[9] The measurements made of gravity at Paris from 1660 to 1800 fell in the range from 980.7 to 981.1 cm/sec^2.[10] Huygens's measurement of 15.096 Paris feet for d_{sec} corresponds to measuring

[2] Blackwell 1986 is a translation into English with an introductory essay by Bos.

[3] George Smith (class lecture notes) has pointed out that the overly small value from Galileo's figures results from ignoring rolling resistance by equating the motion of rolling balls in Galileo's inclined plane experiments with idealized frictionless sliding.

[4] See Koyré 1968, 98–102.

[5] Koyré (1968, 108) gives Riccioli's result as 15 Roman feet. According to Smith (1997), this is about 13⅔ Paris feet.

[6] The equation $p = 2\pi(l/g)^{1/2}$, where p is the period, l is the length, and g is the acceleration of gravity at the surface of the earth, expresses, in modern notation, the relation underlying Huygens's result for idealized pendulums (Blackwell 1986, xxiv).

[7] His theory gives

$$d_{sec} = \pi^2(l_{sec}/2),$$

where d_{sec} is the distance a body dropped near the surface of the earth would freely fall in the first second in the absence of air resistance and l_{sec} is the length of a seconds pendulum.
A seconds pendulum, one which beats seconds, takes one second to swing each way so that its period is two seconds. This makes the equation $p = 2\pi(l/g)^{1/2}$ from note 6 yield

$$g = \pi^2 l_{sec} \quad \text{Paris feet/sec}^2$$

[8] Smith 1997.

[9] As we pointed out above (chpt. 4 note 30), Smith quotes from Aiton the value of 2.7069 cm per Paris inch (see Aiton 1972, note 65 on p. 89). This makes 1.7 Paris lines about 0.38 millimeters.

[10] Smith 1997.

$$g = 30.192 \text{ Paris feet/sec}^2 = 32.176\text{ft/sec}^2 = 980.72\text{cm/sec}^2$$

The modern value for g at Paris is 980.9435cm/sec^2.[11]

Huygens's invention of the pendulum clock gave him the means to accurately determine the length, l_{sec}, of a seconds pendulum.

The most appropriate clocks for this purpose are ones in which each oscillation marks off one second, or one-half second, and which have a pointer to show this. First let such a clock be regulated by observations of the fixed stars over a period of several days, using the method explained in our description of the clock. Then next to it suspend another simple pendulum, that is, one composed of a lead sphere, or some other material of constant weight, tied to a thin string, and set it in motion with a small push. Then increase or decrease the length of the string until its oscillations, over a quarter of an hour or half an hour, occur in unison with the oscillations of the adjusted clock. I said the pendulum should be set in motion with a small push because small oscillations, for example, 5 or 6 degrees, are sufficient to give equal times, but not a large number of degrees. Then measure the distance from the point of suspension to the center of the simple pendulum. (Blackwell 1986, 168)

The small arcs are required because periods of simple pendulums are not independent of arc length. For simple pendulums, large arcs result in significant differences in swing times for pendulums of the same length.[12] Huygens's later remarks indicate that the distance to be measured from the point of suspension would have to be corrected if the difference between the center of the suspended sphere and the center of oscillation is significant.[13]

The passage we have been quoting goes on to propose using seconds pendulums to define a universal standard of length.

For the case in which each oscillation marks off one second, divide this distance into three parts. Each of these is the length of an hour foot, a name we have used above. By doing all this, the hour foot can be established not only in all nations, but can also be reestablished for ages to come. Also, all other measurements of a foot can be expressed once and for all by their proportion to the hour foot, and can thus be known with certainty for posterity. For example, we have already said above that the Parisian foot is related to the hour foot as 864 to 881. In other words, if the

[11] Cited from Smith 1997.

[12] Huygens's investigation of pendulums led to his invention of the cycloidal pendulum, in which having the cord swing against curved cheeks enabled the period to be independent of the amplitude of the arc of swing. His book contains a very elegant investigation of mathematical properties of cycloidal curves (Blackwell 1986, 73–104). See also Yoder 1988, 71–89.

Huygens makes it clear that the above relation of d_{sec} to the period of a pendulum holds accurately for cycloidal pendulums and for small arc simple pendulums, but not for simple pendulums with large arcs of swing.

[13] Much of part IV of Huygens's book (Blackwell 1986, 105–72) is devoted to finding the centers of oscillation for pendulums with variously shaped suspended weights. In proposition 24, Huygens points out that the center of oscillation cannot be determined for cycloidal pendulums (166–7). This provides additional motivation for his using small arc simple pendulums to measure d_{sec}.

Parisian foot is given, then we would say that a simple pendulum, whose oscillations mark off seconds of an hour, has a length equal to three of these feet, plus eight and one-half lines. (Blackwell 1986, 168)

That Huygens made such a proposal shows he was convinced that the impressive stability of the results he had obtained for l_{sec} would continue to hold for seconds pendulums in the future.

As we have seen, Huygens's measured length (l_{sec} = 3 Paris feet and 8½ lines = 3.059 Paris feet) for a seconds pendulum yields the distance (d_{sec} = 15.096 Paris feet) that a body would freely fall in one second. This indirect, *theory-mediated*, measurement differs considerably from the estimates of 12 Paris feet and 13⅔ Paris feet obtained by Mersenne's and Riccioli's attempts to use experiments with falling bodies to establish this distance of the one second's fall by direct observation. Let us consider grounds provided by Huygens for trusting his *theory-mediated* measurement over the much lower estimates resulting from these attempts at direct observation.

Huygens was able to obtain impressively stable results over many repetitions for experimentally determining the length of a seconds pendulum. Given his background assumptions, these stable agreeing measurements of l_{sec} count as stable agreeing measurements of d_{sec}. Huygens's law of the cycloidal pendulum also allows d_{sec} to be measured by pendulums of many different lengths. The agreement of these additional measurements of d_{sec} with those from seconds pendulums counts as a convergent extension of the stability of outcomes for d_{sec} to a more general class of measurements.

Huygens also provided laws allowing d_{sec} to be determined from the period and height of a conical pendulum. A conical pendulum is one where the bob is spun in a circle so that the supporting string forms a cone with its vertex at the point of support.[14] The agreement of the measurements of d_{sec} from conical pendulums counts as convergent additional support for counting the seconds pendulum measurements as accurate.[15] Like those for simple pendulums of different lengths, the agreeing measurements of d_{sec} from conical pendulums can be extended to pendulums of different heights.

Suppose we simply combined these various sorts of estimates of d_{sec}, using Gauss's least-squares weighted-mean formula. The large difference between the direct estimates together with the very great stability exhibited in the *theory-mediated* measurements would make repetitions of the *theory-mediated* measurements quickly overwhelm the direct estimates. Our modern least-squares assessment counts exhibited stability as a

[14] For such a pendulum, $p = \pi(2h/d_{sec})^{1/2}$, where p is the period (the time for the bob to complete a circle) and h is the height of the cone formed by the moving string. This makes $d_{sec} = 2h(\pi^2/p^2)$.

[15] Smith points out that, while Huygens's derivations of both pendulum laws are based on the Galilean hypothesis that the acceleration of gravity is uniform, they differ in other theoretical background assumptions appealed to. The conical pendulum law added the v^2/r measure of what Huygens called the centrifugal force corresponding to the inertia resisting the centripetal acceleration of the circling bob (see note 28 below for Huygens's statement of this result). The key addition for a cycloidal pendulum was that the acceleration is proportional to the sine of the angle of inclination of the constrained motion. This is what simple pendulums can approximate if the arcs are sufficiently small.

measure of reliability. The impressive convergence of the *theory-mediated* measurements gives strong grounds for regarding the results from attempts at direct measurements as unreliable.

Huygens was able to provide a strikingly compelling direct empirical defense of his *theory-mediated* measurement against estimates based on direct observation. He was able to use his new resources for accurately determining pendulum swing times to help carry out his own more accurate direct measurements of d_{sec} by experiments with falling bodies.

This agrees completely with our most accurate experiments, in which the precise point of time at which the fall ends is not identified by the judgment of the eye or the ear. Neither of these is completely satisfactory; but the space crossed during a fall can be determined without any error by another method which we will try to show here.

A half oscillation of a pendulum suspended from a wall or an erect board indicates the lapse of time used by a falling body. In order that the sphere of this pendulum be released at the same moment as the lead weight which is going to fall, each is tied to, and held by, a thin string, which is cut by a flame. But before this, another thin cord has been attached to the lead weight which is to fall. Its length is such that when all of it is stretched out by the falling weight, the pendulum has not yet hit the wall. The other end of this thin cord is attached to a prepared scale made on paper or on a thin membrane. This scale is attached to the wall or board in such a way that it is easy for it to follow the cord which pulls it, and to descend in a straight line along the length of the cord, passing the place where the sphere of the pendulum will hit the wall. When the whole length of the cord has been used up, the upper part of the scale will be pulled downwards by the falling lead weight before the pendulum hits the board. The size of this part is indicated when the sphere, which has been lightly covered with soot, makes its mark on the scale which is sliding by. If the length of the cord is added to this, then one has an exact definition of the space of the fall. (Blackwell 1986, 171–2)

This experiment avoids the main source of error infecting previous attempts, the difficulty of coordinating the times and the places reached by the falling body.[16] Its outcome agrees with the seconds pendulum estimate. This provides direct evidence that the earlier results obtained by Mersenne and Riccioli were not accurate, as well as providing additional convergent support for the accuracy of the *theory-mediated* measurements of d_{sec} from pendulum experiments.

Huygens follows up his description of this experiment by remarking on air resistance.

Even though we have assumed that air resistance has no effect in these matters, the measure we have assigned to falling bodies agrees exactly with experiments. Certainly air resistance is not large

[16] The experiment described here in the *Pendulum Clock* of 1673 differs from the Mersenne-like experiment of 1659 by Huygens which is described by Koyré. Where Mersenne (Koyré 1968, 102), and the earlier Huygens experiment (108–9), use the coincidence of two sounds – that of the falling body hitting the ground and that of the bob of the timing pendulum hitting the wall, the present experiment lets the mark made on the tape by the timing pendulum directly fix the distance fallen by the weight in the time of the pendulum swing. It is worth noting that the possible systematic error introduced by the drag of the string and attached paper scale would be expected to slow it down not speed it up.

enough to make a sensible difference at the heights to which we can ascend, provided that the solid bodies are made of metal...

...What must be primarily noted is that the height could be so great, or if the height is moderate, the lightness of the projected body could be so great, that air resistance would change the acceleration of the motion quite considerably from what we have proven above. (Blackwell 1986, 172)

Another advantage of the indirect seconds pendulum measurements is that they are not as much affected by air resistance as falling body experiments.

Smith points out that a correction for air resistance would add about 0.042 lines for estimates of l_{sec}.[17] This is just less than 0.1 millimeters. He points out that estimates of l_{sec} need an additional correction for air buoyancy. The weight of the air displaced by the bob would have to be added to recover what it would weigh in a vacuum. According to Smith, the corresponding correction would typically add 0.048 lines. If these corrections are applied to Huygens's value of 440.5 lines for l_{sec}, the value of 980.72 cm/sec^2 for g corresponding to the original estimate is transformed to 980.9 cm/sec^2. This is closer to the modern value of 980.9435cm/sec^2 for g at Paris.[18]

2. *Gravity varies with latitude*

We can think of the length of a seconds pendulum as a phenomenon. This length, l_{sec} established by Huygens, is a generalization which accurately fits a large set of data and can be expected to fit additional data of the same sort. As we have seen, this expectation was realized with considerable precision for seconds pendulums in the vicinity of Paris.

Huygens's proposal to use the length of a seconds pendulum as a universal standard of length was undermined when in 1673, the very year his *Horologium* was published, Richer reported that on his expedition to Cayenne[19] he had to shorten the length of a seconds pendulum by 1¼ lines. Richer had obtained 3 feet 8⅗ lines = 440.6 lines at Paris.[20] This gives his result for Cayenne as $l_{sec} = 440.6 - 1.25 = 439.35$ lines. This measures a corresponding one second's fall of $d_{sec} = 15.056$ Paris feet.[21] Cayenne is at latitude 4°55′ north of the equator. This is much closer to the equator than Paris, which is at about 48°50′ north latitude. Richer's measurement for Cayenne is significantly less than the measured value of $d_{sec} = 15.096$ Paris feet (15.099 for Richer) in Paris. This raises the possibility that the acceleration of gravity, g, varies with latitude

[17] Smith (1997) offers this, as what he describes as a typical, correction for l_{sec} for air resistance.

[18] Smith (1997) also gives two corrections for arc length. These are 0.038 for 2-inch and 0.085 for 3-inch arcs. The increase from 2 to 3 inches in arc length more than doubles the needed correction. If we add the correction for 2-inch arcs to the other corrections, the corresponding value for g is 980.986. The larger correction for 3-inch arcs would give a value of 981.09 for g. Comparing these corrected estimates with the modern value of g suggests that Huygens was able to keep his Paris seconds pendulum arcs quite small.

[19] Cayenne is French Guyana.

[20] C&W, 829; Smith 1997.

[21] $d_{sec} = (1/2)\pi^2\, 3.051 = 15.056$ Paris feet.

so that bodies nearer the equator fall with less acceleration than those at the latitude of Paris.

Other expeditions led to results which were not so easily fit into this pattern for the variation of g with latitude. Varin, Deshayes, and De Glos obtained $440\frac{5}{9}$ lines for l_{sec} (15.098 Paris feet for d_{sec}) at Paris and had to shorten it by 2 lines for $438\frac{5}{9}$ lines (d_{sec} = 15.029 Paris feet) for Goree[22] at latitude 14°40′ and shorten it by $2\frac{1}{18}$ lines to 438.46 lines (d_{sec} − 15.026 Paris feet) for Guadaloupe at latitude 15°00′.[23] Picard, who got the same 440.5 lines as Huygens for Paris at latitude 48°50′, found no discernable difference for Tycho's Uraniborg to the north of Paris at latitude 55°54′ and no discernable difference for Cape Cete to the south of Paris at latitude 43°24′.[24] Mouton obtained the somewhat anomalous estimate of 438.3 lines for Lyons at latitude 45°47′.[25] The value of d_{sec} corresponding to Mouton's estimate of l_{sec} is 15.02 Paris feet.

Table 5.1 gives these estimates of the one second's fall together with their latitudes[26] and lengths of seconds pendulums.

Table 5.1

		Latitude	l_{sec} Paris lines	d_{sec} Paris feet
Mouton	(Lyons)	45°47′	438.3	15.02
Varin et al.	(Guadaloupe)	15°	438.5	15.027
Varin et al.	(Goree)	14°40′	438.56	15.029
Richer	(Cayenne)	4°55′	439.35	15.056
Picard	(Cape Cete)	43°24′	440.5	15.096
Picard	(Uraniborg)	55°54′	440.5	15.096
Picard	(Paris)	48°50′	440.5	15.096
Huygens	(Paris)	48°50′	440.5	15.096
Varin et al.	(Paris)	48°50′	440.56	15.098
Richer	(Paris)	48°50′	440.6	15.099

In his *Discourse on the Cause of Gravity*, Huygens reports the following reaction to Richer's report:

There still remains one property, that until now we believed less certain, namely that bodies weigh as much in one place on the Earth as they do in another. Since this has been shown otherwise by some recently made observations, it is worth the trouble to examine its origin and its consequences.

Someone maintains to have found that a seconds pendulum is one and a quarter lines shorter in Cayenne, a country in America only four or five degrees distant from the equator, than it is in

[22] Goree is Senegal. [23] See C&W, 829; Smith 1997.
[24] *GHA 2B*, 24; Smith 1997. [25] Smith 1997.
[26] These cited latitudes are quite rough.

Paris. From this it follows that, if we take pendulums of equal length, the one in Cayenne makes its arc more slowly than the one in Paris. Supposing this is true, we can conjecture only that this would be a sure sign that weighted bodies descend more slowly in Cayenne than in France. And because this difference would cause a completely opposite effect, were it known to be attributable to the tenuity of the air, which is greater in the torrid zone, I do not see that there could be any other reason except that like bodies weigh less at the equator than in regions separated from it. (Huygens 1690, 145–6)[27]

Huygens takes quite seriously the inference from the reported observations in Cayenne to the claim that bodies weigh less at the equator than in regions separated from it. He goes on to point out that this could be caused by the rotation of the earth.

I realized, as quickly as this new phenomenon had been communicated to us, that the cause could be attributed to the daily motion of the Earth. This, being greater in each country the nearer to the equator, must produce a proportional endeavor to push bodies from the center and thus to rid them of a specific part of their gravity. It is easy, from the things explained above, to know what part this must be in bodies that are found at the equator. Having found, as we have seen, that if the Earth turned 17 times faster than it does, the centrifugal force at the equator would be equal to the total gravity of a body, it is necessary that the motion of the Earth, such as it is now, remove one part of the gravity, which would be to the entire gravity as one to the square of 17, which is to say 1/289; because the forces of bodies to move away from the center around which they turn are as the squares of their velocities, according to my Theorem Three in *Vi Centrifuga*.[28] (Huygens 1690, 146)

He then calculates the difference that this effect of the earth's rotation would have on pendulums between Paris and at the equator.

Seeing that the seconds pendulum in Paris is some 3 feet 8½ lines, it follows that the length of a pendulum on an immobile Earth, or at the pole, would be some 3 feet 9⅙ lines. Removing from that 1/289, which makes 1½ lines, we would have the length of the seconds pendulum at the equator as some 3 feet 7⅔ lines. Thus, this pendulum would be shorter than that in Paris by ⅚

[27] Passages from Huygens's *Discourse on the Cause of Gravity* are from a manuscript of a translation by Karen Bailey. This translation contains page references to Huygens's publication of this discourse on gravity as an addition to his *Treatise on Light*, in 1690.

[28] Huygens included statements of his theorems on "centrifugal force" in part V of his *Horologium*. Here are theorems II and III:

> II. If two equal bodies move with equal speed on unequal circumferences, their centrifugal forces will be inversely proportional to the diameters.
>
> III. If two equal bodies move with unequal speed on equal circumferences, but each maintains a constant motion as we wish this to be understood in these theorems, then the centrifugal force of the faster will be related to the centrifugal force of the slower as the squares of the speeds. (Blackwell 1986, 176)

These, together, are equivalent to having what Huygens calls "centrifugal forces" measured by the oppositely directed centripetal accelerations given by v^2/r for uniform circular motion, where v is the tangential velocity and r is the radius.

We don't call this measure of the inertial effect of circular motion a force.

of a line, which is a little less than what has been found at Cayenne by Mr. Richer, namely one and a quarter lines. (Huygens 1690, 149)[29]

He goes on to comment on this difference from the reported observations.

But we cannot entirely trust these first observations, the occurrence of which we do not see as in any way [disquieting] – and we can trust still less, from what I believe, in those that are said to have been made in Guadeloupe, where the shortening of the Paris pendulum has been found to be two lines. We must hope that in time we will be informed exactly of these different lengths, as much at the equator as in other regions;[30] (Huygens 1690, 149–50)

So, Huygens is convinced that l_{sec} and d_{sec} vary with latitude due to the centrifugal effect of the earth's rotation. Moreover, he does not regard the discrepancies between his calculated amounts and the reported observed differences as any sort of serious difficulty. Huygens was aware that failing to keep the swings very small could account for exaggerating the observed amounts by which l_{sec} should be shortened.[31]

3. *The shape of the earth*

Huygens calculated that the centrifugal effect of the earth's rotation would lead a weight on a string at the latitude of Paris to suspend at an angle of some 1/10 of a degree from vertical on a perfectly spherical earth.

This angle, some 1/10 of a degree, is considerable enough to cause us to believe that we would have noticed it, either in astronomical observations or in those which we make with a level. So to speak only of the latter, would it not be necessary that looking from [toward] the direction of the north, the line of the level would drop visibly below the horizon? This has certainly never been noticed, nor surely will it ever occur. And the reason for it, which is another paradox, is that the earth is not entirely spherical, but in the figure of a sphere pulled down toward the two poles, so that it would be close to an ellipse turning on its small axis. (Huygens 1690, 151–2)

[29] From his calculation that the earth would have to rotate 17 times faster than it does to have the centrifugal effect (the centripetal acceleration v^2/r) of its rotation equal the acceleration of gravity, Huygens calculates that a seconds pendulum at the equator would be $1/17^2 = 1/289$ shorter than its length at the pole. He also calculates Δl(lat)$= \sin^2(90°\text{-lat})(1/289)$ (Huygens 1690, 149). Huygens gives the latitude of Paris as $48°51'$, so for Paris $\Delta l = 1/667.4$ he gives 1/668. The length of a seconds pendulum on an immobile earth or at the pole is, therefore, $440.5 + (1/668)440.5 = 441.16$ lines, or about 3 Paris feet 9⅙ lines. This gives $441.16 - (1/289)441.16 = 439.63$ lines or about 3 feet 7⅔ lines for the l_{sec} at the equator. So the shortening from Paris is 8½ − 7⅔ = ⅚ lines.

[30] This quotation continues as follows:

> and certainly it is something that well deserves being researched with care, even if it would only be to correct, according to this theory, the motions of pendulum clocks, in order to make them serve as a measure of longitude at sea. So a clock, for example, which was accurately adjusted in Paris, being transported to some place at the equator, will be slowed down around one minute and 5 seconds in 24 hours, as is easy to calculate following the preceding reasoning. And likewise in proportion for each different degree of latitude. (Huygens 1690, 150)

This illustrates a practical application of research, which Huygens turned into an empirical challenge to Newton's universal gravitation (as shown in Smith's impressive account of 1997).

[31] Smith (1997) points out that a seconds pendulum having a total arc of swing of just over 10 inches, corresponding to 16° total or a mere 8° of swing from vertical on each side, would produce an extra apparent excess shortening of over one line for l_{sec}.

Huygens concludes that the fact that the centrifugal effect of the earth's rotation does not result in observable differences from vertical for suspended weights shows that the earth is oblate rather than perfectly spherical.

Newton also argues, in proposition 19 book 3, that the earth is an oblate spheroid (C&W, 821–4). His ingenious argument is based on universal gravity and the earth's rotation to calculate the oblateness needed to have a channel full of water from the pole to the center of the earth balance the water in a connecting channel from the center to the equator (C&W, 823–4). This yields an oblateness where $^{230}/_{229}$ is the ratio of an equatorial radius to a polar radius (C&W, 824). In proposition 20 book 3, Newton calculated that the increase of weight for bodies on the surface in going from the equator to the poles is very nearly as the square of the sine of the latitude, so that the length l_{sec} at the equator would be 1.087 lines shorter than at Paris (C&W, 828).

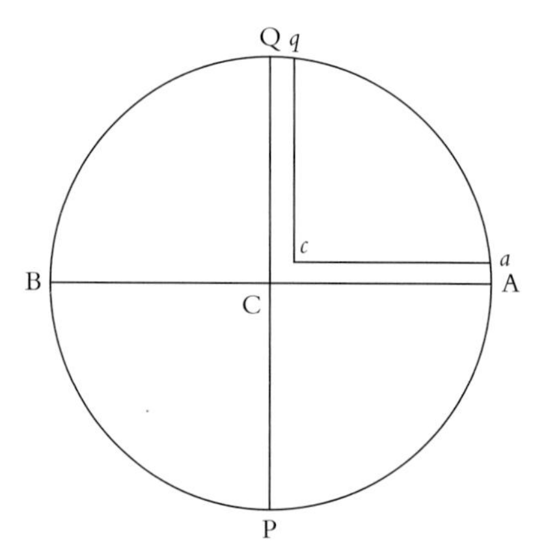

Figure 5.1 Newton's channels

After reading the first edition of Newton's *Principia*, Huygens was able to adapt Newton's oblateness calculation to the assumption that gravity is the same inside the earth as it is at the surface. This gives a ratio of $^{578}/_{577}$ for an equatorial radius to a polar radius (Huygens 1690, 156). Newton's calculation is for an earth of uniform density, for which universal gravitation requires that gravity on bodies inside the earth varies as their distance from the center. The following comment by Huygens, comparing his equatorial excess of $^{1}/_{578}$ of the basic semidiameter *EC* to Newton's,[32] highlights features of universal gravitation to which he objects.

[32] Huygens's *EC* is the equatorial radius *AC* in the above diagram. The above diagram, from Cohen's reader's guide chapter 8 (C&W, 234), is oriented so that the polar axis *QP* is vertical and the equatorial axis *BA* is horizontal, in accord with our usual custom today. Both Newton (C&W, 823) and Huygens (1690, 158) represent the equatorial axis as vertical in their diagrams.

As was found a little before, this excess is only 1/578 of EC, the semi-diameter of the Earth.

Mr. Newton came up with 1/231 of EC, and hence that the figure of the Earth differs much more from the spherical, using a completely different calculation that I will not examine here because I am not especially in agreement with a Principle that he supposes in this calculation and others, namely, that all the small parts that we can imagine in two or more different bodies attract one another or tend to approach each other mutually. This I will not concede, because I believe I see clearly that the cause of such an attraction is not explicable either by any principle of mechanics or by the laws of motion. Nor am I at all persuaded of the necessity of mutual attraction of whole bodies, having shown that, were there no Earth, bodies would not cease to tend toward a center because of what we call their gravity. (Huygens 1690, 159)

Huygens objects to Newton's extension of gravity to include a universal mutual attraction between all the small parts of bodies that we can imagine. This, he claims, is ruled out because he sees clearly that the cause of such an attraction is not explicable by any principle of mechanics or by the Laws of Motion.[33] He, also, is not persuaded of the necessity of the mutual attraction of whole bodies, since his own hypothesis about the cause of gravity would have bodies tend toward a center even if there were no earth.[34]

Huygens's rejection of mutual gravitation between the small parts of bodies led to his smaller calculated oblateness and to his smaller calculated variation in pendulum rates with latitude. For Huygens, only the differing effects of the earth's rotation made pendulum rates (and, therefore, rates of pendulum clocks) vary with latitude, while for Newton these rates were affected by variation of gravity with latitude due to oblateness as well as by rotation effects. Huygens exploited a voyage, designed to test the adequacy of his pendulum clocks for determining longitude at sea, to mount an empirical challenge to Newton's universal gravity. Given the incorrect assumption that his pendulum clock was not interfered with by the rolling of the ship, the initial results obtained on this voyage appeared to support Huygens's smaller calculated corrections of pendulum rates for latitude against Newton's.[35]

The extensions of Picard's measurement of a degree of latitude to longer arcs extending both to the south and to the north of Paris undertaken by D. Cassini and J. Cassini led to another, less informative but far more famous, empirical challenge to Newton. From the results reported in 1720, the southern arc measurement gave a longer estimate for the length of a degree than the northern arc measurement.[36] This

[33] In chapter 9 we will explore this objection to Newton's controversial application of Law 3 to gravity between Jupiter and the sun construed as an interaction at a distance between separated bodies. In section I.4 we quote Huygens's statement of his commitment to the mechanical philosophy and its rejection of non-contact interactions. For Huygens, the Laws of Motion are laws of collision. See Huygens 1669 translated by Wilson in Murray, Harper and Wilson 2011.

[34] See section 4 below.

[35] Smith 1997 provides a very fine account of Huygens's empirical challenge to universal gravity. See also Schliesser and Smith, forthcoming, *Archive for the History of the Exact Sciences*.

[36] See *GHA 2B*, 26. Todhunter offers the following description of this result:

The general result obtained is the following: from the southern arc which extended over nearly 6°19′, the length of a degree was found to be 57097 toises; from the northern arc which extended over rather more than

led them to infer, contrary to both Newton and Huygens, that the earth was a prolate (elongated at the poles) rather than an oblate (flattened at the poles) spheroid. Maupertuis's paper of 1733 promoted expeditions to Lapland and Peru to measure northern and southern meridian degrees.[37] His 1738 account of the result of his Lapland measurement reported that a northern degree was longer.[38] This inspired a redoing of the Paris measurements, which by 1740 had led to a clear consensus that the earth was oblate rather than prolate.[39] This consensus was further confirmed when results of the Peruvian expedition were finally reported in 1749.[40] Though these measurements confirmed oblateness, they, and their successors through at least the 1770s, were far too rough and divergent to give a stable empirically measured value for the amount of oblateness.[41] These results were, therefore, unable to clearly distinguish between Newton's theory and Huygens's proposal for a cause of gravity.

4. *Huygens's hypothesis as to the cause of gravity*

For Huygens, as for other proponents of the Mechanical Philosophy – the dominant approach to Natural Philosophy at this time – applications of principles of mechanics in natural philosophy are restricted to making natural phenomena intelligible by showing how they could be caused by the different magnitudes, figures, and motions of bodies.[42] He explicitly rejects any appeal to a quality or tendencies of bodies to draw near each other.

> In order to find an intelligible cause of gravity we must see how gravity is possible – supposing that we take into account in the case of bodies in nature that are made from a like matter, neither a quality nor a tendency to draw near one another, but only the different magnitudes, figures, and motions. How might it yet be possible that several of these bodies tend directly toward a common center, and are held together around it? This is the most ordinary and important phenomenon of what we call gravity. (Huygens 1690, 129)

Huygens proposes to answer this question by providing such a mechanically intelligible cause for this phenomenon of gravity.

2°12′, the length of a degree was found to be 56960 toises. This was considered to make it sufficiently evident that the length of a degree of the meridian must diminish from the equator to the pole. (Todhunter 1962, 56)

[37] *GHA 2B*, 27.

[38] Ibid.

[39] Ibid.

[40] Ibid.

[41] *GHA 2B*, 27–34.

[42] The confidence in the possibility of confirmations that establish very high probability in Huygens's statement of his hypothetico-deductive method (see chpt. 1 sec. VII) represents a departure from the basic Cartesian version of the mechanical philosophy. As we have seen when discussing Newton's contrasting *vera causa* condition in his Rule 1 (chpt. 4 sec II.2), the aim of the mechanical philosophy as stated by Descartes was to make motion phenomena intelligible by finding qualitative analogies that suggested possible hypotheses that would show how the phenomena could be explained by contact action on bodies. This departure makes Huygens's statement of method more like today's philosophers of science than like Descartes.

He begins with an experiment which shows that such tending of bodies toward a common center about which they are held together can arise from the centrifugal tendencies of other bodies in constrained circular motion about that center.

We can see this effect in an experiment that I have done expressly for the purpose, which surely merits being remarked on because it reveals to the eye a likeness of gravity. I took a cylindrical vessel, around eight or ten inches in diameter, the base of which was clean and solid, with a height only a half or a third of its width. Having filled it with water, I threw in some crushed Spanish beeswax which, being just a bit heavier than the water, goes to the bottom. I then covered it with a piece of glass laid directly on the water, which I attached all the way around with some cement so that nothing could escape. Having arranged it thus, I placed this vessel in the middle of the round table of which I spoke a little earlier and, causing it to turn, I saw immediately that the bits of Spanish wax, which were touching the bottom and following more the motion of the vessel than that of the water, gathered all around the sides, the reason being that they have more force than the water to move away from the center. Having continued to turn the vessel with the table for a little while, causing the water to acquire more and more circular motion, I suddenly stopped the table. At that instant all the Spanish wax fell into the center in a pile, which represented to me the effect of gravity. The reason for this was that the water, notwithstanding the vessel's rest, still maintained its circular motion and consequently its endeavor to move away from the center, whereas the Spanish wax had lost its motion, or very nearly so, on account of touching the bottom of the stopped vessel. I noticed also that this powder returned to the center in spiral lines because the water still carried it a bit. But if we arrange some body in this vessel so that it cannot fully follow the motion of the water but can only go toward the center, it will then be pushed entirely straight. For example, if L is a small ball that can roll freely on the bottom between threads AA, BB, and a third one a little higher, KK, held horizontally through the middle of the vessel, we see immediately that if the motion of the vessel is stopped, the ball will go to the center D. And we must notice that we can make the body L of the same gravity as the water in this last experiment and it will succeed even better; so that, without any difference in the gravity of the bodies that are in the vessel, the motion alone produces the effect here. (Huygens 1690, 132–3)

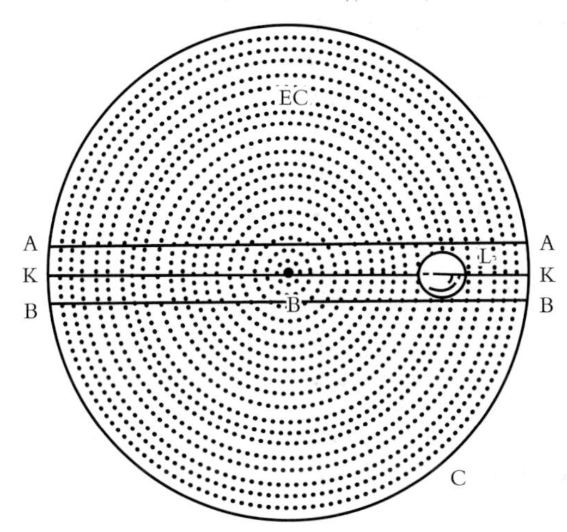

So, if the body is not carried along with the circular motion it will have less centrifugal effect than the water. The constrained circling water will, therefore, drive it to the center.

Huygens makes his hypothesis recover the phenomenon of weight toward a center by having many different very small layers or shells of vortical particles swirling in all different directions.

Thus, in order to explain gravity of the sort that I envisage here, I will suppose that in a spherical space, which includes the Earth and the bodies that surround it up to a great distance, there is a fluid matter that consists of very small parts and that is diversely agitated in all directions with great speed. Since the matter cannot leave this space, which is surrounded by other bodies, I say that its motion must become in part circular around the center; yet not so much that it begins to turn all in the same direction, but in such a way that most of these different motions would be made on spherical surfaces around the center of said space, which in this case would also be the center of the Earth. (Huygens 1690, 135)

The centripetal tendencies imparted by the actions of these very tiny layers of vortical fluid matter on the parts of a body will add together to produce its weight.

It is not difficult now to explain how gravity is produced by this motion. If within the fluid matter that turns in the space that we have supposed, the circular motion encounters some parts much greater than those that compose this matter, or some bodies formed by an amassing of small parts hooked together, these bodies will necessarily be pushed towards the center of motion, since they do not follow the rapid motion of the aforementioned matter; and if there is a sufficient amount, they will form the terrestrial globe, supposing that the Earth was not yet formed. The reason is the same as the one reported in the experiment above, which showed that the Spanish wax amassed in the center of the vessel. This then is in all likelihood what the gravity of bodies truly consists of: we can say that this is the endeavor that causes the fluid matter, which turns circularly around the center of the Earth in all directions, to move away from the center and to push in its place bodies that do not follow this motion.

Now, the reason why heavy bodies that we see descend in the air do not follow the spherical motion of the fluid matter is obvious enough. Since this motion is directed toward all its sides, the impulses that a body receives follow so suddenly one upon the other that less time passes than would be necessary for it to acquire a detectable motion. (Huygens 1690, 137)

So, while the centripetal tendencies add up to produce the weight of a body, the transverse tendencies imparted by the actions of these very tiny layers of vortical particles will cancel out.

As we have noted, one fundamental feature of the centripetal force of terrestrial gravity is that bodies at equal distances from the center of the earth have weights equally proportional to their quantities of matter. This is the property that it generates equal component accelerations on all such equally distant bodies. This is the feature that makes the inverse-square centripetal forces maintaining the planets and moons of Jupiter and Saturn in their respective orbits count as what Stein has called centripetal fields of acceleration. Here is a passage in which Newton describes this feature as a problem for mechanical causes of gravity.

Indeed, this force arises from some cause that penetrates as far as the centers of the sun and planets without any diminution of its power to act, and that acts not in proportion to the quantity of the *surfaces* of the particles on which it acts (as mechanical causes are wont to do) but in proportion to the quantity of *solid* matter, . . . (C&W, 943)

In his classic paper "Newton and the Reality of Force," Andrew Janiak emphasizes this as a reason that Newton could rule out any mechanical cause of gravity while still leaving open possibilities of causes that would avoid action at a distance by local action due to subtle matter.[43]

The "subtle matter" could not be *mechanical* in Newton's sense, *but* would have to flow through material bodies, interacting somehow with their *masses*. Similarly, it would have to exhibit differential densities, or some other feature that renders the *distance* between masses their salient relation. As Newton knew, no physical account of gravity of this period – including those of Huygens and Leibniz – was remotely capable of accounting for these aspects of gravity. (Janiak 2007, 145)[44]

Huygens certainly thought his theory would be able to account for the proportionality of weight to mass. Here are two of the topic headings from his *Discourse on the Cause of Gravity*:

That the matter that causes gravity passes through pores of all known bodies

That the gravities of bodies maintain the same proportion as the quantities of matter that compose them.

The parts of his hypothesized fluid matter need to be small enough to penetrate through bodies to interact with the particles which compose them.

Huygens points out that such extreme smallness is already needed to account for the notable effect that enclosing bodies in containers doesn't change their weight.

The extreme smallness of the parts of our fluid matter is still absolutely necessary to provide a reason for one notable effect of gravity, namely that heavy bodies, enclosed on all sides in a vessel of glass, metal, or any other material that exists, are always found to weigh equally. So it must be that the matter that we have said is the cause of gravity passes very easily through all bodies that we regard to be the most solid, and with the same ease as passing through the air. (Huygens 1690, 139)

He points out that on his account, the weight of a body is proportional to the total volume of the particles which make it up.

[43] We agree with Janiak's main point that Newton is consistent in appealing to the possibility of such causes to claim that his controversial applications of Law 3 to spatially separated bodies do not commit his theory to action at a distance. In chapter 9 we will point out that what would be needed to implement such possibilities corresponding to suggestions made by Newton in his scholium to proposition 69 book 1 seem to be, indeed, possible.

[44] Janiak (2008, 78) repeats this contention.

This matter then passes easily through the interstices of the particles that compose the bodies, but not through the particles themselves; and this causes the various gravities [weights], for example, of rocks, metal, etc. This is because the heavier of these bodies contain more of such particles, not in number but in volume: for only in their place [only in the places unoccupied by the particles] is the fluid matter able to rise. (Huygens 1690, 139)

If the particles are all equally solid (have the same density) then the weights of bodies will correspond to the amounts of matter that comprise them. Contrary to Janiak's suggestion, it appears that such a version of Huygens's hypothesis would be able to recover weight proportional to quantity of matter.

Huygens makes the question of the solidity of the particles an empirical one.

But because one could doubt if these particles, being impenetrable to the fluid matter, are as such entirely solid (because not being solid, or equally being empty, they would still have to cause the same effect, for the reason I just stated), I will show that they have this perfect solidity and that consequently the gravities of bodies correspond precisely to the proportion of matter that composes them. (Huygens 1690, 139–40)

He argues that it is answered by what experience shows about collisions.

To this end, I will point out what occurs during the impact of two bodies when they meet in horizontal motion. It is certain that the resistance that causes bodies to be moved horizontally, as a ball of marble or lead placed on a very level table would be, is not caused by their weight toward the Earth, since the lateral motion does not draw them away from the Earth, and so is not at all contrary to the action of gravity that pushes them down.

There is nothing then in the quantity of matter attached together contained in each body that produces this resistance. So, if two bodies each contain as much matter as the other, they will reflect equally, or both will remain completely motionless, depending on whether they are hard or soft. But experience shows that every time two bodies reflect equally in this way or stop one another, having come to meet with equal velocities, these bodies are of equal gravity. It follows then from this that those bodies that are composed of equal quantities of matter are also of equal gravity. (Huygens 1690, 140)

It is empirically supported by pendulum experiments which confirm the rules given in Huygens's and Wren's 1668–9 reports on laws of collisions.[45] These experiments measure the equal weights of bodies shown to have equal quantities of matter by their equal reactions when having come together horizontally with equal velocities.

If these passages were written before Huygens had read Newton's *Principia*, then these somewhat cruder earlier pendulum experiments cited by Huygens afforded measurements that anticipated the measurements of the equality of ratios of weight to mass of terrestrial bodies afforded by the more sophisticated pendulum experiments Newton reports in proposition 6 book 3.

[45] See Murray, Harper, and Wilson (2011) for translations by Curtis Wilson of these famous reports by Wallis, Wren, and Huygens on laws of collisions, together with comments by Murray, Harper, and Wilson.

Huygens used his measurement of l_{sec} and his theory of circular motion to compute the velocity of the fluid matter needed to cause terrestrial gravity. He calculates this velocity to be about 17 times greater than the velocity corresponding to the rotation of the earth at the equator.[46] In chapter 8 we shall consider Huygens's calculation of the awesome velocity,

[46] Here is his calculation:

> "But I find with my theory of circular motion, which agrees perfectly with experience, that if we claim that the endeavor of a body turning in a circle to move away from the center exactly equals the endeavor of its gravity alone, it must make each turn in as much time as a pendulum of the length of the semidiameter of this circle takes to make two arcs. We must then determine in how much time a pendulum of the length of a semidiameter of the Earth would make these two arcs. This is easy, by the known property of pendulums, and by the length of those that beat seconds, which is 3 feet 8½ lines, measured in Paris. I find it would be necessary for these two vibrations to be 1 hour 24½ minutes, supposing, according to the exact measurement of Mr. Picard, that the semidiameter of the Earth is some 19,615,800 [Paris] feet by the same measure. The velocity then of the fluid matter near the surface of the Earth must be equal to that of a body that would make the circuit of the Earth in this time of 1 hour 24½ minutes. This velocity is very nearly 17 times greater than that of a point below the equator, which makes the same revolution in relation to the fixed stars as we must take here, in 23 hours 56 minutes." (Huygens 1690, 143)

Huygens's measurement of 3 Paris feet 8½ lines for l_{sec} gives 15.096 Paris feet for $d_{sec} = \pi^2(l_{sec}/2)$. His basic theorem of pendulums gives the period p in seconds of a pendulum of any length l in feet corresponding to uniform vertical acceleration $g = 2d_{sec}$ Paris feet/sec^2 of gravity near the surface of the earth

$$p = 2\pi(l/g)^{1/2} = 2\pi(l/2d_{sec})^{1/2}.$$

Taking l to be equal to the 19,615,800 Paris feet for the semidiameter of a spherical earth according to Picard's measurement of the length of a degree of longitude, gives

$$2\pi(19,615,800/30.192)^{1/2} = 5064.5 \text{ seconds},$$

which to the nearest half-minute is 1 hour 24½ minutes, as the period of such a pendulum. Huygens shows that this is the time to complete a revolution of the earth required to have the "centrifugal force" of such motion on a circumference of a circle equal the weight of a body. The velocity corresponding to this traversing of the length of the circumference of the earth in about 1 hour 24.5 minutes is

$$23.933/1.408 = 16.99 \text{ times greater},$$

which is, indeed, very nearly 17 times greater than the velocity which would traverse this circumference in the time, 23 hours 56 minutes, of one revolution of the earth with respect to the fixed stars.

Here is proposition X of Huygens's *Theorems on Centrifugal Force Arising from Circular Motion*:

> If a body moves on a circumference, and if each revolution is completed in the same time in which a pendulum, whose length is the radius of the circle, completes a very small revolution of a conical motion or completes two, very small, lateral oscillations, then it will have a centrifugal force equal to its weight. (Blackwell 1986, 177)

It is the result he appeals to, here.

Huygens's measure of "centrifugal force" for a body moving uniformly on the circumference of a circle is equal in magnitude to the oppositely directed centripetal acceleration

$$v^2/r = (2\pi r/p)^2/r = 4\pi^2 r/p^2$$

corresponding to such uniform circular motion. His calculation is equivalent to setting this centripetal acceleration equal to the acceleration of gravity

$$g = 2d_{sec} \text{ Paris feet/sec}^2 = 30.192 \text{ Paris feet/sec}^2$$

corresponding to his measured one second's fall d_{sec} of 15.096 Paris feet for bodies near the surface of the earth.

49 times greater than what we have found near the Earth, which was already 17 times greater than the velocity of a point at the equator (Huygens 1690, 168),

needed to account for the surface gravity of the sun. These are clear cases of successful measurement from phenomena of parameters of Huygens's proposed cause of gravity.

5. Huygens on Newton's inferences to inverse-square gravity

In a passage quoted above, Huygens stated that he objects to the universal mutual attraction between all the small parts of matter and, also, to such mutual attraction between whole bodies. This passage is followed by a paragraph giving his reasons for agreeing with Newton's inferences to inverse-square centripetal forces, extending at least from the sun to the planets and from the earth to the moon.

I have nothing against *Vis Centripeta*, as Mr. Newton calls it, which causes the planets to weigh (or gravitate) toward the Sun, and the Moon toward the Earth, but here I remain in agreement without difficulty because not only do we know through experience that there is such a manner of attraction or impulse in nature, but also that it is explained by the laws of motion, as we have seen in what I wrote above on gravity. Because nothing hinders the action of this *Vis Centripeta* toward the Sun, it would be similar to what pushes bodies that we call heavy to descend toward the Earth. I thought for a long time that the spherical figure of the Sun could be produced by the same thing that, according to me, produces the sphericity of the Earth, but I had not extended the action of gravity to such great distances as from the Sun to the planets or from the Earth to the Moon, because the vortices of Mr. Descartes, which formerly appeared very likely to me, and which I still had in mind, came to a crossroad [cut across it]. I had not thought at all about the regular diminution of gravity, namely that it is in inverse proportion to the squares of distances from the center. This is a new and quite remarkable property of gravity, the basis of which is well worth the trouble of investigating. But seeing now from the demonstrations of Mr. Newton that, if one supposes such a gravity towards the Sun that diminishes according to said proportion, it counterbalances the centrifugal force of the planets so well and produces exactly the effect of elliptical motion that Kepler had predicted and verified by observations, I can scarcely doubt that these hypotheses concerning gravity would be true, or that the system of Mr. Newton, insofar as it is founded thereupon, would likewise be true. (Huygens 1690, 160–1)

Huygens accepts Newton's attribution of an inverse-square diminution with distance as a new and quite remarkable discovery about gravity. Investigating the basis for such inverse-square gravitation toward the earth and the sun is declared to be well worth doing. Seeing Newton's demonstration that such inverse-square gravity toward the sun would so well balance the "centrifugal force" of the planets and would produce Kepler's elliptical motion convinces him that the hypothesis of such inverse-square gravity is true. Huygens tells us that he can scarcely doubt that Newton's system, insofar as it is founded upon such inferences to inverse-square gravity, would likewise be true.

The paragraph continues by outlining how such inverse-square gravitation is supported by its solution to problems for Descartes's vortex theory, in which the planets are carried around the sun by huge vortices, like chips of wood in a whirlpool.

This should appear so much the more probable as we find in it the solution of several difficulties that are a problem for the vortices supposed by Descartes. We see now how the eccentricities of the planets are able to remain constantly the same; why the planes of the orbits do not join together, but retain their different inclinations with respect to the plane of the ecliptic; and why the planes of all these orbits necessarily pass through the Sun. We see how the motion of planets can accelerate and decelerate to the extents that we observe, which could occur in this way with difficulty if they floated in a vortex around the Sun. Finally, we see how comets can pass through our system. For, while we know that they often enter in the region of the planets, we had some difficulty imagining how they could sometimes go in a motion contrary to that of the vortex that had enough force to carry the planets. But this doubt is also removed with the doctrine of Mr. Newton, since nothing prevents the comets from traveling in elliptical paths around the Sun, like the planets, but in more extended paths, and in a figure more different from circular, so that these bodies have their own periodic revolutions, as certain ancient and modern philosophers and astronomers had imagined. (Huygens 1690, 161–2)

Huygens points out that these features of the motions of solar system bodies, which make Descartes's vortexes problematic, can all be naturally accounted for by an inverse-square gravity toward the sun.

In a later passage, Huygens describes the moon-test as providing additional support for the inverse-square variation of the gravities of the planets toward the sun, as well as for the gravities of satellites toward their planets.

We have seen how in Mr. Newton's system the gravities, as much of the Planets toward the Sun as of the satellites towards their planets, are supposed to be in inverse-square proportion to their distances from the center of their orbits. This is confirmed admirably by what he demonstrates concerning the Moon, namely that its centrifugal force, which its motion gives it, precisely equals its gravity toward the Earth, and so these two contrary forces hold it suspended where it is. Because the distance from here to the moon is 60 semi-diameters of the Earth, and therefore the gravity in its region is 1/3600 of what we feel, it is necessary that the centrifugal force of a body, which is moved like the Moon, would likewise equal 1/3600 of the weight it would have at the surface of the Earth. (Huygens 1690, 165–6)

Huygens takes the moon-test to show that gravity toward the earth varies inversely as the squares of distances from its center. As we shall see below, he infers that this inverse-square variation holds for distances ranging from a little beyond the vicinity of the earth's surface out to at least the distance of the moon. Huygens also takes the moon-test to show that it is this inverse-square gravity toward the earth which maintains the moon in its orbit by balancing what he calls the "centrifugal force" of the circular motion about the earth corresponding to its orbit.

Neither Huygens nor Newton regard the questions they raise about the non-spherical shape of the earth or the variation of surface gravity with latitude as undermining the inference from the moon-test to the inverse-square variation with distance from the center of the earth for gravity above the region of the earth's surface.[47] Both, also, regard the moon-test as decisively establishing that the moon is maintained in its orbit by inverse-square gravity toward the earth. Like Newton, Huygens sees these moon-test inferences and the inferences to inverse-square centripetal forces maintaining planets in their orbits about the sun, as well as to such forces maintaining satellites of Jupiter and Saturn in their orbits of these planets, as mutually supporting one another.

Huygens builds into his explanation of the cause of gravity that the weights toward the center of the earth it produces on bodies are proportional to their quantities of matter. This ensures that, in absence of air resistance, freely falling bodies will be equally accelerated by gravity. He requires that gravity toward the earth be inverse-square with distance from the center of the earth for distances ranging from a little above the vicinity of the surface of the earth out to at least the distance of the moon. This requires that his hypothesized system of rapidly moving fluid spheres recover, for at least that range of distances, Newton's measurable quantities for an inverse-square centripetal force directed toward the center of the earth. In particular, it will produce equal gravitational accelerations on bodies at any equal distances from the center of the earth, in that range, and so will have for locations in this region about the earth accelerative measures that vary inversely as the squares of the distances from the center. So, Huygens's hypothesized system of rapidly moving spherical shells is to be construed so as to produce an inverse-square centripetal acceleration field for distances ranging from a little above the vicinity of the surface of the earth out to at least the distance of the moon.

II Lessons on scientific method

Huygens can be a very useful foil for helping us understand what we can learn from Newton's method. One difference is the way that Huygens's commitment to the mechanical philosophy prevented him from following Newton in the later inferences

[47] In a corollary added to proposition 37 book 3 in the second edition, Newton carries out a moon-test calculation that takes into account the variation of gravity with latitude, the rotation and oblate shape of the earth and the location of Paris on it, and the two-body correction, as well as the basic distortion of a circular orbit that accounts for the variational inequality of the moon's mean motion. This calculation is formulated as an argument for the lunar distance of 60.4 terrestrial semidiameters, specified in its opening sentence. To the extent that this specified distance is compatible with distance estimates from lunar observations, it shows that Newton's moon-test inference continues to hold up when some of the simplifying assumptions of the basic calculation are replaced by more realistic approximation. A detailed account of this more complex moon-test calculation is given in the appendix of chapter 6.

to universal gravity. We shall continue our exploration of Huygens as a foil for Newton's method when we discuss these inferences in our discussion of Newton's arguments for propositions 7 and 8 of book 3 in chapter 9 below. It will be suggested that Newton was as sensitive to the oddness of the idea of action-at-a-distance as Huygens and the other mechanical philosophers, and as eager to provide a mechanical explanation of gravity. His genius was not that he did not share their metaphysical qualms, but rather that he did not let failure to give a mechanical explanation undercut his inferences from phenomena to universal gravity.

Here in this chapter it will be useful to focus on two contrasts. One is the richer notion of empirical success that contrasts Newton's method from Huygens's commitment to the hypothetico-deductive identification of empirical success as prediction. The other is the contrast between Newton's commitment to extend inverse-square gravitation toward the earth to arbitrarily great distances beyond the moon and Huygens's refusal to extend it very much beyond the moon.

1. *Empirical success and resilience*

In chapter 1 we quoted Huygens's statement of hypothetico-deductive method with its clear identification of empirical success with accurate prediction. It was argued that the agreement between the moon-test estimates and the seconds pendulum estimates of the strength of the earth's gravity counts as an example of Newton's stronger conception of empirical success. The agreement of the cruder moon-test estimates empirically backs up the more precise pendulum estimates. One important way it does this is that the moon-test estimates provide more empirical support for resisting large changes than would be provided by the cited pendulum data alone. We illustrated this agreement by combining the cited moon-test estimates and the cited Paris seconds pendulum estimates in a single table. As we pointed out in the introduction, these numbers suggest that, though the moon-test estimates may be irrelevant to small differences from Huygens's estimate, the agreement of the cruder moon-test estimates of the strength of gravity affords increased resistance to large changes from that estimate.

In chapter 4 we argued that an application of modern least-squares assessment can reveal empirical advantages of multiple agreeing measurements that are not made explicit by just listing them. We used Gauss's method of appropriately weighted means to combine estimates of differing reliability to exhibit the increased resiliency afforded by adding the moon-test estimates to the four cited pendulum estimates from experiments at Paris. The resulting graph clearly reinforced the suggestion that the moon-test estimates would afford increased resistance to large differences from Huygens's measurement, while being irrelevant to small differences.

Now let us consider the result of adding the pendulum estimates from other latitudes together with the moon-test estimates.

Table 5.2

Moon-test			Seconds pendulum
Ptolemy	14.271		
Vendelin	15.009		
Huygens	15.009		
		15.02	Mouton (Lyons)
		15.027	Varin et al. (Guadaloupe)
		15.029	Varin et al. (Goree)
		15.055	Richer (Cayenne)
		15.096	Picard (Cape Cete)
		15.096	Picard (Uraniborg)
		15.096	Picard (Paris)
		15.096	Huygens (Paris)
		15.098	Varin et al. (Paris)
		15.099	Richer (Paris)
Copernicus	15.261		
Street	15.311		
Tycho	15.387		

The mean of the ten seconds pendulum estimates is 15.0712 Paris feet with an *sd* of 0.0325 Paris feet.[48] The differences among the estimates from different latitudes are reflected in this *sd* which is twenty-five times more than the 0.0013 Paris feet of the *sd* of the Paris pendulum estimates.[49] These differences among pendulum estimates from different latitudes are, nevertheless, still a very great deal smaller than those among the moon-test estimates. We saw above that the *sd* for the moon-test estimates is about 0.37 Paris feet.[50]

We can carry out the weighted-mean calculation for the result of combining the moon-test estimates with all ten of these cited pendulum estimates from various locations. The result is a weighted-mean of 15.071 Paris feet for the one second's fall. As in Figure 4.1 of chapter 4, here in Figure 5.2, the horizontal straight line represents this result. The straight line of constant slope represents the new mean resulting from adding a hypothetical new pendulum estimate of value x to the pendulum estimates. The smaller slope of this line, compared with that of its counterpart in Figure 4.1, represents increased resiliency afforded by the larger number of pendulum estimates. As before, the curved line represents the weighted-mean resulting

[48] A standard way of representing their clustering about it is by their standard deviation (*sd*), the square root of the average of the squares of their differences from the mean. The *sd* of these pendulum estimates is 0.0325 Paris feet.

[49] See chapter 4 note 42.

[50] See chapter 4 note 43.

from combining the new pendulum estimate with the cited pendulum estimates together with the moon-test estimates.[51] Here again, the range of agreeing overlap shows that for small differences the moon-test data are irrelevant.

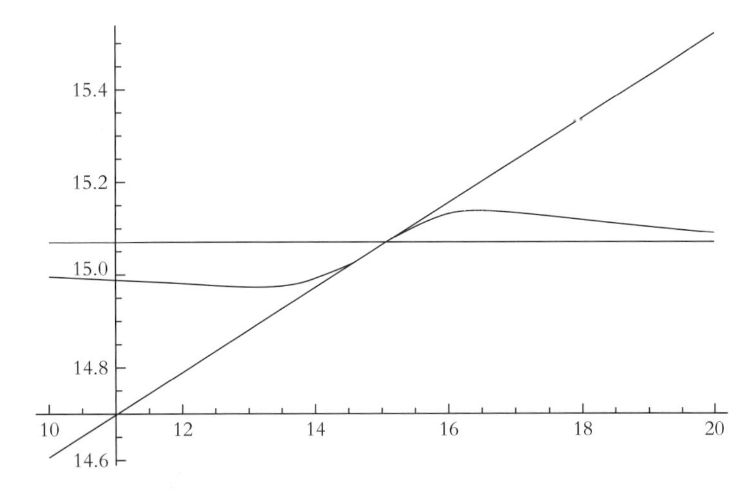

Figure 5.2 Empirical support added to Paris estimate + estimates from other latitudes from moon-test estimates of edition 3

Our discussion has stressed the resiliency resulting from agreeing measurements by diverse phenomena, as exemplified in combining pendulum estimates with the cited moon-test estimates of the strength of gravity. The moon-test estimates increase the resiliency of the resistance to large changes in estimates of parameter values. Making a different phenomenon count as an agreeing measurement of the same theoretical parameter makes available another whole set of data, together with the resources available from researchers working on that other phenomenon for generating new data.

Another advantage, illustrated here, is exploiting divergences as sources of research for refining one's model. A really important advantage of making seconds pendulum measurements at different latitudes, over just making a lot more repetitions of the measurement at Paris, was finding the small but systematic differences that led to research into variation with latitude and then to the investigation of the shape of the earth.

2. *Extending inverse-square gravity*

The phrase "so far as possible" in Newton's Rule 2 reflects a major difference between him and Huygens. Newton takes the inverse-square variation of this acceleration field as the *law* of the variation of gravitation toward the earth with distance to its center. For him, any distance beyond the vicinity of the surface, however far, counts as a distance

[51] See weighted-mean formula in chapter 4 note 46.

to which it is possible to extend these agreeing measurements of inverse-square adjusted acceleration of gravity toward the earth.[52] Huygens accepts the moon-test as showing that this acceleration field extends at least out to the moon, but his hypothesis as to the cause of gravity leads him to conclude that it must end somewhere beyond. Newton's method endorses extending the inverse-square variation out to arbitrarily great distances, unless there are boundaries or regions beyond which such indefinitely continued extension of the measured agreement is not possible.

To see what might count as showing such impossibility of indefinitely continued extension, let us consider the hypothetical extension of such inverse-square variation below the surface of the earth. Huygens and Newton both reject this extension. They would reject it, for example, for a freely falling body in an empty canal from the center to the surface of the sort appealed to in Newton's oblateness calculation. They reject counting the inverse-square adjustment of g measured by the length of a seconds pendulum and the inverse-square adjustment of the centripetal acceleration measured by the lunar orbit as agreeing measurements of the acceleration a freely falling body below the surface would have.[53] According to Newton's universal gravity, these two phenomena count as giving agreeing measurements of acceleration varying directly with distance from the center for distances from the center smaller than the distance to the surface.[54] On Huygens's hypothesis, the alternatives measured by these two phenomena have the acceleration constant and equal to that at the surface for bodies below the surface, so that their measurements of the acceleration of gravity for bodies at the surface give the same measured values for differing distances below the surface.[55]

In the case of extending the inverse-square adjustments as agreeing measurements for distances smaller than r, Huygens and Newton both provide alternatives that appear to do at least as well at counting the two phenomena as agreeing measurements. The evidence cited by Huygens against inverse-square surface gravity corresponding to the actual oblateness of the earth may count as empirical grounds against extending inverse-square variation to distances smaller than r in the corresponding idealized

[52] As we pointed out in the introduction, and will explore in more detail in chapter 7 section II, Newton's third rule for doing Natural Philosophy is applied to explicitly endorse this inference, when he argues in proposition 6 that all bodies, however distant they may be, gravitate toward the earth.

[53] The agreement displayed in the moon-test shows that for any arbitrary distance D, the inverse-square adjustments,

$$g(r/D)^2 \text{ and } (4\pi^2 R/t^2)(R/D)^2,$$

of the acceleration g measured by a seconds pendulum and the centripetal acceleration, $(4\pi^2 R/t^2)$ of the moon in its orbit would give correspondingly agreeing accelerations.

[54] In Newton's theory, the respective alternatives for $D < r$ are

$$g(D/r) \text{ and } (4\pi^2 R/t^2)(R/r)^2(D/r),$$

corresponding to weight varying directly as D for D smaller than the earth's radius.

[55] For Huygens, the surface accelerations g and $(4\pi^2 R/t^2)(r/R)^2$ each give the same measured values for all distances D for bodies below the surface of the earth.

model of the earth as a perfect sphere.[56] Huygens also points out a divergence problem.[57] To extend inverse-square gravity all the way to the center requires weights to approach infinity as the distances from the center approach zero. These suggest that the alternatives provided by Newton and Huygens do better than extending the inverse-square to distances below the surface.

Having an alternative that does better at counting the two phenomena as agreeing measurements of the acceleration of gravity for distances smaller than *r* would undercut extending the inverse-square variation measured by them to such distances. If Huygens could produce such an alternative that would actually do better for distances greater than the moon, then Newton's method would endorse it rather than Newton's extension of the inverse-square. In the absence of an alternative good enough to count as a serious rival for the values measured by these phenomena, or some contrary phenomenon making such extension liable to exceptions, Newton's Rule 4 would recommend extending the agreeing inverse-square measurements indefinitely.[58]

That Huygens's hypothesis as to the cause of gravity would lead one to conclude that the inverse-square variation must end somewhere beyond the moon is not a serious rival, unless it shows sufficient promise for developing contrary alternative measurements or is supported by phenomena requiring exceptions to Newton's extension. An important part of Newton's method is to push the measurements until empirical problems are actually found, and then to exploit the divergences as *theory-mediated* phenomena that can guide further research.

[56] Huygens argues that the inverse-square variation of surface gravity with these varying distances from the center of the earth would lead to an additional variation of pendulum rates with latitude that would double the slowing of these rates as latitude decreases from the poles toward the equator. This would call for even more slowing of clocks and pendulums as latitude decreases toward the equator than Newton's universal gravity. (Huygens 1690, 167)

[57] He points out:

"it would have to be said that gravity, in going toward the center, would increase to infinity, which is not likely." (Huygens 1690, 167)

[58] This is in line with what we saw in our discussion of Rule 4 in chapter 1, and will see in more detail in chapter 7, when we discuss Newton's actual introduction of Rule 4 in his argument to infer inverse-square gravity toward planets without orbiting satellites to measure it.

6

Unification and the Moon-Test: Critical Assessment

Part I investigates exaggeration in the precision of the claimed agreement with Huygens's measurement in Newton's moon-test. In section I.1 we argue that the precision exhibited in the agreement of Newton's moon-test calculation goes beyond what modern least-squares assessment can support from his cited data. A least-squares assessment also shows that the sharp agreement resulting from Newton's dubious application of the correction factor motivated by his precession theorem should not be counted as legitimate support for it. Section I.2 argues that Newton is innocent of Westfall's main accusation of data fudging in the moon-test; but, that Westfall raises questions that prompt a review of Newton's selection of data.

Section I.3 reviews Newton's moon-test argument in his original version of book 3. We include some discussion of details of this version of Newton's moon-test calculation as well as the estimates he cites. In this original version Newton cites five lunar distance estimates of 59 terrestrial radii explicitly attributed to different astronomers instead of the single estimate of 59, which he attributed to Ptolemy and most astronomers in the third edition. It also contains an estimate of $59\frac{1}{3}$ attributed to Flamsteed, an estimate of $62\frac{1}{2}$ attributed to Kircher, and assigns 61 as the correction to Tycho's estimate, instead of the $60\frac{1}{2}$ we saw assigned as the correction to Tycho. The Kircher estimate is sufficiently deviant that a least-squares assessment would suggest that including it might be counted as an illegitimate effort to offset the low estimates.

Section I.4 reviews the revisions for the moon-test in the first and second editions. In the first edition, Flamsteed's estimate is omitted and the five estimates of 59 are collapsed into a single estimate of 59 attributed to most astronomers. In the second edition, Kircher's estimate is dropped and the correction to Tycho is changed to $60\frac{1}{2}$. Section I.5 reviews the correction to Tycho. The mean of the alternative corrections considered is very close to 61. The value $60\frac{1}{2}$, to which it was changed in the second and third editions, would seem to require ignoring some of these alternatives. Section I.6 argues that such features of Newton's treatment as ignoring alternatives by relying on a single assumed value rather than the mean do not violate the standards of Newton's day. The opening remarks of an argument by Thomas Simpson make it clear that even as late as 1755, there was not yet in place a commonly accepted norm for taking the mean of a number of estimates as better than any single one of them.

In Part II we assess the extent to which empirical success is legitimately achieved in Newton's moon-test. Section II.1 argues that Newton's inference does not depend on his dubious precession correction or on his selection of which estimates to include. Newton's basic inference is still supported, even if Kircher's dubious estimate and the dubious second correction to Tycho's are dropped while keeping all the rest. Section II.2 argues that the lunar distance calculated from Huygens's measurement of g could be counted as accurate. It reviews how a new lunar distance measurement, by Le Monnier in 1748, removed doubts by Euler based on smaller estimates by Flamsteed and by Cassini in 1740. Section II.3 argues that a correction for syzygy distances can defend the larger lunar distance Newton assumes in his moon-test of corollary 7 proposition 37. This section also discusses an interesting critique by Thomas Bayes, which points out that Thomas Simpson's argument for relying on the mean assumes that the chances of errors of the same magnitude in excess or in defect are upon average nearly equal. In section II.4, the weighted-mean construction together with the syzygy correction is applied to illustrate that the increased resiliency corresponding to the empirical success of the agreeing measurements does not depend on Newton's dubious precession correction factor or on any of his various selections of which distance estimates to include. Section II.5 concludes that the empirical success of the agreeing measurements supports Newton's inference, and that nothing in Westfall's accusations of data fudging in any way undercuts the cogency of this inference to identify the force that maintains the moon in its orbit with terrestrial gravity.

Appendix 1 discusses the details of Newton's moon-test calculation from corollary 7 of proposition 37 book 3. This calculation takes into account the variation of gravity with latitude, the rotation and oblate shape of the earth and the location of Paris on it, as well as the two-body correction and the basic distortion of a circular orbit that accounts for the variational inequality of the moon's mean motion. It shows that Newton's moon-test inference continues to hold up when the simplifying assumptions of the basic calculation are replaced by more realistic approximations.

I. Exaggerated precision

1. Assessing precision and support for Newton's precession correction factor

The difference between Huygens's 15 feet 1 inch $1\frac{7}{9}$ lines for a one second's fall at the earth's surface and Newton's 15 feet 1 inch $1\frac{4}{9}$ line is only $\frac{3}{9}$ of a line. Since a line is $\frac{1}{12}$ inches, this difference is a mere 0.028 inches. The corresponding cited agreement is, therefore, to better than 3 hundredths of an inch! We now have available techniques for assessing evidential support which show that this precision is far beyond what counts as legitimate to claim on the basis of the evidence cited by Newton.

In chapter 4, we applied least-squares techniques to assess the result of carrying out Newton's basic moon-test calculation for each of the six lunar distance estimates he cited. The resulting estimate for the one second's fall at the surface of the earth was a mean value of 15.041 Paris feet, with a 95% Student's *t*-confidence interval of $\pm$ 0.43 Paris feet. The large size of these 95% error bounds, more than 5 inches, shows that the extreme precision of Newton's cited agreement with Huygens is undercut by the uncertainty introduced by the differences among his cited estimates of the lunar distance.

We can make it even more obvious that the evidence cited by Newton does not support expecting that anything at all close to an agreement within 0.03 inches with Huygens's measurement will hold for future distance estimates, of similar precision to those cited. The 50% Student's *t*-confidence estimate is 15.041 $\pm$ 0.122 Paris feet.[1] The moon-test with the six distance estimates cited by Newton, therefore, gives ground for about 50% confidence that similar future distance estimates would lead to one second's falls at least 0.067 feet = 0.8 inches greater or at least 0.177 feet = 2.1 inches less than Huygens's measurement. Our least-squares assessment, therefore, makes it clear that, even, the more modest agreement, cited to the nearest inch in the first edition,[2] goes beyond the precision that the data cited by Newton can support.

The foregoing least-squares assessment did not take into account Newton's dubious correction, which increases the one second's fall in the ratio (178.725/177.725) = 1.0056 to offset the (1/178.725) = 0.0056 reduction of the centripetal force on the moon that would correspond to setting $c = 0.0056$ in corollary 2 of proposition 45 book 1. Consider the result of applying such a least-squares assessment to the result of applying this correction to each of the one second's falls corresponding to the six cited distance estimates. The resulting 95% Student's *t*-confidence estimate is about

$$15.126 \pm 0.43 \text{ Paris feet}$$ [3]

[1] The Student's *t*-parameter for 50% confidence estimates for five degrees of freedom corresponding to $n = 6$ is 0.73 (Freedman et al. 1978, A71). This gives $\pm$ 0.73*SE* = 0.73(0.167) = 0.122 Paris feet.

[2] Newton 1687, 407.

[3] Here are the results.

Table 6.1

	x	1.0056*d*(*x*)
Ptolemy	59	14.351
Vendelin	60	15.093
Huygens	60	15.093
Copernicus	60⅓	15.346
Street	60.4	15.397
Tycho	60.5	15.474

The mean for the corrected 1.0056*d*(*x*) is 15.126 feet with an *SE* of 0.168 feet, for a 95% Student's *t*-confidence bound 2.57*SE* = 0.432 feet.

The corresponding 50% confidence estimate would be 15.126 ± 0.123 Paris feet.[4] So, even if Newton's correction factor were justified, his cited data would support expecting with 50% confidence that similar future lunar distance estimates would yield corrected one second's falls either at least 1.8 inches greater or at least 1.1 inches smaller than Huygens's measurement.[5]

Such least-squares assessments also show that the more precise fit with Huygens's measurement afforded by Newton's larger correction factor 1.0056, rather than the smaller factor 1.0028, is not significant. These same six estimates together with the correction factor 1.0028 corresponding to the alternative ratio $c = 1/357.45$ yield

$$d_{sec} = 15.084 \pm 0.43 \text{ Paris feet}$$[6]

as a 95% Student's *t*-confidence estimate. The Huygens value agrees with both of these estimates. The differences, 15.125 − 15.096 = +0.029 Paris feet for 1.0056 and 15.084 − 15.096 = −0.012 Paris feet for 1.0028, make it clear that the apparent closer agreement resulting from Newton's 1.0056 is an artifact of his selection of the specific estimate of 60 terrestrial radii as the lunar distance. The extremely sharp agreement with Huygens resulting from his correction factor 1.0056 in his moon-test calculation is, therefore, not to be legitimately counted as support for it.

2. *Westfall on fudge factors in the moon-test*

Richard Westfall has suggested that Newton fudged his data in order to go from the more modest initial correlation with Huygens, cited to the nearest inch in the first edition, to the extraordinarily precise agreement of later editions.[7] Westfall frames his

[4] The result of multiplying the *SE* of 0.168 feet (note 3) by the Student's *t* parameter of 0.73 for 50% confidence estimates for 5 degrees of freedom (note 1) is 0.1226.

[5] 15.126+0.123 = 15.249, which is 0.249−0.096 = 0.153 feet = 1.8 inches greater than Huygens's measurement of 15.096 Paris feet. 15.126−0.123 = 15.003 feet, which is 0.093 feet or 1.1 inches less than Huygens's measurement.

[6] Here are the results.

Table 6.2

	x	$1.0028d(x)$
Ptolemy	59	14.311
Vendelin	60	15.051
Huygens	60	15.051
Copernicus	60⅓	15.303
Street	60.4	15.354
Tycho	60.5	15.43

The 95% Student's *t* estimate for the corrected $d(x)$ is

$$15.083 \pm 0.430 \text{ Paris feet}$$

The 50% Student's *t* estimate is

$$15.083 \pm 0.122 \text{ Paris feet}$$

[7] See Westfall 1973, 754–5.

accusation in a context which characterizes the role of the *Principia* in establishing a new quantitative paradigm of physical science.

> The role of the *Principia* in establishing the quantitative paradigm of physical science extended well beyond its dynamic explication of accepted conclusions. Far more impressive was its success in raising quantitative science to a wholly new level of precision. (Westfall 1973, 751)

After pointing out that Galileo did not attempt to include perturbing factors in his idealized mathematical models, Westfall went on:

> In contrast, Newton enlarged the definition of science to include those very perturbations by which material phenomena diverge from the ideal patterns that had represented the object of science to an earlier age. The *Principia* submitted the perturbations themselves to quantitative analysis, and it proposed the exact correlation of theory with material event as the ultimate criterion of scientific truth. (Westfall 1973, 751)

Newton did introduce a paradigm where perturbations, divergences from idealized patterns, were themselves submitted to quantitative analysis. This, however, is a paradigm of successively better approximations. Westfall's last claim, that in the *Principia* "the exact correlation of theory with material event was proposed as the ultimate criterion of empirical truth," does not do justice to Newton's use of approximations. Newton's method is fundamentally approximative. As we saw in chapter 1 and will see in more detail in chapter 7, Newton's key statement of this method, in Rule 4 of his *Rules for Reasoning in Natural Philosophy*, is explicit in its openness to approximations.

Westfall exploits what he takes to be Newton's proposal of exact correlation as the criterion of truth to give an impressively rhetorical expression of his accusation that Newton was guilty of fudging.

> And having proposed exact correlation as the criterion of truth, it took care to see that exact correlation was presented, whether or not it was properly achieved. Not the least part of the *Principia*'s persuasiveness was its deliberate pretense to a degree of precision quite beyond its legitimate claim. If the *Principia* established the quantitative pattern of modern science, it equally suggested a less sublime truth – that no one can manipulate the fudge factor quite so effectively as the master mathematician himself. (Westfall 1973, 751–2)

Westfall claims that Newton manipulated fudge factors to present a deliberate pretence to a degree of precision beyond what he had established as legitimate to claim.

In support of this suggestion, Westfall remarks upon a calculation that Newton had included in a draft for a proposed scholium to proposition 4 book 3 he had sent to Roger Cotes, the editor of the second edition.

> While the new edition was being prepared, Newton sent the editor, Roger Cotes, a scholium to Proposition IV in which he carried out the correlation substantially as it finally appeared. Since it assumed material from Propositions XIX and XXXVII, Newton eventually broke the scholium up and inserted it in those two propositions, most of it in the latter. In that proposition the character of the finely corrected correlation is clear.

...

Corollary 7, which contains the correlation, is actually a calculation of the mean distance of the moon from the value of *g*. (Westfall 1973, 755)

Here is a more detailed description Westfall gives of this calculation:

Starting with the measured orbit of the moon, Newton calculated the distance that the attraction of gravity causes it to deviate from a rectilinear inertial path during 1 minute. This figure required correction by the amount of the sun's disturbance of the moon's orbit, measured by the motion of the moon's line of apsides, yielding a calculated distance of 14.8538067 feet. A body removed one-sixtieth as far from the earth would fall the same distance in 1 second. Since he had set the distance of the moon at 60⅖ times the maximum radius of the earth, Newton corrected the calculated fall in 1 second accordingly. He took the oblate shape of the earth into account, computing the distance of fall at the mean radius (45° latitude), and adding two-thirds parts of a line to correct for the smaller radius at the latitude of Paris. Hence at Paris, bodies would fall 15 Parisian feet, 1 inch, $4^{25}/_{33}$ lines in 1 second. The rotation of the earth gives rise to centrifugal effects which diminish the fall by 3.267 lines at Paris. Hence, by calculation from the centripetal acceleration of the moon, ½ *g* should equal 15 feet, 1 inch, 1½ lines [3][8]. Beneath its unsettling combination of decimals carried to seven places and fractions of varying complexity, the calculation purports to find a correlation accurate to a fraction of a line, a degree of precision perhaps somewhat better than 1 part in 3000. (Westfall 1973, 752)

This calculation is described as applying a number of corrections, including one for the amount of the sun's disturbance of the lunar orbit measured by the motion of the moon's line of apsides, as well as ones taking into account the oblate shape of the earth and centrifugal effects due to the earth's rotation, to yield 15 feet 1 inch, 1½ lines, an even more precise correlation with Huygens.

The calculation Westfall is describing is the one from corollary 7 of proposition 37 book 3. In a note in the last chapter,[9] we pointed out that this version of the moon-test showed that the correlation was more than just an artifact of simplifying assumptions, such as a concentric circular lunar orbit about a spherical earth taken to be at rest, built into the moon-test calculation of proposition 4. To the extent that the lunar distance argued for in it holds up, this more refined calculation shows that the agreement between values of ½ *g* measured by inverse-square adjusting the centripetal acceleration of the lunar orbit and values of ½ *g* measured by terrestrial pendulum experiments can continue to hold as simplifying assumptions are replaced by assumptions compatible with better approximations to the complexity of the actual earth–moon system.[10]

[8] Here is Westfall's note [3]:

[3] I. Newton, *Principia*, (1, pp. 407–9; 482–3). The computation of the centrifugal effect used in Proposition XXXVII is found in Proposition XIX (1, p. 425). (Westfall 1973, 758)

The "1" refers to his note 1, which specifies the Motte–Cajori translation of Newton's *Principia*.

[9] See chapter 5 note 47.

[10] A more detailed account of this more elaborate moon-test calculation will be given in appendix 1.

Here is Westfall's suggestion of how this correlation calculation supports his accusation that Newton is guilty of fudging data. According to Westfall:

> When he placed it in Corollary 7 to Proposition XXXVII, Newton did not indeed explicitly present the correlation as anything more than an exact calculation of the moon's mean distance. When he thought originally of making it a scholium to Proposition IV, however, he seemed to be trying implicitly to present it as something else. Cotes, who had caught the spirit of the enterprise rather well, saw as much. "In y^{e} Scholium of IVth Proposition," he suggested, "I think the length of y^{e} Pendulum should not be put 3 feet and 8$\frac{2}{5}$ lines: for the descent would then be 15 feet 1 inch 1$\frac{1}{3}$ line. I have considered how to make y^{t} Scholium appear to the best advantage as to y^{e} numbers, & I propose to alter it thus" [19, 20].[11] He went on to select values for the distance corresponding to 1° on the earth's surface and for the latitude of Paris that led to a very precise correlation. Cotes need not have worried. Newton was also considering how to make the correlation appear to the best advantage as to the numbers. Ultimately he settled on different values for the elements of the calculation, but by treating the distance of the moon as terminus ad quem instead of terminus a quo, he reached a correlation quite as satisfactory as Cotes'. In both calculations it was more public relations than science. (Westfall 1973, 755)

Westfall points out that in corollary 7 of proposition 37 this elaborate correlation is presented as a calculation of the lunar distance. He then insinuates that Newton, when he originally intended to include the calculation in a scholium to proposition 4, planned to support a more precise correlation with Huygens by calculating the lunar distance needed to get the agreement. This insinuation is given some support by quoting Cotes's remark,

> I have considered how to make y^{t} Scholium appear to the best advantage as to ye numbers, & I propose to alter it thus,

which certainly does suggest a deliberate effort to select numbers to make for a closer agreement with Huygens.

Westfall's accusation that Newton fudged the moon-test is his suggestion that, in order to arrive at the more precise correlation cited in the second edition, Newton appealed to the calculation that ended up in corollary 7 of proposition 37 to adjust what he took to be the lunar distance; thus, illegitimately, treating the distance of the moon as "*terminus ad quem* instead of *terminus a quo*." Westfall accuses Newton of treating the lunar distance as the end point reached rather than as the assumed starting point of this correlation calculation. The credibility of this accusation is undermined by the fact that Newton did not adjust the lunar distance assumed in his correlation calculation when revising from the first edition to the more precise correlation of the

[11] Here are Westfall's notes:

> 19. Cotes to Newton, 16 February 1711/12 (20, pp. 61–2). It transpired that the fraction 2/5 in the length of the pendulum had been a slip of Newton's pen. He had intended 3/5.
> 20. J. Edleston, Ed., *Correspondence of Sir Isaac Newton and Professor Cotes* (John W. Parker, London, 1850).

second edition. The same lunar distance, of 60 earth radii, assumed in calculating the correlation in the first edition was also the lunar distance assumed in the correlation calculations of the second and third editions.

As we have seen, Newton's more precise correlation follows from his correction factor, (178.725/177.725) = 1.0056, applied to the result of inverse-square adjusting the one minute's fall corresponding to the centripetal acceleration of the moon, together with his assumption of 60 earth radii as the lunar distance. On these assumptions, Newton's cited lunar period and his cited estimate of the circumference of the earth yield his precise value, 15 Paris feet 1 inch and $1\frac{4}{9}$ lines, for the inverse-square adjusted fall in one second at the surface of the earth corresponding to the centripetal acceleration of a concentric circular lunar orbit about a spherical earth.

The only part of the elaborate calculation described by Westfall that plays any role in generating the more precise correlation of the second and third editions is the correction factor, 1.0056. Here is what Westfall had to say about it, in his above description of the more elaborate correlation calculation that got put in corollary 7 of proposition 37.

> This figure required correction by the amount of the sun's disturbance of the moon's orbit, measured by the motion of the moon's line of apsides, yielding a calculated distance of 14.8538067 feet. (Westfall 1973, 752)

So, the only part of the elaborate calculation that plays any role in generating the incredibly precise correlation of the proposition 4 moon-test in the second and third editions is a correction Westfall describes as needed to correct for an amount of the sun's disturbance measured by the motion of the moon's line of apsides. As we have seen, and contrary to what Westfall's remark suggests, Newton's application of this correction factor is dubious. But, contrary to Westfall's accusation of fudging, it is not used to adjust the lunar distance either before or after the moon-test of the first edition.

What about the suggestive remark Westfall quoted from Cotes? Here is the passage from Cotes's letter to Newton from which Westfall quoted this remark.

> In the Scholium to the IVth Proposition I think the length of the Pendulum should not be put 3 feet & $8\frac{2}{5}$ lines; for the descent will then be 15 feet 1 inch $1\frac{1}{3}$ line. I have considered how to make that Scholium appear to the best advantage as to the numbers & I propose to alter it thus. To take 57220 Toises for the measure of a degree instead of 57230; for 57220 is the nearest round number to a mean amongst 57060, 57292, 57303. To take 3 feet $8\frac{10}{19}$ lines for the length of the Pendulum; for ye French sometimes make it $8\frac{1}{2}$, sometimes $8\frac{5}{9}$, & $8\frac{10}{19}$ is a mean betwixt these numbers. (Cotes to Newton February 16, 1711/12; *Corresp.* V, 226)

Here is Newton's response.

> Sr
>
> In the scholium to ye IVth Proposition I should have put the length of ye Pendulum in vacuo 3 feet & $8\frac{3}{5}$ lines. It was by an accidental error that I wrote $8\frac{2}{5}$ lines. The Pendulum must be

something longer in Vacuo then in Aere to vibrate seconds. You may put it either $8\frac{3}{5}$ or $8\frac{10}{19}$ as you shall think fit, the difference being inconsiderable. If you chuse $8\frac{10}{19}$, the numbers computed from thence may stand. (Newton to Cotes February 19, 1711/12; *Corresp.* V, 230)

Finally, here is Cotes's answer to this response.

Sr

I received your last. As I reviewed the XXth Proposition I perceiv'd it was by a slip of the Pen that you had put $8\frac{2}{5}$ instead of $8\frac{3}{5}$ lines in Your former Letter. I choose this number rather than $8\frac{10}{19}$ for the reason which You gave & because the fraction is more simple & already in use amongst the French. (Cotes to Newton 23 February 1711/12; *Corresp.* V, 232)

As both Cotes and Newton acknowledge, Newton's "$8\frac{2}{5}$," which Cotes said ought to be changed, was just a slip of the pen. The 3 feet $8\frac{3}{5}$ lines selected by Cotes is what Richer obtained at Paris for the length of a seconds pendulum.[12] Westfall's accusation is not supported, when Cotes's suggestive remark is seen in its proper context in this exchange of letters with Newton.[13]

The foregoing least-squares assessment does show that Newton's numbers display a precision of agreement with Huygens that, we now know, is far beyond what his cited evidence can support. We have seen that the correction factor Newton applies to achieve this is not justified. In addition, by our standards today such features of Newton's calculation as picking the single one second's fall of 15.009 Paris feet corresponding to a single assumed distance of 60 earth radii over the mean value, 15.041 feet, of the outcomes of the test for all six cited estimates of the lunar distance would be counted as questionable.

In critically assessing such assumptions by Newton we should keep in mind that our norms for evidential assessment today differ from the, rather looser, standards of Newton's day. Clearly, our appeal to least-squares assessment standards for evaluating levels of precision legitimately supported by multiple measurements of a parameter was not available to Newton. Least-squares techniques did not come into very general use until after the pioneering work of Gauss in 1809, nearly a century after the work by Newton and Cotes on revising the *Principia*.[14]

Westfall's additional discussion may suggest deception for which we might more readily hold Newton accountable.

Indeed, Proposition XXXVII serves to indicate that even the modest correlation of the first edition is somewhat deceptive. To be valid, the correlation must start with two quantities that are measured independently, the distance of the moon and *g*. Huygens had provided the latter with high precision. For the former, Newton had a range of varying measurements. Most astronomers, he said, set it at 59 radii of the earth. Vendelin set it at 60, Copernicus at $60\frac{1}{3}$, Kircher at $62\frac{1}{2}$, and Tycho at $56\frac{1}{2}$. Tycho's result was founded on a false theory of refraction, however,

[12] See above, chapter 5 section I.2.

[13] Westfall acknowledges this (in his note 19). See note 11 above.

[14] See *GHA 2B*, 203–7.

and when his observations were corrected they gave a distance of 61. From this set of numbers Newton extracted an average of 60 which yielded in turn a centripetal acceleration that correlated with a value of ½ *g* of 15 feet, 1 inch. It is difficult to believe that the value of ½ *g* did not influence his averaging of the lunar measurements. Significantly, in edition two he omitted Kircher's measure and found that the correction of Tycho's came, not to 61, but to 60½ times the radius of the earth. (Westfall 1973, 755)

Westfall quotes the distance estimates Newton cited in the first edition and suggests that the value 60, assumed by Newton, was extracted as an average. If one counts the 59 attributed to most astronomers as a single estimate, the average of these distance estimates cited in the first edition is 60.57. This would make it misleading to count the value 60 as representing the average of these cited values.

Westfall, also, describes the changes made in the second edition as significant. These changes, omitting Kircher's estimate and changing the correction to Tycho's from 61 to 60½, lead to an average of 59.96, which is considerably closer to the value 60 Newton assumed in his moon-test calculation.

3. *The moon-test in Newton's original version of book 3*

The more diffuse presentation of the moon-test in his earlier version of book 3 gives information about additional data available to Newton and about alternative ways he considered formulating the moon-test calculation. In what Cajori titles section 10 of this earlier essay version of his *System of the World*, Newton uses the moon-test to argue for the inverse-square variation of the circumterrestrial force. In section 9 he had just argued that the circumsolar force, throughout all the regions of the planets, varies as the inverse-square of the distances from the sun.

10. The circumterrestrial force decreases in the doubled ratio [that is, varies inversely as the square] of the distances from the earth. This is proved on the hypothesis that the earth is at rest.

I gather as follows that the circumterrestrial force decreases likewise in the doubled proportion [or varies inversely as the square] of the distances. The moon's mean distance from the center of the earth, as measured in terrestrial semidiameters, is 59 according to Ptolemy, Kepler in his *Ephemerides*, Boulliau, Hevelius, and Riccioli; according to Flamsteed 59⅓; according to Vendelin 60; according to Copernicus 60⅓; according to Kircher 62½; according to Tycho 56½. But Tycho and all those who follow his tables of refractions, by making the refractions of the sun and moon (entirely contrary to the nature of light) greater than those of the fixed stars – and doing so by about four or five minutes – increased the moon's parallax by the same number of minutes, that is, by about a twelfth or fifteenth of the whole parallax. If this error is corrected, the distance will come out about 61 terrestrial semidiameters, about the same as the others have made it.[15]

[15] This translation, which corresponds to the statement and first paragraph of the argument for section 10 on pages 559 and 560 of Cajori, is from a draft translation by I.B. Cohen. It differs from Cajori's translation in minor ways and in one major way. Cajori gives the correction to Tycho's estimate as 60 or 61.

Correct this error, and the distance will become 60 or 61 semidiameters of the earth, nearly agreeing with what the others have determined. (Cajori 1934, 560)

What is here claimed to be shown by the moon-test is the inverse-square variation of the circumterrestrial force. There is, therefore, precedent in this earlier argument for the appeal to the moon-test to add support to the argument for inverse-square variation in proposition 3.

The calculation for the one second's fall at the surface corresponding to the assumed lunar distance is interesting.

> Let us assume a mean distance of 60 semidiameters and a lunar period against the fixed stars completed in $27^{d}7^{h}43^{m}$, as astronomers have determined it; and (by prop. 4, corol. 6) a body revolving in our air near the surface of the earth at rest, by a centripetal force that would be to the same force at the distance of the moon inversely in the doubled ratio [that is, as the square] of the distances from the center of the earth, that is, as 3600 to 1, would complete – with the resistance of the air removed – a revolution in $1^{h}24^{m}27^{s}$. Suppose the circumference of the earth to be 123,249,600 Paris feet (as the French have recently determined by measurements); then the same body, with its circular motion removed and under the action of the same centripetal force as before, would in falling describe 15 1/12 Paris feet in one second. This is gathered by a calculation made by means of prop. 36 of book 1 and it agrees with experience. For by making experiments with pendulums and computing from those, Huygens has demonstrated that bodies descending with all that centripetal force (of whatever kind) by which they are urged near the surface of the earth describe 15 1/12 Paris feet in the time of one second.[16]

Newton uses corollary 6 of proposition 4 book 1 to calculate the period of a hypothetical small moon orbiting just above the surface of the earth from the period of the actual moon together with his assumed lunar distance of 60 earth radii, and the assumption that the forces maintaining those moons in their respective orbits are inversely as the squares of their distances from the center of the earth. This yields $t' = 84.653$ min $= 1^{h}24^{m}39^{s}$ for the orbital period of such a hypothetical small moon orbiting just above the earth.[17] Newton gives this as $1^{h}24^{m}27^{s}$.

This difference makes a difference for the claimed agreement to the nearest inch. Newton's $t' = 1^{h}24^{m}27^{s} = 5067$ seconds. It yields 15.081 Paris feet for the fall in one second corresponding to the centripetal acceleration exhibited by this small moon's orbit.[18] This is 15 feet 0.97 inches. This, to the nearest inch, is clearly 15 feet 1 inch. The correct value of t' is $1^{h}24^{m}39^{s} = 5079$ seconds. It yields 15.009 Paris feet.[19] This is

[16] The quoted passage is, for the most part, from the above cited manuscript translation by I.B. Cohen. I have, however, put as the calculation to which Newton refers proposition 36 book 1 from Cajori 560, which Newton cites above as equivalent to the arc^2/D calculation.

[17] According to corollary 6 proposition 4 book 1, having the forces be as the -2 power of the distance is equivalent to having the periods be as $R^{3/2}$. So where the radius R of the lunar orbit is taken as 60 while the radius of the orbit of the body near the surface of earth is taken as 1, we have

$$t' = t \quad (1/60^{3/2}),$$

where t' is the period of the smaller orbit and t is the cited lunar period $27^{d}7^{h}43^{m} = 39343$ min.

$$39343(1/60^{3/2}) = 84.653.$$

[18] $arc^2/D = (123{,}249{,}600/5067)^2/(123{,}249{,}600/\pi) = 15.081.$

[19] $arc^2/D = (123{,}249{,}600/5079)^2/(123{,}249{,}600/\pi) = 15.009.$

exactly what we calculated as the fall in one second corresponding to the force at the surface equal to inverse-square adjusting the force maintaining the actual moon in its orbit. As we have seen, this is only 15 feet 0.11 inches. To the nearest inch it is 15 feet 0 inches, rather than 15 feet 1 inch.

Here are the ten cited lunar distances $x = R/r$ in earth radii together with the corresponding correct moon-test estimates $d(x)$ of the one second's fall:

Table 6.3

	x	$d(x)$
Ptolemy	59	14.271
Kepler	59	...
Boulliau	59	...
Helvelius	59	...
Riccioli	59	14.271
Flamsteed	59⅓	14.514
Vendelin	60	15.009
Copernicus	60⅓	15.261
Tycho (1)	61	15.772
Kircher	62.5	16.964

The corresponding 95% Student's *t*-confidence estimates for the lunar distance in earth radii and one second's fall in Paris feet are[20]

$$59.82 \pm 0.84r \quad 14.89 \pm 0.64 \text{ Paris feet}$$

The Huygens measurement, 15.096 Paris feet, is well within the 95% Student's *t*-confidence bounds, [14.25, 15.53], afforded by the moon-test estimates of the one second's fall at the surface of the earth corresponding to these ten cited lunar distance estimates.

The result is explicitly described as proved under the hypothesis that the earth is at rest. The next section is entitled,[21]

11. The same is proved under the hypothesis that the earth moves.

In it, the two-body correction is applied. Newton estimated the lunar mass, from its mean apparent diameter of 31½ arc minutes, to be about 1/42 of the mass of the

[20] The 95% Student's *t*-confidence parameter for the nine degrees of freedom corresponding to ten estimates is 2.26 (Freedman et al. 1978, A71).

Table 6.4

	x	$d(x)$
mean	59.817	14.888
SE	0.372	0.285
2.26*SE*	0.841	0.643

[21] The translation is from the Cohen manuscript. See, also, Cajori 1934, 560.

earth.[22] This gives 60.47 or about 60½ as the two-body corrected distance.[23] The 95% Student's *t*-confidence least-squares estimate of the lunar distance corresponding to the above ten estimates Newton cited from astronomers is 59.82 ± 0.84 terrestrial radii. The two-body corrected distance, 60.47, is further from this mean than the assumed one-body distance of 60 to which it corresponds, but both are well within the 95% confidence bounds. This shows that the selection of 60 for the moon-test is not undercut by the two-body correction corresponding to Newton's estimate of the mass of the moon.

The additional estimates of the lunar distance, especially the greater detail about astronomers giving 59 as their estimate, may help illuminate some of the issues raised by Westfall's suggestions about fudging data. In this regard, note that the correction to Tycho's distance estimate is here given as 61, as in the first edition, rather than as the 60½ which was given in the second and third editions.

With five estimates of 59 as well as Flamsteed's 59⅓, and taking the correction to Tycho as 61, the mean of the ten cited distance estimates is 59.82. Newton's assumed value of 60 is quite close to this mean. Our least-squares assessment of these estimates makes Kircher's estimate of the lunar distance diverge from the mean by more than three of the corresponding 95% Student's *t*-confidence bounds.[24] This is, perhaps, sufficiently removed to be dismissed as an outlier. Given the five cited estimates of 59 by major astronomers, one may wonder whether Newton's inclusion of Kircher's estimate in the first place is, perhaps, suggestive of a concern to keep the average close to 60. Without Kircher's estimate, the mean of the nine remaining is only about 59.52.

4. *The moon-test in the first and second editions*

Here are proposition 4 book 3 and the moon-test distances cited in the first edition.

Prop. 4 *The moon gravitates toward the earth and by the force of gravity is always drawn back from rectilinear motion and kept in its orbit.*

The mean distance of the moon from the center of the earth in terrestrial semidiameters is according to most astronomers 59, according to Vendelin 60, according to Copernicus 60⅓, according to Kircher 62½, and according to Tycho is 56½. But Tycho and all those who follow his tables of refractions, by making the refractions of the sun and moon (entirely contrary to the nature of light) be greater than those of the fixed stars – in fact greater by about four or five minutes – have increased the parallax of the moon by that many minutes, that is, by about a twelfth or fifteenth of the whole parallax. Let that error be corrected, and the distance will come to be roughly 61 terrestrial semidiameters, close to the value that has been assigned to the others. (Newton 1687, 406)

[22] See Cajori 1934, 561.

[23] This gives $60\ (43/(43^2 42)^{1/3}) = 60.47$, as the two-body corrected distance R' corresponding to the assumed distance $R = 60$ earth radii. (See chpt. 4 sec. I.3 above for Newton's two-body correction.)

[24] Kircher's estimate of 62.5 is 2.68 semidiameters greater than the mean. This is $(2.68/.84) = 3.2$ of the 95% Student's *t*-confidence bounds beyond the mean.

Here is our least-squares assessment of this first edition version of the moon-test.

Table 6.5

	x	$d(x)$
most astronomers	59	14.271
Vendelin	60	15.009
Copernicus	60⅓	15.261
Tycho (1)	61	15.772
Kircher	62.5	16.964
estimate[25]	$60.57 \pm 1.62r$	15.455 ± 1.25 Paris feet

Huygens's measurement is cited as 15 1/12 Paris feet. This value 15.083 Paris feet is certainly well within the 95% confidence interval, [14.2, 16.7], for these estimates.

In this first edition, by citing both Huygens's measurement and the one second's fall computed from the moon's orbit as 15 1/12 Paris feet, the precision of the claimed agreement is limited to the nearest inch. However, even this more modest agreement is more precise than can be expected to hold for one second's falls corresponding to similar future distance estimates, on the basis of those cited. The 50% estimate corresponding to these five distances is 15.455 ± 0.33 Paris feet.[26] So we can, with 50% confidence, expect future distance estimates of similar precision to those cited to yield one second's falls either less than 15 feet 1.5 inches or greater than 15 feet 9.4 inches.

One change made by Newton in going from the original version of the moon-test to this, of the first published edition, is to collapse the explicit list of five estimates of 59 by specified astronomers to the claim that the lunar distance in terrestrial semidiameters is "according to most astronomers 59." This change has the effect of giving less weight to estimates significantly below 60. The other change is to simply omit Flamsteed's[27] estimate of 59⅓, another estimate which may be regarded as significantly below 60.

[25] The 95% Student's t-confidence parameter for the four degrees of freedom corresponding to five estimates is 2.78 (Freedman et al. 1978, A71). Here are the results for these distances cited in the first edition.

Table 6.6

	x	$d(x)$
mean	60.57	15.455
SE	0.58	0.45
2.78*SE*	1.62	1.25

[26] The 50% Student's t-parameter for the four degrees of freedom corresponding to five estimates is 0.74 (Freedman et al. 1978, A71). The *SE* for the estimate corresponding to the above five distances cited in the first edition is 0.45 (see previous note). The 50% error bound $tSE = 0.74(0.45) = 0.33$ Paris feet.

[27] These revisions in going from Newton's original version of book 3 to that of the first edition of 1687 were made well before Newton's, much discussed, falling out with Flamsteed, which led him to suppress references to Flamsteed in revisions for the second edition of 1713.

The result is a mean of 60.57, which is much less close to 60 than the mean, 59.82, of the ten lunar distances cited in the original version of the moon-test.

With the mean of the distance estimates at 60.57, Newton's two-body correction helps support his choice of 60 as the distance to assume on the assumption that the earth is at rest.

> This calculation is founded on the hypothesis that the earth is at rest. For if the earth and the moon move around the sun and in the meanwhile also revolve around their common center of gravity, then the distance of the centers of the moon and earth from each other will be roughly 60½ terrestrial semidiameters, as will be evident to anyone who computes it (according to prop. 60, bk. 1). (Newton 1687, 407)

In the first edition he gave what he called a *quam proxime* (or a rough) estimate of 1 to 26 as the ratio of the earth's mass to that of the moon.[28] This gives about 60.76 as the two-body corrected lunar distance.[29] In the above passage Newton cites this corrected distance as 60½. These are each quite a bit closer to the mean of this estimate than 60 is.[30] This makes Newton's estimate of the two-body correction help support his use of 60 as the distance in the moon-test, even though the mean of the distances cited is much closer to 60½.

Westfall has pointed out that in the second edition Newton omitted Kircher's estimate and changed the correction to Tycho from 61 to 60½. Here is our least-squares assessment of the moon-test for the distance estimates cited in the second edition.

Table 6.7

	x	$d(x)$
most astronomers	59	14.271
Vendelin	60	15.009
Copernicus	60⅓	15.261
Tycho (2)	60½	15.387
95% estimate[31]	$59.96r \pm 1.07r$	$14.98 \pm .79$ Paris feet

[28] See corollary 4 proposition 37 book 3 in his first edition (Newton 1687, 466). See notes 34–6 in chapter 4 above for more on these estimates.

[29] $60(27/(27^2 26)^{1/3}) = 60.76$, from proposition 60 book 1. See note 23 above.

[30] That these values for the two-body corrected distance are within the 95% least-squares confidence estimate, $60.57r \pm 1.62r$, corresponding to the cited lunar distance estimates by astronomers shows the important weaker claim that Newton's estimate of the mass of the moon is compatible with his assumption of $60r$ as the one-body distance in the moon-test.

[31] The Student's t-parameter for 95% confidence with the three degrees of freedom corresponding to four estimates is 3.18 (Freedman et al. 1978, A71).

Table 6.8

	x	$d(x)$
mean	59.96	14.982
SE	0.34	0.25
tSE	1.07	0.79

Here, in the second edition, the cited lunar distances average to almost exactly $60r$. As we have seen, the 95% confidence estimate ($60.57r \pm 1.62r$) from the estimates cited in the first edition is closer to $60\frac{1}{2}$ than it is to 60. The second edition estimates are also more closely grouped about 60, as indicated by their somewhat narrower confidence interval. The two changes resulting in this increased prominence to Newton's assumed value of 60 are dropping Kircher's estimate and changing the correction to Tycho from 61 to $60\frac{1}{2}$. These are changes that Westfall suggested were significant.

Kircher's lunar distance estimate of $62\frac{1}{2}$ earth radii would be counted as quite deviant by our least-squares assessment of the ten lunar distance estimates cited in Newton's original version of the moon-test. We saw that it is more than three of the corresponding 95% Student's *t*-confidence bounds away from the mean of those estimates.[32] In spite of this great divergence from the other estimates, Newton included this estimate in the first version of the moon-test. He also selected it as one of the five to keep, when making his revisions to the distances cited in the moon-test of the first edition. The least-squares estimate corresponding to those five, $60.57r \pm 1.62r$, makes Kircher's 62.5 difference from the mean greater than one of those 95% Student's *t*-confidence bounds.[33]

Neither this somewhat smaller difference from the mean of these five of the first edition, nor the very great difference from the mean of the original ten cited distances in the unpublished original version of the *System of the World*, counted for Newton as sufficient to exclude Kircher's estimate in those earlier versions of the moon-test. It was dropped in the second edition. This did lead to much closer agreement with Newton's assumed distance of $60r$ among the cited estimates of the second edition than had been exhibited by his cited estimates of the first edition.

Perhaps Cotes was more stringent than Newton had earlier been about which estimates to select. Even without the least-squares assessment available to us today, the great divergence of Kircher's estimate from the others might have provided legitimate reason for counting it as less reliable. Indeed, as we suggested earlier, including it in the first place may well be a better candidate than his later dropping it for Westfall's suggestion that a concern to cite distances close to his assumed value of 60 unduly influenced Newton's selection of estimates to cite.

5. *The correction to Tycho*

Here is the Cajori translation of the correction to Tycho from the original version of book 3.

But *Tycho*, and all who follow his tables of refraction, making the refractions of the sun and moon (altogether against the nature of light) to exceed those of the fixed stars, and that by about four or

[32] See note 24 above.

[33] $1.96r = 1.96/1.62 = 1.19$ of those $1.62r$ confidence bounds.

five minutes in the horizon, did thereby augment the horizontal parallax of the moon by about the like number of minutes; that is by about the 12th or 15th part of the whole parallax. Correct this error, and the distance will become 60 or 61 semidiameters of the earth, nearly agreeing with what others have determined. (Cajori 1934, 560)

It differs from the corresponding passage quoted above from a manuscript translation by I.B. Cohen by citing 60 as well as 61 as permissible corrected values for Tycho's estimate.

Diurnal parallax is the angular difference of the moon's position against the background of fixed stars when viewed directly overhead and when viewed horizontally.

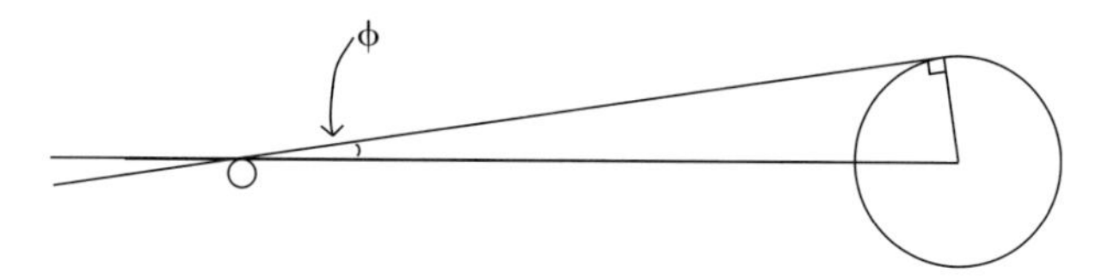

Figure 6.1 Diurnal parallax

Such parallax was long used to estimate the moon's distance from the earth. Where ϕ is the angle of parallax the moon's distance in radii of the earth is $1/\sin\phi$. Tycho's distance estimate of 56½ earth radii corresponds to a parallax of about 60.85 minutes.[34]

As light moves from empty space through the increasingly denser medium of the earth's atmosphere it is bent toward the normal to the surface. This is an instance of the phenomenon of refraction. It makes an object viewed near the horizon appear at a higher angle from the horizon than its actual angular position on the celestial sphere. Refraction will have no effect on observations of an object viewed directly overhead, since light coming from it will already be along the normal to the surface at the viewer's location.

In estimating parallax, the directions are specified by locations, say of a part of an edge of the moon, against fixed stars at very nearly the same position on the celestial sphere. Since the light from these stars will be striking the atmosphere from almost the same direction as the light from the corresponding part of the moon's edge, it will be bent in very nearly the same amount by the transitions to increasingly dense layers of the atmosphere. Refraction will, therefore, have negligible effect on such estimates of diurnal parallax.

[34] Newton's assumed value of 60 corresponds to a parallax of about 57.30 minutes, a reduction of about 3.55 minutes of the parallax corresponding to Tycho's estimate. The value 61 corresponds to a parallax of 56.36, a reduction of about 4.50 minutes. Finally, 60½, cited as the corrected value of Tycho's estimate in the second and third editions, corresponds to a parallax of about 56.825, a reduction of about 4.02 minutes from the parallax corresponding to Tycho's estimate.

According to Newton, Tycho added 4 or 5 minutes of arc, increasing his estimates of parallax by about a 12th or 15th part of the whole parallax. If we subtract 4 minutes from the parallax of 60.85 minutes, corresponding to Tycho's estimate of 56.5 terrestrial semidiameters as the lunar distance, the result is a parallax of 56.85 minutes. This corresponds to a lunar distance of 60.47 terrestrial semidiameters. If we subtract 5 minutes from the parallax of 60.85 minutes corresponding to Tycho's distance estimate we get 55.85 minutes. This corresponds to a distance estimate of about 61.56 semidiameters or radii of the earth. The mean of these two distance estimates is 61.015 such earth radii. So, given what Newton has said about Tycho's estimate, 61 terrestrial semidiameters appears to be a more appropriate correction than the 60½ to which it was changed in the second edition.[35] Newton could get his distance estimate of 60½ semidiameters from the 60.47 corresponding to the correction of subtracting 4 minutes from Tycho's parallax. This, however, would require ignoring his other cited alternative corrections.[36]

6. Standards of Newton's day

The correction to Tycho and the inclusion and dropping of Kircher's estimate are examples, like Newton's selection of the one second's fall corresponding to the single assumed distance of 60 rather than the mean of the one second's fall for all his cited estimates of the lunar distance, which violate the basic idea that the mean of several estimates is the value on which to rely. Even this more basic idea, however, was not commonly accepted as a standard for legitimate practice in Newton's day. As late as the middle years of the eighteenth century, Bradley and Lacaille stood out as unusual for being astronomers whose practice regularly conformed to this idea.[37] The widespread earlier practice, exemplified in work by such eminent astronomers as Helvelius and Flamsteed, was to use the estimate which seemed best without regard to the rest.[38] On April 10, 1755 a letter of Thomas Simpson to the President of the Royal Society on the advantage of taking the mean of a number of observations was read to the society.[39] Simpson began as follows:

It is well known to your Lordship, that the method practised by astronomers, in order to diminish the errors arising from the imperfections of instruments, and of the organs of sense,

[35] In addition to his suggestions that Tycho increased his parallax by 4 or 5 minutes, Newton also describes these as increases of a 12th or 15th part of the whole parallax. A 15th part of 60.85 is 4.06 minutes. Subtracting this would give a parallax of 56.79 minutes, which would correspond to a distance estimate of 60.54*r*. A 12th part of 60.85 minutes is about 5.07 minutes. Subtracting this gives 55.78 minutes, which corresponds to a distance estimate of 61.63.
The mean of these four distance estimates (60.47, 60.54, 61.56, 61.63) is 61.05.

[36] The 60 semidiameters cited by Cajori are even further from the mean. Moreover, the correction of subtracting 3.55 minutes from Tycho's parallax to which it corresponds is not among the alternative corrections to Tycho mentioned by Newton.

[37] See *GHA 2B*, 198.

[38] Ibid.

[39] See *Phil. Trans.* vol. XLIX, 1755, 83–93.

by taking the Mean of several observations, has not been so generally received, but that some persons, of considerable note, have been of the opinion, and even publickly maintained, that one single observation taken with care was as much to be relied on as the Mean of a great number. (*Phil. Trans.* vol XLIX, 82–3)

Simpson's opening remark makes it clear that, even as late as 1755, there was not yet in place a commonly accepted norm for taking the mean of a number of estimates as better than any single one of them.

This suggests that to regard Newton's failures to do justice to the mean as the best representative of several estimates as examples of fudging data may be an illegitimate application to Newton of normative standards that did not come into place until later than his time. Moreover, as we shall see from comments by Thomas Bayes on Simpson's paper, there are serious limitations on the appropriateness of relying on the mean as the best representative for combining several estimates. We have already seen a need for modification of this basic idea of using the mean reflected in Gauss's method for combining estimates of differing degrees of accuracy.

II Empirical success and the moon-test argument

1. Newton's inference does not depend on his dubious correction factor, nor upon his selection of which estimates to cite

Even if the application of the correction for $c = 1/178.725$, the selection of 60½ as the correction to Tycho, and the other selections of numbers giving exaggerated precision of agreement with Huygens did not actually violate what were accepted as norms of practice in Newton's day, such selection should not count as contributing legitimate additional force to his argument. That this was appreciated by Cotes is clear from the following remark from his letter to Newton of February 23, 1711/12.

I am satisfied that these exactnesses, as well here as in other places, are inconsiderable to those who can judge rightly of Your book: but ye generality of Your readers must be gratified with such trifles, upon which they commonly lay ye greatest stress. (*Corresp.* V, 232–3)

To what extent, therefore, is the moon-test argument independent of such excessive *exactnesses*?

Let us carry out the moon-test with all the other estimates cited in his various versions of the argument taken together, while omitting Kircher's very deviant 62½ and the perhaps somewhat dubious 60½, to which the correction to Tycho's estimate was changed in the second edition. Here are the estimates of $x = R/r$ and of the corresponding one second's fall $d(x)$.

The Huygens measurement, 15.096, is still within the resulting 95% t-confidence interval, [14.378, 15.118], for $d(x)$ in Paris feet. Newton's basic inference is still supported, even if Kircher's estimate and the second correction to Tycho's are dropped while keeping all the rest.

Table 6.9

	x	$d(x)$
Ptolemy	59	14.271
Kepler	59	...
Boulliau	59	...
Helvelius	59	...
Riccioli	59	14.271
Flamsteed	59⅓	14.514
Vendelin	60	15.009
Huygens	60	15.009
Copernicus	60⅓	15.261
Street	60.4	15.311
Tycho	61	15.772
95% t estimate[40]	$59.64 \pm 0.49r$	14.748 ± 0.37 Paris feet

With these distance estimates, however, Newton's large two-body correction is somewhat marginal. Given his third edition estimate of 1/39.788 lunar mass in earth masses,[41] the two-body correction of the mean estimate $R/r = 59.64$ is 60.136.[42] This falls just outside the 95% Student's t-confidence interval, [59.15, 60.13], for the lunar distance in earth radii.

2. Can the lunar distance calculated from Huygens's measurement of g *be counted as accurate?*

When explicating Newton's *vera causa* condition in Rule 1, we pointed out that his identification of the centripetal force on the moon with terrestrial gravity committed him to counting any phenomenon measuring the strength of gravity as also measuring the centripetal acceleration of the moon. Just as Picard's measurement of the earth allowed inverse-square adjusting the centripetal acceleration of the lunar orbit to measure surface gravity, so it ought to allow the inverse-square adjustment of Huygens's measurement of gravity to yield an estimate of the lunar distance. Huygens's measured

[40] We have a 95% Student's t-parameter of 2.23 for the ten degrees of freedom corresponding to eleven estimates (Freedman et al. 1978, A71; 1998, A106).

Table 6.10

	x	$d(x)$
mean	59.64	14.748
sd^+	0.73	0.54
SE	0.22	0.164
tSE	0.49	0.37

[41] See corollary 4 of proposition 37 book 3.

[42] The two-body corrected distance $R' = R[(S+P)/[(S+P)^2 S]^{1/3}]$, where R is the one-body distance. (prop. 60 bk. 1 chpt. 4 sec. II.3).

We have R=59.64, S=1 and P=1/39.788.

value, $d = 15.096$ Paris feet, yields $x(d) = 60.12$ earth radii as the lunar distance.[43] Like the above two-body correction, this falls within, but only just within, the 95% Student's t-confidence interval corresponding to these estimates of the lunar distance from astronomers.

The conflict of this value with the smaller estimates by Flamsteed and Cassini led to another of Euler's early objections to inverse-square gravitation. Here is a translation of a passage from a long paper Euler presented to the Berlin Academy of Sciences on June 8, 1747.

> Knowing the quantity of gravity at the surface of the Earth, I have concluded from it the absolute force which ought to act on the Moon, by supposing that it decreased in doubled ratio of the distances. From this force compared to the periodic time of the Moon, I have deduced the mean distance of the Moon to the Earth, and then its horizontal parallax to this same distance. But this parallax has been found a little too small, with the result that the Moon is less distant from us, than according to the theory, and leaving the force, by which the Moon is impelled toward the Earth, smaller, than I had supposed. (Waff 1976, 57)

Euler reported new lunar parallax data that gave reason to remove this discrepancy in a letter to d'Alembert dated September 28, 1748.

> You will remember that I have also alleged, in order to prove that the moon does not follow exactly the theory of attraction, this reason that the observed parallax of the moon surpassed by more than a minute the one which is found by the theory, and you still make the same remark if you envisage the table of parallaxes of Mr. Cassini or of Flamsteed. But the last eclipse of the sun [on 25 July 1748] has quite convinced me that the true parallax of the moon is perfectly in agreement with the theory, and I have seen with the greatest satisfaction that Mr. le Monnier has established the parallax of the moon almost a minute smaller than Mr. Cassini. (Waff 1976, note 10, 57–8)

Here we see later parallax observations supporting the distance estimates calculated from the theory against the smaller distances estimated from the larger parallaxes of Cassini and Flamsteed.

Le Monnier observed the solar eclipse of July 25, 1748 in Scotland.[44] He reports an observed lunar diameter of 29 minutes 47½ seconds, which closely confirmed the sun's apparent diameter of 29 minutes 48½ seconds for the time of the middle of the

[43] Where d is the one second's fall of gravity near the surface of the earth,

$$x(d) = (60^2 d/0.25015)^{1/3}$$

is the corresponding estimate of the lunar distance, R/r, in earth radii. So,

$$x(15.096) = (60^2(15.096)/0.25015)^{1/3} = 60.1157.$$

This follows from our formula

$$d(x) = 0.25015x(x/60)^2,$$

for the one second's fall $d(x)$ in Paris feet resulting from inverse-square adjusting the centripetal acceleration of the lunar orbit for estimate $x = R/r$ of the lunar distance in earth radii and lunar period of 39,343 minutes for a concentric circular orbit about a spherical earth of circumference 123,249,600 Paris feet (see chpt. 4 note 40).

[44] When I asked Curtis Wilson for help finding information about Le Monnier's observations, he found Le Monnier's report to the French Academy and translated the relevant passages from it for me.

eclipse according to Halley's tables. Wilson also made available to me a list of parallaxes from a *Mémoire* by La Lande,[45] from which I have taken the following parallaxes from Halley, Cassini, and Le Monnier.

Table 6.11

	Parallax			Mean R/r[46]
	Max	Min	Mean	
Halley 1719	61′ 7″	53′29″	57′18″	60.0
Cassini 1740	62′11″	54′33″	58′22″	58.9
Le Monnier 1748	61′ 8″	53′29″	57′18″	60.0

The following comment:

I think what the solar eclipse meant to Euler was that Halley's value of the lunar parallax was being narrowly confirmed, and this value was smaller than the values Euler had been previously aware of.

is from the letter Wilson sent me along with the material he made available to me from the Paris Academy *Mémoires* of Le Monnier and La Lande.

3. What about the distance of corollary 7 proposition 37?

In corollary 7 proposition 37 book 3, Newton appeals to a moon-test calculation to defend an estimate of 60.4 earth radii as the lunar distance. When we omitted Kircher's estimate and used only the $61r$ correction of Tycho's estimate, while keeping all five estimates of $59r$ as well as Flamsteed's $59\frac{1}{3}$, we got

$$59.64r \pm 0.49r = [59.15r, \quad 60.13r]$$

as the corresponding 95% Student's t-confidence estimate of the lunar distance. The distance estimate of $60.4r$, which Newton defended with his calculation in corollary 7 proposition 37 falls well outside this 95% Student's t-confidence bound.

Newton points out that the distance estimates were made in syzygies. As we have seen, Newton's variational orbit corresponding to the sun's action on a concentric circular orbit would support a ratio of about (70/69) of the distance at quadratures to that at syzygies. This would lead to a ratio of about $(69\frac{1}{2}/69) = 1.007$ for the distance in the corresponding concentric circular orbit to the distance in the variational orbit at syzygies. When this correction factor is taken into account, we have

[45] From La Lande 1756.

[46] I have computed these mean distances, $R/r = 1/\sin\theta$, from the mean parallaxes.

Table 6.12

	$1.007x$	$d(1.007x)$
Ptolemy	59.413	14.573
Kepler	—	—
Boulliau	—	—
Helvelius	—	—
Riccioli	59.413	14.573
Flamsteed	59.749	14.821
Vendelin	60.42	15.326
Huygens	60.42	15.326
Copernicus	60.756	15.583
Street	60.823	15.635
Tycho	61.427	16.106
estimate[47]	60.060 ± 0.493r	15.060 ± 0.373 Paris feet

The 95% Student's *t*-confidence interval for the lunar distance that would be afforded by these corrected estimates is [59.57, 60.55]. The distance estimate of 60.4 earth semidiameters, which Newton defended in his calculation of corollary 7 proposition 37 book 3, falls within these error bounds. One reason for Newton to have pointed out that the distance estimates were in the syzygies is that it would support applying this correction.[48]

The following remarks are included in a letter of Newton to Flamsteed, dated April 24, 1695.

Sr

I now send you ye Tables I promised. They are accurate enough for computing ye Moons Parallax & thence her longitude & Latitude from Observation. The little Table of the Equation of ye Moons Parallax is founded on ye 28th Prop. of ye 3d Book of my Principles where I shew that the Moons Orb (without regard to her excentricity) is Oval & that her distance in the Quadratures is greater than her distance in ye Octants in ye proportion of 70 to 69. In ye Table of

[47] Here is the least-squares assessment for these eleven corrected estimates of the lunar distance $1.007x$ and the corresponding one second's falls $d(1.007x)$ for bodies near the surface of the earth.

Table 6.13

	$1.007x$	$d(1.007x)$
mean	60.060 earth radii	15.060 Paris feet
sd^+	$0.734r$	0.555ft
SE	$0.221r$	0.167ft
95%t = 2.23*SE*	$0.493r$	0.373 Paris feet

The Student's *t*-parameter for the ten degrees of freedom corresponding to eleven estimates is 2.23 (Freedman et al. 1978, A71).

[48] Huygens's measured one second's fall, 15.096 Paris feet, falls well within [14.69, 15.43], the 95% Student's *t*-confidence bounds that would be afforded by the calculated one second's falls corresponding to these corrected distance estimates. So, the corrected distances, also, add support to the basic moon-test inference.

her horizontal parallaxes, I make her horiz. Parallax in the syzygies less than you make it in your printed Tables by about half a minute, & in ye Quadratures I make it less than you do by about 11/3′. Were the French mensuration of ye earth to be confided in as exact, these parallaxes ought to be still less: but I am unwilling to diminish them any further as yet. (*Corresp*, IV, 105–6)

The last remark about the French measurement of the earth suggests that one source of Newton's smaller parallaxes is an appeal to the moon-test to justify the greater lunar distances. The first part, however, appeals to the sort of correction we have been discussing. This appeal is made more explicit in the following remarks, which accompany the table sent by Newton.

These tables are grounded on the supposition that the mean distance of the Moon in the Octants is 60⅕ semi-diameters of the Earth, and by consequence her horizontal Parallax in that mean distance 57′ 5″ 39‴. And that her mean distance in the syzygies is less in the proportion of 69 to 69½, & in the Quadratures greater in the proportion of 70 to 69½. (*Corresp*, IV, 108)

This makes it clear that the recommended correction factor relating distances in syzygies to distances in octants is the one we have applied.

In his reply of April 27, Flamsteed cited observations of eclipses of the moon that suggested that Newton's parallaxes are too small and, therefore, that the lunar distance is less than that given in the table sent by Newton.[49] The following remarks from Newton's response, of May 4, 1695, may be instructive.

The Table of horizontal Parallaxes was made by such limits as I gathered in Autumn from your two first synopses of Observations. I do not pretend to be accurate in it. But what you object from Lunar Eclypses overthrows it not because these & ye Solar ones disagree. You think to reconcile them by supposing yt ye parallax is greater in the Sun's Perige less in his Apoge: whereas ye contrary is true. The Sun in his Perige *draws ye Moon off from ye earth* & thereby diminishes her parallax in winter & on ye contrary encreases it in summer thô not sensibly. The reason therefore why the Lunar Eclipses make the Parallax greater than ye solar ones do, is to be enquired. One reason you hint, namely yt ye diluteness of ye shadow neare ye limb makes it seem broader then it is. Another may be that all ye Suns light wch passes through ye Atmosphere within 20 or 24 miles of the earth is scattered by ye refraction of ye Atmosphere & goes not to ye edge of the shadow. A third may be some mistakes in your calculation. For you make ye [Moon]s mean Anom. 5s.15°.28′.26″ & thence her horiz. Parallax 59′.57″: you should have said 60′.49″. For 59′ 57″ is ye Parallax agreeing to the mean Anom. 4s. 15°.28′. See therefore if there be not some such mistake in your calculations.

But yet if my Table satisfy you not, you may use your printed one, and only apply to it that little menstrual equation wch I sent you. (*Corresp*, IV, 120–1)

The last sentence suggests that even the overly small distances from Flamsteed's exaggerated parallaxes can be used, provided the correction we have been discussing is applied to convert these distances at syzygies to mean-distances at octants.

[49] See Forbes, E.G. et al. 1997, 588.

This reinforces our main lesson here. The appropriateness of this correction shows that Newton's inference is not an artifact of his selection of which distance estimates to cite in the published editions of *Principia*.

The main part of the above discussion, however, argues that Flamsteed's parallax estimates are too large. This suggests that there may be other systematic errors, in addition to the need for the correction we have been discussing, that infect the five estimates of 59 as well as Flamsteed's estimate of 59⅓ for the lunar distance in terrestrial semidiameters. Systematic errors of this sort would provide reasons to support Newton's selection of which distance estimates to include in the published editions of *Principia*.

It turns out that the, above-mentioned, 1755 defense by Thomas Simpson of using the mean as the best representative supported by a group of estimates was commented upon by Thomas Bayes. Here are some of these comments from a letter Bayes wrote criticizing Simpson's proposal.

> Now that the errors arising from the imperfection of the instruments & the organs of sense shou'd be reduced to nothing or next to nothing only by multiplying the number of observations seems to me extremely incredible. On the contrary the more observations you make with an imperfect instrument the more certain it seems to me that the error in your conclusion will be proportional to the imperfection of the instrument made use of. For were it otherwise there would be little or no advantage in making your observations with a very accurate instrument rather than with a more ordinary one, in those cases where the observation cou'd be very often repeated: & yet this I think is what no one will pretend to say, Hence therefore as I see no mistakes in Mr. Simpson's calculations I will venture to say that there is one in the hypothesis upon which he proceeds. And I think it is manifestly this, when we observe with imperfect instruments or organs; he supposes that the chances for the same errors in excess, or defect are exactly the same, & upon this hypothesis only has he shown the incredible advantage, which he wou'd prove arises from taking the mean of a great many observations, Indeed Mr. Simpson says that if instead of that series of numbers which he uses to express the respective chances for the different errors to which any single observation is subject any other series whatever shou'd be assumed the result will turn out greatly in favor of the method now practiced by taking a mean value. But this I apprehend is only true where the chances of errors of the same magnitude in excess or defect are upon average nearly equal. For if the chances of errors in excess are much greater than those in defect, by taking the mean of many observations I shall only more surely commit a certain error in excess & vice versa. This I think is manifest without any particular calculation. & consequently the errors which arise from the imperfection of the instrument with which you can make a careful observation cannot in many cases be much diminished by repeating the observations ever so often & taking the mean. (Bayes's letter to John Canton[50])

So, according to Bayes, the Simpson defense for relying on the mean rests on the assumption that the chances of errors of the same magnitude in excess or in defect are upon average nearly equal.

[50] Made available by David Bellhouse of the Department of Statistics and Actuarial Science at the University of Western Ontario. I learned about Simpson and Bayes on the idea of using the mean to represent the result of combining multiple observations from an excellent presentation he gave on Bayes.

The point brought up by Bayes is now recognized as a general constraint on legitimate applications of least-squares assessments of inferences. So, if there are systematic errors infecting them, as Newton's discussion suggests, then Newton's treatment of Flamsteed's and the other small estimates of the lunar distance as unrepresentative is in line, even, with what we now take to be reasonable standards of scientific practice.

4. *Empirical success and resilience*

We have argued that the rougher agreeing measurements afforded by Newton's inference from the lunar distances increase the resiliency of Huygens's more precise estimate. Resiliency is resistance to large changes in the estimate of a parameter value. For a sufficiently divergent alternative estimate of the one second's fall *d*, adding estimates from inverse-square adjusting the cited estimates of the lunar distance would provide more evidence against it than would be provided by the cited pendulum estimates alone.

In section IV.1 of chapter 4 we illustrated this by combining moon-test estimates of *d* from the lunar distance estimates of the third edition with the four estimates from seconds pendulum experiments at Paris discussed by Newton and Huygens. In section II.1 of chapter 5 we combined the third edition moon-test estimates with the ten cited seconds pendulum estimates, including those from other latitudes as well as from Paris. Let us now combine all eleven of the above cited corrected moon-test estimates with all ten of the cited seconds pendulum estimates.

As we have noted,[51] Gauss explains that his least-squares principle can be extended to observations of unequal accuracy by introducing weights regarded as reciprocally proportional to the ratios of the expected accuracies.[52] This indicates that, from the very beginning of least-squares methods, it was realized that means needed to be weighted when combining data of differing accuracies.

In a distribution resulting from such a combination of estimates from different sources, each estimate would be weighted by the reciprocal of the variance for estimates from its source. When combining the moon-test estimates with the pendulum estimates we use the standard deviations exhibited by the two given sets of data to estimate the respective variances of the two sources.[53]

In the following plots, the horizontal line $y = 15.071$ Paris feet represents the weighted-mean resulting from combining the above cited eleven syzygy corrected

[51] See chapter 4 note 46.

[52] See Gauss 1963, 250–1.

[53] The appropriate weighted-mean is given by

$$(1/((n/\varsigma^2) + (m/\tau^2)) \times [(n \times w/\varsigma^2) + (m \times z/\tau^2)],$$

where *n* and *m* are respectively the number of pendulum and moon-test estimates, ς and τ are their respective standard deviations, and *w* and *z* are their respective means. Other more conservative weightings would use the sd^+'s or, even more appropriate with such small numbers of estimates, the Student's *t*-confidence bounds.

moon-test estimates of *d* with all ten of the pendulum estimates cited by Huygens and Newton for other locations as well as Paris. The curved line is the plot of the weighted-mean resulting from adding a new pendulum estimate *x* to the combination of these ten with the eleven moon-test estimates. The sloping straight line is the new mean resulting from combining the new pendulum estimate with the ten pendulum estimates taken on their own.

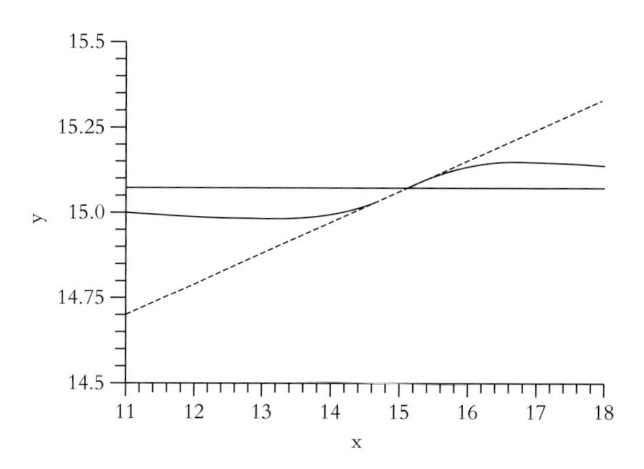

Figure 6.2 Empirical support added to all ten cited pendulum estimates from the above cited eleven syzygy corrected moon-test estimates of *d*

We can see that for new values that are not too divergent, the addition of the moon-test data does not make a difference. The moon-test data are irrelevant to the relatively small differences corresponding to the variation with latitude. For more divergent new estimates *x*, however, the weighted-mean resulting from the combination with the moon-test data provides strong resistance to the larger differences that would result from the pendulum data alone. These plots display the increase in resiliency provided by the agreeing measurement of the one second's fall from the lunar data.

We have been focusing on random errors in the above discussion. This has helped make our main point that agreement among measurements from distinct phenomena counts as a form of empirical success, rather than a merely theoretical or pragmatic advantage. In chapter 5 we pointed out that an additional important advantage of combining different sorts of measurements of the same parameter is exploiting divergences as sources of research for refining one's model. The important advantage of making seconds-pendulum measurements at different latitudes, over just making a lot more repetitions of the measurement at Paris, was making available the small but systematic differences that led to research into variation with latitude and then to the investigation of the shape of the earth.

There is another very important advantage of measurements from diverse phenomena. To the extent that systematic errors of measurements from a distinct phenomenon can be legitimately regarded as independent, the agreeing measurements from that distinct phenomenon can afford additional evidence that the results of the measurements from an initial phenomenon are not merely artifacts of systematic error. In chapter 10 we will see the decisive role agreeing measurements from diverse phenomena played in the recent transformation of dark energy from a dubious hypothesis into part of the accepted background framework guiding empirical research in cosmology. This appears to be an application of Newton's scientific method that exploits this important advantage of Newton's ideal of empirical success as convergent accurate measurements of parameters by diverse phenomena.

5. *Concluding remarks*

We began by pointing out that Newton's classic deductions of inverse-square centripetal forces from Keplerian phenomena are measurements. They realize an ideal of empirical success according to which a theory succeeds by having its theoretical parameters receive convergent accurate measurements from the phenomena it purports to explain. We then showed how the application of Newton's first two rules of reasoning to argue from the moon-test to the identification of the force holding the moon in its orbit with terrestrial gravity is backed up by this ideal of empirical success.

We have seen that on Newton's identification, two phenomena, the length of a seconds pendulum at the surface of the earth and the centripetal acceleration exhibited by the lunar orbit, are found to give agreeing measurements of the strength of the same inverse-square centripetal acceleration field, the earth's gravity. We have also explored how this realizes the ideal of empirical success by making available the data from both phenomena to back up the measured value of this common parameter. Such agreement increases the resiliency – the empirical grounds for resisting large changes – of the estimate of the parameter value.

We have also seen that nothing in Westfall's accusations of data fudging in any way undercuts the cogency of Newton's inference. Attention to details of other related propositions reveals considerably more complexity in the evidential reasoning supporting Newton's moon-test inference than might be suggested in an account based only on the basic text of proposition 4. The syzygy distance correction is an example where these additional complexities provide significant additional evidence in support of the inference.

Chapter 6 Appendix: The moon-test of corollary 7 proposition 37 book 3

Newton's moon-test calculation of proposition 4 appeals to simplifying assumptions of a concentric circular orbit about a spherical earth. One of Newton's major contributions to lunar theory was his account of the action of the sun to perturb a basic concentric circular lunar orbit to make the lunar distance least in the syzygies and greatest in the quadratures. This made a better approximation to the lunar orbit than his simplified model of a concentric circular orbit. Other contributions toward more accurate assumptions were his calculations of the oblate shape of the earth, the effect of the earth's rotation on seconds-pendulum measurements and the variation of gravity with latitude. In corollary 7 of proposition 37 book 3, Newton carries out a moon-test calculation which takes into account such departures from the simplified calculation of proposition 4.

Newton's proposition 37 is stated as the problem to find the force of the moon to move the sea.

> **Proposition 37 Problem 18** *To find the force of the moon to move the sea.*

Its corollary 7 is stated as a defense of a lunar distance estimate of 60⅖ greatest semidiameters of the earth in the octants of the lunar orbit.

1. *Octants and Newton's variational orbit*

This calculation opens with an estimate of the lunar distance it will be offered as support for.

> And the mean distance of the center of the moon from the center of the earth (in the octants of the moon) will be nearly 60⅖ greatest semidiameters of the earth. (C&W, 879)

This lunar distance is specified to be 60⅖ greatest semidiameters of the earth when the moon is in its octants. The octants are when the moon's earth-centered angular position with respect to the sun is 45°, 135°, 225°, or 315°. The octants, therefore, are midway between the syzygies (when the moon and sun are lined up at 0° or 180°) and the quadratures (when the moon and sun are perpendicular at 90° or 270°).

In propositions 26–9 book 3, Newton developed his account of the perturbations which would result from the tidal action of the sun on a concentric circular lunar orbit in the plane of the earth's orbit of the sun.[54] The resulting variational orbit makes the moon's distance least in the syzygies and greatest in the quadratures. In proposition 29 Newton points out that the moon is at its mean-distance from the earth in the octants of such a variational orbit.

> Hitherto we have investigated the variation in a noneccentric orbit, in which the moon in its octants is always at its mean distance from the earth. If the moon, because

[54] Having the moon's motion with respect to the earth coplanar with the motion of the sun with respect to the earth is a simplifying assumption. Newton's treatment of the motion of the nodes and inclination of the lunar orbit in propositions 30–5 of book 3 extends his account of the action of the sun on the motions of the moon to include the result of combining its variational effects with its action on an inclined lunar orbit.

> of its eccentricity, is more distant or less distant from the earth than if it were placed in this orbit, the variation can be a little greater or a little less than according to the rule asserted here; but I leave the excess or deficiency for astronomers to determine from phenomena. (From prop. 29 bk. 3, C&W, 847–8)

These remarks emphasize the approximative nature of Newton's investigation. Extending the account to include variations in the lunar distances due to the eccentricity of its basic orbit is left for astronomers to determine from phenomena.[55] By specifying that the lunar distance be for the moon in its octants, Newton is picking lunar positions where its observed distance, in the deformed variational orbit due to the action of the sun, would agree with the radius of the corresponding concentric circular orbit.

2. *Oblate earth*

Let us return to the calculation in corollary 7. Newton begins his defense of the foregoing distance estimate by computing the corresponding distance in Paris feet.

> For the greatest semidiameter of the earth was 19,658,600 Paris feet, and the mean distance between the centers of the earth and moon, which consists of 60⅖ such semidiameters, is equal to 1,187,379,440 feet. (C&W, 879)

Newton's cited length of the greatest semidiameter of the earth comes from his estimates of the size and oblateness of the earth in proposition 19 book 3. After citing measurements by Picard, Cassini and others of meridian arcs, Newton uses Picard's to give the circumference of the earth and semidiameter of the earth on the hypothesis that it is spherical.[56]

> And from these measures the circumference of the earth is found to be 123,249,600 Paris feet, and its semidiameter 19,615,800 feet on the hypothesis that the earth is spherical. (From prop. 19 bk. 3, C&W, 822)

Newton's cited circumference exactly corresponds to Picard's measurement of 57,060 Paris toises for one degree.[57] Given a spherical earth, the semidiameter is just the radius $r = c/2\pi$. Newton gives 19,615,800 Paris feet as the radius corresponding to this circumference of 123,249,600 Paris feet.[58]

[55] That variation in the lunar distance due to eccentricity of the moon's basic orbit is left for astronomers to determine from phenomena is very much in line with the fact that he has left the detailed treatment of the lunar precession problem to future researchers.

In the very sketchy remarks Newton includes on eccentricity and the precession of the apsides in his scholium to proposition 35 book 3, Newton cites an eccentricity of 0.05505 (C&W, 871). The current mean eccentricity of the orbit of the moon about the earth is given as 0.05490 (*ESAA*, 701). This gives a distance of about 1.055 at apogee (the point on the orbit of greatest distance from the center of the earth) and 1-.055 at perigee (the point of least distance) for an elliptical orbit with 1 as the mean distance equaling its semi-major axis. The sun's action distorts such an elliptical orbit by reducing its distances in syzygies and increasing its distances in quadratures, as it does to distort a concentric circular orbit into Newton's basic variational orbit. See Airy 1969, 64–72, for an accessible qualitative account.

[56] As we have seen (chpt. 4 note 16), the basic moon-test holds up for these other estimates.

[57] One Paris toise is 6 Paris feet. (57,060)(6)(360) = 123,249,600.

[58] Newton's cited 19,615,800 feet is 17 feet longer than 123,249,600/2π feet, which equals 19,615,783. Perhaps he has rounded 783 to 800.

To estimate the oblateness of the earth, Newton supposes a narrow channel full of water going from the pole to the center and from that center out to the equator. On his theory of universal gravity and the assumption of uniform density of the earth, he calculates the oblateness that would be required to make the water in the two legs balance given the centrifugal effect of the earth's rotation at the equator. Here is his result.[59]

> Therefore the diameter of the earth at the equator is to its diameter through the poles as 230 to 229. And thus, since the mean semidiameter of the earth, according to Picard's measurement, is 19,615,800 Paris feet, or 3,923.16 miles (supposing a mile to be 5,000 feet), the earth will be 85,472 feet or 17 1/10 miles higher at the equator than at the poles. And its height at the equator will be about 19,658,600 feet and at the poles will be about 19,573,000 feet. (From prop. 19 bk. 3, C&W, 824)

Newton gives 19,658,600 feet at the equator as the greatest semidiameter corresponding to the resulting flattening of a spherical earth of his basic radius 19,615,800 feet.[60] He gives 1,187,379,440 Paris feet as the mean-distance between the centers of the earth and moon, which consists of 60 2/5 such greatest semidiameters.[61]

3. *Two-body correction*

Newton goes on to correct for the fact that the earth and moon orbit their common center of gravity.

> And this distance (by the preceding corollary) is to the distance of the center of the moon from the common center of gravity of the earth and moon as 40.788 to 39.788; and hence the latter distance is 1,158,268,534 feet. (C&W, 879)

The preceding corollary (corollary 6 of proposition 37) gives 40.788 to 39.788 as the ratio of the distance of the center of the moon from the center of the earth to the distance of the moon from their common center of gravity. This is based on Newton's very erroneous estimate (corollary 4) of 1/39.788 as the ratio of the mass of the moon to the mass of the earth.[62] Newton gives 1,158,268,534 feet as the distance of the center of the moon from this

[59] Newton's ratio of 230 to 229 makes the semidiameter at the equator about 1.0044 times the semidiameter through the pole. The modern value, given by the ratio of 6,378,140 meters for the equatorial radius to 6,356,755 meters for the polar radius is about 1.0034. (*ESAA*, 700)

[60] The flattening of a spherical earth corresponding to Newton's ratio 230/229 = 1.004367 of the equatorial to the polar semidiameters gives 1.0021835 as the ratio of the equatorial semidiameter to the radius of the corresponding unflattened sphere. This gives 1.0021835 (19,615,800) = 19,658,631 Paris feet, which is quite close to Newton's 19,658,600 for the greatest semidiameter of the earth.

[61] 60.4(19,658,600) = 1,187,379,440.

[62] As we have seen (sec. II.3 of chpt. 4 above), the true ratio is about 1/81.3. See also *ESAA*, 710, where this ratio is given as .0123002 = 1/81.29949.

Most of proposition 37 is devoted to generating this estimate from Sturmy's estimate, from observations of the tides in Bristol harbor, of 9 to 5 as the ratio of heights of tides at syzygies (where solar and lunar tidal forces maximally reinforce each other) to heights of tides at quadratures (where solar and lunar tidal forces maximally interfere). The calculation introduces a number of corrections that are hard to estimate and have been criticized by Westfall as additional examples of data fudging. The greatest source of the error, however, in Newton's use of such calculations to compare the masses of the earth and moon comes from his incorrect estimate of the distance to the sun.

common center of gravity calculated by multiplying its mean-distance of 1,187,379,440 feet by (39.788/40.788).[63]

As we have seen, Newton's two-body correction from proposition 60 book 1 gives the ratio

$$R'/R = (M + m)/((M + m)^2 M)^{1/3}$$

where R′ is the two-body distance, R is the corresponding one-body distance, and M and m are the respective masses of the two bodies. This would give

$$R(1 - \cos\theta_m)(1/60)^2(R/r)^2$$

where θ_m is the angle of its orbit the moon would complete in one minute and r is the radius of a spherical earth, as the fall in one second at distance r corresponding to inverse-square adjusting the one minute's fall of a concentric circular orbit of radius R. Where M is the mass of the earth and m is the mass of the moon, $R_0 = R'$ (M/M+m) is the distance of the moon from the common center of gravity of the earth–moon system. Here in corollary 7 of proposition 37, Newton will give

$$R_0(1 - \cos\theta_m)(1/60)^2(R'/r)^2$$

as the fall in one second corresponding to inverse-square adjusting the one minute's fall of the lunar orbit. This is equivalent to the above application of the two-body correction from proposition 60 book 1.[64]

4. *The one minute's fall at orbit*

Newton next calculates the fall in one minute toward the earth exhibited by the centripetal acceleration of this orbital motion.

> And since the moon revolves with respect to the fixed stars in $27^d7^h43^{4/9m}$, the versed sine of the angle that the moon describes in one minute is 12,752,341, the radius being 1,000,000,000,000,000. And the radius is to this versed sine as 1,158,268,534 feet is to 14.7706353 feet. The moon, therefore, falling toward the earth under the action of that force with which it is kept in its orbit, will in the time of one minute describe 14.7706353 feet. (C&W, 879)

[63] 1,187,379,440 feet (39.788/40.788) = 1,158,268,440.69 feet, which is about 93 feet less than Newton's 1,158,268,534 feet.

[64] To see this, note that

$$R(1 - \cos\theta_m)(1/60^2)(R/r)^2 = R_0(1 - \cos\theta_m)(1/60)^2(R'/r)^2,$$

just in case

$$R^3 = R_0(R')^2.$$

These are equal because

$$R^3 = [R'((M + m)^2 M)^{1/3}/(M + m)]^3 = (R')^3(M + m)^2 M/(M + m)^3 = (R')^3(M/(M + m))$$

and

$$R_0(R')^2 = R'(M/(M + m))(R')^2 = (R')^3(M/(M + m)).$$

Newton here gives 39343 + 4/9 minutes as the lunar period.[65] He gives the versed sine,

$$(1 - \cos\theta_m) = .000000012752341,$$

of this angular arc θ_m to an extraordinary fifteen decimal places.[66] Newton's 1,158,268,534 feet for the radius yields 1,158,268,534 (.000000012752341) = 14.7706353 Paris feet for the one minute's fall corresponding to the centripetal acceleration of the lunar orbit.[67]

5. *The precession correction*

Newton next multiplies by his correction ratio 178.725/177.725 = 1.0056 that would offset having c=1/178.725=.0056 in corollary 2 of proposition 45 book 1. As we have seen, this is exactly twice what would be needed to offset a radial solar tidal force that would account for half the lunar precession. A tidal force for which such a correction would be appropriate would, to fairly good approximation, account for the observed precession of the lunar orbit. Newton cites his corollary of proposition 3 as justification for applying this correction.

> And by increasing this force in the ratio of 178 29/40 to 177 29/40, the total force of gravity in the orbit of the moon will be found by prop. 3, corol. [of this book 3]. And falling toward the earth under the action of this force the moon will describe 14.8538067 feet in the time of one minute. (C&W, 879)

Applying this correction to Newton's one minute's fall of 14.7706353 feet yields, to seven decimal places, 14.7706353(178.725/177.725) = 14.8537448 feet. This agrees with Newton's 14.8538067 when each is rounded to four places.[68]

6. *The one second's fall at latitude 45°*

Newton now calculates the fall in one second that would be produced by this inverse-square force on bodies at a location at the mean-distance of its surface from its center.

> And at 1/60 of the distance of the moon from the center of the earth, that is, at a distance of 197,896,573 *[19,789,657.3]* feet from the center of the earth, a heavy

[65] *ESAA*, 708 gives 27.321661 decimal days = $27^d 7^h 43.192^m$ = 39343+1.7/9 minutes as the mean period of the orbit of the moon.

[66] To twelve decimal places, (1-cos(360/39343)) = .000000012752 = (1-cos(360/(39343+4/9)). To fifteen places, (1-cos(360/(39343+4/9))) = .000000012752196 rather than Newton's 0.000000012752341. Perhaps, Newton is using tables for versed sines that are not accurate beyond twelve places.

[67] 1,158,268,534 (.000000012752341) = 14.77063532. Leaving the versed sine to twelve decimal places would give to four places

$$1,158,268,534(.000000012752) = 14.7702 \quad \text{Paris feet,}$$

for the one minute's fall corresponding to the moon in this orbit. Using R_0 = 1,158,268,440.69 Paris feet, which correctly corresponds to Newton's R' and his exaggerated ratio 1/39.788 of the moon's mass to that of the earth as well as leaving the versed sine to twelve places also gives

$$14.7702 \quad \text{Paris feet,}$$

for the one minute's fall to four places.

[68] Applying the correction (178.725/177.725) = 1.0056 to the four place precision estimate 14.7702 Paris feet for the one minute's fall corresponding to the orbit (and Newton's 1/39.788 ratio for the mass of the moon in the two-body correction) would yield 1.0056(14.7702) = 14.8529 Paris feet.

> body – falling in the time of one second – will likewise describe 14.8538067 feet. And so, at a distance of 19,615,800 feet (which is the mean semidiameter of the earth), a heavy body in falling will describe – in the time of one second – 15.11175 feet, or 15 feet 1 inch and 4 1/11 lines. (C&W, 879)

Given that the accelerations are inversely as the squares of the distances, the one minute's fall, 14.8538067 Paris feet, corresponding to the centripetal acceleration Newton assigns to the two-body mean-distance R′ = 1,187,379,440 Paris feet equals the one second's fall corresponding to the acceleration at (1/60)R′ = 19,789,657.3 Paris feet.[69] The fall in one second at the mean semidiameter distance 19,615,800 Paris feet, given that the fall in one second at 19,789,657 feet is 14.8538067 feet in this inverse-square acceleration field, is 14.8538067 times $(19{,}789{,}657/19{,}615{,}800)^2$. Newton gives this as 15.11175 feet, which is 15 feet 1 inch and 4.092 = 4 and 1.01/11 lines or, as he says, 15 feet 1 inch 4 1/11 lines.[70]

7. *The correction for Paris (latitude 48° 50′)*

> This will be the descent of bodies at a latitude of 45 degrees. And by the foregoing table, presented in prop. 20, the descent will be a little greater at the latitude of Paris by about 2/3 of a line. Therefore, by this computation, heavy bodies falling in a vacuum at the latitude of Paris will – in the time of one second – describe approximately 15 Paris feet 1 inch and 4 25/33 lines. (C&W, 879–80)

Newton adds 2/3 of a line to his 15 feet 1 inch 4 1/11 lines for the one second's fall at latitude 45°, to give what he characterizes as, approximately 15 Paris feet 1 inch and 4 25/33 lines as the one second's fall at the latitude of Paris.[71]

In proposition 20 book 3 Newton gives 48°50′ as the latitude of Paris (C&W, 827) and 3 Paris feet 8 5/9 lines = 440.555 lines as the length of a seconds pendulum at Paris, corrected for the weight of the air.

> at the latitude of Paris the length of a seconds pendulum is 3 Paris feet and 8 1/2 lines (or rather, because of the weight of the air, 8 5/9 lines) (C&W, 828)

His table gives 3 feet 8.428 lines = 440.428 lines as the length of a seconds pendulum at latitude 45°.[72] Huygens gives $d_{sec} = (1/2)l\pi^2$ for the fall in one second measured by having l be the length of a seconds pendulum. The ratio $(\frac{1}{2}\pi^2(440.555)/\frac{1}{2}\pi^2(440.428) = 440.555/440.428 = 1.00029$ gives the factor by which to augment the 15.11175 foot fall in one second at 45° latitude Newton computed from the inverse-square adjusted lunar orbit to get the corresponding

[69] Newton's 197,896,573 feet has a misplaced decimal. It should be 19,789,657.3, exactly 1/60 of the 1,187,379,440 feet he gave as the moon's distance in the octants of its orbit.

[70] $14.8538067(19{,}789{,}657/19{,}615{,}800)^2 = 15.118275$ Paris feet. Newton's 15.11175 is about .0065 Paris feet too short. Multiplying $(19{,}789{,}657/19{,}615{,}800)^2$ by the correct four place one second's fall 14.7702 Paris feet for distance R′/60 = 19,789, 657 gives, to four decimal places, 15.0332 Paris feet.

[71] 1/11 + 2/3 = 25/33.

[72] Newton's table is based on the following theorem he states, without proof, in proposition 20 book 3.

> The increase of weight in going from the equator to the poles is very nearly as the versed sine of twice the latitude, or (which is the same) as the square of the sine of the latitude. (C&W, 827)

For a proof of this proposition see the reader's guide section 10.15 (C&W, 350–5). See also Chandrasekhar, section 110, 394–5.

fall in one second at Paris. We have 1.00029(15.11175) = 15.11613 feet, for a difference of 0.00438 feet = 0.63 lines. Newton gives this difference as 2/3 of a line = 0.67 lines.[73]

8. *Effects of rotation*

> And if gravity is diminished by taking away the centrifugal force that arises from the daily motion of the earth at that latitude, heavy bodies there will – in the time of one second – describe 15 feet 1 inch and 1½ lines. And that heavy bodies do fall with this velocity at the latitude of Paris has been shown above in props. 4 and 19 [of this book 3]. (C&W, 880)

In proposition 19 book 3, Newton gives 7.54064 lines as the versed sine, r(1-cosα), where r is the length, 19,615,800 Paris feet, of the semidiameter of the earth at the equator and α is the angle the earth will rotate in one second.

> A body revolving uniformly in a circle at a distance of 19,615,800 feet from the center, making a revolution in a single sidereal day of $23^h56^m4^s$, will describe an arc of 1,433.46 feet in the time of one second, an arc whose versed sine is 0.0523656 feet, or 7.54064 lines. (Proposition 19; C&W, 822)

The period $23^h56^m4^s$ = 86160 seconds. Where r is 19,615,800 feet and α = 360/86160 = .004178°, the versed sine r(1-cosα) = 0.052158 feet = 7.51081 lines.[74]

Newton also gives the ratio of the centrifugal effect of bodies at the equator to the centrifugal effect due to the earth's rotation at the latitude of Paris.

> The centrifugal force of bodies on the earth's equator is to the centrifugal force by which bodies recede rectilinearly from the earth at the latitude of Paris (48°50′10") as the square of the radius to the square of the cosine of that latitude, that is, as 7.54064 to 3.267. (Prop. 19; C&W, 822)

Newton here gives 48°50′10″ = 48°50.167′ = 48.8361° as the latitude of Paris.[75] The square of the cosine of this angle is 0.4332. The result of multiplying his 7.54064 lines for the effect of the earth's rotation at the equator by this factor 0.4332 is 3.26696, which gives his 3.267 lines as the amount the centripetal effect of the earth's rotation will reduce the fall in one second at the latitude of Paris. We have $4^{25}/_{33}$ = 4.7576 and 4.7576 − 3.267 = 1.49. The result of correcting a fall of 15 feet 1 inch and $4^{25}/_{33}$ lines by minus 3.267 lines is, therefore, 15 feet 1 inch 1½ lines.[76]

9. *Aoki on the moon-test of corollary 7*

Shinko Aoki comments as follows on Newton's use of 60⅖ as the lunar distance in this calculation:

[73] The correction for latitude applied to our one second's fall to four decimal places of 15.0332 Paris feet at latitude 45 degrees yields 15.0332(1.00029) = 15.0376 Paris feet.

[74] The arc in one second is 2π(19,615,800/86160) = 1,430.475 feet. Where this arc length is a and the diameter $D = 2(19{,}615{,}800)$ feet, $a^2/D = 0.052158$ feet = 7.51081 lines.

[75] The modern value of the latitude for Paris is 48° 50.2′ (*Astronomical Alamanc*, J10).

[76] Applying this correction, 3.267 lines = 0.0227 feet to our four place one second's fall 15.0376 for Paris gives 15.0603 Paris feet or 15 feet .7336 inches or 15 feet 8.68 lines.

> . . . , in Cor. 7 to Prop. XXVII, he used 60⅖ when he took the ellipticity (flattening) of the Earth's figure into account. The observed mean distance of the lunar orbit should not have been altered, however, even when deducing the surface gravity by a different formula, since the observations are independent of theory. In other words, if the mean distance is changed the result must be considered to have been deliberately chosen[80] in each case. (Aoki 1992, 183–9)

Aoki's note [80] reads

> Or, fabricated. See [24], also compare footnote 9.
>
> The reference [24] is to Westfall's fudge factor paper. Here is Aoki's footnote 9, which gives an *OED* entry for "fudge":
>
> According to the *OED* (second edition, 1989), the meaning of this word is as follows (there was no entry for the adjective, but only for the verb): To fit together or adjust in a clumsy, makeshift, or dishonest manner; to patch or 'fake' up; to 'cook' accounts. Often in schoolboy language: To make (a problem) look as if it had been correctly worked, by altering figures: to conceal the defects of (a map or other drawing) by adjustment of the parts, so that no glaring disproportion is observed; and in other like uses.
> I suppose some similarity to the word 'fabricate'; or "To fudge is to alter the data to bring them in agreement with the theory," following Wilson's opinion [29].

The reference [29] reads:

> Wilson, Curtis, private communication, December 12, (1990)

This reference attributes to Wilson the quoted definition of "To fudge," which Aoki proposes is appropriate to his own use of "fabricate" when he applies that word to Newton's use of 60⅖ in corollary 7.

So, it looks like Aoki is proposing to call Newton's use of 60⅖ as the lunar distance in the corollary 7 proposition 37 calculation, rather than the 60 assumed in the moon-test of proposition 4, a "fudge."

The basis for this accusation is his attribution that Newton is presenting this lunar distance as one generated from the lunar observations. Newton's text makes it clear that, as Westfall also points out (in a passage we have quoted above in chpt. 6 sec. I.2), the calculation of corollary 7 is a defense of the distance estimate,

> And the mean distance of the center of the moon from the center of the earth (in the octants of the moon) will be nearly 60 ⅖ greatest semidiameters of the earth. (C&W, 879)

This distance estimate is followed by the calculation showing its agreement with terrestrial gravity via the theory, when the shape of the lunar orbit, the oblate shape of the earth, and other complications are taken into account. It is not presented as one generated from the lunar distance observations.

This distance estimate is also explicitly specified to be in the octants of the moon. This makes Newton's results on the shape of the lunar orbit in his account of its *variation*[77] inequality

[77] See chapter 2 footnote 42.

relevant. As we have seen (chpt. 6 sec. II.3), these assert that the sun's action makes the lunar mean-distance in the octants 1.007 times its smaller mean-distance in the syzygies. As we have also seen (chpt. 4 sec. II.1), the distance of 60 assumed in the calculation of proposition 4 is explicitly cited as a distance from observations in syzygies. We have, 1.007(60) = 60.42. Therefore, the distance defended by the calculation of terrestrial gravity in corollary 7 is, actually, in fairly good agreement with the 60 assumed in the moon-test of proposition 4. Aoki does offer some critical comments on Newton's treatment of the moon's variation (see pages 183–9 of Aoki 1992); but, his charge of fudging is undercut by his failure to consider the relevance of Newton's treatment of the shape of the lunar orbit to his moon-test calculation.

7

Generalization by Induction (Propositions 5 and 6 Book 3)

Part I is devoted to Newton's argument for proposition 5 and his important Rule 4 for reasoning in natural philosophy. Rule 4 is a very informative characterization of theory acceptance for Newton and for scientific method today. Part II is devoted to Newton's argument for proposition 6 and his important Rule 3. The phenomena cited count as agreeing measurements supporting the proportionality of weight to quantity of matter for all bodies at equal distances from any planet. This supports an interpretation of Rule 3 which informs the role of *theory-mediated* measurements in supporting scientific inferences today.

I Proposition 5

In proposition 5 and its corollaries Newton identifies the inverse-square centripetal acceleration fields toward Jupiter, Saturn, and the sun with gravity toward those bodies and generalizes to also assign such gravities to planets without moons. The basic identification counts all these orbital phenomena as effects of gravity attracting satellites toward primaries. The generalization to planets without moons counts these phenomena as affording agreeing measurements of centripetal direction and inverse-square variation of gravity toward all planets. Section I.1 introduces the basic inference of proposition 5 and its extensions in the corollaries and scholium of that proposition.

The scholium to proposition 5 appeals to Newton's Rule 4 for doing natural philosophy. As we have seen in chapter 1, this rule characterizes his concept of provisional acceptance of theoretical propositions as guides to research. It articulates a commitment to accept theoretical propositions that are counted as gathered from phenomena by induction. This commitment to accept theoretical propositions as premises that make phenomena measure theoretical parameters is central to the practice of Natural Science today. In section I.2 we shall use Newton's specific application of this Rule to inform our interpretation of its characterization of this important conception of theory acceptance. In section I.3 we shall discuss Newton's controversial appeal to his third Law of Motion in corollaries of proposition 5. This will lay groundwork for our detailed examination, in chapter 9, of the extent to which Rule 4 can support this controversial inference.

1. The basic argument

Proposition 5 Theorem 5

The circumjovial planets [or satellites of Jupiter] gravitate toward Jupiter, the circumsaturnian planets [or satellites of Saturn] gravitate toward Saturn, and the circumsolar [or primary] planets gravitate toward the sun, and by the force of their gravity they are always drawn back from rectilinear motions and kept in curvilinear orbits.

For the revolutions of the circumjovial planets about Jupiter, of the circumsaturnian planets about Saturn, and of Mercury and Venus and the other circumsolar planets about the sun are phenomena of the same kind as the revolution of the moon about the earth, and therefore (by rule 2) depend on causes of the same kind, especially since it has been proved that the forces on which those revolutions depend are directed toward the centers of Jupiter, Saturn and the sun, and decrease according to the same ratio and law (in receding from Jupiter, Saturn, and the sun) as the force of gravity (in receding from the earth). (C&W, 805–6)

Here we have an appeal to Rule 2:

Therefore, the causes assigned to natural effects of the same kind must be, so far as possible, the same. (C&W, 795)

by itself.

This inference explains all these phenomena as effects of the same kind of force – gravity attracting a satellite toward its primary. It counts these phenomena as agreeing measurements of general features of this kind of force – a centripetal direction and an inverse-square accelerative measure. Having an inverse-square accelerative measure requires each such force to have its attractions equally proportional to masses for bodies at equal distances. Newton's inference is to identify all these different inverse-square acceleration fields as gravity attracting satellites toward primaries.

Whewell[1] suggested that the problem of deciding when effects are of the same kind renders applications of Rule 2 vacuous.

Are the motions of the planets of the same kind with the motion of a body moving freely in a curvilinear path, or do they not rather resemble the motion of a floating body swept round by a whirling current? The Newtonian and the Cartesian answered this question differently. How then can we apply this rule with any advantage? (Butts 1988, 332)

[1] We have noted (see chpt. 4 sec. II.2) that the agreeing measurements in Newton's moon-test are a realization of what Whewell called "consilience of inductions." It can be argued that Whewell's conception of colligations of facts affords an insightful interpretation of Newton's phenomena (see Harper 1989 for an application to Kepler's Harmonic Rule). Whewell's discussions of these methodological themes and his critiques of Mill's overly simplistic account of induction show him to have significantly grasped important features of Newton's achievement. Nevertheless, his rejection of Newton's basic inferences from phenomena suggests that he failed to appreciate the full power of Newton's scientific method. I suspect he was too much in the grip of the hypothetico-deductive model of scientific inference.

Contrary to Whewell's suggestion, this application of Rule 2 is one that vortex theorists who read Newton's *Principia* were willing to accept.[2] The detailed criteria of similarity Newton points to, centripetal direction and inverse-square variation of the accelerations produced, underwrite the identification. They do this independently of whether the forces are interpreted as direct interactions between the primary and the satellites or as local interactions of the satellites with vortex particles pushing them toward the primary. The agreement of the measurements of the centripetal direction from the Area Rule motion of the orbits and the inverse-square variation from the Harmonic Rule relation among the orbits in each of these systems makes the application of Rule 2 to count them all as gravitation non-vacuous.

As we have noted, Huygens explicitly builds centripetal force on a body proportional to the quantity of matter of that body into what he demands from his proposed cause of gravity.[3] After Newton's *Principia* was published, a main enterprise of vortex theorists was to attempt to give vortex-theoretic accounts of forces that could produce such centripetal acceleration fields by local interactions with vortex particles.[4]

Let us now consider inferences in corollaries which are straightforward extensions of this basic inference. In the first part of corollary 1, Newton extends gravity to planets without moons.

> Corollary 1 *[first part]*. Therefore, there is gravity toward all planets universally. For no one doubts that Venus, Mercury, and the rest [of the planets, primary and secondary] are bodies of the same kind as Jupiter and Saturn . . . (C&W, 806)

For planets without moons there are no centripetal accelerations of bodies to count as effects of the same kind. Rule 2 does not apply. Newton offers the remark that no one doubts that these planets without moons "are bodies of the same kind as Jupiter and Saturn."

Corollary 2 extends the inverse-square variation of gravity to all the planets.

> Corollary 2. The gravity that is directed toward every planet is inversely as the square of the distance of places from the center of the planet. (C&W, 806)

This explicitly extends to planets without moons the inverse-square variation of gravity, which, in the case of the sun and planets with satellites, is measured by the Harmonic Rule and by the absence of orbital precession. Like the planets with moons, each of these other planets is asserted to be the center of an inverse-square centripetal acceleration field of gravity.

[2] As we saw in chapter 5 above.

[3] In chapter 5 section I.4 we described this feature of Huygens's treatment of his hypothesis. As we pointed out, this undermines some of the significance Andrew Janiak (2007, 145; 2008, 78) attributes to Newton's characterization of mechanical causes as acting in proportion to the surfaces of bodies.

[4] See chapter 5 section I.4. See *GHA* 2B, 6–13 and Aiton 1972. For Leibniz see Bertoloni Meli 1993, 126–42 and Aiton 1972, 125–51.

Newton's discussion of proposition 5 ends with the following scholium.[5]

Scholium

Hitherto we have called "centripetal" that force by which celestial bodies are kept in their orbits. It is now established that this force is gravity, and therefore we shall call it gravity from now on. For the cause of the centripetal force by which the moon is kept in its orbit ought to be extended to all the planets, by rules 1, 2, and 4. (C&W, 806)

This third edition scholium is the one explicit appeal to Rule 4 in Newton's argument. It is added to back up appeals to Rules 1 and 2. The inference supported is to identify the centripetal forces by which celestial bodies are kept in their orbits as gravity and to extend such centripetal forces of gravity to all the planets.

2. Regulae Philosophandi: *Rule 4*

We introduced Rule 4 in chapter 1. This rule was first published in the third (1726) edition of *Principia.*[6] Here is the Cohen and Whitman translation of Newton's formulation of this important rule, together with his comment on it, as they appear in his *Principia.*

Rule 4. *In experimental philosophy, propositions gathered from phenomena by induction should be considered either exactly or very nearly true notwithstanding any contrary hypotheses, until yet other phenomena make such propositions either more exact or liable to exceptions.*[7]

This rule should be followed so that arguments based on induction may not be nullified by hypotheses. (C&W, 796)

Newton's characterization of his own philosophy for investigating nature as "experimental philosophy" is an explicit contrast to the mechanical philosophy of his continental critics.[8] Newton's comment suggests that the point of this rule is to defend arguments based on induction from being undercut by mere hypotheses.

[5] We used this scholium to introduce Rule 4 in chapter 1 above.

[6] It appears in Newton's interleaved and annotated copy of the second edition (Koyré and Cohen 1972, 555). We shall see (chpt. 9 sec. IV) that this rule was anticipated in Newton's response to an objection by Cotes to Newton's appeal to Law 3 to count gravitation maintaining a satellite in orbit as an interaction between the satellite and its primary.

[7] Here is the Latin.

Regula IV

In philosophia experimentali, propositiones ex phenomenis per inductionem collectae, non obstantibus contrariis hypothesibus, pro veris aut accurate aut quamproxime haberi debent, donec alia occurrerint phaenomena, per quae aut accuratiores reddantur aut exectionibus obnoxiae. (Koyré and Cohen 1972, 555)

Note that in the above translation "exactly or very nearly true" translates *"veris aut accurate aut quamproxime"* and "make such propositions more exact" translates *"per quae aut accuratiores reddantur"*. This last clause makes it clear that to translate "accurate" as "exact" must be compatible with the idea that something counted as exact can be made even more exact.

[8] See Harper and Smith 1995 for more detail about important features distinguishing Newton's experimental philosophy from the mechanical philosophy. It traces differences from his continental critics, such as Newton's focus on empirically establishing propositions rather than conjecturing hypotheses about causal mechanisms from his responses to objections to his theory of light and colors up through his development of and later defenses of his theory of gravity.

Newton tells us "propositions gathered from phenomena by induction are to be considered either exactly or very nearly true . . . until yet other phenomena make such propositions either more exact or liable to exceptions." One thing we see here is *acceptance subject to empirical correction* rather than just assigning and adjusting probabilities. Acceptance of theoretical propositions is central to Newton's methodology. The inclusion of "very nearly true" as an option, explicitly, makes room for accepting propositions gathered from phenomena by induction as approximations. Propositions can be *accepted as approximations* even if they are not exactly true. This is very much in line with construing such propositions as outcomes of measurements by phenomena, as is the provision for other phenomena making such propositions either more exact or liable to exceptions.

Rule 4 tells us to consider propositions gathered from phenomena by induction as "either exactly or very nearly true" and tells us to maintain this in the face of "any contrary hypotheses." In chapter 1 we appealed to Newton's ideal of empirical success to clarify the difference between what are to count as propositions gathered from phenomena by induction and what are to be counted as mere hypotheses.

As we have seen, Newton's basic inferences to inverse-square centripetal forces are backed up by agreeing measurements of the centripetal direction from the Area Rule motion of the orbits and the inverse-square variation from the Harmonic Rule relation among the orbits from which they are inferred.

These examples Newton counts as propositions gathered from phenomena by induction are all backed up by agreeing measurements of the same parameter by distinct phenomena. For such propositions, this ideal of empirical success as convergent accurate measurement of parameters by phenomena informs the application of Rule 4. Consider an alternative conjecture according to which some planet without moons is assumed to not have such gravity directed toward it. Such an alternative would not treat the agreeing measurements from planets with satellites as extending to count as measurements of the centripetal direction and inverse-square variation of gravity for this planet. Rule 4 will count the claim of such a skeptical challenge as a mere contrary hypothesis to be dismissed, unless such an alternative is given with details that actually deliver on measurement support sufficient to make it a serious rival, or is supported by other phenomena making the proposition inferred by Newton liable to exceptions.

3. *Gravity as mutual interaction*

The second part of corollary 1, unlike the first part, exemplifies what was a prime focus of objections to Newton's theory by vortex theorists.[9] It also prompted a challenge by Roger Cotes, the editor of the second (1713) edition. Cotes pointed out that Newton's application of the third Law of Motion to gravity construed as an interaction between spatially separated bodies appeared to be an assumed hypothesis rather than a deduction

[9] See chapter 5 sec I.3.

from phenomena. In chapter 9 we will see that Newton's Rule 4 was formulated as part of his response to this challenge by Cotes. In that chapter we will contrast Newton's and Huygens's commitments to avoid action at a distance to help clarify how Newton's responses to Cotes illuminate his scientific method. This will build on and extend our interpretation of Rule 4 in this chapter.

Here is this controversial application of the third Law of motion to construe gravity maintaining bodies in orbits as mutual attractions between spatially separated bodies.

> Corollary 1 *[second part]*... And since, by the third Law of Motion, every attraction is mutual, Jupiter will gravitate toward all its satellites, Saturn toward its satellites, and the earth will gravitate toward the moon, and the sun toward all the primary planets. (C&W, 806)

This application of the third Law of Motion to attribute mutual gravitation between primaries and satellites claims that gravity can be treated as a mutual attraction or direct interaction. We shall see that, like Cotes, Howard Stein challenges that this appears to count as an assumed hypothesis.[10]

Huygens and Leibniz did not object to Newton's basic identification of the inverse-square centripetal acceleration fields exhibited by the orbits of moons and planets as gravitations of the satellites toward their respective primaries; but, as vortex theorists, they strongly objected to the claim that these gravitations are mutual interactions between the primary and the satellites. They defended alternative hypotheses according to which each primary is surrounded by a configuration of moving matter which maintains its satellites in their orbits. For them, the relevant equal and opposite reaction would not be the attraction of the primary to the satellite; but, rather, the interaction between the satellite and the vortex corpuscles that deflect it from tangential motion into the curved path of its orbit.

Our moon, as well as the satellites of Jupiter and Saturn, is included among what Newton characterizes as "Venus, Mercury and the rest" of the planets.[11] The above-considered first part of corollary 1 together with corollary 2 attributes inverse-square gravity to all these bodies. Each is inferred to have its own inverse-square acceleration field of gravity directed toward its center. We have seen that a major difference between Newton and Huygens is that Newton extends the inverse-square accelerative measure of gravity toward a planet to arbitrarily great distances.[12] Given Newton's extension of these inverse-square acceleration fields of gravity to such large distances, the gravitations of the primaries toward their satellites follow from the first part of

[10] Stein 1991, 217–19. In chapter 9 below we will find that exploring issues raised by this interesting objection can greatly illuminate differences between Newton's stronger ideal of empirical success for theories and the weaker empiricist standard of empirical success as prediction alone.

[11] Newton's wide use of "planet" is suggestive of the geocentric tradition in which the moon, like the sun and the primary planets, is distinguished as a "planetae" – a wanderer among the fixed stars. See chapter 1 section I.1. See also Kuhn 1957, 45.

[12] Recall chapter 5 section II.2.

corollary 1, even without Newton's appeal to Law 3 to construe these gravitations as mutual attractions between the primaries and their satellites.

Newton's extension of inverse-square acceleration fields toward planets to arbitrarily great distances from their centers leads to gravitation of planets toward one another, of the moon toward the sun, and of the seas of the earth toward the sun and moon. Corollary 3 alludes to empirical evidence for the existence of perturbations between Jupiter and Saturn, as well as perturbation of the lunar motion by the sun and of our sea by both the sun and moon.[13]

Corollary 3. All the planets are heavy toward one another by corols. 1 and 2. And hence Jupiter and Saturn near conjunction, by attracting each other, sensibly perturb each other's motions, the sun perturbs the lunar motions, and the sun and moon perturb our sea, as will be explained in what follows. (C&W, 806)

The allusion to explanations to follow makes it clear that Newton regards his treatments of such perturbations as part of what he counts as his argument for universal gravitation.[14]

The *Principia* does not contain any detailed treatment of perturbations to be ascribed to gravitations of Jupiter and Saturn toward one another.[15] In a memorandum of a conversation with Newton by David Gregory, dated May 4, 1694, we find the suggestion that attractions between Jupiter and Saturn near conjunction did produce sensible perturbations.[16]

37. The mutual interactions of Saturn and Jupiter were made quite clear at their very recent conjunction. For before their conjunction Jupiter was speeded up and Saturn slowed down, while after their conjunction Jupiter was slowed down and Saturn speeded up. Hence corrections of the orbits of Saturn and Jupiter by Halley and Flamsteed, which were afterwards found to be useless and had to be referred to their mutual action. (*Corresp*, III, 318)

Astronomers had long been having considerable difficulty giving accurate tables for these planets.[17] The gravity toward one another following from attributing inverse-square gravity to each would be a natural place to look for a source of these difficulties.

[13] This corollary was added in the second (1713) edition (Koyré and Cohen 1972, 571).

[14] This strongly supports Stein's impressive case for arguing that Newton's "deduction from the phenomena" of universal gravitation is not complete at proposition 7, but includes all his later treatments of such complexities and so includes more or less the whole of book 3. (Stein 1991, 219–20)

[15] In chapter 8 below we will consider remarks Newton offered about the action of Jupiter on Saturn in proposition 13 book 3 in the third edition.

[16] Another of these memoranda contains this remark:

> Flamsteed's observations when compared with those of Tycho and Longomontanus prove the mutual attraction of Jupiter and Saturn at their most recent past conjunction in 1683. (*Corresp.* III, 337)

[17] See Wilson (1985, 36–53) for empirical evidence for and efforts to deal with anomalies in the mean motions of Jupiter and Saturn from Kepler to the efforts by Flamsteed and Halley, which Gregory's memorandum suggests are the corrections alluded to by Newton. See below chapter 9 section IV.2 for more discussion of Halley's proposal.

Moreover, the sort of interaction near conjunction described by Gregory must have seemed a very plausible conjecture suggested by Newton's theory.

We now know, however, that the mutual perturbation that makes the most sizable effects, the great inequality of Jupiter's and Saturn's motions, has a period of nearly 900 years and that perturbations corresponding to conjunctions are too small to produce effects that would have been detectable by astronomers of Newton's day.[18] As we shall discuss further in chapter 9 below, this great inequality was not successfully treated as gravitational perturbation until its solution was announced by Laplace in 1785, nearly one hundred years after Newton's *Principia* was first published.[19]

Newton's successful treatments of the variational orbit and the solar perturbations of the nodes and inclination of the lunar orbit provide very strong evidence that gravitation toward the sun perturbs the orbital motion of the moon about the earth, even if a detailed treatment of the lunar precession as a solar perturbation had to wait until Clairaut's work in 1749.[20] Similarly, Newton's treatment of the actions of gravitation toward the sun and moon to explain the major tidal phenomena counted as very strong evidence that the sun and moon do perturb our sea. Newton's gravitational explanation was far superior to the alternative explanation offered by Galileo and to any of the rival vortex theoretic explanations, even though it was considerably improved upon by Laplace's later gravitational account of the tides.[21]

We shall need to assess the extent to which appreciating the ideal of success exemplified by the accurate measurement of theoretical parameters by phenomena can support the reasonableness of regarding Newton's proposal to treat gravity as a force of interaction between separated bodies as more than a hypothesis. To what extent does Newton's argument allow this critical background assumption to reasonably count as "gathered from phenomena by induction" and so by Rule 4 to count as at least approximately true, even in the face of vortex hypotheses supported by philosophical commitment to the idea that intelligible physical causes must act by contact? We will, however, be in a better position to address this question in chapter 9. Our assessment will exploit resources developed in our discussion of Newton's argument for proposition 6 in the next section of this chapter. It will also build upon our discussion in our next chapter of Newton's arguments for proposition 7 and its application to measure relative masses of sun and planets with satellites in corollary 2 of proposition 8.

[18] See Wilson 1985, 16–17. Wilson (1985, 24–36) gives an excellent account of the main Jupiter–Saturn perturbations which includes a somewhat improved version of Airy's (1834) geometrical representation of them.

[19] See Wilson (1985, 227–85) for a detailed exposition of Laplace's treatment.

[20] See chapter 4 section V above. See also Waff (1976; 1995), Smith G.E. (in C&W, 263–4), and Wilson (2001).

[21] See Aiton 1954 for Galileo's theory of the tides, Aiton 1955a for Newton's contribution, as well as those of Daniel Bernoulli and Euler, to the gravitational theory of the tides, and Aiton 1955b for Descartes's rival vortex theory of the tides. See Bowditch 1832, vol. 2, 526–792 for a translation into English and commentary on Laplace's impressive treatment of the tides.

II Proposition 6: proportionality to mass from agreeing measurements[22]

Proposition 6 backs up Newton's argument for assigning gravitation as inverse-square centripetal acceleration fields to all planets by appealing to diverse phenomena that provide agreeing measurements of the equality of accelerations toward planets for bodies at equal distances from them.

1. *The basic argument*

Proposition 6 Theorem 6

All bodies gravitate toward each of the planets, and at any given distance from the center of any one planet the weight of any body whatever toward that planet is proportional to the quantity of matter which the body contains. (C&W, 806)

1.i *Pendulum experiments*

Others have long since observed that the falling of all heavy bodies toward the earth (at least on making an adjustment for the inequality of the retardation that arises from the very slight resistance of the air) takes place in equal times, and it is possible to discern that equality of the times, to a very high degree of accuracy, by using pendulums. I have tested this with gold, silver, lead, glass, sand, common salt, wood, water, and wheat. I got two wooden boxes, round and equal. I filled one of them with wood, and I suspended the same weight of gold (as exactly as I could) in the center of oscillation of the other. The boxes, hanging by equal eleven-foot cords, made pendulums exactly like each other with respect to their weight, shape, and air resistance. Then, when placed close to each other [and set into vibration], they kept swinging back and forth together with equal oscillations for a very long time. Accordingly, the amount of matter in the gold (by book 2, prop. 24, corols. 1 and 6) was to the amount of matter in the wood as the action of the motive force upon all the gold to the action of the motive force upon all the [added] wood — that is, as the weight of one to the weight of the other. And it was so for the rest of the materials. In these experiments, in bodies of the same weight, a difference of matter that would be even less than a thousandth part of the whole could have been clearly noticed. (C&W, 806–7)

Galileo took uniform acceleration of terrestrial bodies as a fundamental principle and evidence for it had accumulated by Newton's day.[23] As we have seen, Huygens's measurement of the acceleration of gravity with pendulums was a major breakthrough. Newton's pendulum experiment is designed to turn the evidence provided by equal accelerations into measurements establishing the equality of ratios of weight to inertial mass. It measures the equality of the ratio of weight to inertial mass for each of the thirty-six pairs of samples of these nine varied materials. These measurements establish this equality to about one part in a thousand.

[22] This treatment of this argument expands upon earlier treatments in Harper and DiSalle 1996 and Harper 1999.

[23] See, e.g., pp. 89–117 of Koyré 1968.

Newton appeals to corollaries 1 and 6 of proposition 24 book 2.[24] Let us consider proposition 24 itself.

Book 2 Proposition 24 Theorem 19

In simple pendulums whose centers of oscillation are equally distant from the center of suspension, the quantities of matter are in a ratio compounded of the ratio of the weights and the squared ratio of the times of oscillation in a vacuum. (C&W, 700)

This makes $m_1 / m_2 = w_1(t_1)^2 / w_2(t_2)^2$, where m_1 and m_2 are the quantities of matter, w_1 and w_2 are the weights, and t_1 and t_2 are the times of oscillation for a pair of simple pendulums in a vacuum. Using a balance, Newton could adjust the amounts in his samples so that, for example, the weight of his sample of gold is equal (up to quite fine tolerances) to the weight of his sample of wood. Given that weights w_1 and w_2 are equal, to establish the equality of the inertial masses m_1 and m_2 to the cited tolerance of one part in a thousand, it is sufficient to establish that the times do not differ by more than one part in two thousand.[25]

The equality of the periods of such pairs of pendulums counts as a phenomenon – a generalization fitting an open-ended body of data – insofar as the experiment is regarded as repeatable.[26] Establishing this phenomenon to tolerances of about one

[24] Here is corollary 1:

> Corollary 1. And thus if the times are equal, the quantities of matter in the bodies will be as their weights. (C&W, 700)

It is an immediate consequence of proposition 24.

We have seen (chpt. 5 sec. I.1) that Huygens made a major breakthrough when he used the slow motions of seconds pendulums to measure the acceleration of gravity far more accurately than it could be measured by attempting to directly determine how far bodies would fall in one second. The much slower motion, in addition to being easier to accurately measure, greatly reduced the relative effect of air resistance.

The slower motion, while minimizing the effect of air resistance, does not minimize the fact that balances in air compare relative buoyancies with respect to air rather than the weights themselves. Proposition 24 is proved for pendulums in vacuums. Corollary 6,

> Corollary 6. But in a nonresisting medium also, the quantity of matter in the bob of a simple pendulum is as the relative weight and the square of the time directly and the length of the pendulum inversely. For the relative weight is the motive force of a body in any heavy medium, as I have explained above, and thus fulfills the same function in such a nonresisting medium as absolute weight does in a vacuum. (C&W, 701),

addresses this worry.

[25] Another immediate consequence of proposition 24 is

> Corollary 2. If the weights are equal, the quantities of matter will be as the squares of the times. (C&W, 700).

Given that $w_1 = w_2$, we have $m_1/m_2 = (t_1 / t_2)^2$. So, to have $m_1 / m_2 = (1001/1000)$ is to have $t_1 / t_2 = (1001/1000)^{½} \approx 1.0005 = 2001/2000$.

So, given equal weights, we would require being able to detect differences in period of one part in two thousand in order to limit differences in masses to one part in one thousand. Though this difference might be too small to pick up on any single swing, it would mount up to detectable amounts over the large numbers of swings available.

[26] Newton's experiment was redone in 1999 at St. John's College, Annapolis, using lead, glass, and sand, with pendulums and timing devices (e.g. no stopwatches) comparable to those that might have been available to Newton. The results were good enough to establish that the periods agreed to within one part in six

part in two thousand measures the equality of the masses of each pair, of these samples of equal weight, to about one part in a thousand.[27]

That different bodies have equal accelerations at any given distance exhibits that the earth's gravitation is an acceleration field, not just a centripetal force. We count this centripetal force as an acceleration field, if and only if the same proportionality would hold between weight and mass for all attracted bodies at any given distance.

1.ii The moon-test The moon-test measured the agreement between the acceleration of gravity at the surface of the earth and the result of increasing the inverse-square centripetal acceleration of the lunar orbit to obtain what the corresponding acceleration at the surface of the earth would be. This shows, also, that the result of decreasing the acceleration of gravity Huygens had measured at the surface of the earth in accordance with the inverse-square to obtain the corresponding centripetal acceleration at the lunar distance agrees with the centripetal acceleration of the moon in its orbit.

Now, there is no doubt that the nature of gravity toward the planets is the same as toward the earth. For imagine our terrestrial bodies to be raised as far as the orbit of the moon and, together with the moon, deprived of all motion, to be released so as to fall to the earth simultaneously; and by what has already been shown, it is certain that in equal times these falling terrestrial bodies will describe the same spaces as the moon, and therefore that they are to the quantity of matter in the moon as their own weights are to its weight. (C&W, 807)

The equality of these accelerations at any given distance is, therefore, a phenomenon that measures the constancy of the proportionality of weight to mass. The outcome of the moon-test and the equality of the periods of Newton's pendulums are phenomena that give agreeing measurements supporting equality of ratios of inertial mass to inverse-square adjusted weight toward the center of the earth for the moon and for terrestrial bodies.

1.iii The Harmonic Rule Newton next argues that inverse-square acceleration fields toward Jupiter and toward the sun are exhibited by orbits satisfying Kepler's Harmonic Rule. Recall that a system of orbits satisfies the Harmonic Rule just in case the periods are as the $3/2$ power of the mean-distances.[28]

thousand and that the masses in the three bobs agreed to within about one part in three thousand. See Wilson 1999, 73 and note 27 below.

[27] Our contention that Newton could have determined the equality of the weights up to quite fine tolerances is supported by the fact that when the experiment was repeated at St. John's College, Annapolis, the equality of the weights was able to be fixed so that $\delta w/w < 0.000014$. See Wilson (1999, 68–9).

At that Annapolis demonstration George Smith pointed out that, by using cat gut cords, Newton's pendulums would have been less susceptible to stretching than the braided wire cords used in Annapolis. This would have made it even easier for him to control for variation in lengths than it was for Wilson's team in Annapolis.

[28] See chapter 2 section I and section IV.

Further, since the satellites of Jupiter revolve in times that are as the $3/2$ power of their distances from the center of Jupiter, their accelerative gravities toward Jupiter will be inversely as the squares of the distances from the center of Jupiter, and, therefore, at equal distances from Jupiter their accelerative gravities would come out equal. Accordingly, in equal times in falling from equal heights [toward Jupiter] they would describe equal spaces, just as happens with heavy bodies on this earth of ours. (C&W, 807)

The Harmonic Rule for circular orbits is equivalent to the inverse-square diminution of the centripetal accelerations.[29] This requires equal accelerations at equal distances; therefore, for any given distance it requires for all bodies at that distance that the ratios of their motive forces of gravity toward Jupiter to their inertial masses be equal.

Newton also appeals to the Harmonic Rule for the primary planets.[30] Like the Harmonic Rule for its satellites for gravitation toward Jupiter, the Harmonic Rule for the circumsolar planets shows that gravitation toward the sun is an inverse-square acceleration field.[31]

And by the same argument the circumsolar [or primary] planets, let fall from equal distances from the sun, would describe equal spaces in equal times in their descent to the sun. Moreover, the forces by which unequal bodies are equally accelerated are as the bodies; that is, the weights [of the primary planets toward the sun] are as the quantities of matter in the planets. (C&W, 807)

The Harmonic Rule for the primary planets is a phenomenon measuring the equality of ratios of mass to inverse-square adjusted weight toward the sun for the planets.

1.iv Un-polarized orbits Newton next argues from the absence of observable perturbations of the orbits of Jupiter's moons to infer that the moons and Jupiter are equally accelerated toward the sun by solar gravity.

Further, that the weights of Jupiter and its satellites toward the sun are proportional to the quantities of their matter is evident from the extremely regular motion of the satellites, according to book 1, prop. 65, corol. 3. For if some of these were more strongly attracted toward the sun in proportion to the quantity of their matter than the rest, the motions of the satellites (by book 1, prop. 65, corol. 2) would be perturbed by that inequality of attraction.[32] (C&W, 807–8)

[29] See chapter 3 section III.1.ii.

[30] Here we see that Newton includes gravitation toward the sun as well as toward the primary planets. As in proposition 5 above, the word "planetae" in proposition 6 is to be read widely enough to include the sun.

That Newton also includes gravitation toward the earth suggests that "planetae" in proposition 6 is to be construed widely enough to allow both the sun and the earth to count as planets.

[31] See chapter 3 section III.2.ii–iii.

[32] In proposition 65 book 1, Newton discussed orbital systems of more than two bodies attracting each other according to inverse-square accelerative forces.

Book 1 Proposition 65 Theorem 25

More than two bodies whose forces decrease as the squares of the distances from their centers are able to move with respect to one another in ellipses and, by radii drawn to the foci, are able to describe areas proportional to the times very nearly. (C&W, 567)

Given the argument for proposition 5 and its corollaries, we can assume that Jupiter and its moons gravitate toward the sun with inverse-square accelerative forces. This makes the uniform motion concentric circular orbit of each moon carry the information that there are no appreciable differences in the accelerative forces toward the sun on that moon and on Jupiter. If there are no appreciable differences in the accelerative measures of centripetal forces then the motive forces on bodies at equal distances must be equally proportional to their quantities of matter.[33]

Newton goes on to appeal to the outcome of a calculation, which he describes as establishing that such differences in accelerative forces toward the sun are less than one part in a thousand.

If, at equal distances from the sun, some satellite were heavier [or gravitated more] toward the sun in proportion to the quantity of its matter than Jupiter in proportion to the quantity of its own matter, in any given ratio, say *d* to *e*, then the distance between the center of the sun and the center of the orbit of the satellite would always be greater than the distance between the center of the sun and the center of Jupiter and these distances would be to each other very nearly as the square root of *d* to the square root of *e*, as I found by making a certain calculation. And if the satellite were less heavy [or gravitated less] toward the sun in that ratio of *d* to *e*, the distance of

In proving this proposition Newton considers two cases, several lesser bodies revolving around one large one, and systems such as Jupiter and its moons revolving around a much greater one such as the sun. The corollaries are about this second case.

> Corollary 1. In case 2, the closer the greater body approaches to the system of two or more bodies, the more the motions of the parts of the system with respect to one another will be perturbed, because the inclinations to one another of the lines drawn from this great body to those parts are now greater, and the inequality of the proportion is likewise greater. (C&W, 569)
>
> Corollary 2. But these perturbations will be greatest if the accelerative attractions of the parts of the system toward the greater body are not to one another inversely as the squares of the distances from that greater body, especially if the inequality of this proportion is greater than the inequality of the proportion of the distances from the greater body. For if the accelerative force, acting equally and along parallel lines, in no way perturbs the motions of the parts of the system with respect to one another, it will necessarily cause a perturbation to arise when there is an inequality in its action, and such perturbation will be greater or less according as this inequality is greater or less. The excess of the greater impulses acting on some bodies, but not acting on others, will necessarily change the situation of the bodies with respect to one another. And this perturbation, added to the perturbation that arises from the inclination and inequality of the lines, will make the total perturbation greater. (C&W, 569)
>
> Corollary 3. Hence, if the parts of this system – without any significant perturbation – move in ellipses or circles, it is manifest that these parts either are not urged at all (except to a very slight degree indeed) by accelerative forces tending toward other bodies, or are all urged equally and very nearly along parallel lines. (C&W, 570)

[33] This argument is much less often cited than the pendulum experiments, but, as Damour (1987, 143–4) points out, it is especially interesting as Jupiter and its moons are massive enough to count as having non-negligible self-gravitational energy. Kenneth Nordtvedt (1968a) found that in the alternative to general relativity proposed by Brans and Dicke, the gravitational self-energy proportion of the total mass-energy of a body would couple differently than the rest to an ambient field so as to produce polarization of orbits.

The extension of the identification of passive gravitational mass with inertial mass to include such massive bodies is sometimes called the gravitational weak equivalence principle (GWEP) (Will 1993, 184). We shall see (sec. II.6 and Table 3) that laser ranging experiments have established limits on polarization of our moon's orbit that measure limits on violations of GWEP of $(2 \pm 5) \times 10^{-13}$ (Dickey et al. 1994).

the center of the orbit of the satellite from the sun would be less than the distance of the center of Jupiter from the sun in that same ratio of the square root of *d* to the square root of *e*. And so if, at equal distances from the sun, the accelerative gravity of any satellite toward the sun were greater or smaller than the accelerative gravity of Jupiter toward the sun, by only a thousandth of the whole gravity, the distance of the center of the orbit of the satellite from the sun would be greater or smaller than the distance of Jupiter from the sun by 1/2000 of the total distance, that is, by a fifth of the distance of the outermost satellite from the center of Jupiter; and this eccentricity of the orbit would be very sensible indeed. But the orbits of the satellites are concentric with Jupiter, and therefore the accelerative gravities of Jupiter and of the satellites toward the sun are equal to one another. (C&W, 808)

Newton does not give the details of the calculation he refers to. According to his description, the calculation is taken to show that if at equal distances the accelerative force toward the sun on a moon were greater than that on Jupiter, then the orbit of the moon would be polarized away from the sun. He tells us that, according to the calculation he has carried out, establishing the phenomenon of absence of polarization of a moon of Jupiter to tolerances of one part in two thousand of the mean-distance of Jupiter from the sun measures a limit of one part in one thousand on differences between accelerations toward the sun of the moon and Jupiter at equal distances. He goes on to point out that one part in two thousand of Jupiter's distance is one fifth of the distance of the outermost satellite from the center of Jupiter, so that a polarization this large would be quite detectable.

The same argument also applies to the orbits of Saturn's moons, and to the earth and its moon if they also gravitate toward the sun. Newton appeals to corollaries 1 and 3 of proposition 5 to argue that they do, indeed, have weight toward the sun.

And by the same argument the weights [or gravities] of Saturn and its companions toward the sun, at equal distances from the sun, are as the quantities of matter in them; and the weights of the moon and earth toward the sun are either nil or exactly proportional to their masses. But they do have some weight, according to prop. 5, corols. 1 and 3. (C&W, 808)

Newton has now appealed to absence of polarization with respect to the sun of the orbits of Jupiter's moons, Saturn's moons and the earth's moon as phenomena exhibiting equality, at equal distances, between gravitational accelerations toward the sun of planets and their moons. All of these, therefore, are phenomena exhibiting equality, at equal distances, of ratios between weights toward the sun of bodies and their inertial masses.

The polarizations toward the sun of orbits of moons of planets (where moon and planet gravitate toward the sun in different proportions to their inertial masses) are among what are now called Nordtvedt effects. In 1968 the physicist Kenneth Nordtvedt established that such effects could distinguish the alternative Brans–Dicke theory from General Relativity.[34] He also provided a calculation which gives polarization in

[34] See Nordtvedt (1968a). For a vivid and accessible historical account see Will 1986, 139–46.

the opposite direction and of a smaller amount than the outcome of the calculation Newton reports.[35] Even though Newton's calculation is incorrect, his data still put strong bounds on differences between ratios of mass to weight toward the sun at equal distances when the correct Nordtvedt calculation is used.[36]

1.v Parts of planets Newton extends his argument for equal ratios between weight and inertial mass to individual parts of planets.

But further, the weights [or gravities] of the individual parts of each planet toward any other planet are to one another as the matter in the individual parts. For if some parts gravitated more, and others less, than in proportion to their quantity of matter, the whole planet, according to the kind of parts in which it most abounded, would gravitate more or gravitate less than in proportion to the quantity of matter of the whole. But it does not matter whether those parts are external or internal. For if, for example, it is imagined that bodies on our earth are raised to the orbit of the moon and compared with the body of the moon, then, if their weights were to the weights of the external parts of the moon as the quantities of matter in them, but were to the weights of the internal parts in a greater or lesser ratio, they would be to the weight of the whole moon in a greater or lesser ratio, contrary to what has been shown above. (C&W, 808–9)

Here, instead of direct measurements by phenomena, we have a thought experiment which makes salient that it would be very improbable to have parts differing in ratios of weight to inertial mass so exactly proportioned that whole planets had equal ratios. This is made especially implausible by the additional fact that measurements from the moon-test establish agreement in ratios between outer parts of the earth (ordinary terrestrial bodies) and the whole of the moon. This completes Newton's main argument for proposition 6.

2. *Corollaries of proposition 6*

Newton goes on to give five corollaries of proposition 6.

Corollary 1. Hence, the weights of bodies do not depend on their forms and textures. For if the weights could be altered with the forms, they would be, in equal matter, greater or less according to the variety of forms, entirely contrary to experience. (C&W, 809)

If the weights of bodies are proportional to their quantities of matter, then they do not depend on their forms and textures. It thus follows from proposition 6 that one cannot alter the weight of a body merely by changing its shape or texture. To change the weight of a body one would have to add to or subtract from its quantity of matter. In addition to the phenomena explicitly appealed to in the basic argument for proposition 6, this corollary is supported by many chemical experiments as well as by vast amounts of ordinary experience. Squeezing flat a round clay ball does not change its weight.

[35] See Nordtvedt (1968b, 1186–7). Damour (1987, 144) points out that Newton's result is incorrect both in magnitude and sign, and refers to this calculation by Nordtvedt for the correct result. See appendix below.
[36] See section II.6 below. See also Harper et al. 2002.

Neither does melting a block of ice, in a container which keeps all the water, change its weight. Newton's suggestion that violations of this corollary would be entirely contrary to experience appears to appeal to this wide body of additional evidence from experience.

The first part of the next corollary expands on the basic argument for proposition 6. It adds appeal to a rule of reasoning to back up the extension of gravitation toward the earth, inferred from the pendulum experiments and the moon-test, to all bodies universally.

Corollary 2 *[first part]*. All bodies universally that are on or near the earth are heavy [or gravitate] toward the earth, and the weights of all bodies that are equally distant from the center of the earth are as the quantities of matter in them. This is a quality of all bodies on which experiments can be performed and therefore by rule 3 is to be affirmed of all bodies universally . . . (C&W, 809)

What is explicitly affirmed of all bodies universally is gravitation toward the earth with weights, at equal distances from the center of the earth, directly proportional to their quantities of matter. This inference from weights producing equal accelerations toward the earth at equal distances for bodies on or near the earth to such weights toward the earth for all bodies universally, is backed up by Newton's Rule 3.[37]

Rule 3. *Those qualities of bodies that cannot be intended and remitted [i.e., qualities that cannot be increased and diminished] and that belong to all bodies on which experiments can be made should be taken as qualities of all bodies universally.* (C&W, 795)

The quality belonging to all bodies on which experiments can be made, which this rule is here applied to generalize to all bodies universally, is weight toward the earth with equal ratios to inertial mass at equal distances from the center of the earth. For any given distance from the center of the earth, the ratio of weight to inertial mass is counted as a quality of bodies which cannot be increased or diminished, and which belongs to all bodies on which experiments can be made.

Corollary 2 continues with the following remarks suggesting untoward consequences of allowing for differing ratios of weight to mass.

Corollary 2 *[continued]* . . . If the aether or any other body whatever either were entirely devoid of gravity or gravitated less in proportion to the quantity of its matter, then, since (according to the opinion of Aristotle, Descartes, and others) it does not differ from other bodies except in the form of its matter, it could by a change of its form be transmuted by degrees into a body of the same condition as those that gravitate the most in proportion to the quantity of their matter; and, on the other hand, the heaviest bodies, through taking on by degrees the form of the other body, could by degrees lose their gravity. And accordingly the weights would depend on the forms of bodies and could be altered with the forms, contrary to what has been proved in corol.1. (C&W, 809)

[37] This rule was a new addition that first appeared in the second edition of 1713. It replaced what in the first edition had been hypothesis 3. See next note 38 below.

The untoward consequence is that, according to a certain widely held opinion, weight would depend on forms and could be altered by transformations of form alone.[38]

The next corollary argues that all spaces are not equally full.

Corollary 3. All spaces are not equally full. For if all spaces were equally full, the specific gravity of the fluid with which the region of the air would be filled, because of the extreme density of its matter, would not be less than the specific gravity of quicksilver or of gold or of any other body with the greatest density, and therefore neither gold nor any other body could descend in air. For bodies do not ever descend in fluids unless they have a greater specific gravity. But if the quantity of matter in a given space could be diminished by any rarefaction, why should it not be capable of being diminished indefinitely? (C&W, 810)

He goes on, in the following corollary, to offer a hypothetical sufficient condition for the existence of a vacuum.

Corollary 4. If all the solid particles of all bodies have the same density and cannot be rarefied without pores, there must be a vacuum. I say particles have the same density when their respective forces of inertia [or masses] are as their sizes. (C&W, 810)

The hypothesis of this corollary, like what Newton suggests is the widely held opinion about transformation of form which in the first edition he stated as hypothesis 3, is quite plausible given a certain specific sort of atomism according to which all atoms are identical and all differences among gross bodies are accounted for by differing arrangements of their atoms. It is quite clear, however, that these corollaries do not assert this sort of atomism as an established fact.[39]

The final corollary of proposition 6 outlines a number of differences between gravity and magnetic attraction.

[38] There was no appeal to Rule 3 in the first edition. In that edition the discussion of transformation of form included appeal to hypothesis 3.

> Hypothesis 3. Every body can be transformed into a body of any other kind and successively take on all the intermediate degrees of qualities. (C&W, 795, note bb; see also Koyré 1965, 263)

The widely shared opinion, referred to in the above statement from the third edition, appears to be this doctrine of possible transformation of form which Newton formulated as hypothesis 3 in the first edition.

Here in the third edition, this commonly held opinion is not endorsed as something Newton is claiming to count as an established fact.

Zvi Biener and Christopher Smeenk have shown that letters from Cotes arguing that this had not been established led Newton to change this corollary and to replace his hypothesis 3 by Rule 3. (Biener and Smeenk, forthcoming)

[39] A salient aspect of Newton's experimental philosophy, from its earliest expression in his contributions to the debate over his theory of light and colors, right up to his last revisions of the later editions of the *Principia*, is his effort to clearly distinguish between propositions he regards as established facts or principles and those he explores as conjectures, however plausible the latter may be (see Harper and Smith 1995).

As we saw in note 38 above, what Newton here describes as a commonly held opinion is what he counted as hypothesis 3 in his first edition. As we have noted (chpt. 2 and chpt. 4) in that first edition, Newton called phenomena and Rules 1 and 2 *hypotheses*. Biener and Smeenk relate how Cotes' arguments finally convinced Newton that his arguments for this conjecture were not sufficient to count it as an established fact or principle.

Corollary 5. The force of gravity is of a different kind from the magnetic force. For magnetic attraction is not proportional to the [quantity of] matter attracted. Some bodies are attracted [by a magnet] more [than in proportion to their quantity of matter], and others less, while most bodies are not attracted [by a magnet at all]. And the magnetic force in one and the same body can be intended and remitted [i.e., increased and decreased] and is sometimes far greater in proportion to the quantity of matter than the force of gravity; and this force, in receding from the magnet, decreases not as the square but almost as the cube of the distance, as far as I have been able to tell from certain rough observations. (C&W, 810)

Salient among these differences is that magnetic force does not have a constant (unvarying and invariable for each body and the same for all bodies at equal distances) ratio to mass. This is the quality of gravity toward the earth that cannot be intended and remitted in Newton's appeal to Rule 3 in corollary 2.

3. Regulae Philosophandi: *Rule 3: Newton's discussion*

Newton's discussions of his other rules of reasoning are at most a few lines for each rule. For Rule 3 he provides more than a full page of discussion. Here are the opening remarks of this discussion in the third edition.

For the qualities of bodies can be known only through experiments; and therefore qualities that square with experiments universally are to be regarded as universal qualities; and qualities that cannot be diminished cannot be taken away from bodies. Certainly idle fancies ought not to be fabricated recklessly against the evidence of experiments, nor should we depart from the analogy of nature, since nature is always simple and ever consonant with itself. (C&W, 795)

The remark about experiments and the remark about not letting idle fancies undercut the evidence of experiments are in line with the general defense of inferences from phenomena from being undercut by conjectured hypotheses that we saw in his discussion of Rule 4. In this case, the defense is focused on defending universalization. The appeal to the analogy of nature and corresponding ideas of uniformity and simplicity of nature are part of this focus on universalization. The appeal to the idea that qualities that cannot be diminished cannot be taken away exploits the specific criterion of not allowing intension or remission of degree.[40]

Newton next provides a rather elaborate discussion defending inferences extending such universal qualities to small parts of bodies and appealing to an analogy with our evidence for qualities, such as extension, impenetrability, and inertia, which were widely regarded as essential to bodies.

The extension of bodies is known to us only through our senses, and yet there are bodies beyond the range of these senses; but because extension is found in all sensible bodies, it is ascribed to all bodies universally. We know by experience that some bodies are hard. Moreover, because the hardness of the whole arises from the hardness of its parts, we justly infer from this

[40] This is, perhaps also, meant to be suggestive of Newton's earlier appeal to ideas about transformation of form.

not only the hardness of the undivided particles of bodies that are accessible to our senses, but also of all other bodies. That all bodies are impenetrable we gather not by reason but by our senses. We find those bodies that we handle to be impenetrable, and hence we conclude that impenetrability is a property of all bodies universally. That all bodies are movable and persevere in motion or in rest by means of certain forces (which we call forces of inertia) we infer from finding these properties in the bodies that we have seen. The extension, hardness, impenetrability, mobility, and force of inertia of the whole arise from the extension, hardness, impenetrability, mobility, and force of inertia of each of the parts; and thus we conclude that every one of the least parts of all bodies is extended, hard, impenetrable, movable, and endowed with a force of inertia. And this is the foundation of all natural philosophy. (C&W, 795–6)

This passage has been interpreted as an explication of the sort of inference that this rule is designed to justify.[41] The application of it we have been examining, however, is not especially focused on this sort of inference from qualities of macroscopic bodies to the same qualities for their microscopic parts.[42] Newton does claim, here, that we count such commonly accepted essential properties of bodies as extension, hardness, and impenetrability, together with mobility and inertia, as empirically established for the least parts of bodies from their holding for all bodies within reach of our experiments. His last remark even claims that this is the foundation of all natural philosophy.

Newton goes on to discuss a possible application of this third rule to adjudicate the question of whether or not matter is composed of indivisible atoms.

Further, from phenomena we know that the divided, contiguous parts of bodies can be separated from one another, and from mathematics it is certain that the undivided parts can be distinguished into smaller parts by our reason. But it is uncertain whether those parts which have been distinguished in this way and not yet divided can actually be divided and separated from one another by the forces of nature. But if it were established by even a single experiment that in the breaking of a hard and solid body, any undivided particle underwent division, we should conclude by the force of this third rule not only that divided parts are separable but also that undivided parts can be divided indefinitely. (C&W, 796)

This suggested application of Rule 3 does not seem unproblematic.[43] Moreover, it is not clear how the criterion of qualities that cannot be increased or diminished would figure in it.

[41] See McGuire 1970.

[42] We have seen that Newton's basic argument to extend gravitational weight proportional to mass to parts of bodies in proposition 6 is not an inference to accept this as, itself, a measurement outcome or as a result of extending such agreeing measurements to bodies beyond the reach of our experiments. Instead, it is a thought experiment designed to suggest how very improbable it would be to have all these agreeing measurements of the equality of such ratios for whole bodies if such equality did not, also, hold for their parts.

[43] The antecedent clause in the last sentence,

> But if it were established by even a single experiment that in the breaking of a hard and solid body, any undivided particle underwent division,

needs some interpretation, especially the phrase "any undivided particle." Clearly, Newton does not intend undivided particles to be as yet undivided small parts in the ordinary sense of "small part." Perhaps his undivided particles are to be understood as candidates for undividable atoms, if anything is. This would raise the problem of identifying that some small part is such a paradigmatic candidate. Another suggestion might be

The final paragraph in Newton's discussion includes the following summary of evidence for universal gravity and an application of Rule 3 to it.[44]

Finally, if it is universally established by experiments and astronomical observations that all bodies on or near the earth gravitate [*lit.* are heavy] toward the earth, and do so in proportion to the quantity of matter in each body, and that the moon gravitates [is heavy] toward the earth in proportion to the quantity of its matter, and that our sea in turn gravitates [is heavy] toward the moon, and that all planets gravitate [are heavy] toward one another, and that there is a similar gravity [heaviness] of comets toward the sun, it will have to be concluded by this third rule that all bodies gravitate toward one another. (C&W, 796)

He goes on to compare the argument from phenomena for universal gravity with the evidence for impenetrability of bodies.

Indeed, the argument from phenomena will be even stronger for universal gravity than for the impenetrability of bodies, for which, of course, we have not a single experiment, and not even an observation, in the case of the heavenly bodies. (C&W, 796)

Given that observations or experiments to establish impenetrability of heavenly bodies were not available, the case from phenomena for universal gravity was even stronger than that for universal impenetrability of bodies.

The example inferences to attribute extension, hardness, and impenetrability as universal, and indeed perhaps essential, properties of bodies were commonly accepted in Newton's day. These inferences, however, are not as impressively backed up by agreeing measurements as Newton's application of Rule 3 to extend weight toward the earth to bodies at even very great distances from the earth. Consider an inference to attribute hardness to Jupiter, a planet we now know to be a gas giant. Newton's actual application of Rule 3 seems to be better in accord with his statement of the rule and, also, more compelling than some of these examples appealed to in his discussion of the rule.

A long tradition had it that impenetrability should be counted as an essential quality of bodies. Newton offers the following remarks to make it clear that he does not regard his argument, or the application of Rule 3 in it, as establishing gravity as an essential quality of bodies.

Yet I am by no means affirming that gravity is essential to bodies. By inherent force I mean only the force of inertia. This is immutable. Gravity is diminished as bodies recede from the earth. (C&W, 796)

to interpret the "any" as universal quantification over undivided particles so that an experiment of the sort referred to would be taken to show that every level of particle reachable by our experiments would be further divisible. This, however, would raise the problem of how one would identify such an experiment.

[44] Janiak argues that in this summary review of the argument for universal gravity in his comments of Rule 3 Newton does not consider gravity a quality of bodies, but rather a type of interaction (Janiak 2008, 97). It is certainly the case that what is being claimed to be universal here is a force of interaction between bodies. But the fact that Newton cites this summary as an illustration of Rule 3 suggests that, in line with his explicit application of Rule 3 in corollary 2 of proposition 6 book 3, he construes the qualities of bodies (which cannot be intended and remitted) widely enough to include the measured equality of the constant ratios of weight to mass at each distance as supporting the generalization to infer "by this third rule that all bodies gravitate toward one another."

These last qualifying remarks were added in the third edition.[45] They are puzzling. The summary of the evidence for universal gravity he has just given ends with his assertion that it is sufficient to make Rule 3 warrant inference to the claim that all bodies mutually gravitate toward one another. Now, he backs up his claim that he does not regard gravity as essential by suggesting that it does not satisfy the condition of not admitting intension or remission of degree specified in Rule 3, because gravity is diminished as bodies recede from the earth.[46]

4. *Rule 3: Newton's explicit application to gravity*

Let us look more closely at the application of Rule 3 to corollary 2 of proposition 6. Here again is the principle:

Rule 3. *Those qualities of bodies that cannot be intended and remitted [i.e., qualities that cannot be increased and diminished] and that belong to all bodies on which experiments can be made should be taken as qualities of all bodies universally.* (C&W, 795)

[45] See Koyré and Cohen 1972, 555.

[46] Newton's actual application of Rule 3 explicitly construes qualities of bodies in Rule 3 widely enough to include heaviness toward the earth with weights equally proportional to masses of all bodies that are equally distant from the center of the earth. I have taken this explicit application of Rule 3 in corollary 2 proposition 6 and the powerful argument extending it affords to proposition 6 itself as carrying more weight than the rather confusing qualification added in this last remark.

Rule 3 does not, explicitly, claim that qualities which satisfy its stated conditions are essential to bodies. It only claims that they are universal. The fact that the examples with which Newton opens his discussion of this rule are qualities which were widely regarded as essential, the fact that his summary of evidence from phenomena for the universality of gravitation is declared to be even stronger than that for such qualities as even impenetrability, and finally the fact that he attempts to back up his claim that he does not regard gravity as essential by suggesting that it does not satisfy the condition of Rule 3, might suggest that he does, after all, regard the conditions stated in Rule 3 as conditions of qualities that are not merely universal but which are also essential to bodies.

J.E. McGuire (1995, 239–61) argues that Newton *did* construe the criterion of not admitting intension or remission of degree as a criterion of *essential* and not merely of *universal* qualities of matter. He suggests that this essentialist use of the criterion is deeply embedded in Newton's own conception of matter, whose essential features, he thought, had to be able to be revealed to our senses in experiments. According to McGuire, Newton appealed to and transformed to his own use an essentialist tradition going back to Aristotle, as he wrestled with the philosophical difficulties of developing this conception. If this is correct, then it gives a reason why Newton was unable to cut away the essentialist implications of the criterion when he used it in Rule 3 as part of his criterion for picking out universal qualities of matter.

I think Janiak is quite right that Newton takes mass to be as essential to bodies as extension, hardness, impenetrability, and mobility, and that he explicitly does not take gravity to be essential. Janiak also argues that Newton does not take gravity to be a quality of bodies:

> More importantly, Newton suggests that gravity is a type of interaction, rather than a quality, because it decreases with an increase in spatial separation, so it is sensitive to the spatial relations of two or more bodies. In that sense we might characterize it as a kind of spatial interaction between bodies. (Janiak 2008, 97)

We have just seen (note 44 above) that a restriction of Rule 3 to qualities specified narrowly enough to exclude gravity would undermine Newton's illustration of Rule 3 in the summary of his argument he included in his discussion of that rule. It also would undermine Newton's explicit application of Rule 3 in his actual argument for universal gravity.

Unless one can find an interpretation of this last remark that does not put it in this sort of conflict with Newton's intuitively powerful applications of Rule 3 in support of his argument, it seems appropriate to continue to regard it as a puzzle.

Here again is the inference it is applied to justify:

All bodies universally that are on or near the earth are heavy [or gravitate] toward the earth, and the weights of all bodies that are equally distant from the center of the earth are as the quantities of matter in them. This is a quality of all bodies on which experiments can be performed and therefore by rule 3 is to be affirmed of all bodies universally. (C&W, 809)

The quality of bodies which is generalized is weight toward the earth. Newton explicitly tells us that weights for bodies at equal distances from the center of the earth have equal ratios to the quantities of matter in them. For each distance, the ratio of a body's weight toward the earth to its quantity of matter is a constant for all bodies at that distance. The equal accelerations at equal distances which exhibit equal ratios of weight to mass in Newton's pendulum experiments and moon-test are all of bodies relatively near to the earth. The inference to all bodies universally extends this to include bodies at any distances, however great those distances may be. This universalization is a large jump beyond the data immediately exhibited in the moon-test and pendulum experiments. Rule 3 is the principle appealed to in order to back up this jump.

Newton's pendulum experiments measure the equality of the ratio of weight to quantity of matter for bodies at the surface of the earth to within one part in a thousand. For each such body x, let us define

$$Q_e(x) = w_x/m_x,$$

where w_x is its weight (toward the earth) and m_x is its quantity of matter (its inertial mass). For each pair x,y of such bodies let

$$\Delta_e(x, y) = 2|Q_e(x) - Q_e(y)|/|Q_e(x) + Q_e(y)|.$$

Our $\Delta_e(x,y)$ is a parameter, modeled on Eötvös ratios, for representing the magnitudes of differences in ratios of weight to mass.[47] By bounding differences to within one part in a thousand, Newton's pendulum experiments exhibit a bound of about

$$\Delta_e(x, y) < 0.001$$

for the pairs tested. Taking the outcome of this experiment to measure the equality of these ratios for bodies at the surface of the earth to within one part in a thousand is equivalent to taking it as measuring a bound of

$$\Delta_e < 0.001,$$

[47] Where a_1 and a_2 are the gravitational accelerations produced on two bodies at similar distances, the Eötvös ratio is

$$\frac{2\ |a_1 - a_2|}{|a_1 + a_2|}$$

In 1922 Eötvös carried out precise torsion balance experiments that measured bounds limiting this ratio to 5×10^{-9} (Will 1993, 24–7). The component acceleration of each toward the other induced by its weight toward the other is proportional to the product of its own mass and that weight toward the other body. This makes these experiments about gravitational attraction between laboratory bodies put that same bound on our corresponding parameter that would represent the bounds on the differences between the ratios of their weights toward one another to their inertial masses.

where Δ_e is a generalized parameter bounding such differences for all pairs of such terrestrial bodies.

Newton goes on to claim that, if terrestrial bodies were raised to the moon's distance and let go, their fall toward the center of the earth would exhibit the same acceleration as the moon. This reverses the direction of the moon-test of proposition 4, which compares the one second's fall $d_{sec} = 15.096$ Paris feet measured by Huygens with the result of inverse-square adjusting to the mean earth radius the fall exhibited by the lunar orbit. Let us compare Huygens's measurement of the acceleration of surface gravity, $g = 2d_{sec}$ Paris feet/sec^2 with the result of inverse-square adjusting to the mean earth radius the centripetal acceleration exhibited by the lunar orbit.

Consider the moon-test as a null experiment measuring the equality of these accelerations. Let us take Huygens's measurement of $g = 30.192$ Paris feet/sec^2 for a_1. Let us use all the estimates $x = R/r$ of the lunar distance cited in the original version of book 3 or in any of the published editions, leaving out the dubious Kircher estimate and the second correction to Tycho. The resulting 95% Student's *t*-confidence estimate for inverse-square adjusting the centripetal acceleration of the moon to the corresponding acceleration at the surface of the earth is[48]

$$g = 29.497 \pm 0.7313 \quad \text{Paris feet/sec}^2.$$

Taking a_2 as the lower of these bounds, 29.297− 0.7313 = 28.5657 Paris feet/sec^2 yields

$$2 \mid 30.192 - 28.5657 \mid / \mid 30.192 + 28.5657 \mid = 0.055$$

as a corresponding 95% confidence bound on the Eötvös ratio representing the magnitude of the difference between Huygens's measured value of the acceleration of gravity at the surface of the earth and the result of inverse-square adjusting the

[48] Here are the estimates:

Table 7.1

	x	g(x)
Ptolemy	59	28.542
Kepler	59	—
Boulliau	59	—
Helvelius	59	—
Riccioli	59	28.542
Flamsteed	59⅓	29.028
Vendelin	60	30.018
Huygens	60	30.018
Copernicus	60⅓	30.521
Street	60.4	30.622
Tycho (1)	61	31.544

These yield, for $g(x)$, a mean of 29.497, $sd^+ = 1.0876$, $SE = 0.3279$, and tSE (95% $t = 2.23$ for the 10 degrees of freedom corresponding to 11 estimates) of 0.7313 for

$$g(x) = 29.497 \pm 0.7313 \quad \text{Paris feet/sec}^2$$

as a 95% Student's *t*-confidence estimate.

centripetal acceleration exhibited by the moon's orbit to the corresponding acceleration at the surface of the earth.

The limits on these Eötvös ratios measured by the moon-test are limits on the extent to which the result of inverse-square adjusting a body's Q_e ratio to the mean earth radius has been established to approximate the same value for the moon and terrestrial bodies.[49] Given inverse-square variation of gravitation toward the earth, the equality of the periods in Newton's pendulum experiments measure

$$\Delta_e < 0.001,$$

where the generalized parameter Δ_e applies not just to pairs of terrestrial bodies but to the differences in ratios of weight (toward the earth) to inertial mass that any pairs of bodies would have at any equal distances above the surface of the earth. The bounds on Eötvös ratios measured in the moon-test can be counted as a cruder measurement of

$$\Delta_e < 0.055,$$

for this same general parameter. Though one is cruder than the other, both these measurements agree in bounding this generalized difference parameter toward zero.

The application of the third rule to universalize gravitation toward the earth with weight toward the earth equally proportional to quantity of matter at equal distances is backed up by the fact that the phenomena appealed to count as agreeing measurements bounding Δ_e toward zero. This generalization to all bodies above the surface, however far they may be away from the earth, is backed up by the claim that this parameter Δ_e has been found to have agreeing measurements of zero for all bodies within reach of our experiments. The bodies within reach of experiments for weight toward the earth, in Newton's day, were terrestrial bodies and the moon. We can thus understand the role of Newton's criterion, of not admitting intension or remission of degree, as one which endorses generalizing to all such bodies universally by Rule 3, those qualities which count as constant parameter values that have been backed up by agreeing measurements applying to all such bodies within reach of our experiments.

5. More bounds from phenomena

The rest of Newton's main argument for proposition 6 applies the same sort of inferences from phenomena which count as agreeing measurements for corresponding constant parameters characterizing gravitation toward Jupiter and toward the sun.

Consider the Harmonic Rule for Jupiter's moons. The Harmonic Rule for circular orbits is equivalent to the inverse-square diminution of the centripetal accelerations.

[49] Given $f=ma$, as applied to continuously acting forces, we have at each instant $Q(x) = a(x)$, where $a(x)$ is the acceleration of body x produced by $f(x) = w(x)$, its weight toward the earth.

For bodies at or above the surface, multiplying these ratios $Q(x)$ by $(D_x/\mathrm{r})^2$, where D_x is the distance of body x from the center of the earth and r is the mean-distance of the surface from the center, is equivalent to inverse-square adjusting all these centripetal accelerations to their corresponding accelerations at the surface of the earth.

Using the data cited in Newton's table gives agreeing inverse-square adjusted estimates of centripetal acceleration toward Jupiter which exhibit a limit of $\Delta_j < 0.014$ from the Eötvös ratios for gravitation toward Jupiter.[50] Though Newton does not mention it, his data for Saturn's moons give estimates that exhibit 95% confidence limits bounding $\Delta_s < 0.04$ for gravitation toward Saturn.[51]

Consider Newton's appeal to the Harmonic Rule for the primary planets. Using the semi-major axis as our estimate of the mean-distance in an elliptical orbit and the centripetal acceleration of the corresponding circular orbit with the same period as our estimate of the mean centripetal acceleration, the Eötvös ratio becomes an estimate of the equality of inverse-square adjusted accelerations toward the sun (Helios) exhibited by the orbital data for the planets. We compute the centripetal accelerations at the mean-distance for each orbit, inverse-square adjusted to the mean-distance of the earth's orbit. This yields a confidence interval about the mean, which gives a bound of $\Delta_H < 0.005$ from the Eötvös ratios.[52]

[50] Using the mean values of the distance estimates cited in Newton's table for phenomenon 1 we have for moons 1–4, distances of respectively 5.463, 8.612, 13.713, and 24.422 of Jupiter's semidiameters. The periods are respectively 1.769, 3.551, 7.1546, and 16.689 decimal days. For concentric circular orbits we compute the centripetal acceleration of each moon as $V^2/R = ((2\pi R)/t)^2/R$ and we take as our distance unit the distance $R_2 = 8.612$ of Jupiter's semidiameters assigned as the distance of moon two. For each moon i the distance $r_j(i)$ of moon i from the center of Jupiter is R_i/R_2. This gives for each moon i an estimate $(((2\pi R_i)/t_i)^2/R_i)(R_i/R_2)^2$.

The numerical values of these four estimates in semidiameters cubed over decimal days squared are respectively 27.7359, 26.963, 26.820, 27.8378. The mean is 27.3392, with an *sd* of 0.2043 and an sd^+ of $(4/3)^{1/2}sd = 0.2359$. The $SE = (N^{1/2}/N)sd^+ = 0.118$. The 95% confidence Student's *t*-parameter for 3 degrees of freedom is 3.18, which gives error bounds of $\pm$ 3.18 sd^+ or 0.375 semidiameters cubed per decimal day squared about the mean.

Let us take a_1 as the mean value 27.3392 and a_2 as an estimate of the acceleration at the distance R_2 of the second moon, which is at the error bound of 0.375 below the mean. The corresponding limit on the Eötvös ratio is 0.014.

[51] Newton cites from Cassini periods *t* equal to respectively 1.8878, 2.7371, 4.5175, 15.9453, and 79.325 decimal days for the five satellites of Saturn known to him. His observed distances *R* for these satellites in semidiameters of Saturn's ring are respectively 1.95, 2.5, 3.5, 8, and 24. These yield centripetal accelerations, $4\pi^2 Rt^2$, of respectively 21.60, 13.17, 6.77, 1.24, and 0.15 semidiameters of the ring per decimal day squared. Multiplying each by $(R/2.5)^2$ to inverse-square adjust them all to the distance 2.5 semidiameters of the ring for the second satellite yields a 95% Student's *t*-confidence estimate of

$$13.24 \pm 0.52 \quad \text{semidiameters/day}^2.$$

Taking the mean 13.24 as a_1 and the lower bound 13.42−.52 = 12.72 as a_2 yields $2 \mid a_1 - a_2 \mid / \mid a_1 + a_2 \mid = 0.04$ for a 95% confidence bound of $\Delta_s < 0.04$ for Saturn.

[52] Where *a* is the mean distance (the semi-major axis of the elliptical orbit) and t is its period, these yield the following estimates $(4\pi^2 a/t^2)(a^2)$ for the centripetal accelerations inverse-square adjusted to the distance of the earth's orbit. Newton's cited periods and distances from Kepler and Boulliau yield the following estimates for the acceleration toward the sun at the earth's distance in units of AU/t_e^2, where AU is the mean sun–earth distance and t_e is the period of the earth's orbit which Newton sets at 365.2565 decimal days.

Table 7.2

	Mercury	Venus	Earth	Mars	Jupiter	Saturn
Kepler	39.7733	39.6171	39.4784	39.4637	39.3739	39.1319
Boulliau	39.0977	39.6139	—	—	40.0299	39.5280

6. *Bounds from un-polarized orbits*

Newton appeals to the absence of observable polarization of orbits of Jupiter's moons with respect to the sun to argue that those moons and Jupiter are equally accelerated toward the sun by solar gravity at equal distances. Data limiting orbital polarization of Jupiter's moons does put bounds on Δ_H, even though the bounds are not as precise as those according to the calculation described by Newton. We can represent bounds on polarization as bounds on $\delta r/r$, where δr is the difference between the radius vector r of the basic orbit in polar coordinates and the radius vector of the corresponding polarized orbit.

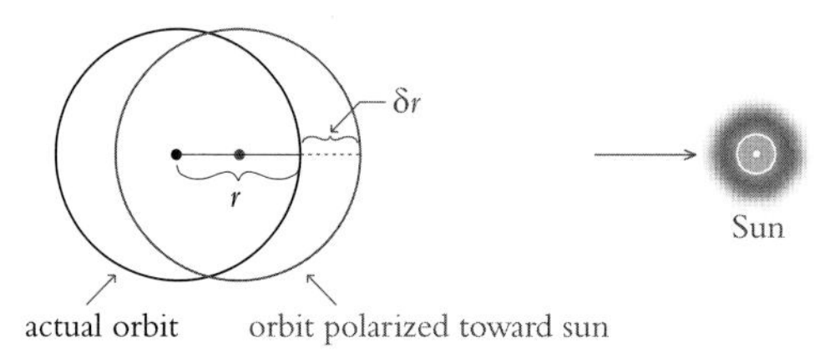

Figure 7.1 Polarized orbit

The calculation described by Newton would make bounding the maximum $\delta r/r < 0.2$ for Jupiter's moon Callisto yield a bound of $\Delta_H(J, C) < 0.001$, where J and C are Jupiter and Callisto at equal distances from the sun. The correct calculation of Nordtvedt would make the bounding maximum $\delta r/r < 0.2$ yield the much weaker bound of $\Delta_H(J, C) < 0.084$ from Callisto's orbit.

Let us see how well data available to Newton could have constrained $\Delta_H(J, C)$ on Nordtvedt's calculation. If we take the mean, 24.42, of the four estimates cited by Newton from other astronomers as r for Callisto in semidiameters of Jupiter and the corresponding 95% Student's t error bound 1.16 semidiameters of Jupiter as our bound on δr, this would bound the maximum $\delta r/r < 0.048$.[53] Nordtvedt's result makes this

The corresponding 95% t-confidence estimate for the centripetal accelerations inverse-square adjusted to the mean distance of the earth is

$$39.5108 \pm 0.199 \quad \text{AU}/t_e^2.$$

Taking as a_1 the 95% upper bound $39.5108 + 0.199 = 39.7098$ AU/t_e^2 and as a_2 the mean 39.5108 AU/t_e^2 yields about

$$2 \mid 39.7098 - 39.5108 \mid / \mid 39.7098 + 39.5108 \mid = 0.005$$

as a 95% t-confidence bound on the Eötvös ratio.

In Harper and DiSalle (1996, S49), the average of Kepler's and Boulliau's estimates were used for the distances and two 95% confidence bounds were used to arrive at 0.008 as a bound on the Eötvös ratio. Harper (1999, 91–3) reports the result of using the averages between Kepler and Boulliau for the distances and one 95% confidence bound to arrive at 0.004 as a bound on the Eötvös ratio. By taking all ten distinct estimates into account and using one 95% confidence bound as the difference between the accelerations, the present calculation gives a better estimate of the 95% confidence bound on the Eotvos ratio corresponding to the data cited by Newton supporting the Harmonic Law for the primary planets.

[53] The four estimates of r for Callisto cited by Newton in his table for phenomenon 1 are respectively 24.67, 24.72, 23, and 25.3 semidiameters of Jupiter. These yield a mean of about 24.42, with an sd^+ of about

Table 7.3 Constraints on Δ

1. Pendulum experiments	Newton (1685)	$\Delta < .001$
	Bessel (1827)*	$\Delta < 2 \times 10^{-5}$
	Eötvös (1922)*	$\Delta < 2 \times 10^{-9}$
	Moscow (1972)*	$\Delta < 10^{-12}$
2. Moon-test	Newton	$\Delta < 0.055$
3. Harmonic Law		
(Jupiter's moons)	Newton	$\Delta < 0.014$
(Saturn's moons)	Newton	$\Delta < 0.04$
4. Harmonic Law		
(Primary planets)	Newton	$\Delta < 0.005$
5. Bounds on polarization of satellite orbits		
Jupiter's moons	Newton	$\Delta < 0.02$
Our moon	Laplace (1825)**	$\Delta < 0.54 \times 10^{-7}$
	Lunar Laser ranging (1994)***	$\Delta < (25) \times 10^{-13}$

* Will, C.M. (1993, 27)
** Damour, T. and Vokrouhlický, D. (1996, 4198–9)
*** Dickey et al. (1994, 485)

bound on $\delta r/r$ for Callisto's orbit limit $\Delta_H(J, C) < 0.02$.[54] Newton's Rule 3 would take this bound on $\Delta_H(J, C)$ as a bound on the general parameter Δ_H, giving differences of ratios of weight toward the sun to inertial mass any pair of bodies would have if they were at any equal distances away from the sun.

The absence of polarization with respect to the sun of the orbits of Saturn's moons and of the orbit of the earth's moon add more phenomena which measure limits on Δ_H. Damour and Vokrouhlický (1996) argue that Laplace achieved calculations able to make such phenomena for the earth's moon limit Δ_H to less than one part in ten to the seventh.[55]

7. *Concluding remark*

Newton has now exhibited phenomena that count as agreeing measurements of zero for Δ_e on gravitation toward the earth, phenomena that measure zero for Δ_j on

0.99, an *SE* of about 0.495 for a Student's 95% *t*-confidence bound of about 1.16 semidiameters. The corresponding bound on $\delta r/r$ is about $1.16/24.42 = 0.048$.

[54] Applying Nordtvedt's basic result to the parameters of Callisto's orbit about Jupiter and Jupiter's about the sun yields maximum

$$\delta r/r = 2.37\Delta.$$

This makes the bound of 0.048 on $\delta r/r$ yield

$$\Delta < 0.02.$$

See appendix of this chapter.

[55] They argue (Damour and Vokroulicki 1996, 4199) that Laplace was able to successfully use data from our moon to bound $\Delta < 0.54\text{x}10^{-7}$.

gravitation toward Jupiter, and that would measure zero for Δ_s on gravitation toward Saturn. He has also provided a remarkable range of phenomena bounding Δ_H toward zero for gravitation toward the sun. These are all the planetary bodies within reach of experiments accessible to Newton. Therefore, Rule 3 would endorse counting these as agreeing measurements bounding toward zero a single universal parameter Δ representing for all planets differences between ratios of mass to weight toward that planet for any pair of bodies at any equal distances from it.

Table 7.3 gives the bounds on Δ by phenomena considered by Newton. It includes the bounds from data available to Newton as well as some of the more precise measurements that have become available since his time. The sharper bounds resulting from these later measurements are examples of what Newton referred to in Rule 4 as making propositions gathered from phenomena by induction more exact. For the equivalence principle represented by $\Delta = 0$, there has not yet been any phenomenon that would make it liable to exceptions.

Chapter 7 Appendix: Polarized satellite orbits as measures of Δ_H

According to Thibault Damour, in his classic comparison of the problem of motion in Newtonian and Einsteinian gravity, Newton's estimate of the orbital polarization that would be generated by differences in ratios of weight toward the sun to inertial mass for a planet and its satellite is incorrect in both magnitude and sign.[56] He cites Nordtvedt's 1968b for a correct calculation of the sort of polarization Newton appeals to. Let us apply Nordtvedt's calculation to one of Jupiter's moons.

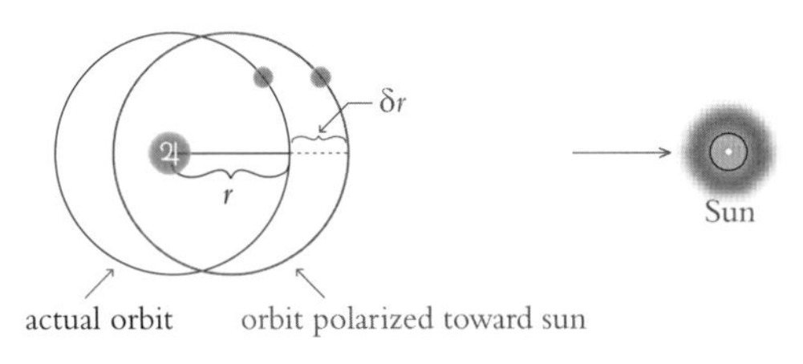

Figure 7.2 Polarized orbit

1. *Nordtvedt's calculation*

Nordtvedt assumes a circular orbit which satisfies the equations of motion

$$d^2r/dt^2 = h^2/r^3 - \mu/r^2 + \delta a \cos\psi \qquad (n1)$$

$$dh/dt = -\delta a\, r \sin\psi \qquad (n2)$$

where r is the moon's radius vector in a coordinate system centered on Jupiter, h is the magnitude of the angular momentum, and δa is the excess acceleration of Jupiter toward the sun over that of the moon. We shall take $\mu = G(m(\text{Jupiter}) + m(\text{moon}))$, where G is the gravitational constant, and the m's are respectively the inertial masses of Jupiter and the moon being considered. Perturbing these equations in (r, h) about a circular orbit of angular momentum h_0 and radius $r_0 = h_0{}^2/\mu$ leads to

$$\delta r = [3\delta a/(\omega_0{}^2 - \Omega^2)] \cos\Omega t, \qquad (1)$$

where ω_0 is the moon's angular frequency, Ω is $\omega_0 - \omega_s$ (where ω_s is the angular frequency of the sun about Jupiter in the Jupiter-centered coordinate system).

[56] Damour is one of the world's leading practitioners of celestial mechanics in the present age. He also takes a keen interest in understanding the role of evidence in the historical development of gravitation theory. See Damour 1987.

This has the effect that $|\delta r|$, the absolute value of δr, is maximal when Ω is 0 or π, i.e. when the moon is directly toward the sun or directly opposite to the sun. It also follows that if δa is negative (i.e. if the acceleration of the moon toward the sun is greater than that of Jupiter toward the sun), the orbit of the moon will be polarized toward the sun. However, Newton describes his calculation as showing that having the ratio for the moon greater than that for Jupiter (i.e. having δa negative) would result in a polarization away from the sun, in contradiction with (1) above.

Let us check the magnitude of δr. If T_0 is the period of a moon about Jupiter and T_j is Jupiter's period about the sun, we have

$$\Omega = \omega_0 - \omega_s = (2\pi/T_0) - (2\pi/T_j) \tag{2}$$

$$\omega_0{}^2 - \Omega^2 \approx 2\omega_0\omega_s = 8\pi^2/(T_0 T_j), \tag{3}$$

since $\omega_s{}^2 \ll \omega_0{}^2$.[57] We now have

$$\delta r = 3\delta a(T_0 T_j/8\pi^2)\cos\Omega t. \tag{4}$$

Defining

$$\Delta := [w_H(Jupiter)/m(Jupiter) - w_H(Callisto)/m(Callisto)]$$

we then have

$$\delta a = \Delta GM/R^2 \tag{5}$$

where w_H is weight toward the sun and m is inertial mass, $G = 6.67 \times 10^{-11}$ meters^{3}kg^{-1}seconds^{-2} is the gravitational constant, $M = 1.99 \times 10^{30}$ kg is the mass of the sun, and $R = 7.78 \times 10^{11}$ meters is Jupiter's mean-distance from the sun. Inserting these values gives

$$\delta a = \Delta\, 2.19 \times 10^{-4}\text{meters seconds}^{-2} \tag{6}$$

and substitution of (6) in (4) yields

$$\delta r = [T_0\ T_j(3\Delta)(2.19 \times 10^{-4})/8\pi^2]\cos\Omega t \tag{7}$$

which expresses δr as a function of the periods T_0 and T_j. Inserting Jupiter's period $T_j = 3.74 \times 10^8$ seconds yields the result

$$\begin{aligned}\delta r &= \Delta\ T_0((3)(2.19)(3.74)/8\pi^2)(10^{-4})(10^8)\cos\Omega t \\ &= \Delta\ T_0(.311)(10^4)\cos\Omega t\ \text{ meters/second}\end{aligned} \tag{8}$$

where T_0 is the period of the moon.

Consider Callisto, the outermost moon of Jupiter that Newton cites data for, as an example. In this case $T_0 = 1.44 \times 10^6$ seconds, and so

$$\delta r = \Delta\,(.448 \times 10^{10})\cos\Omega t \leq 4.48 \times 10^6 \Delta\ \text{km}.$$

[57] Jupiter's period is about 11.8565 Julian years (*ESAA*, 704), which makes $w_s{}^2$ 1.68(10^{-16}) radians^2/sec^2. For Callisto, Jupiter's outermost Galilean satellite, the period is 16.689 days (*ESAA*, 708), making $w_0{}^2 = 6.8$ (10^{-8}) radians^2/sec^2.

The radius r of Callisto's orbit is 1.888×10^6 km, giving

$$\delta r/r = (4.48 \times 10^6)/(1.888 \times 10^6)\,\Delta = 2.37\,\Delta$$

2. *Chandrasekhar on Newton's calculation*

Chandrasekhar (1995, 364–9) offers a reconstruction of Newton's calculation as an application of the method of variation of orbital elements. His proposed reconstruction begins with a characterization of the equation of motion which includes a disturbing function corresponding to differences in ratios of passive gravitational to inertial mass:[58]

$$d^2\boldsymbol{r}/dt^2 = -\mu\boldsymbol{r}/r^3 + GM_\odot\ [-\delta_2\boldsymbol{r}/\rho^3 + \boldsymbol{R}(\delta_2/\rho^3 - \delta_1/R^3)]$$

where $\delta_1 = m_{pg}(\text{planet})/m(\text{planet})$, $\delta 2 = m_{pg}(\text{moon})/m(\text{moon})$, $\boldsymbol{r}$ is the vector from the planet to the moon, $\boldsymbol{R}$ is the vector from the planet to the sun, $\boldsymbol{\rho}$ is the vector from the moon to the sun, $\mu = G(m(\text{planet}) + m(\text{moon}))$, and M is the gravitational mass of the sun.

The disturbing force F is given by the second term on the right-hand side. Chandrasekhar employs a right-handed coordinate system defined by unit vectors along the directions of the angular momentum vector $\boldsymbol{h}$, the radial vector $\boldsymbol{r}$, and a (which is 90 degrees more than the direction of $\boldsymbol{r}$ in the plane of the orbit). We will use bold letters for vectors and the same letters un-bolded for their scalar magnitudes. These scalar magnitudes are respectively $h = |\boldsymbol{h}| = |\boldsymbol{r}\times\boldsymbol{v}|$, where $\mathrm{v} = d\boldsymbol{r}/dt$, $r = |\boldsymbol{r}|$, and $a = hr$. The components of the disturbing force vector $\boldsymbol{F}$ are

$$F_r = (1/2\,GM)/R^3[\delta_2 r(1 + 3\cos 2\psi) + 2(\delta_2 - \delta_1)R\cos\psi]$$
$$F_a = (1/2\,GM)/R^3[-3\delta_2 r\ \sin 2\psi - 2(\delta_2 - \delta_1)R\sin\psi]$$
$$F_h = GM/R^3[-3\delta_2 r\sin U \sin i \cos\psi - (\delta_2 - \delta_1)R\sin U \sin i]$$

In F_h, U is the angle of the sun from the ascending node, i is the inclination of the orbit, and ψ is the angle between the radial vector $\boldsymbol{r}$ and the vector $\boldsymbol{R}$ from the origin to the sun.

Chandrasekhar observes that the terms in δ_2 - δ_1 are one order lower in r/R than the standard tidal terms due to the action of the sun on the planet—moon system. He argues [1995, 366] that if these terms are not to swamp the well-confirmed tidal terms: we must, in fact, require

$$(R/r)(\delta_2 - \delta_1) \ll 1.$$

He then offers summary calculations which give bounds of, respectively, 2.57×10^{-3} and 2.61×10^{-3} on $(\delta_2 - \delta_1)$ for the earth–moon and Jupiter–Callisto systems.

The calculation Chandrasekhar attributes to Newton gives the variation in the semi-major axis a due to $(\delta_2 - \delta_1)$ as a function of the perturbing force component $F\alpha$. By requiring the net change in a due to the combination of tidal effects and the $F\alpha$ component due to (δ_2 - δ_1) to vanish, Chandrasekhar arrives at the result that

If $(\delta_2 - \delta_1) = 1/1000$ then $\delta r/R = 1/2250$.

This is approximately ⅙ the Jupiter–Callisto distance. Newton cites 1⁄2000, which is approximately ⅕ the Jupiter–Callisto distance.

Chandrasekhar's proposed reconstruction, by setting the variation in a to zero, ignores the variation in the angular momentum h. On Chandrasekhar's proposal, Newton's incorrect result would be due to his having ignored the variation of h. The Nordtvedt calculation can be

[58] The respective weights toward the sun of the planet and its moon are proportional to their respective passive gravitational masses $m_{pg}(\text{planet})$ and $m_{pg}(\text{moon})$.

obtained from the variation of Keplerian orbital elements if the variation of h is properly taken into account. The following sketch starts from the same characterization of the disturbing function F used by Chandrasekhar.

Nordtvedt assumed a circular orbit with zero inclination. Since $i=0$, $F_h = 0$. Also, since $r/R \ll 1$, $r/R^3 \rightarrow 0$, the first terms in F_r and F_α involving just δ_2 drop out. Expression of the vectorial equation for d^2r/dt^2 above in terms of its components gives:

$$\begin{aligned} d^2r/dt^2 &= -\mu/r^2 + h^2/r^3 + 1/2((GM/R^3)\,2(\delta_2 - \delta_1))\,R\cos\psi \\ &= -\mu/r^2 + h^2/r^3 + GM/R^2(\delta_2 - \delta_1)\cos\psi \end{aligned}$$

From the other component we obtain for the variation of h

$$\begin{aligned} dh/dt = rF_a &= r(1/2)(GM/R^3)\ \ (-2(\delta_2 - \delta_1)\,R\ \sin\psi) \\ &= -r(GM/R^2)(\delta_2 - \delta_1) = r(GM/R^2)\Delta\sin\psi \end{aligned}$$

Suppose that δ_2 (the ratio for the moon) is greater than δ_1 (the ratio for the planet). This makes Δ (and δa) negative, polarizing the orbit toward the sun, in exact agreement with what Nordtvedt calculated from his equations of motion (n1) and (n2), since he takes δ_2 for the planet and δ_1 for the moon.

3. *Another proposal for Newton's calculation*

In his essay review of Chandrasekhar's book, George Smith (1996) points out that the number 1/2250 reached by Chandrasekhar's reconstruction differs non-trivially from Newton's 1/2000. Smith points out that David Gregory, who was a close associate of Newton, offered a less elaborate but more obvious candidate for Newton's calculation in his (unpublished) commentary on the *Principia*. Smith kindly made available to me relevant passages from Gregory's Latin manuscript. Curtis Wilson generously took the time to translate these passages into English for me. The following interesting suggestion by Gregory is from this translation provided to me by Wilson.

> [Uti calculis quibusdam initis inveni: As I found by some calculations begun] I wish that the author would indicate the basis for them. Perhaps the following is not dissimilar to them. If a satellite were heavier (for its quantity of matter) toward the Sun than toward Jupiter, it would be necessary that the center of the satellite's orbit be beyond Jupiter with respect to the Sun, so that, compensating by this greater distance for the greater gravity, the orbit could subsist. If Jupiter's distance from the Sun were 1000, and its gravity also 1000, and if the center of the satellite's orbit were distant from the Sun by 1001, its gravity (if it were the same as Jupiter's gravity) would be 1000−2, namely decreasing in the duplicate ratio of the increased distance (for 1000, 1001, 1002 are arithmetically proportional, and therefore geometrically proportional to a near approximation).

Suppose Jupiter is in a concentric circular orbit about the sun while Callisto is in a concentric circular orbit about Jupiter. We can think of the center of Callisto's orbit and Jupiter as two bodies that need to have equal accelerations toward the sun. The basic idea here is that if one of these bodies has a larger $w_H/m = \delta_1$ it must be further from the sun in order to have its acceleration a_1 toward the sun equal the acceleration a_2 toward the sun of the other body which has the smaller $w_H/m = \delta_2$. Having the weights toward the sun inversely as the squares of the distances makes

$$|a_1|/|a_2| \approx (\delta_1/\delta_2)(R_2^2/R_1^2).$$

To have

$$|a_1|/|a_2| = 1$$

would require having

$$(R_1^2/R_2^2) \approx (\delta_1/\delta_2)$$

so that having

$$(\delta_1/\delta_2) = (1001/1000)$$

would make

$$R_1/R_2 \approx (1001/1000)^{1/2} \approx (2001/2000)$$

just as Newton claims. Perhaps this is what Newton did.

Gregory, however, goes on to point out that if one takes the centripetal acceleration of the circular orbit of the planet about the sun and of the orbit about the sun of the center of the moon's orbit about Jupiter, the ratio of the *R*'s goes as the cube root rather than as the square root of the ratio of the δ's.

> But on account of the greater circle described in the same time as Jupiter, the centrifugal [!] force will be greater (by Prop. 4 of Bk. I) in the ratio of the increased distance from the center. Therefore for this reason [caput] the gravity of the satellite will be 1000–1. Hence for both reasons it will be 1000–3; that is, gravity decreases in the triplicate ratio of the increased distance. Whence the distance is to be increased in the subtriplicate ratio of the increased gravity of the satellite over the gravity of Jupiter; that is, in the subtriplicate ratio of *d* to *e*; in order, indeed, that the satellite's orbit should subsist and not fail. For as the satellite is found in every point in its orbit successively, it may be conceived as if it were a solid ring subsisting around Jupiter. The same reasoning is to be used if the satellite is less heavy toward the sun than Jupiter. But by this reasoning, it is not the halved [subduplicate] ratio that is to be applied, as the Author does, but the subtriplicate.

8

Gravity as a Universal Force of Interaction (Propositions 7–13 Book 3)

In Part I we review Newton's argument for proposition 7, its corollaries, and the conceptual transition from gravity as centripetal forces of attraction toward planets to gravity as a universal force of pair-wise interaction between bodies. Section I.1 emphasizes the role of Newton's application of Law 3 in his argument. Further details are offered in appendix 1. Section I.2 points out that Newton's inferences to inverse-square attraction toward particles is backed up by systematic dependencies that make the inverse-square variation of gravitation toward a whole sphere measure the inverse-square variation for gravitations toward the particles of which it is composed. This is backed up and further illustrated in section 2 of appendix 2 and section 2 of appendix 3. Section I.3 argues that the conceptual transition from centripetal forces of attraction to a universal force of pair-wise interaction is an example of the methodology of seeking successively more accurate approximations.

Part II reviews Newton's application of theorems about attractive forces toward spherical bodies in proposition 8 and its corollaries. Section II.1 reviews Newton's basic argument for proposition 8 and its motivation by the problem of extending the inverse-square variation of gravity toward a planet right down to the surface of that planet. Section II.2 reviews Newton's application of inverse-square gravity toward planets to compare the surface gravities of the sun, Jupiter, Saturn, and the earth. Passages from Huygens testify to his great admiration for this achievement of Newton's. Section II.3 reviews Newton's application of Law 3 to measure the masses of the sun, Jupiter, Saturn, and the earth from orbits about them. This achievement is one that Huygens does not accept, because he objects to the application of Law 3 on which it depends.

Part III reviews Newton's application of his theory of gravity to the motions of solar system bodies. Section III.1 examines his application of measurements of masses from orbits to arrive at his surprising resolution of the two chief world systems problem. Evidence available to Newton to sufficiently limit the masses of planets without moons is explored. Section III.2 reviews Newton's project of applying universal gravity to account for motions of solar system bodies by successively more accurate approximations.

Appendix 1 gives an exposition of Newton's proof of proposition 69 book 1. This is the proposition appealed to in Newton's application of Law 3 to argue that gravitation toward a planet is proportional to the mass of that planet. Readers will find that Newton's reasoning in this important inference is quite accessible. It also adds illumination to our exposition in chapter 8 section I.1 of his argument to extend this result to parts of planets.

Appendix 2 reviews Newton's theorems about attractive forces toward spherical bodies. Given Newton's basic result about attraction toward spherical shells, his inferences to his other propositions can be quite accessibly explained. Section 1 explains these propositions from book 1 that Newton appeals to in his argument for proposition 8 of book 3 and its applications to cosmology. Section 2 backs up Newton's inference to inverse-square attraction to particles. It extends an integral that Chandrasekhar argues is equivalent to Newton's proposition 80 of book 1 to show that Newton's inference to inverse-square attraction to particles from inverse-square attraction toward whole solid spheres is backed up by systematic dependencies. This is illustrated by a plot comparing variation of attraction with distance toward a whole spherical solid for the inverse-square power law, with alternatives given by alternative power laws for the attractions toward the particles which sum to make it up.

Appendix 3 gives details of Newton's proofs about attraction toward spherical shells. Section 1 expounds Newton's basic arguments and the lemma on first and last ratios to which they appeal. Section 2 extends Newton's own proof to support measuring inverse-square attraction toward particles by inverse-square attraction toward a spherical shell made up of those attracting particles.

Appendix 4 reviews details of Newton's measurements of properties of planets from orbits in the corollaries of proposition 8 book 3. Section 1 reviews and assesses Newton's measurements of comparative surface gravities. Section 2 reviews and assesses Newton's measurements of comparative masses and densities of the sun and planets from orbits about them.

I Proposition 7

Newton's proposition 7 is the culmination of his basic argument for universal gravity. Here is the proposition and its proof.

Proposition 7 Theorem 7

Gravity exists in all bodies universally and is proportional to the quantity of matter in each.

We have already proved that all planets are heavy [or gravitate] toward one another and also that the gravity toward any one planet, taken by itself, is inversely as the square of the distance of places from the center of the planet. And it follows (by book 1, prop. 69 and its corollaries) that the gravity toward all the planets is proportional to the matter in them.

Further, since all the parts of any planet A are heavy [or gravitate] toward any planet B, and since the gravity of each part is to the gravity of the whole as the matter of that part to the matter of the whole, and since to every action (by the third Law of Motion) there is an equal reaction, it follows that planet B will gravitate in turn toward all the parts of planet A, and its gravity toward any one part will be to its gravity toward the whole of the planet as the matter of that part to the matter of the whole. Q.E.D. (C&W, 810–11)

The two paragraphs correspond to two parts of the proof. The first paragraph argues that the gravity toward each planet is proportional to the quantity of matter of that planet.[1] This argument appeals to proposition 69 book 1 and its corollaries. Proposition 69 book 1 applies the third Law of Motion to argue that if bodies attract each other by inverse-square accelerative forces, then the attraction toward each will be as its quantity of matter or mass.[2] The second paragraph argues to extend this result to all the parts of planets. This second argument appeals directly to Law 3. In both arguments the forces are interpreted as interactions between the bodies in such a way as to allow Law 3 to apply, by counting the attraction of each toward the other as action and equal and opposite reaction.

Given these applications of Law 3, all planets and parts of planets count as centers toward which all bodies gravitate; so, this short, two-paragraph, argument leads right to Newton's revolutionary construal of gravity as a universal force of interaction between bodies.[3]

[1] In proposition 6 it was shown that all bodies gravitate toward each planet, and that at any given distance from the center of any one planet the weight of any body toward that planet is proportional to the mass of that body being attracted. As we noted above (chpt. 1 note 86), some physicists today (e.g. Damour 1987, 143) would put this as an identification of passive gravitational mass with inertial mass. Here, we have gravitation toward a planet proportional to the mass of the planet itself, as the body doing the attracting. This, together with the extension to construe all bodies as such centers of gravitation proportional to their masses, extends this identification to include, also, what such physicists today would put as the identification of active gravitational mass with inertial mass.

[2] See appendix 1 for an exposition of Newton's illuminating argument.

[3] To the extent that all bodies universally are counted as either planets or parts of planets, this argument leads directly to universal gravitation. As Newton's discussion in proposition 6 suggests, any ordinary body can become a part of a planet just by falling on it.

On the traditional use of "planet," the fixed stars and the earth are not planets. Comets, irregular wanderers, may or may not have counted as planets; but, Newton's impressive application of gravitation toward the sun to explain their motions, in propositions 40–2 book 3, will clearly bring them into the range of bodies directly covered by this argument.

The earth, of course, has gravity toward it. Moreover, given his application of the third law to gravitation, Newton's argument for proposition 7 directly shows that gravity toward the earth is proportional to the mass of the earth. To the extent that the fixed stars count as bodies, proposition 7 could be extended to them by Rule 3.

In his admirable book *Newton's Gift: How Sir Isaac Newton Unlocked the System of the World*, David Berlinski remarks:

> With this proposition, western science passed at once from its inception to full maturity. For the first and only time in the *Principia*, Newton concluded **Proposition VII** with the letters QED – *quad erat demonstrandum*. Thus it is demonstrated. (Berlinski 2000, 141)

The association with Euclid's use of these letters to indicate that a proof he has offered establishes a proposition as a demonstrated theorem of geometry (e.g. Heath, 317 for the Pythagorean theorem), may very well indicate Newton's high regard for the strength of this very significant argument. The propositions he counts as theorems in books 1 and 2 are ones he regularly concludes with "QED." This suggests that he wants us to regard this central claim in his argument for universal gravity as having similar warrant to these theorems that directly follow from his Laws of Motion.

1. The argument for proposition 7

Here, again, is the first paragraph of the argument for proposition 7 book 3.

We have already proved that all planets are heavy [or gravitate] toward one another and also that the gravity toward any one planet, taken by itself, is inversely as the square of the distance of places from the center of the planet. And it follows (by book 1, prop. 69 and its corollaries) that the gravity toward all the planets is proportional to the matter in them. (C&W, 810)

As we have seen, proposition 6 book 3 shows that planets gravitate toward each other, and that for any given distance from the center of any given planet the weight toward that planet of any body at that distance from it is proportional to the mass of that attracted body. Corollary 2 of proposition 5 shows that gravitation toward any planet varies inversely with the squares of distances from its center. These features asserted in the first sentence make gravity toward planets satisfy conditions explicitly appealed to in Newton's proof of proposition 69 book 1. This makes that proof show that if gravitation is an *interaction*, such that the equal and opposite reaction to the gravitational attraction of a planet on a body demanded by Law 3 is the gravitational attraction of that body on the planet, then the gravitational attraction toward each planet is proportional to the mass of that planet. See appendix 1 for details.

Let us now consider Newton's extension of this result to *parts* of planets. Here again is the second paragraph of his argument for proposition 7.

Further, since all the parts of any planet A are heavy [or gravitate] toward any planet B, and since the gravity of each part is to the gravity of the whole as the matter of that part to the matter of the whole, and since to every action (by the third Law of Motion) there is an equal reaction, it follows that planet B will gravitate in turn toward all the parts of planet A, and its gravity toward any one part will be to its gravity toward the whole of the planet as the matter of that part to the matter of the whole. Q.E.D. (C&W, 810–11)

For any planets *A* and *B*, each part *a* of planet *A* is itself a body being accelerated toward planet *B*. Newton's supposition

since the gravity of each part is to the gravity of the whole as the matter of that part to the matter of the whole

follows from what he established in proposition 6.[4]

As Berlinski's remarks suggest, none of the arguments for the other propositions we have considered in book 3 are followed by these letters. On the basis of my search through book 3, propositions 30 and 31 together with lemmas 1, 2, 10, and 11 have "QED" after their arguments. In these cases, it appears that the claims in question are both quite solidly established from their assumptions together with the Laws of Motion and that Newton's arguments for them (especially, in the cases of propositions 30 and 31 on the motion of the nodes of the moon) are ones of which he would have had significant reason to be proud.

[4] According to proposition 6, we have the gravitations of part *a* and whole planet *A* toward planet *B* proportional to the masses of part *a* and of whole planet *A*. We can write this proportionality to mass as follows:

As in the proof of proposition 69, the third Law of Motion is applied to yield the conclusion

> and since to every action (by the third Law of Motion) there is an equal reaction, it follows that planet B will gravitate in turn toward all the parts of planet A, and its gravity toward any one part will be to its gravity toward the whole of the planet as the matter of that part to the matter of the whole. Q.E.D.

According to this application of Law 3, planet *B*'s gravitation toward part *a* is the equal and opposite reaction to the gravitation of part *a* toward planet *B*. So, since the ratio of the gravitation of part *a* toward planet *B* to the gravitation of the whole of planet *A* toward planet *B* is as the mass of part *a* to the mass of the whole planet *A*, this leads directly to the conclusion that the ratio of planet *B*'s gravitation toward part *a* to its gravitation to the whole of planet *A* is equal to the ratio of the mass of part *a* to the mass of the whole planet *A*.[5]

The heart of this proof is the application of Law 3 to construe gravitation of planet *B* toward each part *a* of planet *A* as the equal and oppositely directed reaction to the gravitation of that part *a* toward planet *B*. This requires that the total acceleration of planet *B* toward planet *A* be the vector sum of its component accelerations toward the parts of planet *A*.

2. *Corollaries to proposition 7*

The first part of Newton's first corollary to proposition 7 expands on his discussion of gravitation toward a whole planet as arising from and compounded of gravitation toward its individual parts. He offers magnetic and electrical attractions as other examples where attraction toward a whole arises from attractions toward its parts.[6]

$$F_B(a)/F_B(A) = Mass(a)/Mass(A), \qquad (1a)$$

where $F_B(a)$ and $F_B(A)$ are the magnitudes of the motive forces of gravitation on part *a* and on planet *A* toward planet *B*, and where *Mass*(*a*) and *Mass*(*A*) are their respective masses.

[5] This application of the third Law of Motion yields

$$F_a(B)/F_A(B) = F_B(a)/F_B(A), \qquad (2a)$$

where $F_a(B)$ and $F_B(a)$ are the equal and opposite gravitation of planet *B* to part *a* and of part *a* toward planet *B*, while $F_B(A)$ and $F_A(B)$ are the equal and opposite gravitations of the planets toward one another. This yields

$$F_a(B)/F_A(B) = Mass(a)/Mass(A), \qquad Conclusion(a)$$

since we already have from proposition 6 that $F_B(a)/F_B(A)$ = Mass(*a*)/Mass(*A*), as in (1*a*) from note 4 above.

[6] In corollary 5 to, the immediately preceding, proposition 6, Newton made it clear that he is aware that magnetic attraction differs from gravity in that its motive measure – the attraction it exerts on a given body – need not be proportional to the mass of that body being acted upon (see above chpt. 7.II.2). Here, he suggests that magnetic attraction is compounded of attractions to individual parts of the attracting magnet.

Magnetic attraction and the attractions exhibited in static electric phenomena are examples, known in Newton's time, where there are attractions between bodies that are not in direct contact with one another. Though these phenomena may suggest some sort of analogy with his treatment of gravitational attraction, Newton's appeal to such examples to motivate the idea of resolving such attractions toward bodies into attractions toward their parts is not backed up by any detailed account of how to carry out such an analysis.

Corollary 1. *[first part]* Therefore the gravity toward the whole planet arises from and is compounded of the gravity toward the individual parts. We have examples of this in magnetic and electric attractions. For every attraction toward a whole arises from the attractions toward the individual parts. This will be understood in the case of gravity by thinking of several smaller planets coming together into one globe and composing a larger planet. For the force of the whole will have to arise from the forces of the component parts. (C&W, 811)

The example of several smaller planets coming together into one globe does help illuminate how the gravity toward the whole would have to arise from the forces toward these smaller planets.

This corollary goes on to answer the objection that gravitations between terrestrial bodies are not detected by our senses.

If anyone objects that by this law all bodies on our earth would have to gravitate toward one another, even though gravity of this kind is by no means detected by our senses, my answer is that gravity toward these bodies is far smaller than what our senses could detect, since such gravity is to the gravity toward the whole earth as [the quantity of matter in each of] these bodies to the [quantity of matter in the] whole earth. (Corollary 1 continued; C&W, 811)

The tininess of the masses of everyday bodies compared to the mass of the whole earth reinforces Newton's claim that our not detecting such attractions by our senses does not count against his theory.[7]

Newton's second corollary asserts the inverse-square variation of gravitation toward particles.

Corollary 2. The gravitation toward each of the individual equal particles of a body is inversely as the square of the distance of places from those particles. This is evident by book 1, prop. 74, corol. 3. (C&W, 811)

The combination of this corollary with proposition 7 provides Newton's most explicit statement of his theory of universal gravitation. He tells us that it is supported by corollary 3 of proposition 74 book 1. Here is his statement of corollary 3 proposition 74 book 1.

Corollary 3. *[of prop. 74, book 1]*. If a corpuscle placed outside a homogeneous sphere is attracted by a force inversely proportional to the square of the distance of the corpuscle from the

[7] In his original *System of the World*, the earlier version of book 3 composed "in popular form," Newton provided the following estimate of the tiny magnitudes the accelerations produced by such attractions might be expected to have.

> Hence a sphere of one foot in diameter, and of a like nature to the earth, would attract a small body placed near its surface with a force 20,000,000 times less than the earth would do if placed near its surface; but so small a force could produce no sensible effect. If two such spheres were distant but by 1/4 of an inch, they would not, even in spaces void of resistance, come together by the force of their mutual attraction in less than a month's time; and lesser spheres would come together at a rate yet slower, namely, in the proportion of their diameters. (Cajori 1934, 569–70; see also Newton 1728, 40–1)

center of the sphere, and the sphere consists of attracting particles, the force of each particle will decrease in the squared ratio of the distance from the particle. (C&W, 594)

Proposition 74 is in section 12 of book 1, which is on the attractive forces of spherical bodies.

We have seen that Newton's basic inferences from phenomena are backed up by systematic dependencies that make the phenomena into measurements of the causal parameter values being inferred. To what extent does the inverse-square variation of gravitation toward a whole sphere measure the inverse-square law for gravitations toward the particles of which it is composed?

In appendix 3 we extend Newton's proof of proposition 71 to allow for alternative powers of attraction to particles. We show that, for the attraction of a corpuscle toward a spherical shell made up of equally attracting particles, the inverse-square variation (with respect to distance from the center of the shell) of the total force on the corpuscle outside being attracted toward it carries the information that the component attractions toward the particles which sum to make it up vary inversely with the squares of its distances from them. Any difference from the inverse-square law for attraction toward the parts would produce a corresponding difference from the inverse-square for the law of attraction toward the center resulting from summing the attractions toward the parts. This suggests that integrating over such nested spheres will afford systematic dependencies to back up the inference from inverse-square attraction toward whole spheres to inverse-square attractions toward the particles whose attractions are being summed to give the attraction toward the whole spheres. In section 2 of appendix 2 we show that this suggestion is realized in an integral provided by Chandrasekhar to represent results obtained by Newton.

3. *From inverse-square fields of acceleration toward planets to gravity as a universal force of pair-wise attraction between bodies*

The argument in proposition 7 transforms the conception of gravity from inverse-square attractions toward planets into a universal force of pair-wise interaction between bodies. In his earlier *System of the World*, Newton offered an extensive discussion devoted to clarifying his new conception of a force of pair-wise interaction between bodies.

Since the action of the centripetal force upon the bodies attracted, is, at equal distances, proportional to the quantities of matter in those bodies; reason requires that it should be also proportional to the quantity of matter in the body attracting.

For all action is mutual, and makes the bodies mutually to approach one to the other, and therefore must be the same in both bodies. (Newton 1728, 38)

Here we see the proportionality of the actions of centripetal forces on bodies at equal distances to the quantities of matter of those attracted bodies cited as a reason to count this action as also proportional to the quantity of matter in the attracting body.

He supports the inference by the claim that all action is mutual. He further supports it by claiming that the attraction makes the bodies mutually approach one another, so that it must be the same in both bodies.

Newton concludes this paragraph by distinguishing our conception of the central body as attracting and the orbiting body as attracted as "more mathematical than natural"; because the attraction is really mutual and, so, must be of the same kind in both.

> It is true that we may consider one body as attracting, another as attracted. But this distinction is more mathematical than natural. The attraction is really common of either to other, and therefore of the same kind in both. (Newton 1728, 38)

We have seen that Newton sharply distinguishes his mathematical characterization of sorts of forces in book 1 from his empirical argument to centripetal forces maintaining moons and planets in their orbits in our solar system. Here he argues that those centripetal forces we have found in nature really are pair-wise mutual interactions. The concept of the action of a centripetal force as an attraction to maintain a body in its orbit is a one-body idealization that needs to be transformed into a mutual interaction. Such a mutual interaction is really common of each toward the other. It must, therefore, be of the same kind in both.

The next paragraph argues that this concept of attraction as a mutual interaction characterizes the attractions that have been argued for.

> And hence it is that the attractive force is found in both. The Sun attracts Jupiter and the other planets. Jupiter attracts its satellites. And for the same reason, the satellites act as well one upon another as upon Jupiter, and all the planets mutually one upon another. (Newton 1728, 38)

It also argues that for this same reason there are mutual interactions among all solar system bodies.

This theme of clarifying the two-body interaction conception of attraction as mutual is extended at some length in the next paragraph.[8]

> And though the mutual actions of two planets may be distinguished and considered as two, by which each attracts the other; yet as those actions are intermediate, they don't make two, but one operation between two terms. Two bodies may be mutually attracted, each to the other, by the contraction of a cord interposed. There is a double cause of action, to wit, the disposition of both bodies, as well as a double action in so far as the action is considered as upon two bodies. But as betwixt two bodies it is but one single one. (Newton 1728, 38–9)

In chapter 3 we saw that Newton appealed to a horse drawing a stone tied to a rope as an example to illustrate Law 3. We noted that this afforded some motivation for

[8] Stein offers an excellent discussion of this passage together with the following related passage (1991, 218). This discussion is expanded upon in his (2002, 287–8).

applying Law 3 to attractions. Here we see Newton exploit the analogy with the mutual attraction of two bodies by the contraction of a cord between them.

Newton goes on to apply the analogy to the attraction which maintains Jupiter in its orbit about the sun.

> 'Tis not one action by which the sun attracts Jupiter, and another by which Jupiter attracts the Sun. But it is one action by which the Sun and Jupiter mutually endeavour to approach each the other. By the action with which the Sun attracts Jupiter, Jupiter and the Sun endeavour to come nearer together, and by the action with which Jupiter attracts the Sun, likewise Jupiter and the Sun endeavour to come nearer together. But the Sun is not attracted towards Jupiter by a two-fold action, nor Jupiter by a two-fold action towards the Sun: but 'tis one single intermediate action, by which both approach nearer together. (Newton 1728, 39)

If the force which maintains Jupiter in its orbit really does satisfy this conception of attraction as a mutual interaction between them, then Law 3 certainly does apply to count the sun's gravitation toward Jupiter as the equal and opposite reaction to Jupiter's gravitation toward the sun.

The application of Law 3 in the first paragraph of Newton's argument for proposition 7 transforms the inverse-square centripetal forces of gravity toward planets into pair-wise interactions between those planets and the bodies which orbit them. The additional application of Law 3 in the second paragraph transforms the conception of gravity into a universal force of pair-wise interaction between bodies.

We have followed Howard Stein in interpreting Newton's inverse-square centripetal forces toward planets as acceleration fields. Clearly, the theoretical transformations undertaken in proposition 7 require revision of this theoretical concept of gravity toward planets. In chapter 1 we pointed out that the original concept is recovered as an approximation. This is very much in line with the Area Rule and Harmonic Rule phenomena. Universal gravity introduces interactions that require corrections to these phenomena that were assumed as premises at the beginning of the argument. We have argued that Newton's scientific method is a method of successive approximations. The deviations from motion in accord with the basic acceleration fields are exploited to provide information about interactions. Such information is then exploited to generate a more accurate revised model.

Newton's center of mass resolution of the system of the world argues that the center of mass of the interacting solar system bodies recovers their true motions. In this center of mass frame the separate centripetal acceleration fields toward solar system bodies are combined into a single system, where each body undergoes an acceleration toward each of the others proportional to its mass and inversely proportional to the square of the distance between them. This recovers appropriate centripetal acceleration components corresponding to the separate inverse-square acceleration fields toward the sun and toward each planet.

II The attractive forces of spherical bodies applied to planets

In the theorems from book 1 that have been appealed to in the basic inferences to inverse-square centripetal forces, bodies have been treated as point masses. A main result of Newton's theorems on the attractive forces of spherical bodies is that inverse-square attractions toward spherical bodies, compounded of spherically symmetric inverse-square attractions toward particles making them up, can be treated as inverse-square attractions toward point masses. Each such spherical body can be represented by a point mass equal to the mass of that whole body concentrated at its center. Another important result is that mutual attractions among such bodies can be treated as interactions among such point masses. Newton's discussion in proposition 8 book 3 is specifically addressed to the worry that inverse-square attraction toward such a sphere might be in error for bodies very near its surface.

1. *Proposition 8 book 3*

Proposition 8 Theorem 8

If two globes gravitate toward each other, and their matter is homogeneous on all sides in regions that are equally distant from their centers, then the weight of either globe toward the other will be inversely as the square of the distance between the centers.

After I had found that the gravity toward a whole planet arises from and is compounded of the gravities toward the parts and that toward each of the individual parts it is inversely proportional to the squares of the distances from the parts, I was still not certain whether that proportion of the inverse square obtained exactly in a total force compounded of a number of forces, or only nearly so. For it could happen that a proportion which holds exactly enough at very great distances might be markedly in error near the surface of the planet, because there the distances of the particles may be unequal and their situations dissimilar. But at length, by means of book 1, props. 75 and 76 and their corollaries, I discerned the truth of the proposition dealt with here. (C&W, 811)

Newton claims that he discerned the truth of proposition 8 by means of book 1, propositions 75 and 76, and their corollaries. He cites these results as removing the worry raised by the question of whether attraction toward a whole sphere made up of compounding inverse-square gravities toward its parts would remain inverse-square at small distances from the surface of that sphere. The fact that the relative differences of distances from an outside body being attracted to the different parts of the sphere increases as it approaches the surface suggests that the inverse-square might be markedly in error near the surface of the planet, even though it holds exactly enough at very great distances.

That Newton explicitly raises this worry, strongly suggests that he sees the argument of propositions 1–7 of book 3 as in need of supplementation by the results which he cites as removing it. This need for such supplementation is reinforced by noting that the worry is one that might be especially relevant to the moon-test argument of proposition 4.

Indeed, as we exploit in section 2 of appendix 2 below, the inverse-square power law and the harmonic oscillator power law (where the force is directly as the distance) are special in that they can be extended right down to the surface of a sphere. These are the two power laws Newton singles out for specific treatment. For other power laws for attractions to the individual particles the variation with distance of the resulting attraction to the whole sphere will not remain the same as that of the attraction to the parts as the distance from the surface becomes small.

2. Measuring surface gravities: corollary 1 of proposition 8

Newton uses orbits of planets and moons to measure the relative weights (of equal bodies at equal distances) toward the sun and toward the planets which those moons orbit. This lets him also measure from these orbits the relative weights equal bodies would have at the respective surfaces of the sun, Jupiter, Saturn, and the earth.

The fundamental property that (for bodies at or above its surface) gravity toward a planet is an inverse-square acceleration field makes the ratio of accelerations toward planets at any given equal distances the same as the ratio of weights toward those planets equal bodies would have at any other equal distances. The ratio of weights of equal bodies at equal distances measures the ratio of the strengths (Newton's absolute quantities of centripetal forces) for these gravities toward the sun and planets. The ratio of the corresponding inverse square adjustments of gravitational accelerations toward planets measures the ratio of the weight a given body would have toward a given planet at a given distance to the weight an equal body would have toward another planet at a different specified distance. This allows orbits together with data about sizes to measure ratios of surface gravities of planets.

Corollary 1. Hence the weights of bodies toward different planets can be found and compared one with another. For the weights of equal bodies revolving in circles around planets are (by book 1, prop. 4, corol. 2) as the diameters of the circles directly and the squares of the periodic times inversely, and weights at the surfaces of the planets or at any other distances from the center are greater or smaller (by the same proposition) as the inverse squares of the distances. I compared the periodic times of Venus around the sun (224 days and 16¾ hours), of the outermost circumjovial satellite around Jupiter (16 days and 16⁸⁄₁₅ hours), of Huygens's satellite around Saturn (15 days and 22⅔ hours), and of the moon around the earth (27 days, 7 hours, 43 minutes) respectively with the mean distance of Venus from the sun, and with the greatest heliocentric elongations of the outermost circumjovial satellite from the center of Jupiter (8′16″), of Huygens's satellite from the center of Saturn (3′4″), and of the moon from the center of the earth (10′33″). In this way I found by computation that the weights of bodies which are equal and equally distant from the center of the sun, of Jupiter, of Saturn, and of the earth are respectively toward the sun, Jupiter, Saturn, and the earth as 1, 1/1,067, 1/3,021, and 1/169,282. And when the distances are increased or decreased, the weights are decreased or increased as the squares of the distances. The weights of equal bodies toward the sun, Jupiter, Saturn, and the earth at distances of 10,000, 997, 791, and 109 respectively from their centers (and hence their weights on the surfaces) will be as 10,000, 943, 529, and 435. What the weights of bodies are on the surface of the moon will be shown below. (C&W, 812–13)

This corollary exploits the above relationships to make orbits measure the relative weights equal bodies would have (at equal distance from their respective centers) toward the sun, Jupiter, Saturn, and the earth. It also uses those orbits, together with information about relative sizes, to measure the relative surface gravities (relative weights equal bodies would have at the respective surfaces) of the sun, Jupiter, Saturn, and the earth. See appendix 4 for details.

In his *Discourse on the Cause of Gravity* of 1690, Huygens made the following comments on the corresponding measurements of surface gravities Newton had provided in the first edition of *Principia* of 1687.

Something remains for me to remark on concerning his System, which has pleased me greatly, namely, that he finds means, supposing the distance from here to the Sun to be known, to define what gravity the inhabitants of Jupiter and Saturn would feel, compared to what we feel here on the Earth, and also what its measure is at the surface of the Sun. These are things which previously seemed quite removed from our knowledge, and which nevertheless are consequences of the principles that I reported a little earlier.

This determination occurs for the planets that have one or several satellites since the periodic times of those and their distances from the planets that they accompany must enter into the calculation. Through this Mr. Newton finds the gravities at the surfaces of the Sun, Jupiter, Saturn, and the Earth in proportion to these numbers: 10000, 804½, 536, 805½. It is true that there is some uncertainty because of the distance of the Sun, which is not sufficiently well known, and which has been taken in these calculations to be around 5,000 Earth diameters, instead of following Mr. Cassini's dimension of around 10,000, which tolerably approaches what I have found previously through similar reasoning in my *System of Saturn*, namely 12,000. I also disagree somewhat with the diameters of the planets. So, by my calculation the gravity in Jupiter, in relation to what we have here on Earth, is found to be 13 to 10, instead of Mr. Newton making them equal, or insensibly different. But the gravity in the Sun, which by the numbers that we have seen, would be around 12 times greater than ours on Earth, I find 26 times greater. (Huygens 1690, 167–8)

Huygens shows great appreciation for Newton's achievement in providing resources for investigations that can bring comparisons of surface gravities of planets within the scope of human knowledge. He also criticizes the very small distance from the earth to the sun which Newton had supposed, which he points out is only around 5,000 earth diameters. He suggests that solar distances more in line with Cassini's dimension of 10,000 earth diameters or his own, even greater, dimension of 12,000 would yield more reasonable estimates. In particular, they would make the ratio of Jupiter's surface gravity to that of the earth about 13 to 10 rather than very nearly equal as Newton had made them in the first edition.

In the first edition, Newton assumed a solar parallax of 20 arc seconds corresponding to taking the solar distance at about 10,000 radii or 5,000 diameters of the earth.[9] In this he was going against a growing consensus to accept Cassini's and Flamsteed's agreeing estimates limiting solar parallax to 10 arc seconds.[10] The correspondingly small estimate of the solar distance, in the first edition, gave 1/1100, 1/2360, and 1/28700 for the ratios of gravity at equal distances toward the sun to that for Jupiter, Saturn, and the earth and 804½, 536, and 805½ respectively as the surface gravities of Jupiter, Saturn, and the earth corresponding to setting 10,000 as the surface gravity of the sun.[11]

Newton was quite impressed with the above quoted comments by Huygens. In a letter to Leibniz of October 16, 1693, responding after a considerable delay (which he attributed to having mislaid Leibniz's letter of March 7) to a request by Leibniz that he comment on Huygens's remarks on the cause of light and gravity,[12] Newton remarks:

> Huygens is a master, and his remarks on my discoveries are brilliant. The parallax of the Sun is less than I had concluded it to be; (*Corresp.* III, 287)

In the second edition Newton adopted the smaller solar parallax of 10 arc seconds of the Cassini-Flamsteed measure.[13] In the third edition he adopted 10½ arc seconds as the solar parallax.[14]

The remarks by Huygens continue with a speculation about the light of the sun suggested by the very great speed that would be required of Huygens's spherical shells to have his hypothesis about the cause of gravity support the much greater surface gravity for the sun.[15]

> From this it follows, in explaining gravity in the manner that I have, that the fluid matter close to the Sun must have a velocity 49 times greater than what we have found near the Earth, which

[9] The solar parallax is the angle between the position of the sun (directly overhead) as it would be seen from the center of the earth and the position as it would be seen (horizontally) from a point on the surface of the earth. For solar parallax θ, the solar distance in radii of the earth is $1/\sin\theta \approx 1/\tan\theta$ for small θ. $1/\sin 20'' = 10{,}313.24033$; $1/\tan 20'' = 10{,}313.24028$.

[10] He was probably influenced by Halley, who had long resisted the growing consensus for smaller solar parallax estimates. See Van Helden (1985, 144–7; and *GHA* 2B, 153–68).

[11] See Newton 1687, 413–14; Koyré and Cohen 1972, 578–9; C&W, 812; as well as Garisto 1991, 44.

[12] Here is the request by Leibniz from his letter to Newton.

> I do not doubt that you have weighed what Christiaan Huygens, that other supreme mathematician, has remarked in the appendix to his book, about the cause of light and gravity. I would like your opinion in reply: for it is by the friendly collaboration of you eminent specialists in this field that the truth can best be unearthed. (*Corresp.* III, 258)

[13] Van Helden (1985, 154); Koyré and Cohen (1972, 581).

[14] Van Helden (1985, 154–5) points out that in 1719 Pound and Bradley obtained a measured parallax of Mars, which would support a solar parallax of between 9″ and 12″, and that Newton's 10½″ is an average between these bounds.

[15] See chapter 5 section 4 for an account of Huygens's corresponding calculation, that to account for the gravity at the surface of the earth his vortical shells would need to revolve with a velocity 17 times that of the earth.

was already 17 times greater than the velocity of a point at the equator. This then is an awesome speed, which has made me wonder whether it would not be sufficient to be the cause of the brilliant light of the Sun, supposing that the light were produced as I have explained in what I have written here; namely that the solar particles, swimming in a more subtle and extremely agitated matter, knock against the particles of the ether that surround it. For, if the agitation of any such matter, with the motion that it has here on Earth, could cause the light of the flame of a candle or of ignited camphor, how much greater would this light be with a motion 49 times more swift and more violent? (Huygens 1690, 168)

Huygens, here, extends the measurements of relative velocity of vortical shells to compute what he describes as the awesome speed that would be required to explain the surface gravity of the sun.

These measurements of the speeds of his hypothesized shells are examples of measurement of causal parameters by the phenomena which they are purported to explain. This is close to the ideal of empirical success exemplified in Newton's argument: convergent accurate measurement of causal parameters by the phenomena which they are purported to explain. This speed is so great that it suggests a speculation about the cause of the brilliant light of the sun. Here we have a second phenomenon, the brilliant light of the sun, which this same causal parameter might also be purported to explain. What we don't have are systematic dependencies that would turn the brightness of the light of the sun into an agreeing measurement of this speed.

Here is Newton's response to Leibniz's request that he comment on Huygens's hypothesis as to the cause of gravity.

For since celestial motions are more regular than if they arose from vortices and observe other laws, so much so that vortices contribute not to the regulation but to the disturbance of the motions of the planets and comets; and since all phenomena of the heavens and of the sea follow precisely, so far as I am aware, from nothing but gravity acting in accordance with the laws described by me; and since nature is very simple, I have myself concluded that all other causes are to be rejected and that the heavens are to be stripped as far as may be of all matter, lest the motions of planets and comets be hindered or rendered irregular. But if, meanwhile, someone explains gravity along with its laws by the action of some subtle matter, and shows that the motion of planets and comets will not be disturbed by this matter, I shall be far from objecting. (Newton to Leibniz October 16, 1693; *Corresp.* III, 287)

Newton's demand that an explanation of gravity explain all its laws can be interpreted to include a demand that it recover the relation making possible the measurement of relative masses of planets from the relative strengths of their inverse-square fields of gravity.

3. *Measuring relative masses: corollary 2 of proposition 8*

Newton's next corollary infers ratios of masses of planets from the ratios of forces at equal distances.

Corollary 2. The quantity of matter in the individual planets can also be found. For the quantities of matter in the planets are as their forces at equal distances from their centers; that is, in the sun, Jupiter, Saturn, and the earth, they are as 1, 1/1,067, 1/3,021, and 1/169,282 respectively. If the parallax of the sun is taken as greater or less than $10'' 30'''$, the quantity of matter in the earth will have to be increased or decreased in the cubed ratio. (C&W, 813)

Newton's inference from ratios of weights of equal bodies at equal distances to ratios of masses of the planets to which those bodies are attracted is based on the claim argued for in the first part of Newton's proof of proposition 7. This is the claim that the absolute measures of gravities toward planets are proportional to the quantities of matter of those attracting planets. Newton's argument for this claim appeals to his application of Law 3 to count the equal and opposite reaction to gravity of a body toward a planet to be the gravity of the planet toward the body it attracts.

This inference to measure masses of planets is one Huygens would not grant. Right after objecting to gravitational attractions between all the small parts of bodies, Huygens went on to object to Newton's claims to have established mutual attraction between whole bodies.[16]

Nor am I at all persuaded of the necessity of mutual attraction of whole bodies, having shown that, were there no Earth, bodies would not cease to tend toward a center because of what we call their gravity. (Huygens 1690, 159)

On Huygens's hypothesis as to the cause of gravity, his supposed rapidly whirling shells of vortex matter would draw ordinary bodies toward their center, whether or not there was any ordinary matter already concentrated there. Indeed, he might regard his explanation of gravity toward that center as an explanation of how the earth came to be formed by the gathering together of ordinary matter attracted toward this center.

So, even though Huygens can exploit the new resources Newton has provided (in corollary 1) for measuring ratios of weights toward planets of equal bodies at equal distances from information about orbits, he cannot follow Newton to the further inferences (here in corollary 2) to count these as measurements of relative masses of the sun and planets.

III The two chief world systems problem

Newton's phenomena are relative motions, with a distinct separate reference frame for each orbital system. The inverse-square centripetal acceleration fields Newton infers from these phenomena for each orbital system, e.g. the sun, Jupiter, Saturn, and the earth, leave open the problem of their consistent combination into a single system. Newton's argument for proposition 7 provides resources for dealing with this problem of combining these separate inverse-square acceleration fields.

[16] See section I.3 of chapter 5 above.

1. *The system of the world*

Newton exploits his measurements of relative masses of the sun and planets to apply his argument to the question that dominated seventeenth-century natural philosophy. Galileo's *Dialogue Concerning the Two Chief World Systems* was an attempt to provide a compelling argument for the Copernican answer to this question.[17]

Let us consider how Newton applies this argument to provide his decisive and surprising resolution of this two chief world systems problem. Newton opens his discussion with a claim that he explicitly identifies as a hypothesis.

Hypothesis 1

The center of the system of the world is at rest.

No one doubts this, although some argue that the earth, others that the sun, is at rest in the center of the system. Let us see what follows from this hypothesis. (C&W, 816)

As we have remarked in chapter 1, Newton will reject both the claim by Ptolemy and Tycho that the earth is at rest and, also, the alternative claim by Copernicus and Kepler that the sun is at rest. By calling it a hypothesis, Newton is refraining from endorsing, as an established fact, even the more general claim that the center of the system of the world is at rest.[18]

Let us take up Newton's suggestion to see what follows from the hypothesis that the center of the world system is at rest. Given this hypothesis, we have a proposition.

Proposition 11 Theorem 11

The common center of gravity of the earth, the sun, and all the planets is at rest.

For that center (by corol. 4 of the Laws) either will be at rest or will move uniformly straight forward. But if that center always moves forward, the center of the universe will also move, contrary to the hypothesis. (C&W, 816)

Here again is corollary 4 of the Laws of Motion.

Corollary 4, Laws of Motion

The common center of gravity of two or more bodies does not change its state whether of motion or of rest as a result of the actions of the bodies upon one another; and therefore the common center of gravity of all bodies acting upon one another (excluding external actions or impediments) either is at rest or moves uniformly straight forward. (C&W, 421)

Given the assumption that external actions and impediments can be ignored when considering the motions of the earth, sun, and the planets, corollary 4 tells us that the common center of gravity of these bodies is either at rest or moves uniformly straight

[17] Galileo's preface claims that he was not arguing "absolutely" for the Copernican side, but rather striving to present it as superior against the arguments of some professed Peripatetics, so that foreign nations would know that it is not from failing to take account of what others have asserted that Italians had yielded to asserting that the earth is motionless. (See Galilei 1970, 5–6)

[18] See chapter 3 section II.1.ii.

forward. Given the hypothesis that the center of this world system is at rest, this center of gravity of the bodies in it will have to be counted as at rest.

The great distance of the nearest stars from our sun renders gravitational attraction to them irrelevant to sensible perturbations of the motions among themselves of solar system bodies.[19] In proposition 10 Newton appealed to his treatment of resistance forces in book 2 to argue that the planets and comets encounter no sensible resistance in moving through the spaces of the heavens. Newton's treatment of resistance forces was far less successful than his treatment of gravity.[20] Even without any appeal to Newton's detailed treatment of resistance, the regular motions of solar system bodies in accord with his theory of gravity without resistance forces afford strong grounds for ignoring hypotheses about such resistance forces.

And against filling the Heavens with fluid Mediums, unless they be exceedingly rare, a great Objection arises from the regular and very lasting Motions of the Planets and Comets in all manner of Courses through the Heavens. For thence it is manifest, that the Heavens are void of all sensible Resistance, and by consequence of all sensible Matter. (Query 28, *Optics*, Newton 1979, 364–5)

This quote is one of many where Newton appeals to the regularity of such motions to dismiss resistance forces in the heavens. We saw similar remarks in the above quoted passage from his response to Leibniz. Such a refusal to count the hypothesis of resistance forces as any sort of serious objection to universal gravity is very much in line with the methodology of Rule 4.

The next proposition is Newton's surprising resolution of the two chief world systems problem.

Proposition 12 Theorem 12

The sun is engaged in continual motion but never recedes far from the common center of gravity of all the planets.

For since (by prop. 8, corol. 2) the matter in the sun is to the matter in Jupiter as 1,067 to 1, and the distance of Jupiter from the sun is to the semidiameter of the sun in a slightly greater ratio, the common center of gravity of Jupiter and the sun will fall upon a point a little outside the surface of the sun. By the same argument, since the matter in the sun is to the matter in Saturn as 3,021 to 1, and the distance of Saturn from the sun is to the semidiameter of the sun in a slightly smaller ratio, the common center of gravity of Saturn and the sun will fall upon a point a little within the surface of the sun. And continuing the same kind of calculation, if the earth and all the planets were to lie on one side of the sun, the distance of the common center of gravity of them all from the center of the sun would scarcely be a whole diameter of the sun. In other cases the distance between those two centers is always less. And therefore, since that center of gravity is continually at rest, the sun will move in one direction or another, according to the various configurations of the planets, but will never recede far from that center. (C&W, 816–17)

[19] In corollary 2 of proposition 14 book 3 (C&W, 819), Newton cites the absence of sensible parallax arising from the annual motion of the earth as evidence of the immense distance of the stars from us.

[20] See Truesdell 1970; Smith G.E. 2000 and 2001a.

Newton's surprising resolution is that neither the sun nor the earth can be counted as a fixed center relative to which the true motions of solar system bodies can be specified. It is made possible by the measurements of masses of the sun and the planets with moons from orbits about them.

Newton points out that the ratio 1/1067 of the mass of Jupiter to that of the sun, which he obtained from comparing the orbit of Venus about the sun to that of Callisto about Jupiter, would put the common center of gravity of the sun and Jupiter at a point a little outside the sun. The data he cites put the sun's radius at about 0.0047AU, where the mean-distance of the earth's orbit from the center of the sun is counted as one AU.[21] His data put the common center of gravity of Jupiter and the sun at about 0.0049AU from the center of the sun.[22] The corresponding ratio for Saturn and the sun would put their common center of gravity a little within the sun. Newton's cited ratio 1/3021 for the mass of Saturn to that of the sun gives about 0.0032AU from the center of the sun as the distance to their common center of gravity.[23]

Newton goes on to argue that if the earth and all the planets were to lie on one side of the sun, the distance of the common center of gravity of all of them would scarcely be a whole diameter of the sun.

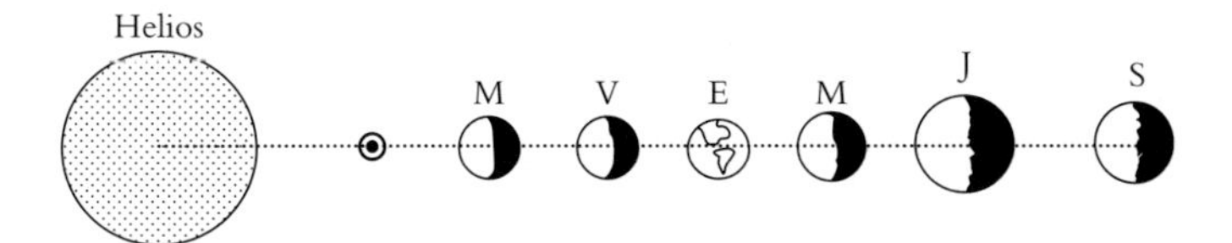

Figure 8.1 All the planets lined up on the same side of the sun

[21] The AU or astronomical unit is defined by the two-body orbit of the earth used by Gauss. See note 31 of this chapter for more detail.

Newton gives 10,000/997 as the ratio of the sun's radius r_H to the radius r_J of Jupiter. On the basis of his data from Pound, Newton sets the diameter of Jupiter as viewed from its mean-distance from the earth at 37.25″. This mean-distance from the earth is taken to be the same as Jupiter's mean-distance from the sun. Taking this distance to be the mean, 5.21AU, of Kepler's and Boulliau's estimates of the mean-distance of Jupiter's orbit from the sun yields 5.21AU×sin18.625″ = 0.00047AU for Jupiter's semidiameter (Jupiter's semidiameter is half its diameter). This yields

$$r_{\mathrm{H}} = (10000/997)(0.00047) = 0.0047Au,$$

for the semidiameter of the sun.

[22] The mean-distance, 5.21AU, of Jupiter from the sun, together with Newton's ratio 1/1,067 for the mass of Jupiter to that of the sun yields

$$5.21/1067 = 0.00488 \text{ or about } 0.0049Au$$

as the distance of the center of mass of Jupiter and the sun from the center of the sun.

[23] Taking the mean, 9.526AU, of Kepler's and Boulliau's estimates as Saturn's mean-distance yields

$$9.526/3021 = 0.003153Au$$

for the distance from the center of the sun to the common center of mass of the sun and Saturn.

What are Newton's grounds for his claim that even this worst-case configuration results in a center of mass quite close to the sun?

The problem is that he had no known satellite orbits to measure masses for Mercury, Venus, or Mars. His measurement for Saturn gives its mass ratio as 1/3021, which is considerably less than that of Jupiter. Even his rather exaggerated mass ratio, 1/169282 (rather than the correct 1/332946),[24] for the earth shows its mass to be a great deal less than that of Saturn. If he could assume that each of Mercury, Venus, and Mars have masses no greater than that of Jupiter, this, together with his measured results for the others, would give as a worst case that the center of mass is no greater than about 0.01AU from the center of the sun.[25] This would put this worst case at only a little over two solar radii from the center of the sun.

Newton had good evidence that Mars, Venus, and Mercury are all very much smaller in size than Jupiter. In the first edition Newton cited 8″, 28″, and 20″ as the respective angles corresponding to the diameters of Mars, Venus, and Mercury viewed from the sun. Using 1.5235AU, 0.724AU, 0.3870AU corresponding to the means of Kepler's and Boulliau's estimates for the mean-distances of these planets from the sun yields .00003AU, .000049AU, and .000019AU for their respective semidiameters.[26] These are all very much smaller than the 0.00047AU Newton's data give for the semidiameter of Jupiter.

The very small sizes of these planets with respect to Jupiter would require very large densities to have their masses be even as great as that of Jupiter. The corresponding densities for Mars, Venus, and Mercury would have to be respectively more than 3,800 times, 880 times, and 15,000 times the density of Jupiter.[27] Our above result shows that even these large densities would put the worst-case center of gravity only a little over one solar diameter from the sun.[28] The extreme density that would be required to have

[24] See note 63 in appendix 4 below.

[25] Taking the means of Kepler and Boulliau, 0.3870, 0.7240, 1, 1.5235, 5.21, and 9.526 as the distances of Mercury, Venus, Earth, Mars, Jupiter, and Saturn and assigning 1/1067, 1/1067, 1/169286, 1/1067, 1/1067, and 1/3021 for the ratios of their masses to that of the sun yields

$$(0.3870/1067) + (0.724/1067) + (1/169286) + (1.5235/1067) + (5.21/1067) + (9.526/3021) = .01Au,$$

as the distance of the center of mass from the center of the sun. This is (.01/.0047)= 2.24 radii or 1.12 diameters of the sun.

[26] The corresponding semidiameters or radii are half the diameters; this yields $1.5235 \times \sin 4'' = 0.0000295$, $0.724 \times \sin 14'' = 0.0000491$, and $0.3870 \times \sin 10'' = 0.00001876$ as the respective semidiameters in AU. In the first edition Newton gives the diameter of Jupiter as 39.5″. This would yield $5.21 \times \sin 19.75''$, or about 0.0005 AU for the semidiameter of Jupiter. In the third edition Newton appealed to Pound's data to adjust the diameter of Jupiter down to 37.25″(C&W, 798). This yields 0.00047AU for Jupiter's semidiameter.

[27] Let us represent the density of Jupiter by $(1067 / 0.00047^3)$, its mass ratio divided by the cube of its semidiameter. The corresponding densities for Mars, Venus, and Mercury at this same mass would be $(1067 / 0.00003^3)$, $(1067 / 0.000049^3)$ and $(1067 / 0.000019^3)$. This yields for Mars a density which would be $(0.00047^3 / 0.00003^3) = 3845$ times that of Jupiter, for Venus $(0.00047^3 / 0.000049^3) = 882$, and for Mercury $(0.00047^3 / 0.000019^3) = 15137$ times the density of Jupiter.

[28] Newton computed the density of the earth to be a little over four times the density of Jupiter, and speculated that the density of Mercury is greater than that of the earth corresponding to its greater exposure to

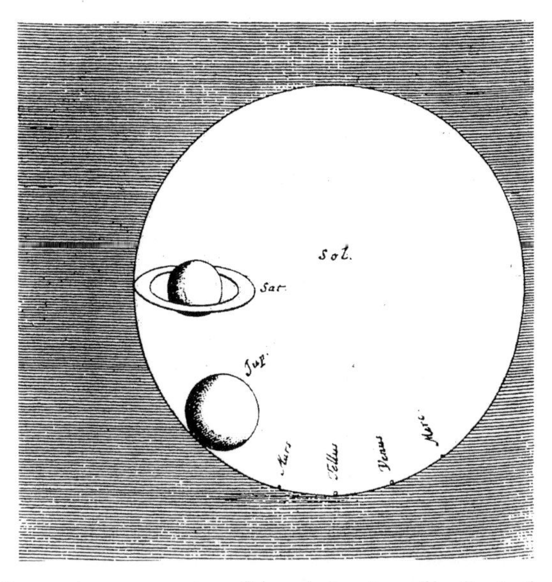

Figure 8.2 Huygens's representation of the relative sizes of bodies in the solar system

the mass of Mercury (or of Venus or Mars) as great as that of Jupiter far exceeds any reasonable bounds on speculations about densities of planets.

As George Smith has pointed out,[29] Newton also had evidence from the absence of data corresponding to significant two-body corrections of the orbits of Mercury, Venus, and Mars. Taking into account this two-body correction makes the Harmonic Rule for a planet measure the sum of the mass of the sun and the mass of that planet.[30]

$$R^3/t^2 \propto M(1+m) \qquad \text{Two-body Harmonic Rule}$$

the heat of the sun (corol. 4 prop. 8 bk. 3: C&W, 814–15). The 15,000 times greater density than that of Jupiter, required to get its mass up to that of Jupiter, is far beyond anything that might be supported by any such speculation.

[29] See Smith G.E. 1999a, 48–9.

[30] In the two-body orbit both bodies orbit their common center of mass. In the corresponding one-body orbit this center of mass just is the center of sun. So where the mass of the sun is set at 1 and m is the mass of the planet in solar masses we have the two-body distance $R'=R(1+m)$, where R is the distance in the corresponding one-body orbit.

The corresponding relation between the two-body and one body versions of the harmonic law for orbits of the sun is captured by Newton's two-body correction in proposition 60 book 1. According to proposition 60 applied to this case,

$$R' = R(1+m)/[(1+m)^2]^{1/3}.$$

This gives

$$R'^3 = R^3(1+m)^3/(1+m)^2 = R^3(1+m),$$

so that

$$R'^3/t^2 = (R^3/t^2)(1+m).$$

where M is the mass of the sun and m is the ratio of the mass of the planet to the mass of the sun.[31] Taking the Harmonic Rule ratios as data for one-body orbits makes them measure the mass of the sun.

$$R^3/t^2 \propto M \qquad \text{One-body Harmonic Rule}$$

We can use Newton's data to estimate confidence bounds limiting two-body corrections.[32] The results limit the mass of Mercury to about 1/556 of the mass of the sun.[33] The corresponding limit for Venus is 1/156 solar masses.[34] For Mars the

[31] In modern Newtonian perturbation theory, the basic model for constructing an elliptical planetary orbit of semi-major axis a and period t is the two-body system, where the sun and the planet orbit their common center of mass. This leads to the following corrected version of the Harmonic Rule:

$$a^3/t^2 = (k^2/4\pi^2)(1 + m),$$

where 1 is the mass of the sun, m is the mass of the planet (together with all its moons) in solar masses, and k is Gauss's constant (Moulton 1970, 151–4). Gauss computed k from the orbit of the earth by setting $a = 1$, $t = 365.2563835$ mean solar days, and $m = 1/354710$. The resulting value was

$$k = 0.01720209895 \quad \text{(Moulton 1970, 153).}$$

The most accurate values of t and m for the earth have since been revised, and t is now known to change over time. It was found convenient, however, to keep the value of Gauss's constant fixed. This has the effect that the astronomical unit, AU (which is defined to correspond to Gauss's original values of t and m), no longer exactly equals the semi-major axis of the earth's orbit (Danby 1988, 146).

A recent value for the mean-distance of the earth from the sun is about 1.0000010178 AU (*ESAA*, 700).

[32] Taking as distances the means between the estimates Newton cites from Kepler and Boulliau would give about

$$1.0006 \pm 0.0035 \quad Au^3/t^2$$

as a 95% Student's t-confidence estimate for the Harmonic Rule ratio corresponding to the mass of the sun.

We take Newton's cited periods in units, 365.2565 decimal days, which Newton cites as the period t_e of the earth, and give as distances the means between those he cites from Kepler and Boulliau in units AU of the mean-distance of the earth.

Table 8.1

	Mercury	Venus	Earth	Mars	Jupiter	Saturn
t	0.24084	0.61496	1	1.88081	11.8616	29.4568
R	0.38696	0.72399	1	1.5235	5.21085	9.52599
R^3/t^2	0.9989	1.0035	1	0.9996	1.0056	0.9962

These give a mean of 1.0006 and a 95% t-confidence bound of 0.0035 for R^3/t^2.

Let us take the lower bound, 1.0006 - 0.0035 = 0.9971 AU^3/t^2, of these estimates as a limit on two-body corrections.

[33] For Mercury, the Harmonic Rule ratio corresponding to taking the mean of Kepler's and Boulliau's distances is 0.9989 AU^3/t^2. The corresponding limit, 0.9989 - 0.9971 = 0.0018, gives a limiting ratio, 1.0006 / 0.0018, which would limit the mass of Mercury to about 1/556 of the mass of the sun.

[34] For Venus, the Harmonic Rule ratio is 1.0035 AU^3/t_e^2. The corresponding limit, 1.0035 − 0.9971 = 0.0064, gives a ratio, 1.0006 / 0.0064, which limits the mass of Venus to about 1/156 of the mass of the sun.

In a brilliant investigation carried out in 1638, Jeremiah Horrocks (1618–41) was able to show that Kepler's own data required assigning a smaller eccentricity to the orbit of the earth than Kepler had done and then to exploit this more accurate eccentricity for the earth to more accurately determine the mean-distance of Venus from observations. The resulting mean-distance for Venus was 0.72333AU. See Wilson 1989a, VI, 252–3.

The Harmonic Rule ratio corresponding to this distance for Venus is $0.72333^3/0.61496^2 = 1.0007$ AU^3/t_e^2. This almost exactly agrees with the above mean value, 1.0006, for R^3/t^2; thereby suggesting that the ratio of the mass of Venus to that of the sun is so small that the corresponding two-body correction is negligible.

The modern value for Venus is about 1/408523.5 solar masses.

resulting limit is 1/400 solar masses.[35] Even with mass ratios for Mercury, Venus, and Mars that are as great as these limits allow, together with Newton's measured values for the mass ratios of the other planets, the center of gravity of the sun and planets in the worst-case configuration is still less than two solar diameters from the center of the sun.[36]

Newton's solution to this two chief world systems problem does not depend on his hypothesis that the center of the system of the world is at rest. Given the exclusion of external actions or impediments, corollary 4 of the Laws counts the common center of gravity of the solar system bodies as an inertial center. It thus counts as an appropriate center for fixing what are to be counted as the true motions of these bodies among themselves. This makes it count as an appropriate center for combining the separate inverse-square acceleration fields of gravitation toward the earth, the sun, and all the planets. Motions fixed relative to their center of mass will recover the ratios among the accelerations of earth, sun, and planets toward one another required by their separate inverse-square acceleration fields. Corollary 5 of the Laws makes it clear that any alternative center that is at rest or in uniform rectilinear motion with respect to this center of mass would do equally well for specifying what can be counted as the true motions of these solar system bodies among themselves. The motions relative to any such center will preserve the ratios of the accelerations toward one another required by combining their separate inverse-square centripetal acceleration fields. Any such center will preserve the features Newton argues for. His surprising solution to the two chief world systems problem is recovered so long as the center of mass can be counted as an inertial center appropriate for specifying the relative accelerations among solar system bodies. Newton's solution, therefore, does not depend on his hypothesis that the center of mass of the system is at rest. As we have seen in chapter 3, corollary 6 extends this to allow for acceleration of the center of mass. So long as the system is sufficiently isolated from external actions and impediments, its center of mass can be treated as an inertial reference frame. Motions that count as accelerations in such a frame are ones that carry information about forces acting to produce relative motions among solar system bodies.

[35] For Mars, the Harmonic Law ratio is 0.9996, which gives a limit, 0.9996 − .9971 = 0.0025, for a mass ratio, 1.0006 /0.0025, which would put the mass of Mars at about 1/400 of the mass of the sun.

[36] Taking the means of Kepler and Boulliau, 0.3870, 0.7240, 1, 1.5235, 5.21, and 9.526 as the distances of Mercury, Venus, Earth, Mars, Jupiter, and Saturn and assigning 1/208, 1/156, 1/169286, 1/400, 1/1067, and 1/3021 for the ratios of their masses to that of the sun yields

$$(0.3870/556) + (0.724/156) + (1/169286) + (1.5235/400) + \\ (5.21/1067) + (9.526/3021) = 0.017Au,$$

as the distance of the center of mass from the center of the sun. This is (0.017/0.0047) = 3.66 radii or about 1.83 diameters of the sun.

2. *Take stable Keplerian orbits as a first approximation*

Newton's specification of the center of mass is sufficiently exact to make it clear that the changes of motion of the sun with respect to it are quite small, while those of the earth are very much greater. Their motions with respect to the sun are reasonably good approximations of what can be counted as true motions of the planets, while their motions with respect to the earth are not at all appropriate as approximations for specifying what can be counted as true motions.

Corollary. Hence the common center of gravity of the earth, the sun, and all the planets is to be considered the center of the universe. For since the earth, sun, and all the planets gravitate toward one another and therefore, in proportion to the force of the gravity of each of them, are constantly put in motion according to the laws of motion, it is clear that their mobile centers cannot be considered the center of the universe, which is at rest. If that body toward which all bodies gravitate most had to be placed in the center (as is the commonly held opinion), that privilege would have to be conceded to the sun. But since the sun itself moves, an immobile point will have to be chosen for that center from which the center of the sun moves away as little as possible and from which it would move away still less, supposing that the sun were denser and larger, in which case it would move less. (C&W, 817)

The common center of gravity of the earth, sun, and all the planets is to be considered the center for fixing what count as the true motions of these bodies, given the hypothesis that the center of this system of the world is at rest. This center is one with respect to which each of these bodies is constantly put into motion and with respect to which these changes of motion measure the forces of gravity of these bodies toward one another.

If (as in the commonly held opinion) that body to which all bodies gravitated most had to be placed in the center, that privilege would have to be conceded to the sun. The accelerations exhibited by the earth with respect to this center of mass are so great that it is very far from counting as any sort of reasonable approximation to a center appropriate for specifying what are to be counted as true motions of solar system bodies. The changes of motion of the sun with respect to the common center of mass, on the other hand, are sufficiently small that it can count as a reasonable first approximation toward a center appropriate for specifying true motions of solar system bodies. That it does so move requires that motions with respect to it will need corrections to better approximate true motions.

Newton recommends that the alternative which is counted as the appropriate center should be one from which the sun moves away as little as possible. This appears to be a recommendation that we begin with what would be orbits in accord with inverse-square gravitation toward the sun and then take observed deviations from such motion as phenomena to be explained, insofar as can be, by gravitations toward other bodies in the system. As Newton's remark suggests, such deviations would be still smaller if the sun were denser and larger relative to the other bodies in the system.

Newton's argument to universal gravity has led to his discovery of this important force of nature from the cited phenomena of orbital motions. This is a successful realization of the first part of the aim of natural philosophy described in our epigram from Newton's preface,

to discover the forces of nature from the phenomena of motions.

Newton is now ready to enter into the second part of the endeavor described in our epigram,

and then to demonstrate other phenomena from these forces.

His center of mass resolution of the problem of specifying true motions has set the stage. He can now demonstrate for the planets Kepler's elliptical orbits, describing areas proportional to the times of description by radii to the center of the sun from his account of the force of gravity acting according to the Laws he has set out.

Proposition 13 Theorem 13

The planets move in ellipses that have a focus in the center of the sun, and by radii drawn to that center they describe areas proportional to the times.

We have already discussed these motions from the phenomena. Now that the principles of motions have been found, we deduce the celestial motions from these principles a priori. Since the weights of the planets toward the sun are inversely as the squares of the distances from the center of the sun, it follows (from book 1, props. 1 and 11, and prop. 13, corol. 1) that if the sun were at rest and the remaining planets did not act upon one another, their orbits would be elliptical, having the sun in their common focus, and they would describe areas proportional to the times. The actions of the planets upon one another, however, are so very small that they can be ignored, and they perturb the motions of the planets in ellipses about the mobile sun less (by book 1, prop. 66) than if those motions were being performed about the sun at rest. (C&W, 817–18)

Newton's demonstration consists in pointing out that Kepler's orbits would follow exactly (from book 1, propositions 1 and 11, and proposition 13, corollary 1)[37] for an idealized model in which the planets did not act on one another and each is accelerated only by its inverse-square varying weight toward a stationary sun. His discussion of the center of mass afforded by the measurements of relative masses of solar system bodies makes clear that the sun is sufficiently more massive than the rest, and that the actions of

[37] We have seen that according to proposition 1, if the force maintaining a body in an orbit were directed toward the center, the rate at which areas are described by radii to that center would be uniform. According to proposition 11 book 1, if a body revolves in an ellipse with the force directed toward a focus, that force will have to vary inversely as the square of the distance (C&W, 462–3). According to corollary 1 of proposition 13 book 1, a body acted upon by a centripetal force that is inversely proportional to the square of the distance of places from the center will move in some one of the conics, having a focus in the center of forces and conversely. (C&W, 467)

So a body moving according to an inverse-square force directed toward a center will move in either a hyperbola, or a parabola, or an ellipse with respect to that center. A hyperbolic trajectory would after closest approach to the sun carry the body off into space. A parabolic trajectory would crash into the sun. The only conic trajectory corresponding to a body maintaining an orbit about the sun is Kepler's ellipse.

the planets upon one another are so small that this idealization actually does hold to fairly good approximation.

Newton's discussion makes clear that the sense in which this proposition attributing these basic sun-centered orbits to planets is to count as gathered from phenomena by induction according to Rule 4 is to be considered very nearly rather than exactly true. Here is more of it.

> Yet the action of Jupiter upon Saturn is not to be ignored entirely. For the gravity toward Jupiter is to the gravity toward the sun (at equal distances) as 1 to 1,067; and so in the conjunction of Jupiter and Saturn, since the distance of Saturn from Jupiter is to the distance of Saturn from the sun almost as 4 to 9, the gravity of Saturn toward Jupiter will be to the gravity of Saturn toward the sun as 81 to 16 x 1,067, or roughly as 1 to 211. And hence arises a perturbation of the orbit of Saturn in every conjunction of this planet with Jupiter so sensible that astronomers have been at a loss concerning it. According to the different situations of the planet Saturn in these conjunctions, its eccentricity is sometimes increased and at other times diminished, the aphelion sometimes is moved forward and at other times perchance drawn back, and the mean motion is alternately accelerated and retarded. (C&W, 818)

He also assesses the smaller power of Saturn to perturb the orbit of Jupiter. (C&W, 818)

In chapter 9, we shall see that there is indeed a sensible perturbation of the orbits of Jupiter and Saturn due to their gravitational interaction. Finding the detailed account of the empirical correction needed and the Newtonian perturbation explaining it was a triumph for Newton's theory that afforded very high-quality evidence in support of it. We shall also see that these successes were not obtained until 1785, when Laplace solved the great inequality of Jupiter-Saturn as a nearly 900-year periodic mutual perturbation.

We shall also see that the role of accepting Kepler's orbits as idealizations corresponding to motions the planets would have if the only force acting on them were their weights toward the sun is that deviations from them are to be sought for as *theory-mediated* phenomena, to yield information about interactions that can be exploited to yield perturbation-corrected orbits. Deviations from these corrected orbits are in turn to be sought for as further *theory-mediated* phenomena to carry information about further interactions to be exploited in a sequence of progressively more accurate approximations. George Smith has argued that the successes of such applications from the time of Newton through that of Laplace and beyond yielded a long-term accumulation of very high-quality evidence in support of gravity as a universal force of interaction between bodies acting according to the law specified.[38]

[38] See Smith G.E. 2002a, 160–7. Smith further develops this important theme in a forthcoming paper (see Smith, forthcoming).

Chapter 8 Appendix 1: Newton's proof of proposition 69 book 1

Book 1 Proposition 69 Theorem 29

> *If, in a system of several bodies A, B, C, D, . . . , some body A attracts all the others, B, C, D, . . . , by accelerative forces that are inversely as the squares of the distances from the attracting body; and if another body B also attracts the rest of the bodies A, C, D, . . . , by forces that are inversely as the squares of the distances from the attracting body; then the absolute forces of the attracting bodies A and B will be to each other in the same ratio as the bodies [i.e., the masses] A and B themselves to which those forces belong.* (C&W, 587)

Newton is claiming that under the assumption that body *A* attracts all the others (including body *B*) with inverse-square accelerative forces and the assumption that body *B*, similarly, attracts all the others (including *A*), then the absolute force of *A* will be to the absolute force of *B* as the mass of *A* is to the mass of *B*.

Newton's proof begins by pointing out that the supposition that each body attracts all the rest with inverse-square accelerative forces requires that the centripetal force toward each body will produce equal accelerations of other bodies toward it at equal distances.

> For, at equal distances, the accelerative attractions of all the bodies B, C, D, . . . toward A are equal to one another by hypothesis; and similarly, at equal distances, the accelerative attractions of all the bodies toward B are equal to one another. (C&W, 587)

The supposition, thus, makes each body, of the pair (*A*, *B*), the center of a centripetal acceleration field.

The ratios of accelerations produced by such forces at equal distances are independent of distance. This makes the acceleration produced at any given distance measure the absolute quantity of such a centripetal force.

> Moreover, at equal distances, the absolute attractive force of body A is to the absolute attractive force of body B as the accelerative attraction of all the bodies toward A is to the accelerative attraction of all the bodies toward B at equal distances; (C&W, 587)

Newton goes on to exploit the fact that the distance of *A* from *B* and the distance of *B* from *A* are equal. The equality of these distances makes the ratio of *B*'s acceleration toward *A* to *A*'s acceleration toward *B* equal the ratio of the absolute attractive force toward *A* to the absolute attractive force toward *B*.

> and the accelerative attraction of body B toward A is also in the same proportion to the accelerative attraction of body A toward B. (C&W, 587)

Consider the following diagram.

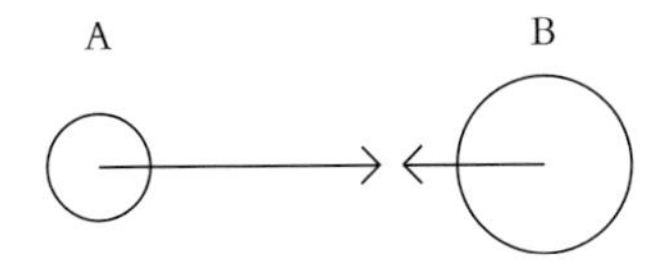

The acceleration of B with respect to A and the acceleration of A with respect to B are the indicated vectors. The ratio of the scalar magnitudes of these vectors equals the ratio of the absolute attractive force toward A to the absolute attractive force toward B. These absolute attractive forces are what Newton specified, in Definition 6, as the absolute quantities or measures of the respective centripetal forces directed toward A and toward B.[39]

The key step in Newton's proof is an application of his third Law of Motion to the motive force attracting B toward A and the motive force attracting A toward B.

> But the accelerative attraction of body B toward A is to the accelerative attraction of body A toward B as the mass of body A is to the mass of body B, because the motive forces – which (by defs. 2, 7, and 8) are as the accelerative forces and the attracted bodies jointly – are in this case (by the third Law of Motion) equal to each other. (C&W, 587)

According to the definitions Newton cites, the motive measure of a centripetal force is proportional to the mass of the body moved times the acceleration generated on it.[40] Let us recall Newton's formulation of the third Law of Motion.

> Law 3
>
> *To any action there is always an opposite and equal reaction; in other words, the actions of two bodies upon each other are always equal and always opposite in direction.* (C&W, 417)

The assumption of proposition 69 is that A attracts B, and all the rest, and that B also attracts A, and all the rest. This assumption is construed so as to interpret the motive force on B toward A as an action of A on B, and to interpret its equal and oppositely directed reaction as the motive force on A toward B. This application of the third Law of Motion construes this attraction as a mutual interaction between A and B. Given this application of Law 3, Newton's conclusion

> Therefore, the absolute attractive force of body A is to the absolute attractive force of body B as the mass of body A is to the mass of body B. Q.E.D. (C&W, 587)

follows.[41]

[39] Let $absF_A$ and $absF_B$ be respectively the absolute measures of the inverse-square centripetal forces toward A and toward B, while $acc_A(B)$ and $acc_B(A)$ are respectively the magnitudes of the acceleration of B with respect to A and the acceleration of A with respect to B. This first step in the argument is that

$$absF_A/absF_B = acc_A(B)/acc_B(A) \qquad (1A)$$

[40] Where $F_A(B)$ and $F_B(A)$ are respectively the motive force attracting B toward A and the motive force attracting A toward B, and $Mass(A)$ and $Mass(B)$ are the respective masses of A and B, we have

$$F_A(B)/F_B(A) = (Mass(B) \times acc_A(B))/(Mass(A) \times acc_B(A)), \qquad (2A)$$

where, as in footnote 39 above, $acc_A(B)$ and $acc_B(A)$ are respectively the magnitudes of the acceleration of B with respect to A and the acceleration of A with respect to B.

[41] Having the motive force attracting B toward A be the equal and opposite reaction to the motive force attracting A toward B, makes the ratio, $F_A(B)/F_B(A)$, of their magnitudes equal one. We have, therefore, from (2A) in footnote 40 above,

$$F_A(B)/F_B(A) = 1 = (Mass(B) \times acc_A(B))/(Mass(A) \times acc_B(A)). \qquad (3A)$$

This, together with (1A) from footnote 39 above gives Newton's conclusion:

$$Mass(A)/Mass(B) = acc_A(B)/acc_B(A) = absF_A/absF_B \qquad (Conclusion\ A)$$

This proof concentrated on the two-body interaction between bodies *A* and *B*, but the same argument applies to each pair of the bodies in the system, when they all interact together.

> Corollary 1. Hence if each of the individual bodies of the system A, B, C, D, . . . , considered separately, attracts all the others by accelerative forces that are inversely as the squares of the distances from the attracting body, the absolute forces of all those bodies will be to one another as the ratios of the bodies [i.e., the masses] themselves. (C&W, 587)

Given the supposition, each body will have an acceleration toward each of the others that is directly proportional to the mass of that other body and inversely as the square of its distance from that body.

The supposition of proposition 69 was that each of the bodies attracts all the rest with inverse-square accelerative forces. The proof, however, does not depend on the inverse-square property of the accelerations produced. Whatever function of distance these accelerations might be, the same proof will go through, so long as ratios of accelerations at equal distances from the attracting bodies are independent of those distances. It is this more general property of being similarly varying acceleration fields, rather than the particular inverse-square variation with distance of the accelerations produced, that counts.

> Corollary 2. By the same argument, if each of the individual bodies of the system A, B, C, D, . . . , considered separately, attracts all the others by accelerative forces that are either inversely or directly as any powers whatever of the distances from the attracting body, or that are defined in terms of the distances from each one of the attracting bodies according to any law common to all these bodies; then it is evident that the absolute forces of those bodies are as the bodies [i.e., the masses]. (C&W, 587–8)

As Newton points out, this corollary follows by the same basic argument as that for proposition 69.

In a last corollary to this proposition Newton claims that if motions in a system of bodies interacting by inverse-square attractions are to closely approximate ellipses with the Area Rule about a common focus in the center of a greatest body, as the planets and sun do, then the ratios of the attractions toward those bodies must approximate the ratios of their masses.

> Corollary 3. If, in a system of bodies whose forces decrease in the squared ratio of the distances [i.e., vary inversely as the squares of the distances], the lesser bodies revolve about the greatest one in ellipses as exact as they can be, having their common focus in the center of that greatest body, and – by radii drawn to the greatest body – describe areas as nearly as possible proportional to the times, then the absolute forces of those bodies will be to one another, either exactly or very nearly, as the bodies, and conversely. This is clear from the corollary of prop. 68 compared with corol. 1 of this proposition. (C&W, 588)

This corollary reinforces Newton's application of Law 3, by showing that the resulting proportionality to masses of the attracting bodies is robust with respect to approximations.[42]

[42] It would be very much worth exploring in detail how significantly this result adds support to Newton's controversial application of Law 3.

Chapter 8 Appendix 2: The attractive forces of spherical bodies

1. *Basic propositions from book 1*

Newton's scholium to proposition 69 opens with the suggestion that cases of attraction toward bodies which, like gravity, are proportional to the quantity of matter in the attracting body should be "reckoned by assigning proper forces to their individual particles and then taking the sums of these forces."[43] The corollaries of proposition 7 tell us that "the gravity toward the whole planet arises from and is compounded of the gravity toward the individual parts," and that "The gravitation toward each of the individual equal particles of a body is inversely as the square of the distance of places from those particles." To have these claims consistent with inverse-square attraction to whole planets requires being able resolve these inverse-square attractions to spherical bodies as sums of inverse-square attractions to the particles making them up.

Newton begins by showing that the sum of inverse-square attractions toward the parts of a spherical surface will be zero for any point inside the sphere.

Book 1 Proposition 70 Theorem 30

If toward each of the separate points of a spherical surface there tend equal centripetal forces decreasing as the squares of the distances from the point, I say that a corpuscle placed inside the surface will not be attracted by these forces in any direction. (C&W, 590)

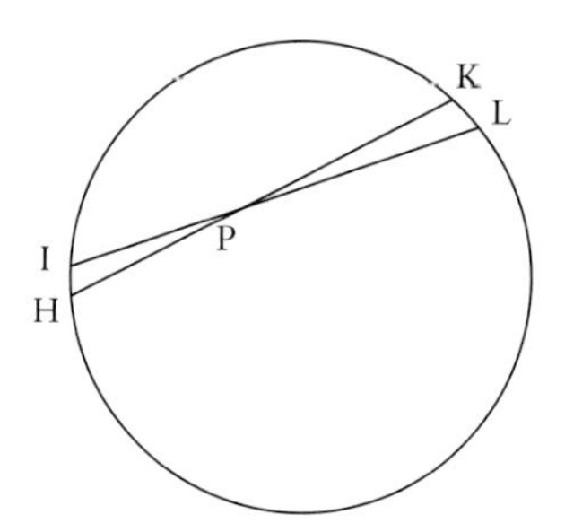

The idea of Newton's proof of this proposition is clear from his diagram. The attractions toward the areas on opposite sides circumscribed by rotating the lines are inversely proportional to the squares of the distances.

As the angles are made small, these traced-out curvilinear areas become directly proportional to the squares of the distances so that these oppositely directed attractions cancel each other out.

Newton shows that the attraction exerted by such a spherical surface on a corpuscle outside this sphere is directed toward the center of the sphere and is inversely as the square of the distance of the corpuscle from the center of the sphere.

[43] This scholium to proposition 69 book 1 is informative about Newton's methodology. We will discuss it in chapter 9 section I below.

Book 1 Proposition 71 Theorem 31

With the same conditions being supposed as in prop. 70, I say that a corpuscle placed outside the spherical surface is attracted to the center of the sphere by a force inversely proportional to the square of its distance from that same center. (C&W, 590)

His proof of this fundamental result appeals to an interesting application of his geometrical treatment of limiting ratios. The details are given in appendix 3 below.

Newton next begins his investigation of attractions toward solid spherical bodies.

Book 1 Proposition 72 Theorem 32

If toward each of the separate points of any sphere there tend equal centripetal forces, decreasing in the squared ratio of the distances from those points, and there are given both the density of the sphere and the ratio of the diameter of the sphere to the distance of the corpuscle from the center of the sphere, I say that the force by which the corpuscle is attracted will be proportional to the semidiameter of the sphere.

For imagine that two corpuscles are attracted separately by two spheres, one corpuscle by one sphere, and the other corpuscle by the other sphere, and that their distances from the centers of the spheres are respectively proportional to the diameters of the spheres, and that the two spheres are resolved into particles that are similar and similarly placed with respect to the corpuscles. Then the attractions of the first corpuscle, made toward each of the separate particles of the first sphere, will be to the attractions of the second toward as many analogous particles of the second sphere in a ratio compounded of the direct ratio of the particles and the inverse squared ratio of the distances [i.e., the attractions will be to one another as the particles directly and the squares of the distances inversely]. But the particles are as the spheres, that is, they are in the cubed ratio of the diameters, and the distances are as the diameters; and thus the first of these ratios directly combined with the second taken twice inversely becomes the ratio of diameter to diameter. Q.E.D. (C&W, 592)

Newton assumes a case where the numbers of the attracting particles in the two spheres are proportional to their volumes and, therefore, to the cubes of their diameters. He also assumes that the distances of the corpuscles from the centers of their respective spheres are proportional to the diameters of those spheres. This makes the distances of the two corpuscles from correspondingly situated attracting particles in their respective spheres, also, proportional to the diameters of those spheres. Where F_1 is the magnitude of the compounded attractions of the first corpuscle toward the particles of the first sphere and F_2 is the magnitude of the compounded attractions of the second corpuscle toward the second sphere and the respective diameters of those spheres are D_1 and D_2,

$$F_1/F_2 = (D_1{}^3 \times D_1{}^{-2})/(D_2{}^3 \times D_2{}^{-2}) = D_1/D_2.$$

The compounded attractions are as the diameters of the attracting spheres and therefore as their radii or semidiameters.

This result is applied, together with proposition 70, to give the attractions compounded of inverse-square attractions toward particles of a sphere on a corpuscle placed inside it.

Book 1 Proposition 73 Theorem 33

If toward each of the separate points of any given sphere there tend equal centripetal forces decreasing in the squared ratio of the distances from those points, I say that a corpuscle placed inside the sphere is attracted by a force proportional to the distance of the corpuscle from the center of the sphere.

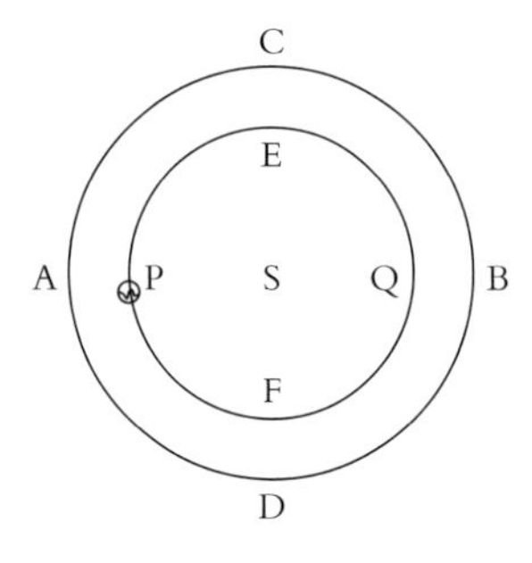

Let a corpuscle P be placed inside the sphere ABCD, described about center S; and about the same center S with radius SP, suppose that an inner sphere PEQF is described. It is manifest (by prop. 70) that the concentric spherical surfaces of which the difference AEBF of the spheres is composed do not act at all upon body P, their attractions having been annulled by opposite attractions. There remains only the attraction of the inner sphere PEQF. And (by prop. 72) this is as the distance PS. Q.E.D. (C&W, 593)

By proposition 70, the concentric spherical shells out of which the difference between the whole sphere and the inner sphere defined by the distance of the corpuscle *P* from the center *S* is made up do not act on *P*. Therefore, the only attraction is that toward this inner sphere. By proposition 72, this is proportional to the diameter of this sphere and, thus, to its radius PS.

Newton's scholium to this proposition applies to the spherical surfaces considered in propositions 70 and 71, as well as to those composing the solid spheres in propositions 72–4.

> Scholium. The surfaces of which the solids are composed are here not purely mathematical, but orbs [or spherical shells] so extremely thin that their thickness is as null: namely, evanescent orbs of which the sphere ultimately consists when the number of those orbs is increased and their thickness diminished indefinitely. Similarly, when lines, surfaces, and solids are said to be composed of points, such points are to be understood as equal particles of a magnitude so small that it can be ignored. (C&W, 593)

Think of the body as divided up into a large but finite number of small but finite parts, each of which has a small but finite mass. By increasing their number while decreasing their size, their finite sums can approximate, to any tolerance we care to specify, the results of integration over continuous mass distributions.

Newton now addresses attraction of corpuscles placed outside such spheres uniformly composed of equally inverse-square attracting particles.

Book 1 Proposition 74 Theorem 34

> *With the same things being supposed as in prop. 73, I say that a corpuscle placed outside a sphere is attracted by a force inversely proportional to the square of the distance of the corpuscle from the center of the sphere.*
>
> For let the sphere be divided into innumerable concentric spherical surfaces; then the attractions of the corpuscle that arise from each of the surfaces will be inversely proportional to the square of the distance of the corpuscle from the center (by prop. 71). And by composition [or componendo] the sum of the attractions (that is, the attraction of the corpuscle toward the total sphere) will come out in the same ratio. Q.E.D. (C&W, 593)

Since the attraction to each spherical surface is inversely proportional to the square of the distance of the corpuscle from the center of the sphere and the attraction toward the total sphere is just the sum of its attractions to these spherical surfaces, this total attraction is also inversely proportional to the square of that distance.

In corollary 3 of proposition 74, Newton infers inverse-square attraction toward the particles making it up from the inverse-square attraction toward a whole sphere. We will consider this corollary in section 2, where we will see that Newton is able to back up this inference by systematic dependencies.

In proposition 75, Newton extends this result for inverse-square attraction of corpuscles to inverse-square attraction of other whole spheres.

Book 1 Proposition 75 Theorem 35

If toward each of the points of a given sphere there tend equal centripetal forces decreasing in the squared ratio of the distances from the points, I say that this sphere will attract any other homogeneous sphere with a force inversely proportional to the square of the distance between the centers.

For the attraction of any particle is inversely as the square of its distance from the center of the attracting sphere (by prop. 74), and therefore is the same as if the total attracting force emanated from one single corpuscle situated in the center of this sphere. Moreover, this attraction is as great as the attraction of the same corpuscle would be if, in turn, it were attracted by each of the individual particles of the attracted sphere with the same force by which it attracts them. And that attraction of the corpuscle (by prop. 74) would be inversely proportional to the square of its distance from the center of the sphere; and therefore the sphere's attraction, which is equal to the attraction of the corpuscle, is in the same ratio. Q.E.D. (C&W, 594)

For each particle of the attracted sphere, its total attraction toward the attracting sphere is equal to a single inverse-square attraction toward one corpuscle located at the center of this attracting sphere. Suppose that for each particle of the attracted sphere this single attracting corpuscle undergoes an attraction toward that particle equal and opposite to the attraction it exerts on that particle. By proposition 74, the result of combining these inverse-square attractions of this corpuscle at the center of the first sphere toward the particles of the second sphere is a total attraction toward the center of that second sphere that is inversely as the square of the distance of the corpuscle from that center. By our supposition, this is equal and opposite to the total attraction of the whole second sphere toward the whole first sphere. Therefore, this attraction of the whole second sphere toward the whole first sphere is directed toward the center of the first and is inversely proportional to the square of the distance between the centers of the two spheres.

Several of Newton's corollaries of this proposition are of special interest to us. The first,

> Corollary 1. The attractions of spheres toward other homogeneous spheres are as the attracting spheres [i.e., as the masses of the attracting spheres] divided by the squares of the distances of their own centers from the centers of those that they attract. (C&W, 594),

illustrates that the proportionality of attraction to the mass of an attracting sphere follows from the supposition that attraction toward that whole homogeneous sphere results from combining equal inverse-square attractions toward the particles which make it up. This argument does not depend on assuming, as a premise, that Law 3 applies to these attractions.

The next corollary explores the case when it is assumed that the attracted body also attracts the attracting body so that Law 3 applies.

> Corollary 2. The same is true when the attracted sphere also attracts. For its individual points will attract the individual points of the other with the same force by which they are in turn attracted by them; and thus, since in every attraction the attracting point is as much urged (by law 3) as the attracted point, the force of the mutual attraction will be duplicated, the proportions remaining the same. (C&W, 595)

Here, with Law 3 assumed to hold for the attraction between spheres, we will have

$$F_1(2) = F_2(1) \text{ proportional to } (M_1 \times M_2)/D^2,$$

where $F_1(2)$ is the attraction of sphere 2 toward sphere 1, $F_2(1)$ is the equal and opposite attraction of sphere 1 toward sphere 2, M_1 and M_2 are the respective masses of those spheres, and D is the distance between their centers.

Newton points out that his results about orbital motions with respect to the focus for conics hold for bodies moving outside an attracting sphere when that sphere is placed in the focus.

> Corollary 3. Everything that has been demonstrated above concerning the motion of bodies about the focus of conics is valid when an attracting sphere is placed in the focus and the bodies move outside the sphere. (C&W, 595)

According to proposition 11 book 1, the centripetal force toward a focus of an ellipse must be inverse-square to maintain a body in that ellipse as an orbit.[44] This is especially relevant to application of such results to orbital motion about planets construed as attracting homogeneous spherical bodies.

In the next proposition Newton goes on to consider a better approximation to the nonhomogeneous densities of actual planets.

Book 1 Proposition 76 Theorem 36

> *If spheres are in any way nonhomogeneous (as to the density of their matter and their attractive force) going from the center to the circumference, but are uniform throughout in every spherical shell at any given distance from the center, and the attractive force of each point decreases in the squared ratio of the distance of the attracted body, I say that the total force by which one sphere of this sort attracts another is inversely proportional to the square of the distance between their centers.* (C&W, 595)

Chandrasekhar offers a very nice exposition of a proof of such a generalization of inverse-square attraction between spheres to ones with nonhomogeneous spherically symmetric densities, based on successive applications of Newton's proposition 75 to nested spheres.[45]

In his classic treatment of Newtonian Mechanics, A.P. French discusses this result for solid spheres of nonhomogeneous but spherically symmetric density, right after his treatment of Newton's basic result for attraction toward homogeneous spherical shells. In his treatment of this basic result, French offered an elegant application of the calculus to show that

[44] According to proposition 10 of book 1, a centripetal force maintaining a body in an elliptical orbit and directed to the center of that ellipse would vary directly as the distance from that center. This harmonic oscillator force law and the inverse-square force law are the only two force laws that would maintain bodies in stable elliptical orbits. Newton alludes to the fact that this harmonic oscillator force law is the law of attraction for bodies inside a sphere made up of equal inverse-square attractions toward the particles making it up.

> Corollary 4. (prop. 75) And whatever concerns the motion of bodies around the center of conics applies when the motions are performed inside the sphere. (C&W, 595)

If a body could move freely inside a homogeneous sphere of inverse-square attracting particles, the resulting force on it would maintain it in a conic with the center rather than a focus located at the center of the attracting sphere.

[45] See Chandrasekhar (1995, 280–1). Newton's statement of the theorem allows both the density and the attractive force to vary. This, together with remarks about addition of non-attracting matter in his proof, suggest that Newton intends to allow different spherical shells to differ in attractive force per unit mass as well as density. It should not be difficult to extend Chandrasekhar's proof to Newton's more general case.

$$F_r = -GMm/r^2$$

is the attraction of a particle of mass m at a distance r (greater than the radius) from the center of a spherical shell of negligible thickness and mass M resulting from integrating the inverse-square attractions toward the particles of the shell.[46] Here are his follow-up remarks:

> Once we have Eq. (8–14), the total effect of a solid sphere follows at once. Regardless of the particular way in which the density varies between the center and the surface (provided that it depends only on R) the complete sphere does indeed act as though its total mass were concentrated at its center. It does not matter how close the attracted particle P is to the surface of the sphere, as long as it is in fact outside. Take a moment to consider what a truly remarkable result this is. Ask yourself: Is it obvious that an object a few feet above the apparently flat ground should be attracted as though the whole mass of the earth (all 6,000,000,000,000,000,000,000 tons of it!) were concentrated at a point (the earth's center) 4000 miles down? It is about as far from obvious as could be, and there can be little doubt that Newton had to convince himself of this result before he could establish, to his own satisfaction, the grand connection between terrestrial gravity and the motion of the moon and other celestial objects. (French 1971, 265)

This splendid passage makes vividly clear that this inverse-square result for solid spheres of nonhomogeneous but spherically symmetrical density is truly remarkable, that Newton had good reason to want it to back up his moon-test, and, also, that it follows from an application of the corresponding result for spherical shells.

Newton offers nine corollaries outlining implications of proposition 76. The following are especially important for his argument to universal gravity.

> Corollary 3. And the motive attractions, or the weights of spheres toward other spheres, will – at equal distances from the centers – be as the attracting and attracted spheres jointly, that is, as the products produced by multiplying the spheres by each other.
>
> Corollary 4. And at unequal distances, as those products directly and the squares of the distances between the centers inversely.
>
> Corollary 5. These results are valid when the attraction arises from each sphere's force of attraction being mutually exerted upon the other sphere. For the attraction is duplicated by both forces acting, the proportion remaining the same. (C&W, 596)

So are the last two, which are directed to orbital motion around the foci of conics.

> Corollary 8. Everything that has been demonstrated above about the motion of bodies around the foci of conics when the attracting sphere, of any form or condition that has already been described, is placed in the focus.
>
> Corollary 9. As also when the bodies revolving in orbit are also attracting spheres of any condition that has already been described. (C&W, 597)

These extend to the more general case of nonhomogeneous but spherically symmetric densities the important results of corollaries 1 and 2 of proposition 4. Corollary 8 recovers the orbital

[46] French 1971, 261–5. Equation (8–14) is on page 265.

results for attracted test bodies while corollary 9 extends them to mutual attraction where the third Law of Motion applies to the interaction between primary and satellite.

Propositions 77 and 78 of book 1 are directed toward attractions toward spheres composed of particles with attracting forces directly proportional to the distance from them. These, harmonic oscillator law, attracting forces also result in composited attractions to whole spheres that have the same distance law right down to the surface of the attracting sphere.

2. *Inverse-square attraction to particles from inverse-square attraction to spheres using an integral given by Chandrasekhar*

Newton follows up the above propositions with a scholium.

> Scholium. I have now given explanations of the two major cases of attractions, namely, when the centripetal forces decrease in the squared ratio of the distances or increase in the simple ratio of the distances, causing bodies in both cases to revolve in conics, and composing centripetal forces of spherical bodies that decrease or increase in proportion to the distance from the center according to the same law – which is worthy of note. It would be tedious to go one by one through the other cases which lead to less elegant conclusions. I prefer to comprehend and determine all the cases simultaneously under a general method as follows. (C&W, 599)

He develops this general method for attractions on bodies located outside of attracting spheres in lemma 29 together with propositions 79–81 of book 1.

Chandrasekhar points out that these results by Newton are an "analytical *tour de force*." He argues that Newton's method is equivalent to an integration to give the attraction on an external particle by a homogeneous sphere made up of particles toward which there tend equal centripetal forces according to any specified law of attraction.

> In the adjoining diagram, S is the centre of a sphere of radius a; P is the external particle at a distance $R(=PS)$ from the centre; EE' is a chord that cuts the axis at right angles at D; and z $(=PD)$ is the distance of P from D. The range of the variable z is clearly $(R-a) \leq z \leq (R+a)$.

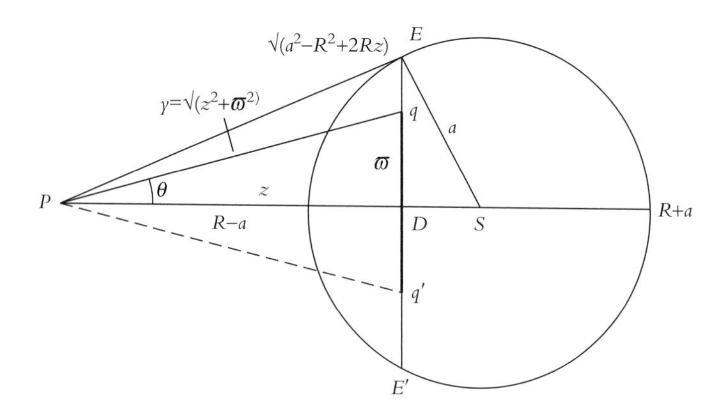

> We determine the attraction of P by the sphere by summing over the contributions by circular discs, EDE', of the same thickness dz, at various distances $z(=PD)$ from P;

> and determine the attraction by the disc EDE', in turn, by summing over the circular rings, qDq' of radii $\bar{\omega}$ in the range $0 \leq \bar{\omega} \leq DE$ (in the manner Newton describes later in Proposition XC). (Chandrasekhar 1995, 287)

This gives the following specification of the force F on a particle P at distance R from the center of the sphere resulting from integrating attractions to particles of the function *f(y)* of distance y from them.

$$F = \pi \int_{R-a}^{R+a} f(y) \left[y^2 - \frac{(y^2 - a^2 + R^2)^2}{4R^2} \right] dy$$

Chandrasekhar gives this integral as equation 9 on page 289 of his 1995 book. He argues that the expression for the force given in this integral is equivalent to Newton's proposition 80 book 1.[47]

If we put $1/y^{2+\delta}$ for *f(y)* and r for distance from the center, we have an expression for the attraction toward homogeneous solid spheres for corpuscles located outside their surfaces

$$f[r.a.\delta] = \pi \int_{r-a}^{r+a} \frac{1}{y^{2+\delta}} (y^2 - \frac{(y^2 - a^2 + r^2)^2}{4r^2}) dy$$

corresponding to alternative power laws (inversely as y to the $2+\delta$) for attractions toward the particles making up the spheres. The inverse-square case is given by $\delta = 0$. This results in

$$F = (4/3)a^3/r^2,$$

which shows that inverse-square attraction toward the particles composing it gives the inverse-square attraction toward the whole sphere. So, proposition 74 follows directly from the more general relation when one solves for $\delta = 0$.[48]

Putting in alternative power laws for attraction toward particles illustrates that Newton's appeal to corollary 3 of proposition 74 to infer inverse-square attraction toward the particles from inverse-square attraction toward whole spheres is backed up by systematic dependencies. The figure on page 326 shows the attractions toward a solid sphere resulting from respectively an inverse 1.5 power law, the inverse-square, and an inverse 2.5 power law for attractions toward the particles making it up.[49]

This illustrates that the inverse 1.5 and 2.5 power laws for attractions toward the particles result in non-uniform ratios of attraction to distance for the corresponding attractions toward a whole sphere on a corpuscle located outside it. For corpuscles located sufficiently far out (say beyond 1.5 distance units in this graphical representation) we have just the sort of dependencies that

[47] Chandrasekhar 1995, 289.

[48] Similarly, when one solves for $\delta = -3$, one can solve for the harmonic oscillator force (directly as y) for attraction toward particles. This results in

$$F = (4/3)a^3 r,$$

which shows that attraction directly as distance from particles results in attraction directly as distance from the center for attraction toward solid spheres made up of such particles.

[49] These diagrams were plotted by Wayne Myrvold.

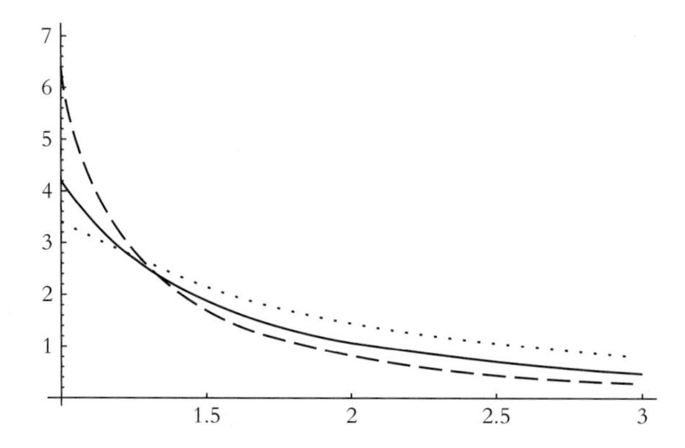

make Newton's basic inferences from phenomena into measurements. Having the attraction toward particles fall off less fast with distance than the inverse-square corresponds to attraction toward the whole sphere that falls off less fast then the inverse-square, while having attraction toward particles fall off faster corresponds to attraction toward the whole sphere that also falls off faster.

The non-uniformities of the ratio of attraction to distance from the center of the attracting sphere for these alternative power laws for attractions toward particles make the evidential situation somewhat more complex. At the distance where its graph crosses that of the inverse-square, an alternative law of attraction toward particles would give the same attraction toward the whole sphere as would the inverse-square. For sufficiently smaller distances the indication relations for alternatives to the inverse-square would be reversed from those corresponding to large distances.

In propositions 85–91 book 1, Newton investigates attractive forces toward non-spherical bodies. In proposition 92 he sums up his results, extending propositions 79–81 for attractions toward spheres to other shaped bodies.[50]

[50] The discussion of this proposition offers an interesting contrast to a comment by Chandrasekhar speculating on Newton's reasons for having his newly published third edition open before him to pages clearly displaying the illustrations for proposition 81 when he posed for a portrait at the age of 83.

> Proposition LXXXI could not have had for Newton any particular 'use in philosophical inquiries'. But the proposition does display Newton's familiarity with and craftsmanship in the use of the integral calculus: inversion of the order of the integrations for reducing a double integral to a single one, and integration by parts to simplify an integral. Was Newton displaying to his critics (if they can understand) that already, in 1685 when he was writing the *Principia*, he was a virtuoso in the Art of the Calculus? (1995, 302; see plates 2 and 3, Chandrasekhar 1995, 300–1, for the portrait and the pages of the original published version of the third edition of the *Principia*)

Newton's discussion in proposition 92 makes it clear that he saw his result in proposition 81 of particular use in inferring the inverse-square law of attraction toward particles making them up from the inverse-square law for attractions toward solid spheres.

Book 1 Proposition 92 Problem 46

> *Given an attracting body, it is required to find the ratio by which the centripetal forces tending toward each of its individual points decrease [i.e., decrease as a function of distance].*
>
> From the given body a sphere or cylinder or other regular figure is to be formed, whose law of attraction – corresponding to any ratio of decrease [in relation to distance] – can be found by props. 80, 81, and 91. Then, by making experiments, the force of attraction at different distances is to be found: and the law of attraction toward the whole that is thus revealed will give the ratio of the decrease of the forces of the individual parts, which was required to be found. (C&W, 618)

Newton makes it clear that his result in proposition 81 (and propositions 80 and 91) can be used to infer the inverse-square law for the attractions towards the particles being summed to make it up from the empirically established inverse-square variation of the attraction toward a whole body.[51]

The harmonic rule relation among the orbits of the planets and the absence of un-accounted for precession in each of those orbits support the inference to the inverse-square law for the attractions toward the particles that sum together to make up the inverse-square centripetal attraction toward the whole sphere of the sun.

[51] The need for such an appeal to experiments finding the force of attraction at differing distances reflects the more complex evidential relations required to rule out alternative force laws toward particles. These complexities can themselves be exploited to help back up the inference from the inverse-square law for attraction toward whole spheres to the inverse-square law for the attractions toward the particles making them up. The only power laws for attraction towards the particles that would have the same power law for attraction toward a whole sphere right down to its surface are the inverse-square and the harmonic oscillator law where the force is directly as the distance. This helps reinforce the weight of evidence for the inverse-square law for gravitation toward particles arising from the evidence for the same inverse-square variation within each of the differing distance ranges explored by the several planets, and the evidence for the inverse-square relation among the accelerations induced at those differing distance ranges, together with the moon-test supporting the inverse-square relation between the centripetal acceleration of the moon and the acceleration of gravity right at the surface of the earth.

Chapter 8 Appendix 3: Propositions 70 and 71 book 1

1. *Newton's proofs of propositions 70 and 71 book 1*

Book 1 Proposition 70 Theorem 30

If toward each of the separate points of a spherical surface there tend equal centripetal forces decreasing as the squares of the distances from the point, I say that a corpuscle placed inside the surface will not be attracted by these forces in any direction.

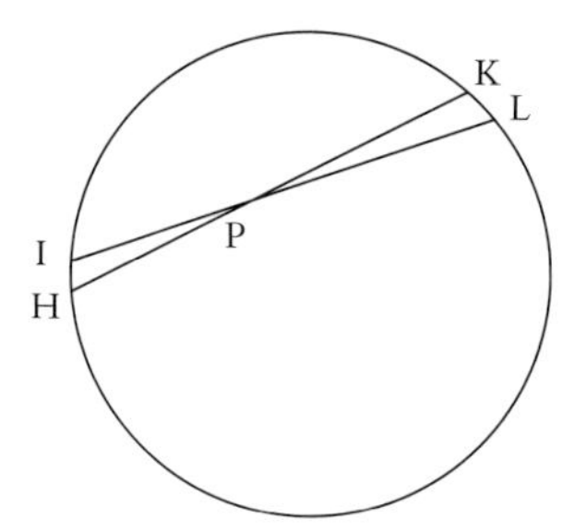

We noted in appendix 2 above that the idea of Newton's proof of this proposition is clear from his diagram. The attractions toward the areas on opposite sides circumscribed by rotating the lines are inversely proportional to the squares of the distances. As the angles are made small, these traced-out curvilinear areas become directly proportional to the squares of the distances, so that these oppositely directed attractions cancel each other out.

Here is Newton's proof that the attraction exerted by such a spherical surface on a corpuscle outside this sphere is directed toward the center of the sphere and is inversely as the square of the distance of the corpuscle from the center of the sphere.

Book 1 Proposition 71 Theorem 31

With the same conditions being supposed as in prop. 70, I say that a corpuscle placed outside the spherical surface is attracted to the center of the sphere by a force inversely proportional to the square of its distance from that same center.

Let AHKB and *ahkb* be two equal spherical surfaces, described about centers *S* and *s* with diameters AB and *ab*, and let P and *p* be corpuscles located outside those spheres in those diameters produced. From the corpuscles draw lines PHK, PIL, *phk*, and *pil*, so as to cut off from the great circles AHB and *ahb* the equal arcs HK and *hk*, and IL and *il*. And onto these lines drop perpendiculars SD and *sd*, SE and *se*, IR and *ir*, of which SD and *sd* cut PL and *pl* at F and *f*. Also drop perpendiculars IQ and *iq* onto the diameters.

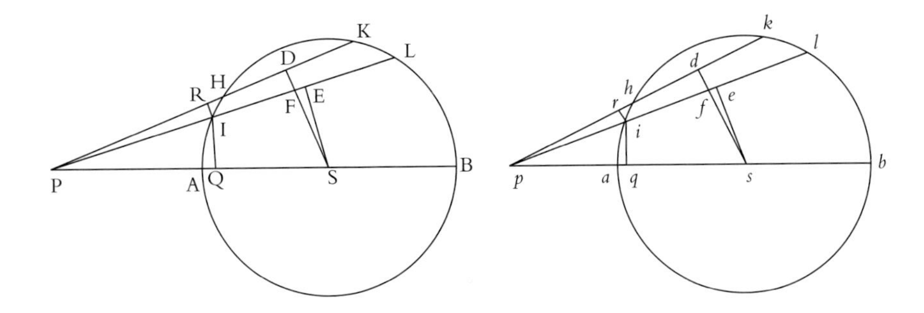

Let angles DPE and *dpe* vanish; then, because DS and *ds*, ES and *es* are equal, lines PE, PF and *pe, pf* and the line-elements DF and *df* may be considered to be equal, inasmuch as their ultimate ratio, when angles DPE and *dpe* vanish simultaneously, is the ratio of equality. (C&W, 590–2)

In Newton's diagram we have equalities of ratios of line segments following from similarity of triangles.
The equality PI/PF = RI/DF follows from the similarity of triangles PRI and PDF, while $pf/pi = df/ri$ follows from the similarity of triangles *pri* and *pdf*. Given the equality of these ratios we have the first equality of (1) by multiplying terms. The second and third equalities follow by a lemma which we shall prove shortly.

$$(\mathrm{PI} \times pf)/(\mathrm{PF} \times pi) = (\mathrm{RI} \times df)/(\mathrm{DF} \times ri) = \mathrm{RI}/ri = \mathrm{IH}/ih \qquad (1)$$

According to this lemma, in the limit, as angles RPI and *rpi* are made to simultaneously approach zero, *df* approaches equality with DF and the ratio RI/*ri* approaches equality with the ratio IH/*ih*, where IH and *ih* are the curved line segments indicated in the diagram.

Right triangle PQI is similar to right triangle PES, so PI/ PS = IQ/ SE. Right triangle *pqi* is similar to right triangle *pes*, so $ps / pi = se / iq$. The first equality of 2 follows from these by multiplication.

$$(\mathrm{PI} \times ps)/(\mathrm{PS} \times pi) = (\mathrm{IQ} \times se)/(\mathrm{SE} \times iq) = \mathrm{IQ}/iq \qquad (2)$$

The second equality of (2) follows from Newton's construction which gives the equality $SE = se$. Let us now multiply equations (1) and (2) together. This yields

$$(\mathrm{PI}^2 \times pf \times ps)/(pi^2 \times \mathrm{PF} \times \mathrm{PS}) = (\mathrm{IH} \times \mathrm{IQ})/(ih \times iq) \qquad (3)$$

Let F be the total force attracting the corpuscle at P toward the curved surface generated by rotating curve IH about the line PS and let f be the total force attracting the corpuscle at p toward the curved surface generated by rotating curve *ih* about the line *ps*. These forces are respectively proportional to the areas (IH × IQ × 2π) and ($ih \times iq \times 2\pi$) of the generated surfaces and inversely proportional to the squares PI^2 and pi^2 of the distances PI and *pi*. This gives (4).

$$F/f = ((\mathrm{IH} \times \mathrm{IQ} \times 2\pi)/\mathrm{PI}^2)/((ih \times iq \times 2\pi)/pi^2) \qquad (4)$$

From (4), together with (3), we have

$$F/f = (pf \times ps)/(\mathrm{PF} \times \mathrm{PS}) \tag{5}$$

Let F_S be the total of the force components toward *S* from attraction toward the curved surface generated by rotating IH, and f_s be the corresponding total force toward *s* from attraction toward the surface generated by rotating *ih*.

$$F_S/f_s = F(\mathrm{PQ}/\mathrm{PI})/f(pq/pi) = F(\mathrm{PF}/\mathrm{PS})/f(pf/ps) \tag{6}$$

The component toward *S* on *P* of the force toward I is the force toward I times the cosine (PQ/PI) of angle IPQ. This factor, cosine of angle IPQ, remains the same as I is rotated about Q, so F_S is just the total force *F* toward the surface times this cosine factor. The same holds for f_s, *f* and its cosine factor (pq/pi). We also have PQ/PI = PF/PS and $pq/pi = pf/ps$ in the limit as triangles PIQ and *piq* become respectively similar to triangles PSF and *pif*.

Combining (5) and (6) yields

$$F_S/f_s = (pf \times ps \times (\mathrm{PF}/\mathrm{PS}))/(\mathrm{PF} \times \mathrm{PS} \times (pf/ps)) = ps^2/\mathrm{PS}^2 \tag{7}$$

For each rotated strip, the total force on a corpuscle at P toward the center S is to the total force on a corpuscle at *p* toward the center *s* for its corresponding rotated strip inversely as the squares of the distances PS and *ps*.

For each point I the attraction on a corpuscle at P toward I can be resolved into a component toward center S and a perpendicular component directly away from the line PS. These perpendicular components cancel out, as they are exactly equal in strength and opposite in direction at opposite sides of the rotated strip. So the total forces toward the respective strips are toward the centers and inversely as the squares of the distances from those centers. The total forces to the spherical surfaces share these properties, since they are just what result from summing over such strips.

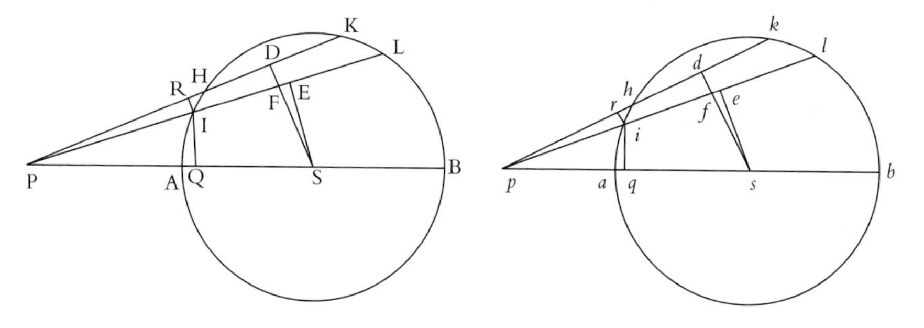

Let us now take up the lemma we assumed in our exposition of the main proof.

Lemma. As angles HPI and *hpi* are simultaneously decreased toward zero by having *I* approach *H* while *i* approaches *h*,

i) *df* becomes equal to DF

and

ii) the ratio RI/*ri* of straight line segments RI and *ri* becomes equal to the ratio IH/*ih* of curved line segments IH and *ih*.

To prove i, note that DF = DS-SF and *df* = *ds*-*sf*. By Newton's construction, DS=*ds*, SE=*se*, the angles HPI and *hpi* are respectively equal to the angles DPF and *dpf*, and the angles PDS,

PES, *pds* and *pes* are all right angles. As the angles DPF and *dpf* simultaneously vanish, SF and *sf* become respectively equal to SE and *se*. Therefore, since SE=*se* by construction, DF becomes equal to *df*.

To prove ii, construct tangents HN at H and *hn* at *h*. Angle NHK = Angle *nhk*, since line

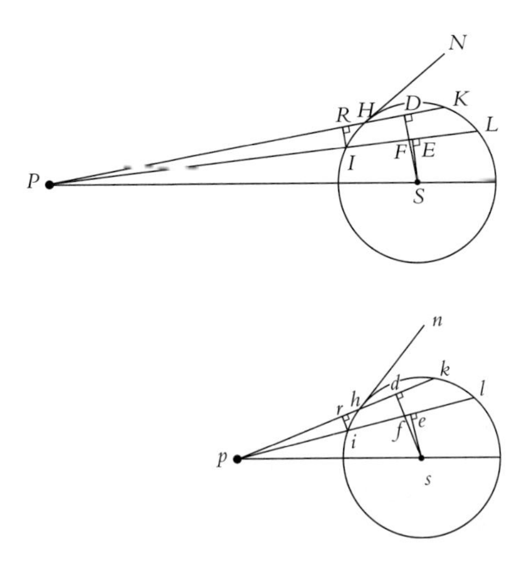

segments HK and *hk* as well as their corresponding circles are equal by Newton's construction. As I approaches H angle RHI becomes ultimately equal to angle NHK, and as *i* approaches *h* angle *rhi* becomes ultimately equal to angle *nhk*. This holds because, by corollary 3 of Newton's lemma 7 on first and ultimate ratios, as I approaches H and *i* approaches *h* chords IH and *ih* become ultimately tangents at H and *h*. The right triangles HRI and *hri*, therefore, become ultimately similar. This makes the ratios RI/*ri* of line segments ultimately equal the ratio IH/*ih* of chords. But, also by corollary 3 of lemma 7, as I approaches H chord IH becomes ultimately equal to curve IH and as *i* approaches *h* chord *ih* becomes ultimately equal to curve *ih*. Therefore, as *I* approaches H and *i* approaches *h*, ratio RI/*ri* becomes ultimately equal to ratio IH/*ih* of curve segments.

2. Extending Newton's proof of proposition 71 to support measuring inverse-square attraction toward particles from inverse-square attraction toward a spherical shell made of those attracting particles

As justification of corollary 2 of proposition 7, Newton cites corollary 3 of proposition 74 book 1.

> Corollary 3. If a corpuscle placed outside a homogeneous sphere is attracted by a force inversely proportional to the square of the distance of the corpuscle from the center of the sphere, and the sphere consists of attracting particles, the force of each particle will decrease in the squared ratio of the distance from the particle. (C&W, 594)

Newton does not provide a proof; however, his proof of proposition 74 was just to consider the solid sphere as made up of innumerable spherical shells and to appeal to proposition 71. Let us, therefore, consider the case where corpuscles outside a spherical shell are attracted toward the center, and where the shell consists of attracting particles.

We want the same law of attraction toward each particle in the shell. Let us show that for the sum to be inverse-square from the center, the individual attractions must be inverse-square from each particle. Suppose the law is inversely as some power r of the distance. Here is the corresponding analogue of (4) from our reconstruction of Newton's proof of proposition 71.

$$F/f = ((\mathrm{IH} \times \mathrm{IQ} \times 2\pi)/\mathrm{PI}^r)/((ih \times iq \times 2\pi)/pi^r) \quad (1')$$
$$= (pi^r \times \mathrm{IH} \times \mathrm{IQ})/(\mathrm{PI}^r \times ih \times iq) \quad (2')$$

Line 3 of our exposition of the proof of proposition 71 is

$$(\mathrm{PI}^2 \times pf \times ps)/(pi^2 \times \mathrm{PF} \times \mathrm{PS}) = (\mathrm{IH} \times \mathrm{IQ})/(ih \times iq) \quad (3')$$

Combining this with (2′) yields

$$F/f = (pi/\mathrm{PI})^r(\mathrm{PI}/pi)^2[(pf \times ps)/(\mathrm{PF} \times \mathrm{PS})] \quad (4')$$

Line 6 of our exposition of the proof of proposition 71 gives

$$F_\mathrm{S}/f_s = F(\mathrm{PF}/\mathrm{PS})/f(pf/ps), \quad (5')$$

where FS is the total of the force components toward S from attraction toward the curved surface generated by rotating IH, and f_s is the corresponding total force toward s from attraction toward the surface generated by rotating ih. Combining (4′) with (5′) yields

$$F_\mathrm{S}/f_s = (pi/\mathrm{PI})^r/(\mathrm{PI}/pi)^2(ps^2/\mathrm{PS}^2), \quad (6')$$

where the factor $[(pi/\mathrm{PI})^r(\mathrm{PI}/pi)^2]$ modifies the basic inverse-square variation, ps^2/PS^2, of the total force toward the center with distance from the center. In the construction, the corpuscle at P is farther from the center S of its sphere than the corpuscle at p is from the center of its sphere, so that PS > ps and PI > pi. Therefore, having $r > 2$ is equivalent to having the factor $[(pi/\mathrm{PI})^r(\mathrm{PI}/pi)^2] < 1$, so that the total force falls off even faster than the inverse-square with distance, while having $r < 2$ is equivalent to having this factor greater than 1, so that the total force falls off less quickly than the inverse-square with distance.

We see here that, just as is the case with Newton's classic inferences from phenomena, the inference in this corollary from inverse-square variation of the total force toward the center with respect to distance from the center to the inverse-square variation of the component attractions toward parts is backed up by systematic dependencies. Any difference from the inverse-square law for attraction toward the parts would produce a corresponding difference from the inverse-square for the law of attraction toward the center resulting from summing the attractions toward the parts. These dependencies make phenomena measuring inverse-square variation of attraction toward the whole count as measurements of inverse-square variation of the law of attraction toward the parts.[52]

[52] Our above proof of Newton's corollary 3 proposition 74 gives systematic dependencies that, given the assumption that attractions toward all the parts obey a common law, makes empirical determinations of the law of attraction toward the whole spherical shell measure the law of attraction to the parts.

In section 2 of appendix 2 above, we showed that the integral provided by Chandrasekhar to represent Newton's results in lemma 29 together with propositions 79–81 of book 1 extends these systematic dependencies to solid spheres made up of spherically symmetric mass distributions.

Chapter 8 Appendix 4: Measuring planetary properties from orbits: details of Newton's calculations

1. *Measuring surface gravities: corollary 1 proposition 8*

> Corollary 1. Hence the weights of bodies toward different planets can be found and compared one with another. For the weights of equal bodies revolving in circles around planets are (by book 1, prop. 4, corol. 2) as the diameters of the circles directly and the squares of the periodic times inversely, and weights at the surfaces of the planets or at any other distances from the center are greater or smaller (by the same proposition) as the inverse squares of the distances. I compared the periodic times of Venus around the sun (224 days and 16¾ hours), of the outermost circumjovial satellite around Jupiter (16 days and 16⁸⁄₁₅ hours), of Huygens's satellite around Saturn (15 days and 22⅔ hours), and of the moon around the earth (27 days, 7 hours, 43 minutes) respectively with the mean distance of Venus from the sun, and with the greatest heliocentric elongations of the outermost circumjovial satellite from the center of Jupiter (8′16″), of Huygens's satellite from the center of Saturn (3′4″), and of the moon from the center of the earth (10′33″). In this way I found by computation that the weights of bodies which are equal and equally distant from the center of the sun, of Jupiter, of Saturn, and of the earth are respectively toward the sun, Jupiter, Saturn, and the earth as 1, 1/1,067, 1/3,021, and 1/169,282. And when the distances are increased or decreased, the weights are decreased or increased as the squares of the distances. The weights of equal bodies toward the sun, Jupiter, Saturn, and the earth at distances of 10,000, 997, 791, and 109 respectively from their centers (and hence their weights on the surfaces) will be as 10,000, 943, 529, and 425. What the weights of bodies are on the surface of the moon will be shown below. (C&W, 812–13)

Newton cites corollary 2 of proposition 4 book 1.[53]

$$f_1 / f_2 = (r_1/t_1^2)/(r_2/t_2^2) \qquad \text{Corol. 2, prop. 4, bk. 1}$$

Proposition 4 of book 1 is about ratios of centripetal forces maintaining bodies in uniform motion on concentric circular orbits. As we have pointed out before, these calculations apply to inverse-square centripetal accelerations for bodies at the mean-distances specified in elliptical orbits. The mean-distance for such an orbit is taken to be equal to the length of the semi-major axis of the ellipse. We have seen that in these elliptical orbits the centripetal acceleration of a body at this mean-distance exactly equals the centripetal acceleration of uniform motion on a concentric circular orbit with the same period and with the radius equal to the semi-major axis of the ellipse.[54] Newton's computations for concentric circular orbits, therefore, are accurate estimates of the inverse-square centripetal accelerations exhibited at the mean-distances specified in the corresponding elliptical orbits.

Let us consider the ratio of the weight toward Jupiter to the weight of an *equal* body toward the sun when both are evaluated at distances from their respective centers equal to the mean-distance R_V

[53] See chapter 3 appendix 2.4.
[54] See chapter 3 footnote 48.

of Venus from the sun. Let W_J be the weight toward Jupiter of a body at a distance from Jupiter equal to the mean-distance R_V of Venus from the sun, and W_H be the weight toward the sun (Helios) that an equal body would have at that same distance.

Given corollary 2 above, we have

$$W_J/W_H = (R_C/t_C^2)(R_C/R_V)^2/(R_V/t_V)^2 = (R_C/R_V)^3(t_V/t_C)^2$$

where $(R_C/t_C^2)(R_C/R_V)^2$ is the inverse-square adjustment to distance R_V of the magnitude (R_C/t_C^2) representing the acceleration toward Jupiter exhibited by the orbit of Callisto, the outermost Galilean satellite of Jupiter with mean-distance R_C from Jupiter and period t_C, and (R_V/t_V^2) represents the acceleration toward the sun exhibited by the orbit of Venus with mean-distance R_V and period t_V. The ratio of these centripetal forces on bodies at equal distances is measured by the ratio of the respective Harmonic Rule ratios R^3/t^2 of corresponding concentric circular orbits.

Newton cites $8'16''$ as the maximum elongation, θ_C, of Callisto from the center of Jupiter. This angle is taken to be adjusted to the mean-distance R_J of Jupiter from the sun. Let us take $R_J = 5.211$AU in units, AU, equal to the mean-distance of the earth from the sun.[55] This makes $R_C = 5.211\sin\theta_C$AU $= 0.01253$AU. Newton cites Callisto's period, t_C, as 16 days 16⁸⁄₁₅ hours. For the mean-distance, R_V, of Venus from the sun let us take 0.724AU.[56] In the above passage Newton cites the period t_V, of Venus as 224 days 16¾ hours. These values for R_V, t_V, R_C, and t_C yield

$$(R_C/R_V)^3/(t_V/t_C)^2 = 1/1064$$

for W_J/W_H. This is the ratio of weights toward Jupiter to weights toward the sun for any equal bodies at any equal distances, so long as the distances are as great or greater than the distance from the center to the surface of the sun. Newton cites this ratio as 1/1067.[57] The modern value for this ratio is about 1/1047.[58]

The corresponding calculation for Saturn using the above cited values for R_V and t_V, together with Newton's cited values for the maximum elongation $\theta_T = 3'4''$ and period $t_T = 15$ days 22⅔ hours of the orbit of Saturn's moon Titan with mean-distance $R_T = 0.0085$AU and $R_S = 9.526$AU for Saturn's mean-distance, yields for equal bodies at equal distances.

$$W_S/W_H = (R_T/R_V)^3/(t_V/t_T)^2 = (.0085/.724)^3(224.698/15.944)^2 = 1/3114$$

Newton gives this ratio W_S/W_H as 1/3021.[59] The modern value is about 1/3499.[60]

[55] In units AU equal to the mean earth–sun distance, 5.211AU is the mean (to three decimal place precision) between the value for Jupiter's distance Newton cites from Kepler (5.1965AU) and the value he cites from Boulliau (5.2252AU). See Newton's phenomenon 4 (C&W, 800).

[56] Newton cites 0.724AU from Kepler and 0.72398AU from Boulliau (C&W, 800).

[57] Robert Garisto points out that Newton's 1/1067 equals 1/1064 to within rounding error.

> For example, using tan $8'16'' = 0.0024$ instead of 0.002405, and $T_{\text{Venus}} = 225$ instead of 224.7, gives a value of 1/1067 exactly. (Garisto 1991, 45)

[58] See *ESAA*, 697.

[59] Garisto also points out that the disagreement between Newton's cited ratio 1/3021 and the ratio 1/3114, calculated from the values for periods and mean-distances available to him, can be explained by approximations similar to those used in the case of Jupiter.

> For example, if one uses 0.00090 for the tangent of $3'4''$, instead of 0.0008921, and uses 225 days for the period of Venus instead of 224.7 days, and 9.53AU instead of 9.526AU for the orbital radius of Saturn, one then obtains 1/3021 exactly. (Garisto 1991, 46)

[60] The ratio W_S/W_H is equal to the ratio of the mass of Saturn to the mass of the sun, for which the modern value is 1/3498.5 (*ESAA*, 697). If we use modern cited values for the mean-distance of

For the earth, Newton's cited maximum elongation $\theta_m = 10'33''$ and period t_m of 27 days 7 hours 43 minutes for the orbit of the moon yield

$$(R_m/R_V)^3/(t_V/t_m)^2 = (.003069/.724)^3(224.698/27.3215)^2 = 1/194000$$

for W_E/W_H for equal bodies at equal distances. Newton gives this ratio as 1/169282. This very large difference suggests a significant error in Newton's calculation from his cited values. The modern value for this ratio is about 1/332946.[61] The great difference between this and the correct calculation from Newton's assumed values suggests significant error in one or more of the assumed values.

The last part of this corollary uses estimates of the relative sizes of the sun, Jupiter, Saturn, and the earth to inverse-square adjust the relative weights at equal distances to give the relative weights equal bodies would have at their respective surfaces. Given his cited ratio 10,000 to 997 for the sizes of the sun and Jupiter, inverse-square adjusting his ratio 1/1067 for the weights at equal distances yields

$$(1/1067)(1/0.0997)^2 = 0.0943,$$

which equals the ratio 943 to 10,000 that Newton gives as that of the surface gravity of Jupiter to the surface gravity of the sun. Similarly, for Saturn, $(1/3021)(1/0.0791)^2 = 0.0529$ equals the ratio Newton gives for its surface gravity to that of the sun. As Garisto points out,[62] the result

$$(1/194000)(1/0.0109)^2 = 0.0434$$

of applying the adjustment corresponding to his size ratio 109 to 10,000 for the earth and sun agrees much better with his claimed ratio 435 to 10,000 of its surface gravity to that of the sun than does the result

$$(1/169282)(1/0.0109)^2 = 0.0497$$

of applying that adjustment to the ratio Newton actually gives for the ratio of gravities toward the earth and sun at equal distances.

2. *Measuring planetary masses and densities: corollaries 2 and 3 of proposition 8*

Newton's next corollary (of proposition 8 book 3) infers ratios of masses of planets from the ratios of forces at equal distances.

> Corollary 2. The quantity of matter in the individual planets can also be found. For the quantities of matter in the planets are as their forces at equal distances from their centers; that is, in the sun, Jupiter, Saturn, and the earth, they are as 1, 1/1,067, 1/3,021, and 1/169,282 respectively. If the parallax of the sun is taken as greater or less than

and period of Titan's orbit about Saturn at the present time (for epoch J2000.0), we get $R_T = 1221.83 \times 10^3$ km $= .008167$AU and $t_T = 15.9454$d (*ESAA*, 708 and 700 for AU in meters), together with Newton's values for R_V and t_V, we get

$$W_S/W_H = (0.008167/724)^3(224.698/15.9454)^2 = 1/3508,$$

which is not far from the modern value.

[61] *ESAA*, 697. Garisto gives this as 1/329000 (1991, 46). According to *ESAA*, 697, 1/328900.5 is the ratio of the mass of the earth + the moon to that of the sun. The ratio for the earth alone is 1/332946.

[62] Garisto 1991, 45.

> 10″ 30‴, the quantity of matter in the earth will have to be increased or decreased in the cubed ratio. (C&W, 813)

The ratio between weights of equal bodies at equal distances toward the earth and sun (and so of the mass of the earth to that of the sun) depends on the solar parallax. Newton's last remark tells us to adjust this ratio for alternative parallaxes by multiplying by the cube of the ratio of that alternative parallax to his assumed parallax of 10½ seconds.[63]

Measuring relative masses provides resources that can be exploited to measure relative densities as well. Newton does this in the next corollary.

> Corollary 3. The densities of the planets can also be found. For the weights of equal and homogeneous bodies toward homogeneous spheres are, on the surfaces of the spheres, as the diameters of the spheres, by book 1, prop. 72; and therefore the densities of heterogeneous spheres are as those weights divided by the diameters of the spheres. Now, the true diameters of the sun, Jupiter, Saturn, and the earth were found to be to one another as 10,000, 997, 791, and 109, and the weights toward them are as 10,000, 943, 529, and 435 respectively, and therefore the densities are as 100, 94½, 67, and 400. The density of the earth that results from this computation does not depend on the parallax of the sun but is determined by the parallax of the moon and therefore is determined correctly here. Therefore the sun is a little denser than Jupiter, and Jupiter denser than Saturn, and the earth four times denser than the sun. For the sun is rarefied by its great heat. And the moon is denser than the earth, as will be evident from what follows [i.e., prop. 37, corol. 3]. (C&W, 814)

[63] Garisto uses this adjustment to illuminate the puzzle about the large error in Newton's assignment of the ratio 1/169,282 for the earth. He points out that the current accepted value of the solar parallax is about 8.80″. Applying the corresponding adjustment factor $(8.80''/10.5'')^3$ to the correct ratio 1/194000 resulting from Newton's cited values for the maximum elongation of the moon and its period, together with his cited mean-distance and period for Venus, yields

$$(1/194000)(8.8''/10.5'')^3 = 1/329550$$

This is quite close to the modern value of 1/332946 for the ratio of the mass of the earth to the mass of the sun.

ESAA, 697 cites 332946 as the ratio of the mass of the sun to that of the earth. According to Garisto,

> The currently accepted value for the mass of the Earth relative to the Sun is about 1/329000. (Garisto 1991, 46)

This is almost exactly in line with 328900.5 cited as the ratio of the mass of the sun to the combined mass of earth + moon.

The corresponding corrected value for Newton's 1/169,282 is about 1/288000. As Garisto points out, this is much too far off to result from mere rounding errors (Garisto 1991, 46). He also gives a very interesting, and quite plausible, reconstruction of how Newton's erroneous ratio might have resulted from copying errors in the course of the revisions between the second and third editions (Garisto 1991, 47).

In the translation by Donahue of this corollary (Densmore 1995, 385), the ratio is given as 1/196,282 with the following note:

> 18. In the third edition, this number was printed as 1/169,282, which is not consistent with his stated heliocentric elongation of the moon. On the conjecture that two digits might have become transposed somewhere in the publishing process, the translator ventures the present correction. It is consistent with a heliocentric elongation of 10′30″.

Newton uses the relative surface gravities divided by the diameters to give the relative densities. We have calculated W_P/W_H as the ratio of the weight toward a planet a body W_P would have at a distance from it equal to the distance R_V of Venus from the sun to the weight W_H toward the sun any equal body would have at that same distance R_V. For any planet and the sun this gives

$$D_P/D_H = (SG_P/r_P)/(SG_H/r_H) = (W_P(R_V/r_P)^2(1/r_P))/(W_H(R_V/r_H)^2(1/r_H))$$
$$= (W_P/W_H)(r_H/r_P)^3 = (M_P/M_H)(r_H/r_P)^3$$

Note that we, also, have

$$D_P/D_H = [M_P/(4/3)\pi r_P{}^3]/[M_H/(4/3)\pi r_H{}^3 = (M_P/M_H)(r_H/r_P)^3$$

Newton's result agrees with what one would expect for ratios of densities of spheres from the fact that density equals mass/volume and the formula $(4/3)\pi r^3$ for the volume of a sphere.

Newton has pointed out that the ratio (M_e/M_H) of the mass of the earth to the mass of the sun is proportional to the cube of the solar parallax. The solar parallax angle is proportional to its tangent r_e/R_e, the radius r_e of the earth divided by the mean-distance R_e of its orbit from the sun. This makes

$$(M_e/M_H) \propto (r_e/R_e)^3$$

The ratio of the densities, $D_e/D_H = (M_e/M_H)(r_H/r_e)^3$, however, is proportional, not to the cube of the solar parallax, but to the cube of the ratio (r_H/R_e) of the solar radius rather than the earth's radius to the mean-distance of the earth's orbit.

$$D_e/D_H \propto (r_H/r_e)^3(r_e/R_e)^3 = (r_H/R_e)^3$$

The ratio (r_H/R_e) of the solar radius to the solar distance is far less difficult to measure than the solar parallax, which depends on the ratio of the earth's radius to the solar distance.

Newton gives for the sun, Jupiter, Saturn, and the earth densities that are as respectively 100, 94½, 67, and 400.

The corresponding modern values for these comparative densities are respectively about 100, 94, 50, and 391.[64]

Even with the terrible value (1/169282 instead of the modern value 1/332946) for the ratio of the mass of the earth to the mass of the sun, Newton's relative density is not all that far off the modern value.[65]

[64] According to *ESAA*, the mean densities of the sun, Jupiter, Saturn, and the earth are respectively 1.41g/cm^3, 1.33g/cm^3, 0.70g/cm^3, and 5.515g/cm^3. The corresponding relative densities, setting the density of the sun at 100, are for Jupiter (1.33/1.41)(100) = 94.33, for Saturn (.70/1.41)(100)= 49.65, and for the earth (5.515/1.41)(100) = 391.135.

[65] If we put in the correct mass ratio (1/194000) calculated from Newton's assumed numbers for distances and periods of Venus and the moon, his assumed relative sizes (r_H/ r_e)= (10,000/109)= 91.74, we get

$$D_e/D_H = (1/194000)(10,000/109)^3 = 3.98,$$

giving 398 as the density of the earth corresponding to taking 100 as the density of the sun. This is a little, but only a little, closer to the corresponding modern value of 391.

9

Beyond Hypotheses: Newton's Methodology vs. Hypothetico-Deductive Methodology

A challenge to Newton's crucial application of Law 3 to count gravity as a pair-wise interaction affords an opportunity to argue in support of Newton's richer notion of empirical success. Part I reviews some of Newton's remarks on method. Section I.1 reviews important methodological remarks in Newton's scholium to proposition 69 book 1. These motivate reckoning centripetal forces by properly assigning and then summing attracting forces toward the individual particles of the central body. They indicate that Newton construes gravity as a pair-wise interaction between solar system bodies as compatible with a variety of possible locally acting causes. They conclude by reiterating Newton's mathematical treatment of forces that allows phenomena to empirically establish which conditions of forces apply to each kind of attracting bodies. Empirically establishing these conditions from phenomena is to precede and inform any further investigations into the physical causes of these forces. Commitment to such an order of investigation reflects a long-standing difference sharply distinguishing the methodology of Newton's experimental philosophy from the methodology of the mechanical philosophy. Section I.2 discusses the first part of the paragraph in which Newton introduces his famous *hypotheses non fingo* passage. Newton begins by announcing he has explained the phenomena of the heavens and of our sea by the force of gravity, even though he has not yet assigned a cause to gravity. He then characterizes features which, he claims, these applications empirically establish that any cause of gravity would need to have. This context illuminates Newton's *hypotheses non fingo* passage by reinforcing the order of investigation he defended in the scholium to proposition 69.

Part II reviews a challenge to Newton's application of Law 3 and Newton's initial response to the letter in which it was raised by his editor Roger Cotes. Section II.1 reviews an articulation by Cotes, and a later articulation by Howard Stein, of the challenge that Newton's crucial application of Law 3 is to be counted as an assumed hypothesis rather than as an inference that ought to be counted as empirically established by the phenomena cited in support of it. Section II.2.i reviews the first part of

Newton's response. Newton characterized his experimental philosophy as one in which propositions are deduced from the phenomena and made general by induction. He claimed that the Laws of Motion and the law of gravity are established by this same method, and that he reserves the term hypothesis for only such a proposition that is assumed or supposed without any experimental proof. Section II.2.ii reviews Newton's appeal to Law 1 to defend the application of Law 3 to attractions. This includes an examination of a thought experiment argument and an empirical experiment with a magnet and piece of iron described in Newton's scholium to the Laws of Motion. Section II.2.iii examines Newton's argument that gravity between the earth and its outer parts counts as such a mutual attraction.

Part III examines the extent to which such arguments can be extended to count gravity between solar system bodies as such mutual attractions. Section III.1 reviews the empirical constraints on combining the acceleration fields toward Jupiter and toward the sun. Section III.2 argues that for such a two-body system one can recover all the same empirical predictions, while allowing for violations of Newton's application of Law 3. This result is extended to include all the *n*-body gravitational interactions among solar system bodies. Section III.3 reviews the case that can be made for counting the challenged application of Law 3 as a making general by induction of measurements by phenomena.

Part IV argues that Newton's Rule 4 can back up the challenged application of Law 3. Section IV.1 points out that Newton's articulation of the methodology of Rule 4 appears to have been developed as part of his response to Cotes's challenge. On the alternatives that make the same predictions without the challenged application of Law 3, weights and masses of solar system bodies are so entangled that neither can be measured. This makes it appropriate to count such alternatives as mere hypotheses to be dismissed, rather than as rivals to be taken seriously. Section IV.2 reviews the role of accumulating empirical success in supporting the acceptance of Newton's application of Law 3 against vortex-theoretic rivals motivated by a commitment to avoid action at a distance. Section IV.3 argues against turning Newton's ideal of empirical success as *theory-mediated* measurement into a necessary criterion for counting a proposition as gathered from phenomena by induction.

I Newton on method

1. *Newton's scholium to proposition 69 book 1*

Proposition 69 is appealed to in Newton's crucial argument to construe gravity as a force of interaction between bodies in proposition 7 of book 3.[1] It is the last proposition in section 11 of book 1, which is titled "the motion of bodies drawn to

[1] See chapter 8 section I.1. Chapter 8 appendix 1 is an account of Newton's proof of proposition 69 book 1.

one another by centripetal forces." The scholium to proposition 69 functions as Newton's transition to the next section of book 1, which is devoted to the attractive forces of spherical bodies. It contains quite significant statements by Newton about his methodology. Here is the first paragraph of this scholium.

Scholium. By these propositions we are directed to the analogy between centripetal forces and the central bodies toward which those forces tend. For it is reasonable that forces directed toward bodies depend on the nature and the quantity of matter[2] of such bodies, as happens in the case of magnetic bodies. And whenever cases of this sort occur, the attractions of the bodies must be reckoned by assigning proper forces to their individual particles and then taking the sums of these forces. (C&W, 588)

This passage suggests that, whenever cases of it occur,[3] this sort of attraction must be reckoned by assigning proper forces toward their individual particles and then taking the vector-sums of these forces. The mass of a whole body is the sum of the masses of its particles; therefore, given that the gravitational attraction of a corpuscle toward a central body is proportional to the total quantity of matter composing that central body, one might expect to be able to find an assignment of forces directed toward the particles of that central body which, when summed as vectors, would give the total force attracting the corpuscle toward that central body.

The next paragraph suggests that Newton intended his use of "attraction" in proposition 69 to be general enough to be compatible with a very wide range of possible causes.

I use the word "attraction" here in a general sense for any endeavor whatever of bodies to approach one another, whether that endeavor occurs as a result of the action of the bodies either drawn toward one another or acting on one another by means of spirits emitted or whether it arises from the action of aether or of air or of any medium whatsoever – whether corporeal or incorporeal – in any way impelling toward one another the bodies floating therein. I use the word "impulse" in the same general sense, considering in this treatise not the species of forces and

[2] The words "of matter" are added by the translators. The Latin of the clause translated "depends on the nature and the quantity of matter" is "pendeant ab eorundem natura & quantitate" (Koyré and Cohen 1972, 298). The addition of "of matter" in the translation can be defended by appeal to Newton's remark in his discussion of "quantity of matter" in definition 1:

> Furthermore, I mean this quantity whenever I use the term "body" or "mass" in the following pages. (C&W, 404)

There may, however, be at least a little less tension with the remark, "as happens in the case of magnetic bodies", if the explicit specification "of matter" is not added to Newton's text. See next note.

[3] Here, as in corollary 1 of proposition 7 book 3, Newton appeals to magnetic attraction as an example of this sort of centripetal force. This reference to magnetic attraction is somewhat puzzling given the disanalogies, some of which Newton pointed out in corollary 5 proposition 6 book 3, between magnetic attraction and gravitation. (See above chpt. 7 sec. II.2.)

their physical qualities but their quantities and mathematical proportions, as I have explained in the definitions. (C&W, 588)

This passage adds significantly to the explanation of the methodological implications of Newton's mathematical treatment of forces, which he gave in his discussion of his definitions of centripetal force and its three measures.[4]

The attraction between bodies *A* and *B*, in proposition 69, must be construed so that the attraction of *A* toward *B* is the equal and opposite reaction to the attraction of *B* toward *A*.[5] Where the bodies *A* and *B* are the sun and a planet, this sort of attraction is incompatible with the vortex theories of Descartes, Huygens, and Leibniz, according to which the interaction explaining the planet's motion is with the vortical particles pushing it and not with the sun. Though it is incompatible with these vortex theories, construing the force maintaining a planet in its orbit about the sun as a mutual attraction satisfying the assumption of proposition 69 is compatible with a number of alternative hypotheses about possible causes. Newton mentions ones which are consistent both with local causation and with the "impelling toward one another" required by the mutual interaction represented by the application of Law 3 to construe the attraction of the sun toward a planet as the equal and opposite reaction to the attraction of that planet toward the sun.

For example, consider "the action of aether or of air or of any medium whatsoever – whether corporeal or incorporeal – in any way impelling toward one another the bodies floating therein." If we construe the action of an aether or other medium as mimicking appropriately contracting rubber bands or cords between the bodies, the application of Law 3 would be quite like the example of the tension in the rope in Newton's discussion of the action of a horse on a stone it is attached to by a rope, and the equal and opposite reaction of the stone to impede the motion of the horse.[6] Even if the physical cause were a pushing together of the bodies by the medium, the application of Law 3 could still be supported, so long as the pushes were coordinated to result in having the acceleration of each body toward the other proportional to the mass of the other.[7] As Newton suggests, what matters for the applications of forces in his argument are their quantities and mathematical proportions.

[4] See chapter 3 section I above.

[5] See above appendix 1 chapter 8.

[6] See his discussion of Law 3 quoted in chapter 3 section II above. Strictly, the equal and opposite reaction to the horse's tug on the rope is the rope's tug on the horse, while that to the rope's tug on the stone is the stone's tug on the rope. Suppose the rope and stone are accelerated by the action of the horse, so that horse and stone are not just in a static equilibrium via the tension in the rope. It is, then, only to the extent that the stretched rope can be treated as rigid and its mass can be treated as negligible (as compared to the masses of the horse and stone) that the forces at the ends of the rope can be taken as "practically" equal in magnitude and the forces on the horse and stone count as "practically" equal and opposite.

[7] Stein (private communication) has suggested having the impulses pushing the bodies together act as though they were produced by the two jaws of a set of tweezers being squeezed together.

Even though the direct applications of Law 3 would be between the bodies and the separate impulses pushing them toward one another, being able to count on having these impulses so coordinated would support the additional application of Law 3 to the motive forces of the two bodies.

We have seen that Huygens's argument for constant ratios of weight to mass was based on experiments supporting the equality of the acceleration of gravity on terrestrial bodies. The experiments served to place as a demand on any adequate causal account that it be consistent with these equalities.[8] In the next section we shall examine Newton's argument from phenomena for establishing the application of Law 3 to attractions as a demand on what could be counted as any viable proposal for a cause of such an attraction.

The scholium to proposition 69 ends with Newton reiterating his commitment to, what he counts as, the appropriate order of investigation. This order of investigation sharply distinguishes his experimental philosophy from the mechanical philosophy advocated by his continental critics.

> Mathematics requires an investigation of those quantities of forces and their proportions that follow from any conditions that may be supposed. Then, coming down to physics, these proportions must be compared with the phenomena, so that it may be found out which conditions [or laws] of forces apply to each kind of attracting bodies. And then, finally, it will be possible to argue more securely concerning the physical species, physical causes, and physical proportions of these forces. (C&W, 588–9)

The mechanical philosophy had an *a priori* commitment to the possibility of explanation by forces as contact pushes between bodies as its chief guide to understanding planetary motion. It was this commitment, more than any of its *empirical* successes, that provided the main motivation for acceptance of vortex theories. Newton's general mathematical treatment of forces provides the resources to reason from phenomena to laws of forces between bodies. According to his experimental philosophy, results of such investigations – such as propositions 1–7 of book 3 – rather than any *a priori* commitment to a particular type of explanation, are to take precedence over (and, indeed, are to count as the chief guide to) arguments concerning how to understand the causes of these forces.

The order of investigation and precedence advocated here is very much in line with Newton's long-standing advocacy of methods of investigation which distinguish his experimental philosophy from the precedence given to hypotheses about explanation in the mechanical philosophy. Here is an informative passage on methodology from a reply Newton wrote in 1672 to questions raised by Ignatius Pardies about his investigation of light and colors.[9]

[8] As we have seen in chapter 5 section I.4, contrary to Janiak (2007, 145; 2008, 78), Huygens has offered grounds for accepting that there could be versions of his proposed cause of gravity that would recover the proportionality of weight to mass.

[9] The publication of Newton's letter on light and colors in the *Philosophical Transactions* of the Royal Society in February 1671/2 (*Phil Trans* number 80; Cohen 1958, 47–59) resulted in a substantial debate carried on in publications of that journal (Cohen 1958, 27–235). Ignatius Pardies, a French Jesuit and professor of mathematics, raised objections, some based on misunderstanding the crucial role played by having the prism oriented at minimum deviation in Newton's main experiment, in a Latin letter published (together with Newton's Latin reply and with translations into English of the letter and reply) in *Phil Trans*

For the best and safest method of philosophizing seems to be, first to inquire diligently into the properties of things, and establishing those properties by experiments and then to proceed more slowly to hypotheses for the explanation of them. For hypotheses should be subservient only in explaining the properties of things, but not assumed in determining them; unless so far as they may furnish experiments. For if the possibility of hypotheses is to be the test of the truth and reality of things, I see not how certainty can be obtained in any science; since numerous hypotheses may be devised, which shall seem to overcome new difficulties. (Cohen 1958, 106)

Newton's contrast between propositions accepted as experimentally established properties of things and mere conjectures was already a central feature of his methodology in his investigation of light and colors. His theory of light rays was designed to allow properties of them to be established experimentally without making any more detailed commitments about their physical constitution.[10] The opposition of this methodology to the mechanical philosophy is quite similar to that of the methodology exhibited in his argument to empirically establish properties of gravity, without regard to hypotheses about how it might be physically caused.[11] Newton suggested to Pardies that the enterprise of discovering interesting propositions that can count as empirically established would be undermined if the possibility of hypotheses were to be counted as a test of truth. In section III below, we shall see that this suggestion can be strikingly supported by an example where we construct an alternative hypothesis that would combine the inverse-square acceleration fields of the sun and Jupiter into a single system without Newton's application of Law 3.

2. *The* hypotheses non fingo *paragraph*

The opening sentence is an announcement.

Thus far I have explained the phenomena of the heavens and of our sea by the force of gravity, but I have not yet assigned a cause to gravity.

number 84 (Cohen 1958, 79–92). Pardies responded with a second letter which was also published, together with Newton's reply and with translations of both, in *Phil Trans* number 85 (Cohen 1958, 97–109). The quoted passage is from the published translation of Newton's reply to this second letter by Pardies.

This reply by Newton to Pardies' second letter begins with the following comment by Newton on Pardies.

> In the observations of the Rev. F. Pardies, one can hardly determine whether there is more of humanity and candor, in allowing my arguments their due weight, or penetration and genius in starting objections. And doubtless these are very proper qualifications in researches after truth. (Cohen 1958, 106)

The contrast between the appreciative attitude displayed here and the growing impatience displayed in Newton's responses to Hook (Cohen 1958, 116–35), Huygens (Cohen 1958, 137–42) and, especially, to Linius (Cohen 1958, 157–62) and Lucas (Cohen 1958, 169–76) is rather striking.

[10] Alan Shapiro (Shapiro 2002) gives an excellent account of the contrast between Newton's treatment of his experimentally established properties of light and colors and his treatment of the corpuscular theory of light. Newton considered his corpuscular theory to be only a probable hypothesis rather than a proposition to be accepted as an experimentally established property of things.

[11] See Harper and Smith 1995 for more on the parallel between this contrast with the mechanical philosophy exhibited in Newton's earlier work on light and colors and the rejection of hypotheses advocated in his later defense of his argument for universal gravity.

It suggests that the achievement of these explanations is not undercut by his having not yet assigned a cause to gravity. It also suggests that efforts to assign a cause are to be guided by these explanations of the phenomena of our heavens and of our sea by the force of gravity. Both of these suggestions are in accordance with the order of investigation we saw endorsed in the scholium to proposition 69 book 1.

The second sentence reports outcomes of such an investigation afforded by the role of the force of gravity in explaining such phenomena.

> Indeed, this force arises from some cause that penetrates as far as the centers of the sun and planets without any diminution of its power to act, and that acts not in proportion to the quantity of the *surfaces* of the particles on which it acts (as mechanical causes are wont to do) but in proportion to the quantity of *solid* matter, and whose action is extended everywhere to immense distances, always decreasing as the squares of the distances.

We have seen that the need for a cause of the force of gravity to act in proportion to the quantity of solid matter was established by agreeing measurements of the equality of ratios of the weights of bodies toward any planet to their quantities of matter for all bodies at any equal distances from the center of that planet. As we saw above in chapter 7, this argument for proposition 6 counts as a very impressive realization of Newton's ideal of empirical success as agreeing measurements of a parameter from diverse phenomena.

The requirement that the action of such a cause extend everywhere to immense distances, always decreasing as the squares of the distances, is also established by agreeing measurements from phenomena. The last part of the next sentence explicitly cites the agreeing measurements of the inverse-square variation of gravity toward the sun from the absence of orbital precession.

> Gravity toward the sun is compounded of the gravities toward the individual particles of the sun, and at increasing distances from the sun decreases exactly as the squares of the distances as far out as the orbit of Saturn, as is manifest from the fact that the aphelia of the planets are at rest, and even as far as the farthest aphelia of the comets, provided those aphelia are at rest. (C&W, 943)

Clairaut's celebrated success in predicting the return of Halley's comet was a particularly striking later vindication of extending the inverse-square law for an acceleration field directed toward the sun to distances beyond those explored by planetary orbits.[12]

That gravity toward the sun is compounded of gravities toward the individual particles of the sun and the resulting requirement that any cause of gravity must penetrate as far as the center of the sun follow from Newton's controversial applications of Law 3 in proposition 7. The property that any such cause must penetrate as far as the centers of the planets without any diminution of its power to act and that it acts in

[12] In 1705 Halley proposed elements of an orbit for this comet we have named after him that give an aphelion distance of 17.86AU, well beyond the vicinity of distances explored by the orbit of Saturn with a mean-distance of about 9.54AU cited by Boulliau. See above chapter 3 section III.3.

proportion to the quantity of *solid* matter follows from the basic result established by the measurements of proposition 6. That such a cause acts not in proportion to the *surfaces* of the particles on which it acts (as mechanical causes are wont to do) appears to be counted as a consequence of this basic proportionality to quantity of solid matter.[13]

With this background, let us now return to further consider Newton's famous *hypotheses non fingo* passage.

> I have not as yet been able to deduce from phenomena the reason for these properties of gravity, and I do not feign hypotheses. For whatever is not deduced from the phenomena must be called a *hypothesis*; and hypotheses, whether metaphysical or physical, or based on occult qualities, or mechanical, have no place in *experimental philosophy*. In this experimental philosophy, propositions are deduced from the phenomena and are made general by induction. The impenetrability, mobility, and impetus of bodies, and the laws of motion and the law of gravity have been found by this method. And it is enough that gravity really exists and acts according to the laws that we have set forth and is sufficient to explain all the motions of the heavenly bodies and of our sea. (C&W, 943; see note 17 below, and Koyré and Cohen 1972, 764, for italics)

Newton has just outlined features of a cause of gravity that he counts himself as having been able to deduce from phenomena. They include the features deduced from his controversial application of Law 3 to count the strength of gravity toward the sun proportional to its mass. They also include corresponding features of the action of such a cause on planets that are very solidly established by convergent agreeing measurements from phenomena. He tells us that these are as far as he has yet gotten. He has not yet been able to deduce from phenomena the reason for these properties.

He tells us that he does not feign hypotheses. He identifies hypotheses with whatever is not deduced from the phenomena. He claims that hypotheses have no place in *experimental philosophy* and he characterizes this experimental philosophy as one in which propositions are deduced from the phenomena and made general by

[13] According to Janiak (2007, 143; 2008, 75), this proportionality to the surfaces of the particles on which it acts is a technical characterization of the action of mechanical causes directed at Leibniz. He sees Newton as accusing Leibniz of conflating local with surface action. He argues that Newton's own commitment to a cause of gravity that acts locally to recover the controversial application of Law 3 is a commitment to a locally acting cause that is not mechanical.

In chapter 5 we criticized Janiak's (2007, 145; 2008, 78) claim that this remark of Newton's shows that Huygens's proposed cause of gravity cannot account for the proportionality of weight to quantity of solid matter. We argued that an examination of details of Huygens's hypothesis suggests that, contrary to Janiak, Huygens has provided fairly good grounds for claiming some version of his proposed cause of gravity would be able to recover this fundamental proportionality of weight to quantity of solid matter. Proportionality to quantity of solid matter appears to be a less significant limitation on hypotheses about mechanical causes than Janiak has suggested.

Perhaps pushes so coordinated as to recover proportionality to quantity of matter needed to recover these applications of Law 3 would be able to be counted as mechanical. Such a possibility is at least suggested by Newton's characterization of the very wide range of hypothetical causes allowed for by his treatment of attraction as mathematical in the scholium to proposition 69 book 1. Newton explicitly includes "corporeal" in his characterization of possible mediums "impelling toward one another bodies floating in them" among the possible causes his mathematical treatment of attraction is not ruling out.

induction. We will see that these characterizations of hypotheses and experimental philosophy were developed in response to Cotes' objection to his controversial application of Law 3 to gravitation between separated bodies. This will afford an opportunity for further discussion of the contrast between this methodology and that of the mechanical philosophy of Huygens.

For now we note that, in contrast to Huygens, Newton regards the possibility of a locally acting cause that could recover his controversial application of Law 3 as very much open. We shall see that this allows him to keep his commitment to avoid action at a distance separated from his assessment of the empirical support for the challenged inference. The properties that any such cause must penetrate as far as the centers of the planets without any diminution of its power to act and that it acts in proportion to the quantity of *solid* matter raise similar difficulties as those raised by the corresponding property for the cause of gravity with respect to the sun. Unlike the inferences to these properties for the sun, which depend on the controversial application of Law 3, the inferences to these similarly problematic properties for the planets are already clearly established by convergent agreeing measurements from phenomena.

The appeal to the empirical successes of the actual and claimed expected additional successes at explaining the motions of heavenly bodies and our sea reinforces Newton's claim that his not having found a cause of gravity does not undercut his inferences to the existence of gravity acting according to the laws he has set forth. This further reinforces the order of investigation, which sharply distinguishes his experimental philosophy from the mechanical philosophy of his critics. According to this order of investigation, which we saw strongly endorsed in the scholium to proposition 69 book 1, the laws and properties of forces are to be established from phenomena before seeking their physical causes. The laws and properties established from phenomena are what any cause of the force of gravity must account for. In this way, he told us in that scholium, it will be possible to argue more securely concerning the physical causes and physical proportions of these forces.[14]

II Beyond hypotheses?

We have seen that the scientific method exhibited in Newton's argument is richer than the basic hypothetico-deductive model of scientific inference. Newton's inferences from phenomena are guided by a richer conception of empirical success. This requires not just accurate prediction of phenomena. It requires, in addition, accurate measurement of parameters by the predicted phenomena. A challenge to Newton's

[14] Though I have disagreed with his dismissive objection to Huygens's hypothesis for a cause of gravity, I am very impressed with Janiak's insight into the extent to which the first part of this paragraph counts as an investigation into the causes of gravity in accord with the order endorsed by the mathematical treatment of forces in the scholium to proposition 69.

application of Law 3 to count gravity as an interaction affords a compelling illustration of advantages realized by using Newton's richer conception of empirical success to guide scientific inference.[15]

1. *The challenge*

Here is the challenge by Roger Cotes, the wonderful editor who did so much to improve the second (1713) edition of the *Principia*. The following passage is from a letter Cotes sent to Newton on March 18, 1713.

> But in the first Corollary of the 5th I meet with a difficulty, it lyes in these words *Et cum Attractio omnis mutua sit* I am persuaded that they are then true when the Attraction may properly be so call'd, otherwise they may be false. You will understand my meaning by an Example. Suppose two Globes *A* & *B* placed at a distance from each other upon a Table, & that whilst *A* remains at rest *B* is moved towards it by an invisible Hand. A by-stander who observes this motion but not the cause of it, will say that *B* does certainly tend to the centre of *A*, & thereupon he may call the force of the invisible Hand the Centripetal force of *B*, or the Attraction of *A* since ye effect appears the same as if it did truly proceed from a proper & real Attraction of *A*. But then I think he cannot by virtue of the Axiom [Attractio omnis mutua est] conclude contrary to his Sense & Observation, that the Globe *A* does also move towards the Globe *B* & will meet it at the common centre of Gravity of both Bodies. (*Corresp*. V, 392)

Cotes goes on to say that he takes this to be a difficulty for the argument to proposition 7 book 3.

> This is what stops me in the train of reasoning by which as I said I would make out in a popular way the 7th Prop. Lib. III. I shall be glad to have your resolution of the difficulty, for such I take it to be. If it appeares so to You also; I think it should be obviated in the last sheet of Your Book which is not yet printed off, or by an Addendum to be printed with ye Errata Table. For 'till this Objection be cleared I would not undertake to answer anyone who should assert You do *Hypothesim fingere* I think You seem tacitly to make this Supposition that the Attractive force resides in the Central Body. (*Corresp*. V, 392)

He points out that what seems to be Newton's tacit assumption that the attractive force resides in the central body appears to count as an assumed hypothesis on which the argument is based, rather than as a conclusion supported by the evidence adduced.

Howard Stein also argues that Newton's appeal to the third Law of Motion to construe the gravitation of a primary toward its satellite as the equal and opposite reaction of the gravitation of that satellite toward the primary is an assumed hypothesis.[16]

[15] The material in the next sections of this chapter is developed from my earlier treatment in Harper 2002. That paper was my contribution to the Stein Fest, a wonderful conference held at the University of Chicago in May of 1999 honoring Howard Stein.

[16] In earlier papers (1967, 179–80 and 1970, 269), Stein, also, mentions difficulties with Newton's appeal to Law 3 to argue that gravitation is an interaction.

Dana Densmore (1999, 104–11) has, also independently, objected to this appeal by Newton to Law 3.

The third law of motion does not tell us that whenever one body is urged by a force directed towards a second, the second body experiences an equal force towards the first; it tells us, rather, that whenever one body is acted upon *by* a second, the second body is subject to a force of equal magnitude and opposite direction. Therefore – putting the point in proper generality – what we may legitimately conclude, from the proposition that each planet is a center of gravitational force acting upon all bodies, is that for each body B there must be some body (or system of bodies) B' which, exerting this force on B, is subject to the required equal and opposite reaction. (Stein 1991, 217)

Imagine, in analogy with the example used by Cotes, that the moon is maintained in its orbit of the earth by invisible hands pushing it toward the earth's center. In this case it would be the pushing hands, rather than the earth, which would be subject to the equal and opposite reaction to the force accelerating the moon toward the center of the earth.

Stein goes on to point out that the leeway implied by the above formulation would not have counted as far-fetched to Newton's readers.

It must not be thought that the leeway implied by this formulation is one merely of far-fetched possibilities – that the only *plausible* subject of the reaction to gravitational force towards a planet is the planet itself. On the contrary, the very widespread view of Newton's time that one body can act on another only by contact – a view that is well known to have had a powerful influence on Newton himself – makes for precisely the opposite assessment: that it is far-fetched to apply the third law in the way Newton does. (Stein 1991, 217)

The widespread view was the fundamental commitment of the mechanical philosophy – that to make motion phenomena intelligible one had to produce a hypothesis that would show how such motions could result from contact pushes among bodies. On vortex theories, which were directly motivated by this commitment, the subject of the equal and opposite reaction to the gravitation of the moon toward the earth would be the vortical particles – invisible hands pushing it toward the earth.

2. *Newton's initial response*

2.i Hypotheses vs. deductions from phenomena Newton responded to this challenge in a letter sent to Cotes on Saturday, March 28, 1713.

I had yours of Feb 18th, & the Difficulty you mention wch lies in these words [Et cum Attractio omnis mutua sit] is removed by considering that as in Geometry the word Hypothesis is not taken in so large a sense as to include the Axiomes & Postulates, so in experimental Philosophy it is not to be taken in so large a sense as to include the first Principles or Axiomes wch I call the laws of motion. These Principles are deduced from Phaenomena & made general by Induction: wch is the highest evidence that a Proposition can have in this philosophy. And the word Hypothesis is here used by me to signify only such a Proposition as is not a Phaenomenon nor deduced from any Phaenomena but assumed or supposed without any experimental proof. (*Corresp.* V, 396–7)

Newton points out that in his experimental philosophy he does not use the word "hypothesis" in so large a sense as to include those first principles he calls the laws of motion. These laws of motion, he tells Cotes, have the highest evidence a proposition can have in his philosophy. Such evidence, he claims, results from deducing propositions from phenomena and making them general by induction. In contrast, he says that he is using "hypothesis" neither for phenomena nor for propositions deduced from phenomena but only for assumptions supposed without any experimental proof.

The letter we are examining instructs Cotes to add remarks to the new edition to clarify these points for readers of the *Principia*. These instructions are to follow up the famous *hypotheses non fingo* passage,

> I have not as yet been able to deduce from phenomena the reason for these properties of gravity, and I do not feign hypotheses. (C&W, 943).

in the General Scholium being added to book 3, with the following remarks:

> For whatever is not deduced from the phenomena must be called a hypothesis; and hypotheses, whether metaphysical or physical, or based on occult qualities, or mechanical, have no place in experimental philosophy. In this experimental philosophy, propositions are deduced from the phenomena and are made general by induction. The impenetrability, mobility, and impetus of bodies, and the Laws of Motion and the law of gravity have been found by this method. And it is enough that gravity really exists and acts according to the laws that we have set forth and is sufficient to explain all the motions of the heavenly bodies and of our sea. (C&W, 943)

A main revision is to add the positive characterization of his experimental philosophy as one in which "propositions are deduced from the phenomena and are made general by induction."[17] So, it turns out that this very significant positive characterization of Newton's method was a direct response to the challenge from Cotes.

[17] Here is a translation of the original passage Newton had earlier sent to Cotes:

> Indeed, I have not yet been able to deduce the reason [or cause] of these properties of gravity from phenomena, and I do not feign hypotheses. For whatever is not deduced from phenomena is to be called a hypothesis; and I do not follow *hypotheses*, whether metaphysical or physical, whether of occult qualities or mechanical. It is enough that gravity should really exist and act according to the laws expounded by us, and should suffice for all the motions of the celestial bodies and of our sea. (C&W, 276 [readers guide])

Here is Newton's instruction with the Latin of his proposed revision.

> And for preventing exceptions against the use of the word Hypothesis I desire you to conclude the next paragraph in this manner
>
> Quicquid enim ex phaenomenis non deducitor Hypothesis vocanda est, et ejusmodi Hypotheses seu Metaphysicae seu Physicae use Qualitatum occultarum sue Mechanicae in Philosophia experimentali locum non habent. In hac Philosophia Propositions deducunter ex phaenomenis & reddunter generales per Inductionem. Sic impenetrabilitas mobilitas & impetus corporum & leges motuum & gravitatis innotuere. Et satis est quod Gravitas corporum revera existat & agat secundum leges a nobis expositas & ad corporum caelestium et maris nostri motis omnes sufficiat. (*Corresp.* V, 397)

As we have seen, Newton's basic inferences from phenomena are backed up by systematic dependencies that make the propositions inferred count as parameter values measured by the phenomena from which they are inferred. His inferences to inverse-square gravity toward the sun and planets with moons are good examples of what we might consider *direct deductions from the phenomena*, according to Newton's methodology. Newton's inference to extend inverse-square gravity to planets without satellite orbits to measure it is a good example which Newton counts as *making general by induction* such propositions. It extends such parameters, with values found constant for the sun and all planets with satellite orbits to measure them, to planets without orbital phenomena accessible to us. The orbital phenomena measuring the centripetal direction and inverse-square variation of gravitation are taken to measure these properties of gravitation toward planets generally.

Cotes's challenge motivates an investigation of the extent to which Newton's appeal to Law 3 to count gravity as an interaction between bodies can be supported by this sort of reasoning from phenomena.

2.ii Appeal to Law 1 After his initial remarks to Cotes, sharply distinguishing deductions from phenomena and mere hypotheses, Newton went on to appeal to passages defending applying the third Law of Motion to attractions.

Now the mutual & mutually equal attraction of bodies is a branch of the third Law of motion & how this branch is deduced from Phaenomena you may see in the end of the Corollaries of ye Laws of Motion, pag. 22. If a body attracts another body contiguous to it & is not mutually attracted by the other: the attracted body will drive the other before it & both will go away together wth an accelerated motion in infinitum, as it were by a self moving principle, contrary to ye first law of motion, whereas there is no such phaenomenon in all nature. (*Corresp.* V, 397)

He argues that a failure of this application of the third Law of Motion to attractions would violate the first Law of Motion. This argument, he claims, counts as a deduction from phenomena of the challenged application of the third Law of Motion to construe attractions as mutual interactions between bodies.

The passage referred to is the argument for extending Law 3 to attractions given in Newton's Scholium to the Laws. We put off considering this argument in our earlier discussion of this scholium.[18] Let us now take up this task. Here is the argument.

I demonstrate the third Law of Motion for attractions briefly as follows. Suppose that between any two bodies A and B that attract each other any obstacle is interposed so as to impede their coming together. If one body A is more attracted toward the other body B than that other body

The printed Latin replaces [, et ejusmodi] in line 2 with [; &]. It also italicizes *hypothesis* in line 2 and *philosophia experimentali* in line 4. (Koré and Cohen 1972, 764)

[18] See chapter 3 section II.3.

B is attracted toward the first body A, then the obstacle will be more strongly pressed by body A than by body B and accordingly will not remain in equilibrium. The stronger pressure will prevail and will make the system of the two bodies and the obstacle move straight forward in the direction from A toward B and, in empty space, go on indefinitely with a motion that is always accelerated, which is absurd and contrary to the first Law of Motion. For according to the first law, the system will have to persevere in its state of resting or of moving uniformly straight forward, and accordingly the bodies will urge the obstacle equally and on that account will be equally attracted to each other. (C&W, 427–8)

This argument appeals to the first Law of Motion, together with the (already established)[19] application of the third Law of Motion to contact pushes, to extend Law 3 to attractions.[20] Having the obstacle and bodies meet in empty space ensures that, to the extent that the system consisting of the two bodies and the obstacle can be treated as a body,[21] the first Law of Motion would be violated if the pressures on the obstacle were not equal and opposite.

Newton goes on to outline an actual experiment in which there is such an attraction between two bodies.

I have tested this with a lodestone and iron. If these are placed in separate vessels that touch each other and float side by side in still water, neither one will drive the other forward, but because of the equality of the attraction in both directions they will sustain their mutual endeavors toward each other, and at last, having attained equilibrium, they will be at rest. (C&W, 428)

Here, in reasonably close approximation to the thought experiment in empty space, even small differences in attraction would generate motion with respect to the water in the direction of the stronger. Still water provides no directionally specific resistance to motion of vessels floating in it; therefore, any invisible hands preventing such motion would have to be coordinated with the direction of the attraction.

[19] As we have seen (chpt. 3 sec. II.3), Newton had just cited work by Wren, Wallis, Huygens, and Mariotte together with a description of his own careful pendulum experiments supporting the application of Law 3 to collisions (C&W, 424–7). Murray, Harper, and Wilson 2011 is an account of the reports by Huygens, Wren, and Wallis together with Newton's more sophisticated versions of experiments extending the results by Wren and Huygens. It includes translations by Wilson of the reports.

[20] The immediate applications of Law 3 to pushes between body A and the obstacle and body B and the obstacle make the equilibrium of the system support the further application of Law 3 to the attractions between A and B themselves.

[21] Another version of the experiment with magnet and iron would have both fastened to a single floating block (Knudsen and Hjorth 1996, 29). Here the relations among the magnet, block, and iron are, rather obviously, sufficiently isolated from disturbance for it to be intuitive to count the system as a body.

In Newton's thought experiment the system, though not actually fastened together, is sufficiently isolated from disturbance of relations among its parts to count as a body for purposes of the experiment. Indeed, as we saw in his letter to Cotes, part of Newton's evidence for Law 1 is that "there is no such phaenomenon in all nature" as having unequal attraction between bodies make the system of both "go away together wth an accelerated motion in infinitum, as it were by a self moving principle."

Magnetic attraction exhibits general regularities – phenomena – that make it count as an attraction. You can make either lodestone or iron move toward the other by holding the other still. You can feel the pull on the lodestone toward the iron, just as you can feel the pull on the iron toward the lodestone. Moreover, the directions of these pulls toward one another are independent of orientation with respect to the still water on which the vessels containing the lodestone and iron float. In 1666 John Wallis appealed to the following comment on the phenomena of magnetic attraction at a distance

> ...it is harder to shew How they have, than That they have it. That the Load-stone and Iron have somewhat equivalent to a Tye; though we see it not, yet by the effects we know... (*Phil Trans.* vol. 1, 282)

to support his conjecture of a similar attraction between the earth and moon.[22]

Even when they are touching one another, the lodestone and iron attract with no visible mechanism pushing them together. This makes such attraction, like the phenomenon of weight of heavy bodies, hard to reconcile with the commitment of the mechanical philosophy to make motion phenomena intelligible by showing how they could be produced by having bodies push on one another.[23] The aim of this mechanical philosophy would motivate attempting to suppose some sort of "ethereal" or other locally acting cause whereby the iron and lodestone would be pushed toward each other by invisible particles.

The above quoted passage from Newton's scholium to proposition 69 book 1 makes it clear that his use of attraction is not intended to exclude such locally acting causes. Newton is, thus, using "attraction" in a wider sense than may be customary. The generality in this usage of "attraction," however, does not make every example of a force directing one body toward another count as an "attraction" toward that other body. Cotes's example with the invisible hand pushing ball *B* toward ball *A* is not an endeavor of these bodies to approach one another.

In contrast to Cotes's example, Newton's thought experiment and his actual experiment with lodestone and iron are cases in which there is such a mutual endeavor. For such cases, Newton argues that unless the endeavors of those bodies to approach one

[22] Wallis's appeal to magnetic attraction was addressed to the following objection to his conjectured motion of earth and moon about a common center of gravity.

> it appears not how two bodies, that have no tye, can have one common center of gravity (*Phil. Trans.*, vol. 1, 282).

Gilbert's 1600 book *De magnete* inspired Kepler's proposals for a celestial physics based on forces modeled on magnetic attraction and repulsion, as well as later proposals such as Wallis's appeal to a "tye" between the earth and moon to explain the tides (*GHA 2A*, chpt. 4 and 5).

[23] In chapter 5 section I.4 we saw Huygens's commitment to contact pushes as a requirement for any hypothesis for a physical cause that would make the phenomena of gravity intelligible.

another are equal and opposite, they would lead to violation of the first Law of Motion for the system construed as a body. This makes satisfaction of equal quantity, needed to count these oppositely directed endeavors as action and reaction according to the third Law of Motion, a mathematical proportion that must be satisfied by the quantities of motion toward one another produced by the cause of such an attraction between bodies – whatever that cause might be.

However implausible they may seem, the mechanical philosophy must be able to provide for coordinated pushes that conserve momentum separately amongst the visible bodies, if it is to accommodate the phenomena of magnetic attraction. To account for the attractions in Newton's experiment by particles acting as invisible pushing hands would require two invisible hands coordinated like tweezers' ends pushing the bodies together.

Such coordination of invisible hands – like a contracting string drawing two bodies together – would make the resulting attraction fit Newton's concept of a force of interaction in which both bodies enter symmetrically.[24] If Newton's defense of his application of Law 3 to attractions succeeds in supporting extension to gravitation, then any mechanical account of the cause of gravity will have to be compatible with having gravitation between Jupiter and the sun satisfy his concept of a force of interaction between bodies.

2.iii Gravity as attraction Newton follows up his discussion of the experiment with lodestone and iron with the following argument for the claim that gravity is such a mutual attraction between the earth and its outer parts.

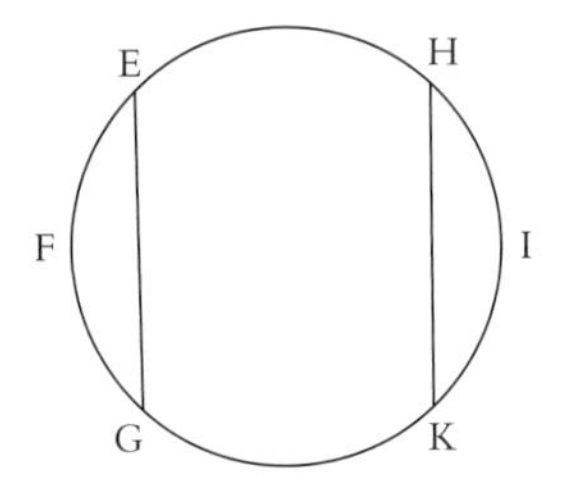

In the same way gravity is mutual between the earth and its parts. Let the earth FI be cut by any plane EG into two parts EGF and EGI; then their weights toward each other will be equal. For if the greater part EGI is cut into two parts EGKH and HKI by another plane HK parallel to the first plane EG, in such a way that HKI is equal to the part EFG that has been cut off earlier, it is manifest that the middle part EGKH will not preponderate toward either of the outer parts but will, so to speak, be suspended in equilibrium between both and will be at rest. Moreover, the outer part HKI will press upon the middle with all its weight and will urge it toward the other outer part EGF, and therefore the force by which EGI, the sum of the parts HKI and EGKH, tends toward the third part EGF is equal to the weight of the part HKI, that is, equal to the weight of the third part EGF. And therefore the weights of the two parts EGI and EGF toward each other are equal, as I set out to demonstrate. And if these weights were not equal, the whole earth, floating in an aether free of resistance, would yield to the greater weight and in receding from it would go off indefinitely. (C&W, 428)

[24] See chapter 8 section I.3.

Given that the parts EGF and EGI have weights toward one another, one might appeal to Law 1 to infer that the earth is not being accelerated by this interaction among its parts, and so conclude that the weights of the parts EGF and EGI toward each other will be equal. This is how the argument is made in Newton's earlier version of book 3, where the diagram has the earth cut into two parts by a single plane and where the analogy with the lodestone and iron is stressed.[25]

Such an argument assumes that there is attraction between the parts EFG and EGI. This assumption of mutual attraction between these parts of the earth might seem to be exactly the sort of mutual attraction of bodies toward one another that Huygens would dismiss out of hand.

The above passage, however, includes an argument which generates the pressing of the parts EGF and EGI upon one another – as well as the equality of those oppositely directed pressings – from the assumption that weights of outer parts of the earth are directed toward its center of gravity and distributed about it so as to be in equilibrium. This assumption is one that mechanical hypotheses about the cause of terrestrial gravity, including Huygens's own proposal,[26] were designed to accommodate.

The outer part EGF will be made up of terrestrial material bodies which are impenetrable to one another and to the central parts of the earth. The weights toward the center of the earth of all these smaller parts will make the outer piece EGF singled out by Newton's diagram press upon the rest of the earth EGI. In the present argument, the larger part EGI is cut by an additional plane HK parallel to the original plane EG, which cuts off another outer part HKI on the opposite side of the earth that is equal to the original outer part EGF. By Newton's construction, the total force by which this outer part HKI presses the middle part EGKH is equal to the oppositely directed total force by which the original outer part EGF presses that middle part from the other side.[27] The middle part EGKH – being, "so to speak . . . suspended in equilibrium" between both outer parts – will not exert a net force toward either.[28] This makes the total force of the larger part EGI pressing on the first outer part EGF equal the total force exerted by the second outer part HKI to drive the middle part EGKH toward that first part. By the construction of HKI, this total force by which the larger part EGI presses the outer part EGF equals the oppositely directed force by which that outer part presses it.

[25] See Cajori 1934, 570 or Newton 1728, 42.

[26] See chapter 5 section 4.

[27] As long as the weights of the outer parts form an equilibrium with respect to the center, this construction can be carried out – even if those weights are not uniformly distributed.

[28] This is in accord with Huygens's hypothesis in which the only force directed by the middle part on either outer part would be that resulting from its resistance to being pushed away from its centered location with respect to the center of the whirling spherical shells.

The middle part, EGKH, of the earth plays the same role in this equilibrium argument as the obstacle plays in Newton's basic equilibrium thought experiment.

The third Law of Motion, thus, applies to gravitation between the earth and its outer parts. As any body lying on the earth can count as an outer part, this argument shows that the cases of gravity as weight toward the center of the earth count as attractions, in Newton's sense, between the earth and terrestrial bodies.

III Gravitation as attraction between solar system bodies?

Let us examine the extent to which such arguments, as the foregoing gravitational equilibrium argument or Newton's appeal to Law 1, can be applied to extend Law 3 to count gravitation of Jupiter toward the sun as an attraction between them.

1. *Combining acceleration fields*

We have seen that the ten distinct distance estimates Newton cites for the planets yield ten inverse-square adjusted estimates of the acceleration toward the sun that would be produced by the inverse-square centripetal acceleration field directed toward the sun on any body at the distance 5.21AU we assumed for Jupiter. These yield

$$1.456 \pm 0.007\ Au/t_e^2$$

as their combined estimate of this centripetal acceleration toward the sun at distance 5.21AU from its center.[29]

Similarly, the inverse-square adjusted centripetal accelerations of Jupiter's moons measure the strength of an inverse-square centripetal acceleration field centered on Jupiter. If, in accordance with Rule 3, we extend the inverse-square acceleration field towards Jupiter to the distance of the sun, we can extend our inverse-square adjusted estimates from Pound's data for Jupiter's four moons to 5.21AU, which we are assuming as the length of the semi-major axis of Jupiter's orbit. This leads to an estimate

$$0.001373 \pm 0.00003\ Au/t_e^2$$

[29] The periods Newton gives together with the estimates of mean-distances he cites from Kepler and Boulliau give the following inverse-square adjusted – $4\pi^2 a/t^2\ (a/5.21)^2$ – estimates in AU/t^2 for the centripetal acceleration toward the sun at the distance 5.21AU we are assuming for Jupiter when its distance from the sun equals the semi-major axis of its orbit:

1.466	1.459	1.454	1.454	1.451	1.442
1.441	1.459			1.475	1.456

in units AU/t_e^2.

These yield a mean 1.456 $sd = \sqrt{(\text{avg}\delta^2)} = 0.00965$ together with an $sd^+ = (10/9)^{1/2} sd = 0.01018$ and an $SE = ((10)^{1/2}/10)\ sd^+ = 0.0032$, for a 95% t-confidence bound $2.26SE = 0.007$.

of the acceleration towards Jupiter the sun should have at the distance of 5.21AU from Jupiter.[30]

We can't use the sun-centered reference frame to represent this, since the sun is motionless in that frame. We, also, can't use a Jupiter-centered frame to represent this. In a Jupiter-centered frame, Jupiter is at rest and the centripetal acceleration of the sun toward Jupiter is 1.456 AU/t_e^2 not 0.00137 AU/t_e^2. We need to combine the inverse-square centripetal acceleration field toward Jupiter with the inverse-square centripetal acceleration field toward the sun. This requires finding a reference frame that can assign appropriate accelerations to each body.

The ratio, 1.456/0.00137 = 1063/1, of the acceleration toward the sun (according to the sun's centripetal acceleration field at the distance we are assuming for Jupiter from the center of the sun) to the acceleration toward Jupiter (at the same distance according to Jupiter's centripetal acceleration field) is a measure of the ratio of the strengths (the absolute quantities) of these two acceleration fields.[31]

George Smith[32] has suggested that Newton could have developed his solution to the problem of combining the acceleration fields of the sun and Jupiter before he had fully developed the concept of mass as quantity of matter. The basic Harmonic Rule ratios $[a^3/t^2]_J$ and $[a^3/t^2]_H$ giving the strengths of the two acceleration fields can do the job. Consider a point c along the line connecting Jupiter with the sun, where r_J and r_H are the respective distances from c to Jupiter and the sun. Let us suppose $r_J/r_H = [a^3/t^2]_H/[a^3/t^2]_J = 1063/1$ so that the sun and Jupiter orbit about c as a common center with circular orbits of respective radii $r_J = 5.21(1063/1064) = 5.205$AU and $r_H = 5.21(1/1064) = 0.0049$AU with a common period $t = 11.8616t_e$. The respective orbits are described by the two

[30] Pound's angular measures θ of maximum elongations of Jupiter's Galilean satellite orbits, at the mean earth–Jupiter distance 5.21AU = the mean Jupiter–sun distance, lead to estimates 5.21sinθ of orbital radii in AU of

r	0.00280	0.00447	0.00712	0.01253

Combining them with the periods

t	0.00484	0.00972	0.01959	0.04569

yields the following results

0.001363	0.001375	0.00138	0.00137

for $4\pi^2 r/t^2\,(r/5.21)^2$ in AU/t_e^2 – inverse-square adjusting Pound's data for the accelerations of Jupiter's moons to the assumed distance, 5.21AU, of the sun from Jupiter.

These, in turn, lead to a mean of 0.001373, an sd of 0.000015, an sd^+ of $(\sqrt{(4/3)})sd = 0.000018$, an SE of the mean of $((\sqrt{4})/4)sd^+$ of 0.000009 for a 95% t-confidence bound for 4−1 = 3 degrees of freedom of 3.18 $SE = 0.00003$.

Of Pound's estimates, only the last two are clearly estimates from cited data about distances. If we restrict our least-squares estimate to those two estimates alone, we arrive at a mean of 0.001375 with a 95% t-confidence bound of 0.0009.

[31] Using the orbit of Venus for the Harmonic Law ratio for orbits about the sun and that of Callisto for the Harmonic Law ratio for orbits about Jupiter, Newton estimates this ratio as 1067/1 (C&W, 812–13). See above chapter 8 section II.2 and appendix 4.

[32] See pp. 46–9 of Smith 1999a.

ends of the line connecting Jupiter and the sun which rotates about center *c*, as in the following diagram from Smith.[33]

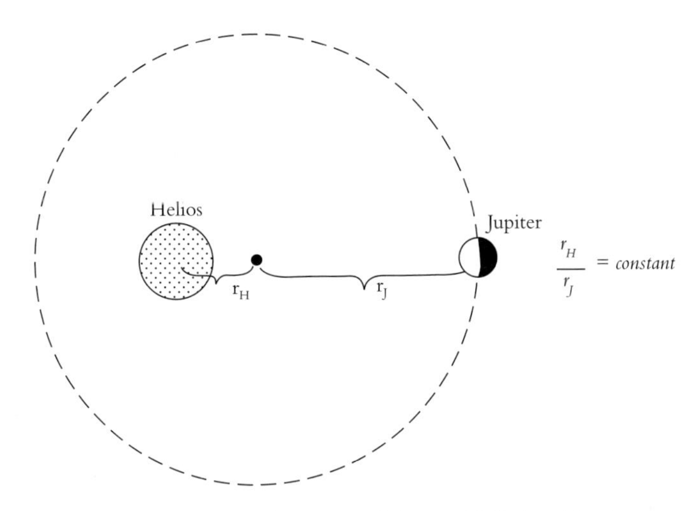

Figure 9.1 A two-body system of the sun (Helios) and Jupiter

The respective centripetal accelerations of these two circular orbits about *c* will be $(4\pi^2/t^2)r_J = 1.46\text{AU}/t_e^2$ for Jupiter and $(4\pi^2/t^2)r_h = 0.0013\text{AU}/t_e^2$ for the sun. They will, thus, be in the ratio $r_J/r_H = 1063/1 = [a^3/t^2]_H/[a^3/t^2]_J$ needed to combine these two inverse-square acceleration fields.

These orbits will satisfy the two-body correction of the Harmonic Rule, where the radius $r_J = 5.205\text{AU}$ of the corresponding one-body orbit for Jupiter is $[a^3/t^2]_H/([a^3/t^2]_H + [a^3/t^2]_J]) = (1063/1064)$ times the two-body distance 5.21AU between Jupiter and the sun. The two-body correction to the Harmonic Rule is, therefore, recovered by Smith's use of the Harmonic Rule ratios to combine these acceleration fields.

We can think of this as a *kinematical* solution to the problem of combining these two acceleration fields. Newton's definition of motive measure (definition 8) of a centripetal force supports an application of the second Law of Motion in the form $f = ma$ familiar to us. The centripetal acceleration of the sun toward *c* times the mass of the sun counts as a motive measure $W_J(H)$ of its weight toward Jupiter, while the centripetal acceleration of Jupiter toward *c* times the mass of Jupiter counts as the motive measure $W_H(J)$ of its weight toward the sun.

One result of combining these acceleration fields is that each has weight toward the other. These weights maintain their oppositely directed orientations toward one another as the sun and Jupiter orbit about *c*. They, therefore, fulfill one major

[33] These motions are described from a reference frame with center at *c* and fixed directions toward the stars.

criterion distinguishing what Newton counts as attraction from Cotes's invisible hand pushing one body toward another. To have these oppositely directed weights count as a single endeavor of these bodies to approach one another requires, in addition, that they be equal so that they satisfy Law 3. To have these oppositely directed endeavors equal is equivalent to having the location of the center *c* be at the center of mass of the sun–Jupiter system. So, this system counts as one where Law 3 applies to the attraction between the sun and Jupiter just in case *c* is located at its center of mass.

2. *Law 3 and Law 1 for a sun–Jupiter system: H-D method is not enough*

Do the orbital motions of Jupiter and the sun about *c*, resulting from combining their acceleration fields into a single system, force Newton's application of Law 3 to attraction between them? Let us consider this a dynamically isolated system, so that to have Law 3 apply at all would require the application Newton wants. Will violations of Law 3 here lead to absurdities comparable to the resulting violation of Law 1 in Newton's thought experiment?

Let us suppose a reference frame with point *c* as origin and a direction fixed for specifying angles is counted as inertial.[34] At time zero let the position vectors r_J and r_H of Jupiter and the sun point in respectively the +y and −y directions along the y-axis. The orbital motion will be in the x–y plane with constant angular rates. Now suppose that the oppositely directed endeavors of Jupiter and the sun toward one another are not equal. That is to suppose that the ratio $W_H(J)/W_J(H)$ of Jupiter's weight toward the sun to that of the sun toward Jupiter is some number *k* different from one.

We can still have the ratio $[(4\pi^2/t^2)r_J]/[(4\pi^2/t^2)r_H]$ of Jupiter's acceleration toward the sun to the sun's acceleration toward Jupiter be 1063/1, as required for combining the two acceleration fields. We have $W_H(J)/m(J) = (4\pi^2/t^2)r_J$ and $W_J(H)/m(H) = (4\pi^2/t^2)r_H$. Therefore, having the ratio of Jupiter's mass to the sun's mass be $k/1063$ instead of 1/1063 will exactly offset having the force ratio $W_H(J)/W_J(H)$ be *k* instead of 1.

If *k* differs from 1, the center of mass c_m will be at a different place than *c* on the line connecting the centers of Jupiter and the sun. At time zero this will be on the y-axis and as time goes by it will move in a uniform circle around point *c*. The centripetal acceleration of this orbit of the center of mass c_m is the analogue to the motion of the system as a whole that Newton counts as a violation of Law 1 in his discussion of the application of Law 3 to attractions.

Unlike rectilinear acceleration of the center of mass, which (insofar as the system as a whole counts as a body) would clearly violate the first Law of Motion, the uniform

[34] We are explicitly allowing for violations of Law 3, so that we can explore whether or not the orbits of Jupiter and the sun about *c* measure the equality of their weights toward one another.

circular motion of the center of mass is not obviously something that ought to count as a violation of the first Law of Motion. Recall Newton's statement with his elucidating discussion for Law 1.

Law 1. *Every body perseveres in its state of being at rest or of moving uniformly straight forward except insofar as it is compelled to change its state by forces impressed.*

Projectiles persevere in their motions, except in so far as they are retarded by the resistance of the air and are impelled downward by the force of gravity. A spinning hoop which has parts that by their cohesion continually draw one another back from rectilinear motions, does not cease to rotate, except insofar as it is retarded by the air. And larger bodies – planets and comets – preserve for a longer time both their progressive and their circular motions, which take place in spaces having less resistance. (C&W, 416)

The elucidating discussion suggests that uniform revolution of the center of mass about *c* might count as a state of motion that would be preserved according to Law 1, rather than as a violation of that Law.

This suggests that, unlike Newton's thought experiment, failure to apply Law 3 to Jupiter and the sun would not result in any comparably clear violation of Law 1 for the system of both construed as a body.

Cotes's challenge that Newton's appeal to Law 3 appears to be an assumed hypothesis rather than a deduction from phenomena is supported insofar as the phenomena measuring the acceleration fields directed toward Jupiter and the sun do not directly measure the equality of their oppositely directed weights toward each other.[35] Without Newton's appeal to Law 3, even the two-body correction fails to carry the information that these weights are equal.

We can continue this construction adding as many bodies as we want, consistently with our information about the relative strengths of their acceleration fields. For a system of *n* gravitationally interacting bodies, the acceleration for each body *i* at time *t* is given by

$$d^2 r_i/dt^2 = \sum_{j \neq i} GM_j (r_j - r_i)/(r_{ij})^3, \tag{1}$$

where the vectors r_i, r_j are defined relative to some inertial frame, r_{ij} (the scalar length of vector $(r_j - r_i)$) is the distance between bodies *i* and *j*, M_j is the mass of body *j*, and *G* is the gravitational constant. This is a standard formulation for the accelerations

[35] That the orbital phenomena cited by Newton do not provide direct measurements of the equality of the weights toward one another of Jupiter and the sun is the problem which has limited the precision to which we have yet been able to measure the gravitational constant.

> The gravitational force between masses of planetary size is not so weak, but this is of no help in determining *G* because only the combination *GM* (where *M* is the mass of the attracting body) appears in the equations of motion of bodies with purely gravitational interactions; hence, planetary observations cannot determine separate values for *G* and *M*. (Ohanian and Ruffini 1994, 3)

The precision to which *G* has been measured is limited to experiments using laboratory bodies, because orbital phenomena don't separate *G* and *M*.

corresponding to the gravitational interactions among n bodies.[36] Now take the same n initial position vectors r_i, r_j, but let the accelerations be given by

$$d^2r_i/dt^2 = \sum_{j \neq i} G'_j M'_j (r_j - r_i)/(r_{ij})^3, \tag{2}$$

where each body is allowed to have its own separate gravitational constant G'_j and correspondingly adjusted mass M'_j. So long as for each body j, the product $G'_j\ M'_j = G\ M_j$, equation 2 will recover exactly the same accelerations as equation 1, even though it need not count the center of mass of the system as inertial.[37] This shows that, so long as they do not result in collisions, the motion phenomena resulting from gravitational interactions among these bodies will not put any bounds on the ratios among these G's, nor among the M's, if Law 3 does not apply to these interactions.

3. Is Newton's application of Law 3 a deduction from the phenomena?

It is clear that the equality of the weights of the sun and Jupiter toward one another is not directly measured by orbital phenomena. Newton's application of Law 3 is, therefore, not what according to Newton's methodology would count as a *direct deduction* from orbital phenomena. To count it as, nevertheless, a *deduction from the phenomena*, Newton would have to construe it as resulting from a legitimate *making general by induction* of some measurements by phenomena.

Let us review the measurements he could appeal to. Consider his thought experiment. It may, perhaps, be counted as, itself, a phenomenon – a general null outcome of many experiments – that "there is no such phaenomenon in all nature" as the violations of Law 1 that would result if Law 3 did not apply to attractions of the sort considered in Newton's thought experiment. Each null outcome of the many examples of such attractions would, thus, measure the equality of the oppositely directed endeavors. The repeatable outcomes of Newton's actual experiments with lodestone and iron are clear examples of phenomena measuring such equalities. Newton's demonstration of such equal attractions between the earth and its outer parts from the assumption that the weights of outer parts of the earth form an equilibrium about its center may go some way toward supporting extending this measured equality to construe gravity as such an attraction between bodies.

Finally, and perhaps most compellingly, we have the requirement that the separate acceleration fields measured by orbital phenomena be combined into a single system. The oppositely directed weights toward one another resulting from combining these acceleration fields is maintained as the bodies move about. This does distinguish gravitation of the

[36] This is the Newtonian part of the n-body point mass equation used as a basis for the orbital ephemerides for the sun, moon, and planets (*ESAA*, 281).

[37] In this case the motion of the center of mass would violate Law 1, if that center of mass were counted as a body.

Katherine Brading has offered an insightful discussion of how composit systems and their parts mutually illuminate the interpretation of bodies on what she identifies as Newton's law-constitutive approach. See Brading 2011.

sun and Jupiter toward one another from acceleration of one body toward another resulting, merely, from an independent push in that direction. Perhaps, Newton regarded fulfilling this criterion for attraction as sufficient to legitimate counting the additionally required equality of those weights toward one another as making general by induction the foregoing measured equalities of attractions between bodies.

As we have seen, additional orbital phenomena corresponding to interactions do not directly measure the equality of the relevant oppositely directed weights. These do, however, provide additional empirical support for the conclusion that gravitation satisfies the crucial criterion – that these bodies maintain forces toward one another as they move about – which distinguishes what Newton counts as attraction from Cotes's example.

Perhaps this would allow the equalities of the oppositely directed forces in attractions between bodies cited in Newton's arguments to extend Law 3 to attractions to count as constant qualities that belong to all bodies on which experiments can be made. If so, then Newton's third Rule for doing natural philosophy,

Rule 3 *Those qualities of bodies that cannot be intended and remitted [i.e., qualities that cannot be increased and diminished] and that belong to all bodies on which experiments can be made should be taken as qualities of all bodies universally.* (C&W, 795)

may well endorse his application of his Law 3 to construe gravity as a universal force of attraction between bodies. This would be a making general by induction that counted the cited phenomena measuring the equalities of oppositely directed forces in attractions between bodies as measuring the equalities of the oppositely directed weights between bodies generally.

IV Beyond hypotheses: yes

1. *Empirical success and Rule 4: Newton's second thought?*

Newton's letter to Cotes ends with the remark

I have not time to finish this Letter but intend to write to you again on Tuesday. (*Corresp.* V, 397)

His Tuesday letter, dated March 31, 1713, opens with the following remarks.

Sr

On saturday last I wrote to you, representing that Experimental philosophy proceeds only upon Phenomena & deduces general Propositions from them only by Induction. And such is the proof of mutual attraction. And the arguments for ye impenetrability, mobility & force of all bodies & for the laws of motion are no better. And he that in experimental Philosophy would except against any of these must draw his objection from some experiment or phaenomenon & not from a mere Hypothesis, if the Induction be of any force. (*Corresp.* V, 400)

The last sentence, which rejects objections from mere hypotheses while endorsing objections drawn from phenomena, is a clear anticipation of the method advocated in

Rule 4, which was first printed in the third edition of 1726. Here again is that Rule characterizing the commitment of his experimental philosophy to provisionally accept propositions gathered from phenomena as guides to research.

Rule 4 *In experimental philosophy, propositions gathered from phenomena by induction should be considered either exactly or very nearly true notwithstanding any contrary hypotheses, until yet other phenomena make such propositions either more exact or liable to exceptions.*

This rule should be followed so that arguments based on induction may not be nullified by hypotheses. (C&W, 796)

It appears that the methodology characterized in this important rule for reasoning in natural philosophy was first introduced as an additional response to Cotes's challenge.[38]

We have argued that mere contrary hypotheses are alternatives that fail to realize Newton's rich conception of empirical success sufficiently to count as serious rivals to propositions counted as gathered from phenomena by induction. Newton's conception of empirical success, as accurate measurement of parameters by the phenomena they are taken to explain, offers far more resources for rejecting alternatives than are provided by the hypothetico-deductive model of scientific inference.

We have seen that in an isolated two-body system, without the assumption that Law 3 applies, the forces and masses cannot be disentangled sufficiently to be measured.[39] Without Newton's application of Law 3, the assumed orbital phenomena are compatible, even, with having Jupiter's mass very much greater than that of the sun. For example, if k is set at $1063(1063) = 1{,}129{,}969$, then the very same orbits about c would be predicted even though Jupiter's mass would be 1063 times that of the sun. The assumed orbital phenomena cannot disentangle weights and masses

[38] Newton had included a somewhat longer statement in an unsent draft of this letter.

Sr

> On Saturday last I wrote to you representing that Experimental philosophy proce[e]ds only upon Phenomena & makes Propositions general by Induction from them. In this Philosophy neither Explications nor Objections are to be heard unless taken from phaenomena. Nor are Propositions here made general by arguments a priore by [*read* but] only by Induction without exception. And upon such an Induction the mutuall and mutually equal Attraction is founded. One may suppose that there may be bodies penetrable or immoveable or destitute of force, or with attraction mutually unequal, but such suppositions without any instance in Phaenomena are mere hypotheses & have no place in experi[ment]al Philosophy: & to introduce them into it would be to overthrow the Arguments from Induction upon wch all the general Propositions in this Philosophy are built. (*Corresp.* V, 401)

This somewhat more extended discussion anticipates Newton's comment on Rule 4,

> This rule should be followed so that arguments based on induction may not be nullified by hypotheses. (C&W, 796),

as well as the method advocated in that Rule.

Newton's long draft of his Saturday letter to Cotes (*Corresp.* V, 398–9) also includes significant anticipation of Rule 4 and of connections between such considerations and Rule 3.

[39] This would make them behave like *gorce* and *morce* in an example Clark Glymour used to illustrate the advantage of measurement over mere prediction as a criterion of empirical success. See Glymour 1980a, 357.

sufficiently to measure either, because they put no bounds on the ratios of the separate *G*'s.[40]

As we have seen, the striking new phenomena corresponding to increasingly precise successive approximations made possible by taking into account gravitational interactions – such phenomena as the two-body correction to the Harmonic Rule for Jupiter's orbit or corrections of Saturn's motion corresponding to perturbations by Jupiter – are all consequences of any such multiple *G* alternative theory. No such phenomenon, therefore, can support Newton's application of Law 3 to gravitation by directly refuting differing *G*'s. We have seen that such phenomena do empirically back up the existence of oppositely directed weights of solar system bodies toward one another that follows from combining their inverse-square acceleration fields. They, thus, add empirical support for attributing to gravity that first main criterion distinguishing what Newton calls "attraction" from examples like Cotes's ball being pushed toward another. They do not, however, measure the equalities of those oppositely directed weights required by Newton's application of Law 3 to count gravity as such a pair-wise attraction between bodies.

Newton's application of Law 3 clearly outstrips our alternative hypotheses of separate *G*'s in realizing his ideal of empirical success. Instead of up to as many differing gravitational constants as there are bodies – resulting in masses too entangled with weights to be measured – Newton's application of Law 3 to gravity as pair-wise interactions among visible bodies in the solar system leads to convergent agreeing measurements of relative masses among them.[41] These measurements of relative masses are clear examples of a realization of Newton's stronger ideal of empirical success that flows from his challenged application of Law 3.[42] This endorses applying Rule 4 to count these alternatives as mere contrary hypotheses that are not to be

[40] Any way of fixing the ratio G_H/G_J will allow the relative acceleration phenomena to measure the weights and masses of these bodies. Moreover, any way of fixing these ratios will predict exactly the same relative accelerations among these bodies. This might lead one to say that differing specifications of the ratios among the *G's* are just alternative specifications of the same theory.

Even though these alternatives would predict the same actual history of relative accelerations among these bodies, they would not agree on all possible predictions. For example, suppose the tangential velocities of the sun and Jupiter with respect to *c* were destroyed so that they fell together. Before their collision the motions toward one another would be the same in all these alternative theories, but the different theories would give differing predictions about the motions after they collided. Theories where G_H/G_J is set at 1,129,969 so that the inertial mass of Jupiter is 1063 times that of the sun would make very different post-collision predictions from those where G_H/G_J is set at 1 so that it is the sun's inertial mass which is more than a thousand times that of Jupiter.

[41] We have seen that in corollary 2 of proposition 8, Newton applies proposition 7 to use Harmonic Law ratios to measure the masses of the sun and planets with moons. Eric Schliesser has pointed out that Adam Smith (1723–90) singles out Newton's ability to calculate the masses of the sun and planets for special praise. See Schliesser 2005, 720.

[42] The resolution of the problem of combining the separate inverse-square acceleration fields is, also, an important achievement that goes beyond successful prediction of the separate phenomena. This achievement is so important that, according to Michael Friedman (1992, 155), Kant appealed to it as a transcendental deduction of Newton's application of Law 3 to construe gravity as universal force of interaction between bodies.

allowed to undercut the application of Law 3 to count gravity as a mutual interaction between bodies.

2. *Acceptance and accumulating support*

Consider the role of accumulating success in supporting Newton's inference over vortex-theoretic alternatives such as Huygens's proposal. On vortex theories it would be the changes in motion of invisible vortical particles resulting from their pushing the planets into orbital motion – rather than the gravitation of the sun toward the planet – that counted as the equal and opposite reaction to the weight of a planet toward the sun. On Newton's Rule 4, the agreement among the measurements of relative inertial masses among solar system bodies provided by orbital phenomena counts as evidence supporting Newton's application of Law 3 to count gravitation among solar system bodies as pair-wise attractions between them. The agreeing measurements of the mass of the sun afforded from the orbits of its six known planets, of the mass of Jupiter from the orbits of its four known moons, of the mass of Saturn from its five known satellites, and of the mass of the earth from the orbit of its moon and from pendulums measuring its surface gravity, are all realizations of Newton's ideal of success as convergent accurate measurements by phenomena. None of these is available to Huygens's alternative proposal.

Huygens resisted Newton's arguments to combine the separately argued for inverse-square centripetal acceleration fields into a single system. He did not accept Newton's center of mass resolution of the two chief world systems problem, on which Kepler's orbits are good starting approximations from which to generate more accurate accounts by correcting for perturbations due to gravitation toward other planets. Huygens was convinced by Newton's arguments to inverse-square attraction of gravity toward the sun that Kepler's elliptical orbits with force toward the sun at a focus were exact descriptions of the motions of the planets. Empirical establishment of perturbation corrections of the basic Keplerian elliptical orbits are clear empirical counterexamples to Huygens's alternative. It turns out, however, that serious work on vortex theories had largely died out in favor of the explanatory power of the Newtonian system by the late 1740s, well before any such perturbation corrections had been empirically established.[43]

As we have seen, however, combining the separate acceleration fields into a single system can be achieved without Newton's application of Law 3. Any of the alternatives with distinct separate *G*'s will do the job. What these do not do is to realize Newton's stronger ideal of empirical success well enough to be counted as serious rivals.

[43] Aiton (*GHA 2B*, 20) points out that the Paris Royal Academy prize of 1740 for an essay on the tides was divided between four competitors, three of whom – MacLaurin, Euler, and Daniel Bernoulli – based their explanations on the Newtonian system. The other, the Jesuit Cavalleri, gave a Cartesian justification of Newton's results. Here is his excellent overall summary of the transition to the Newtonian system:

> It was their recognition of the explanatory power of the Newtonian system that led the ablest Cartesians to combine Newton's mathematical theory with physical vortices. Then when they lost faith in vortices, remarks such as those of Maupertuis concerning the equal unintelligibility of

It was a surprisingly long time before applications of Newton's theory led to significant, empirically established, corrections of Kepler's orbits to take into account perturbations.[44] Curtis Wilson offers the following remarks on the astronomical tables of Cassini II, published in 1740, and those of Halley, published posthumously in 1749:

> These tables, the most highly respected at the time of their publication, and still in use in the 1780s, were purely Keplerian in principle, except for Halley's inclusion in his tables of an anomalous acceleration of the mean motion of Jupiter and an anomalous deceleration in the mean motion of Saturn. (Wilson 1985, 16)

Halley, of course, was aware of and in support of Newton's contention that on his theory of universal gravity one would expect Jupiter and Saturn to mutually perturb one another's orbital motions. The problem was how to give a detailed account of the empirical correction needed as well as of the Newtonian perturbation explaining it.

Wilson plots the dominant Jupiter–Saturn perturbation, the great inequality, along with the next two largest perturbational inequalities of each.[45] This shows the great inequality to have a period of about 900 years, with Jupiter speeding up and Saturn slowing down for half the cycle and Jupiter slowing down and Saturn speeding up for the other half.[46] It also suggests that the pattern reverses in the mid 1700s, so that Halley's correction becomes increasingly inaccurate after the 1760s.[47] In 1773 Lambert showed that Halley's supposition had become empirically untenable.[48]

Finally, on November 23, 1785 Laplace announced to the Paris Academy that the anomalies in the mean motions of Jupiter and Saturn could be accounted for on the assumption of universal gravitation. Wilson describes the enormous import of Laplace's achievement.[49]

> ... in the wake of LAPLACE's "Théorie de Jupiter et de Saturne" and primarily as a result of it, the practice of predictive astronomy had been transformed. On the basis of the new procedures introduced by LAPLACE, the way appeared open to a marked reduction in the gap between tables and observations, and a new period of advance, both theoretical and observational, was entered upon. (Wilson 1985, 23)

This new period of advance was the extraordinarily successful research program that, from the work of Laplace at the turn of the nineteenth century and up through the

> impulsion and attraction (of which there are hints in Bouguer's essay) could help them to abandon the vortices with a clear conscience and accept universal gravitation as a physical axiom. In the second edition of his prize essay (published in 1748), Bouguer himself, having in the meantime observed the deflection of a plumb-line in the neighbourhood of a mountain in Peru, announced his conversion to Newtonianism, though with a modified attraction law. (*GHA 2B*, 21)

[44] The following material from Wilson, its implications for arguing for Newton's stronger ideal of empirical success, as well as the argument against turning theory-mediated measurement into a necessary condition for accepting propositions as gathered from phenomena by induction, are also treated in my forthcoming paper in the Leiden volume edited by Janiak and Schliesser.

[45] Wilson 1985, 35. [46] Ibid. [47] Ibid.

[48] Wilson 1985, 20. [49] Wilson 1985, 22–3.

work of Simon Newcomb at the turn of the twentieth century, led to increasingly accurate perturbation-corrected orbits fitting increasingly precise data and affording increasingly accurate measurements of the masses of solar system bodies.

The solar tables of Lacaille of 1758 were the first to include perturbations of a planet.[50] They included perturbations of the motion of the earth due to the moon, Venus, and Jupiter. They also were the first tables to take account of the aberration of light and the nutation of the earth's axis, two effects that had prevented advances in telescopes and clocks from achieving precision far exceeding the best naked eye observations. Bradley announced his discovery of aberration of light in 1729 and of the nutation of the earth's axis in 1748. The successful theoretical treatment of nutation as a Newtonian perturbation was by d'Alembert and Euler. D'Alembert took his treatment of nutation as affording striking confirmation of attraction of the earth towards the moon. Wilson quotes the following passage from d'Alembert's memoir of 1749.

> The nutation of the terrestrial axis, confirmed by both the observations and the theory, furnishes, it seems to me, the most complete demonstration of the gravitation of the Earth toward the Moon, and hence of the principal planets toward their satellites. Previously this tendency had not appeared manifest except in the ocean tides, a phenomenon perhaps too complicated and too little susceptible to a rigorous calculation to reduce to silence the adversaries of reciprocal gravitation. (Wilson *GHA* 2B, 48)

Euler refined d'Alembert's treatment and was inspired by it to develop the first treatment of mechanics of rigid bodies in 1750.[51]

Euler, however, continued to object to Newton's inference to gravity proportional to the mass of an attracting body. He did so in a letter to Mayer, who was extending the work of Euler and Clairaut on applications of Newton's theory to the lunar precession to develop the first really accurate lunar tables.[52] Here is his negative assessment of the empirical evidence supporting the proposition that gravitational attraction toward heavenly bodies is proportional to their masses in his letter to Mayer dated December 25, 1751.

> ... it is still not decided by any single phenomenon that the attractive forces of heavenly bodies are proportional to their masses. On the contrary, Newton tried to determine the masses on this basis since there is no other way of specifying them. As soon as one now places the statement that the attractive forces are proportional to the masses (which is founded on a crude hypothesis) in doubt, this objection against my idea is completely eliminated. (Forbes 1971, 44)

Euler wanted to avoid action at a distance. He continued to look for a cause of gravity that would avoid action at a distance and he continued to look for phenomena that would afford evidence of an aether that would require modification of Newton's

[50] Wilson 1985, 19.
[51] Wilson (*GHA 2B*, 53; 1987, 253)
[52] Forbes and Wilson *GHA 2B*, 62–8.

theory, even as he developed fundamental contributions to the analytic treatment of perturbations within Newton's theory. Unlike Newton, who regards it as a requirement that any adequate proposal for a cause of gravity must meet, Euler regards the proportionality of attraction to mass as not sufficiently established.

Like Euler, Newton wanted to avoid commitment to action at a distance. As he makes clear in his recently much cited letter to Bentley,[53]

> That gravity should be innate inherent & essential to matter, so yt one body may act upon another at a distance through a vacuum, wthout the mediation of anything else, by and through wch their action or force may be conveyed from one to another is to me so great an absurdity that I believe no man who has in philosophical matters any competent faculty of thinking can ever fall into it. (*Corresp.* III, 254)

Janiak argues that for Newton, the inconceivability of action at a distance by brute matter was an unrevisable conceptual commitment.[54] This passage suggests that Newton shared a sensitivity to the absurdness of action at a distance with Huygens and Euler. Like Huygens and Euler, he appears to have been committed to avoid action at a distance between bodies. Like Euler, Newton did not let the apparent clash between avoiding action at a distance and his contested application of Law 3 to construe gravity as an interaction between separated bodies undermine his efforts to develop the theory of universal gravity and its application to the solar system. Unlike both Huygens and Euler, Newton did not let any philosophical commitment to avoid action at a distance undermine his methodological commitment to make *theory-mediated* measurements afford empirical answers to questions about the force of gravity and the masses, interactions, and motions among solar system bodies.

We have seen that Newton regards the possibility of locally acting causes that could recover his applications of Law 3 as not ruled out. Janiak suggests that this helps make

[53] Here are relevant remarks which precede the quoted passage in Newton's letter to Bentley:

> Tis unconceivable that inanimate brute matter should (without ye mediation of something else wch is not material) operate upon & affect other matter without mutual contact; as it must be if gravitation in the sense of Epicurus be essential & inherent in it. And this is one reason why I desired you would not ascribe innate gravity to me. (*Corresp.* III, 253–4)

Here are remarks which follow the quoted passage:

> Gravity must be caused by an agent acting constantly according to certain laws, but whether this agent be material or immaterial is a question I have left to ye consideration of my readers. (*Corresp.* III, 254)

Eric Schliesser, forthcoming, argues that these, less often quoted remarks, support a more subtle account of Newton's interpretation of gravity as a force of pair-wise interaction between bodies.

[54] See Janiak 2008, 34–49. Harper 1977 provides extended conditional belief representations that can model an agent accepting a proposition as a conceptual commitment and not merely as a knowledge claim. If we represent Newton's commitment to avoid action at a distance in such a model, his successful applications of Law 3 would carry no weight to support action at a distance for him.

These models can be extended to represent rational conceptual changes. As was pointed out in chapter 3 note 22, Harper 1979 applies such a model to represent the transition from Galilean invariance of classical kinematics to the invariance of the Poincaré group of special relativity as a rational conceptual change for a Hertzian physicist who had initially maintained Galilean invariance as a conceptual commitment.

his separation of his philosophical commitment to avoid action at a distance from his methodological commitment to make *theory-mediated* measurements afford empirical answers to questions about the force of gravity sharper than Euler's. On this interpretation, the interactions generated by the proposed applications of Law 3 do not carry any weight to support action at a distance for Newton.[55]

We have seen that Newton's initial argument for universal gravity explicitly leaves open a number of problems to be dealt with later. It is clear that the solutions to problems of demonstrating deviations from his basic model of the lunar orbit due to action of the sun on the earth–moon system generated by Newton in his account of the variational orbit in propositions 25–30 of book 3, and the much later account of the lunar precession by Clairaut, contribute additional evidence for Newton's theory. Insofar as such developments make good on explicit promissory notes, they can be taken as further developments required to complete the initial argument. The important lesson of Rule 4, however, is that such developments increase the resiliency of the commitment to accept universal gravity as a guide to empirical investigation of the motions of solar system bodies. They do this by raising the standard for what would be required of an alternative to count as a serious rival rather than be appropriately dismissed as a mere contrary hypothesis.

One important aim of Newton's argument was to establish the methodology of his experimental philosophy in place of the alternative methodology of the mechanical philosophy. The review of the *Principia* in the *Journal des Scavans* of August 1688 described the work as "a mechanics, the most perfect that one could imagine," but dismissed Newton's argument in book 3 with the sarcastic comment

> In order to make an *opus* as perfect as possible, M. Newton has only to give us a Physics as exact as his Mechanics. He will give it when he substitutes true motions for those that he has supposed. (Feingold 2004, 32)[56]

Newton's task was not just trying to produce evidence for his theory. He was arguing for a new standard by which theories are to be judged. This review is a good illustration of the sort of difficulty he had to face.[57]

3. *An* ideal *of empirical success, not a necessary criterion for acceptance*

I want to argue against turning Newton's ideal of empirical success as *theory-mediated* measurement into a necessary criterion for counting a proposition as gathered from

[55] Apparently in contrast to Newton, the empirical support for Newton's theory came to be taken by most who developed its application to the solar system as evidence for action at a distance. By Kant's day, the empirical support for gravity as a universal force of mutual interaction between bodies was so great that it was taken to empirically settle the causal question in favor of action at a distance.

[56] Feingold suggests that the author was probably an orthodox Cartesian named Régis.

[57] Koffi Maglo (2003) has provided an interesting discussion of the reception of Newton's *Principia*. He suggests that a reinterpretation of the concept of attraction by Maupertuis on which it became congruent with mechanistic semantics played an important role, opening the way for more positive interest in Newton's theory.

phenomena by induction. The first direct measurement establishing the proportionality of gravitational attraction to the mass of an attracting body was not achieved until Cavendish's laboratory measurement of the gravitational constant in 1798.[58] Before Cavendish there was no direct measurement to support the application of Law 3 to count gravity as a pair-wise attraction between separated bodies. This might suggest that Newton's application of Law 3 to count gravity as a pair-wise attraction should not have been counted as a proposition gathered from phenomena by induction before Cavendish completed his experiment.

I have suggested that, on Newton's methodology as explicated in Rule 4, Newton's convergent agreeing measurements of relative masses of solar system bodies in corollary 2 of proposition 8 book 3 may well put his theory sufficiently far ahead to count Huygens's alternative as a mere hypothesis. Newton's own responses to Cotes suggest that he would appeal to the whole of the *Principia.* Cotes, in his preface, focuses on the tides and precession of the equinoxes as affording phenomena that testify to attraction of our sea and earth toward the moon. Even if one rejected Newton's own treatments of the tides and precession of the equinoxes as insufficient to count the attraction of the earth by the moon's gravity as acceptable as an approximation, d'Alembert's and Euler's treatment of nutation and Lacaille's tables of 1758 with perturbations of the earth by the moon would each afford clear empirical support for attraction of the earth toward the moon. According to the methodology of Newton's Rule 4, by the time of Laplace's solution to the great inequality of the Jupiter–Saturn mutual perturbation the indirect support afforded by convergent agreeing measurements of parameters that had accumulated to back up universal gravity was sufficiently great that it is no surprise that few, if any, regarded Cavendish's measurement as removing any serious obstacle to accepting Newton's theory.

There is a long history of philosophers turning powerful sufficient conditions for some apparently clear cases of knowledge into necessary conditions, which then lead to skeptical arguments aimed to undercut other commonly accepted cases. Descartes's example of knowledge by perception limited to subjective contents, that not even a *malin génie* could deceive one about, got turned into a strict subjectivist empiricism according to which knowledge of external bodies is problematic. I suggest that turning the Newtonian ideal of convergent accurate measurement of parameters from diverse phenomena into a necessary condition for acceptance of theoretical propositions in natural philosophy would be another example of this misguided practice that promotes unwarranted skepticism.

Ernan McMullin has argued that Newton's impressive achievement was too idiosyncratic from an epistemic standpoint to serve as a model for the natural sciences

[58] George Smith pointed this out in a vivid, masterful presentation he gave at a conference in Leiden. As I suspect he intended, it was a surprise to many that direct measurement of the mutuality of gravitation between separated bodies was so late in coming.

generally.[59] He points out that Kant, taking his interpretation of Newton's achievement as the paradigm for natural science generally, dismissed chemistry as "systematic art rather than science."[60] The following remarks sum up his interesting discussion of the impact of Newton's *Principia* on philosophers Berkeley, Hume, Reid, and Kant in the century following its publication.

> This "Newtonian" interlude in the history of the philosophy of science would today be accounted on the whole a byway. The *Principia*, despite its enormous achievement in shaping subsequent work in mechanics, was from the beginning too idiosyncratic from an epistemic standpoint to serve as a model for the natural sciences generally. (McMullin 2001, 279)

He suggests that Newton's method is more or less limited to mechanics. On McMullin's assessment Newton is doubly irrelevant today. Not only has his theory been superseded, we are now encouraged to recognize that his idiosyncratic methodology was too limited from the beginning. McMullin suggests that the hypothetico-deductive methodology excluded by Newton's rejection of hypotheses is in fact the scientific method that characterizes most sciences today.

We have been arguing that the features by which Newton's method goes beyond hypothetico-deductive methodology were very effectively applied by such successors as Laplace in generating very high-quality evidence for universal gravity from its application to the solar system. In the next chapter we will see that this richer method of Newton's endorses the transition to Einstein's theory and is very successfully employed in the development and application of testing frameworks for relativistic theories of gravity. We shall also see that Newton's account of theory acceptance, guided by his ideal of empirical success as convergent accurate agreeing measurements from diverse phenomena, appears to be strikingly realized in cosmology today. Newton's scientific method is far richer, more effective, and more widely used today than McMullin's discussion would suggest.

We have not, however, argued that this rich scientific method of Newton's should be made into a criterion for what counts as science. One does not have to reject hypothetico-deductive inference as unscientific in order to understand the features that make Newton's inferences from phenomena more effective in investigations where they can be realized. For one thing, the methodology of Newton's Rule 4 can be quite context sensitive. As long as there is sufficient empirical success to distinguish propositions counted as deduced from the phenomena from alternatives to be dismissed as mere hypotheses, Rule 4 can be applied to characterize an appropriate role for provisional acceptance as guides to research in a scientific investigation. This can be achieved by hypothetico-deductive success in the sorts of empirical research guided by

[59] McMullin 2001, 279.

[60] McMullin quotes this dismissal from Kant's *Metaphysical Foundations of Science* (Kant [1786] 2004, 4). See McMullin 2001, 307.

clinical trials.[61] I believe it is also achieved by other sorts of empirical success sufficient to support acceptance in evolutionary biology. Moreover, the informativeness of Rule 4 does not require that we turn it into a definition of scientific investigation. That would be exactly the sort of destructive philosopher's mistake we have been arguing against.

[61] Mayo and Spanos 2010 gives papers from a conference that very informatively explore applications of error-testing approaches to scientific inference.

10

Newton's Methodology and the Practice of Science

This chapter argues that the features that make Newton's scientific method richer than the hypothetico-deductive model of scientific inference inform the applications of Newton's theory of gravity to solar system bodies by Newton and his successors. We will argue that it informs the radical conceptual transition from Newton's theory to Einstein's. This methodology of Newton's continues to inform the scientific practice exhibited in the development and application of testing frameworks for relativistic theories of gravity. It also appears to be at work in cosmology today.

Part I is a brief review of the contribution of Newton's richer method to inform theory acceptance and empirical support in applications of Newton's theory to solar system phenomena. Section I.1 reviews the features which Newton's method adds to the hypothetico-deductive method. Section I.2 points out the role of seeking successively more accurate approximations in Newton's argument and in subsequent applications by his successors. Section I.3 remarks on the increased support accorded to Newton's theory by its success in these applications. It calls attention to the work of Simon Newcomb, who explicitly made accurate measurement of relative masses a central goal of his method for generating accurate predictive tables of motions of solar system bodies.

Part II argues that the Mercury perihelion problem illustrates Newton's rich method before and after Einstein. Section II.1 reviews Asaph Hall's proposal to revise the inverse-square power law to account for the excess precession. It then reviews Ernest Brown's use of his refinements of our moon's motion to measure the inverse-square precisely enough to rule out Hall's hypothesis. Section II.2 notes the capacity of Einstein's theory to account for the excess precession, together with its Newtonian limit which allowed it to take over the successful measurements of Newton's theory. These gave Einstein good grounds to expect that his theory would be able to beat Newton's theory on Newton's own standard of empirical success. Section II.3 reviews Newton's rich method at work in the application of testing formalisms for relativistic gravity theories to resolve a challenge to General Relativity involving Mercury's perihelion precession.

Part III argues that Newton's scientific method does not require or endorse scientific progress as progress toward Laplace's ideal limit of a final theory of everything. Part IV argues that the more modest Newtonian goal of delivering successively more accurate theoretical models that afford an increasingly detailed and refined understanding of nature is not undermined by Larry Laudan's confutation of convergent realism.

Part V is a postscript that highlights quotes from a leading investigator in supernova cosmology.[1] These describe the decisive role agreeing measurements from diverse phenomena played in the recent transformation of dark energy from a dubious hypothesis into part of the accepted background framework guiding empirical research in cosmology. This appears to be a striking example of Newton's scientific method in action today.

I Empirical success, theory acceptance, and empirical support

1. Newton's scientific method adds features that significantly enrich the basic hypothetico-deductive model of scientific method

On the Cartesian-inspired mechanical philosophy, the aim of natural philosophy is to render motion phenomena intelligible by devising mechanical hypotheses that would show how they could be caused by bodies pushing on one another. In this original Cartesian version it was enough to devise a merely possible hypothetical mechanical cause by contact action. Huygens went beyond this basic demand for devising possible hypotheses by advocating a hypothetico-deductive model of confirmation that would allow such hypotheses to be counted as empirically supported by successful prediction of observable consequences. Here again, from the 1690 preface to his *Treatise on Light*, is Huygens's interesting statement of this model for confirming hypotheses by successful predictions inferred from them.

> One finds in this subject a kind of demonstration which does not carry with it so high a degree of certainty as that employed in geometry; and which differs distinctly from the method employed by geometers in that they prove their propositions by well established and incontrovertible principles, while here principles are tested by the inferences which are derivable from them. The nature of the subject permits of no other treatment. It is possible, however, in this way to establish a probability which is little short of certainty. This is the case when the consequences of the assumed principles are in perfect accord with the observed phenomena, and especially when these verifications are numerous; but, above all when one employs the hypothesis to predict new phenomena and finds his expectations realized. (Huygens 1690; trans. from Matthews 1989, 126–7)

[1] Robert Kirshner was a member of the High-z supernova search team. See Kirshner 2004, pp. 190–3 for his account of the formation of the High-z team, pp. 221–33 for his account of their discovery of cosmic acceleration.

Here we see the two basic features of the hypothetico-deductive model of scientific methodology.

Hypotheses are verified by the conclusions to be drawn from them

and

Accurate prediction counts as empirical success.

Huygens is quite positive about the prospects for this sort of confirmation to establish quite high probability for the assumed hypotheses.

The cases he cites are all ones where a Bayesian model of scientific inference might well be expected to result in high posterior probability for the hypothesis after successful predictions of the sort specified.[2] Abner Shimony has developed interesting models of scientific inference based on what he calls tempered Bayesianism.[3] Jon Dorling has shown that Bayesians can recover interesting appropriate judgments in real cases of scientific inference.[4] Wayne Myrvold has shown that Bayesians will assign higher posteriors to unifying hypotheses that make distinct data relevant to one another.[5] These successes, impressive as they may be, are not yet enough for Bayesians to recover all the advantages of Newton's method. I suggest that in order to recover the features that we have seen to make Newton's method so successful in physics and cosmology, the Bayesian framework needs to be enriched with an account of theory acceptance informed by Newton's conception of empirical success.[6]

[2] The basic Bayesian model is updating by conditionalizing on the observed outcomes. Where P_O (H) is your initial epistemic probability for proposition H,

$$P_O(H/E) = P(H\&E)/P(E)$$

is your conditional degree of belief in H given the assumption that E obtains. Conditionalization is to shift your epistemic probability upon having observed E to obtain; so that you revise your epistemic probability for H from your prior assignment $P_O(H)$ to a posterior assignment

$$P_1(H) = P_O(H/E).$$

[3] See his "Scientific Inference" (Shimony [1970] 1993, 183–273) and "Reconsiderations on inductive inference" (Shimony 1993, 274–300).

[4] See Dorling 1979.

[5] See Myrvold 2003.

[6] Wayne Myrvold has convinced me that what is needed is an account of acceptance that can accommodate accepting propositions for which one's degree of belief is significantly less than certainty.

Jason Stanley (2005) has emphasized the role of personal interests at stake in decisions to be based on it to inform how good the evidence has to be to support a claim that an agent knows a proposition in a given context.

I want to suggest that an important part of counting something as a scientific inference is a normative commitment (at least in these investigations of gravity and cosmology) that a decision to accept it should be able to be defended by appeal to empirical success. I have made a beginning of extending the Bayesian model to accommodate Newton's inferences in Harper 2007b.

I used Kyburg's (1990, 245–50) (risk/gain) ratio account of full belief to model Huygens's decision to reject Newton's application of Law 3. I modeled Huygens as guiding his decision by epistemic utilities. I modeled the initial decision to reject when Huygens was confronted with Newton's inference in proposition 7. I then modeled his decision to continue to reject after being confronted with the agreeing measurements of relative masses of sun and planets afforded by Newton's inference, and finally modeled his continuing to reject after being confronted with all the successes Newton was able to produce in the entire *Principia*. On the assignments of epistemic value I gave to the Huygens I modeled, he could rationally continue to reject Newton's inference, even after reading the whole *Principia*.

I then compared the same sequence of decisions evaluated according to Newton's rich conception of empirical success, rather than the epistemic utilities I had attributed to Huygens. On Newton's conception of empirical success, the agreeing measurements of relative masses of solar system bodies would be enough to make Huygens's alternative to fail to count as a serious rival.

Here again are the features we have cited that go beyond the basic H-D model to make Newton's scientific method so much more informative in his argument for universal gravity. One of these is the richer ideal of empirical success that is exemplified in Newton's classic inferences from phenomena.

> A Richer *Ideal* of Empirical Success:
> This requires not just accurate prediction of phenomena; but, in addition, accurate measurement of parameters by the predicted phenomena.

We have seen how the convergent accurate measurements of relative masses of solar system bodies strikingly realize this richer ideal of empirical success. The comparison with Huygens's vortex theory illustrated the stronger resources Newton's ideal of empirical success affords for preventing alternative hypotheses from undermining what are to be counted as propositions gathered from phenomena by induction.

Another feature that Newton's scientific method adds to the basic hypothetico-deductive model is its use of *theory-mediated* measurements.

> *Theory-mediated* measurements:
> Newton's methodology exploits, insofar as possible, *theory-mediated* measurements to give empirical answers to theoretical questions.

We have seen that this exploitation of *theory-mediated* measurements is explicitly endorsed by Newton's Rule 3 for doing natural philosophy. The agreeing measurements supporting equal ratios of weight toward planets to masses for all bodies at equal distance from them are an application that clearly counts as very high-quality evidence.

The third feature is acceptance of theoretical propositions as provisional guides to research.

> *Theory Acceptance*:
> Newton's methodology provisionally accepts theoretical propositions as guides to research.

We have seen that Newton's Rule 4 for doing natural philosophy articulates the nature of the acceptance his method accords to propositions inferred from phenomena.

2. *Successive approximations*

All three of these features, which go beyond the basic hypothetico-deductive model described by Huygens, come together in a method of successive approximations that informs applications of universal gravity to motions of solar system bodies. On this method, deviations from the model developed so far count as new *theory-mediated* phenomena carrying information to be exploited in developing a more accurate successor.

One striking feature of Newton's argument is that its conclusion, his theory of universal gravity, is actually incompatible with the Keplerian phenomena assumed as premises in his argument for it. The application of Law 3 to the sun and planets leads to two-body corrections to the Harmonic Rule. Newton expected that interactions between Jupiter and Saturn would sensibly disturb their orbital motions. The treatment of such deviations by Newton and his successors exemplifies the method of successive approximations that informs applications of his theory to motions of solar system bodies. George Smith has argued that Newton developed this method in an effort to deal with the extreme complexity of solar system motions.[7] Here again is the striking passage Smith has dubbed "the Copernican scholium."

By reason of the deviation of the Sun from the centre of gravity, the centripetal force does not always tend to that immobile centre, and hence the planets neither move exactly in ellipses nor revolve twice in the same orbit. There are as many orbits of a planet as it has revolutions, as in the motion of the Moon, and the orbit of any one planet depends on the combined motion of all the planets, not to mention the action of all these on each other. But to consider simultaneously all these causes of motion and to define these motions by exact laws admitting of easy calculation exceeds, if I am not mistaken, the force of any human mind. (Wilson 1989b, 253)

We have suggested that Newton's response to this daunting complexity problem was a method of successive approximations that continued to inform the work of his successors in developing their improved models for applying his theory to the motions of solar system bodies. At each stage, deviations from motion in accord with the model developed so far would count as *theory-mediated* phenomena to be exploited as carrying information about interactions to aid in developing a more accurate successor. These successive corrections led to increasingly precise specifications of perturbation-corrected solar system motion phenomena backed up by increasingly precise specifications of details of the interactions which explain them and increasingly precise measurements of the masses of the gravitationally interacting solar system bodies.[8]

3. Accumulating support

The eventual acceptance of Newton's theory was evidentially backed up by an extraordinary accumulation of successful resolutions of open problems left from Newton's original argument in the hands of such successors as d'Alembert, Clairaut, and Euler. It was also evidentially supported by the accuracy of the predictive models backed up by accurate measurements of parameters resulting from these developments, as well as other new successful applications of his theory of gravity. These long-term developments led to the entrenchment of his methodology as a guide to investigating nature. This was a pivotal development in the transformation of natural philosophy from the sort of investigation advocated by the mechanical philosophers into natural

[7] See Smith G.E. (2002a, 153–67). See chapter 1 section VII.3.

[8] See *GHA 2B* and a number of Curtis Wilson's papers, e.g. 1980, 1985, 1987, 1995. The volumes by Laplace in his *Celestial Mechanics* (1798–1825) and by Simon Newcomb (1895) give fairly detailed accounts of the states of the art that had been achieved by their respective times. See Smith G.E., forthcoming, for a wonderful account of the very high-quality evidence afforded by these developments.

science as we know it today. This transformation, in which Newton's *Principia* played such an important role, is what modernists called *the* scientific revolution.

It is important to note that all three features that make Newton's scientific method richer than the hypothetico-deductive model are exemplified in these developments. The provisional acceptance of models and the role of deviations as *theory-mediated* phenomena carrying information to be exploited in developing a more accurate successor model are quite obviously exemplified in these impressive applications of perturbation theory. The emphasis on measurement as a method of affording empirical answers to theoretical questions and the stronger criterion of empirical success are exemplified in the practice, even when practitioners describe the empirical success they achieve as accurate prediction.

One notable instance of an important researcher explicitly advocating Newton's emphasis on measuring causal parameters was Simon Newcomb. Where earlier researchers such as Le Verrier had used models with different assignments of relative masses in constructing their tables of the celestial motions for different planets, Newcomb made the assignment and empirical refinement of a single consistent set of relative masses for solar system bodies a central goal of his method for generating such tables for the *American Ephemeris*.[9] His efforts led to the dominance of the U.S. Naval Observatory as the most successful source for producing accurate tables representing the motions of solar system bodies.

It is important that the increased support for Newton's basic theory afforded by this success at developing successively more accurate models is not exhausted by the increases in epistemic probability that would correspond to successively updating by Bayesian conditioning on the increasingly precise sets of data that would be fit with increasing accuracy by the perturbation-corrected phenomena. George Smith has argued that these predictive empirical successes of the sort cited by Duhem[10] do not

[9] The tables of predicted motions are called "ephemerides." The following quotation is from the preface of Newcomb's 1895 volume outlining the work on which his new tables were based.

> The diversity of the adopted values of the elements and constants of astronomy is productive of inconvenience to all who are engaged in investigations based upon these quantities, and injurious to the precision and symmetry of much of our astronomical work. If any cases exist in which uniform and consistent values of all these quantities are embodied in an extended series of astronomical results, whether in the form of ephemerides or results of observations, they are the exception rather than the rule. The longer this diversity continues the greater the difficulties which astronomers of the future will meet in utilizing the work of our time.
>
> On taking charge of the work of preparing the *American Ephemeris* in 1877 the writer was so strongly impressed with the inconvenience arising from this source that he deemed it advisable to devote all the force which he could spare to the work of deriving improved values of the fundamental elements and embodying them in new tables of the celestial motions. (Newcomb 1895, III)

Salient among these fundamental elements are the relative masses of solar system bodies:

> Until this plan was mapped out, and work well in progress, it was not noticed that the planetary masses adopted in LEVERRIER's tables were so diverse that corrections to reduce the geocentric places to a uniform system of masses would be necessary. (Newcomb 1895, 7)

[10] See above chapter 3 section IV.3.

do justice to the far higher quality evidence afforded by the increasingly accurate specifications of the details of the perturbational interactions accounting for the corrections needed.[11] On Newton's Rule 4 these developments put increasing demands on what a successful rival would have to account for. Among these developments affording increased support for Newton's theory are the measurements of relative masses on which we have focused in our story.[12]

II Mercury's perihelion[13]

The famous Mercury perihelion problem, which contributed to the overturning of Newton's theory of gravity, illustrates the relevance of Newton's scientific method to the transition from his theory to Einstein's and the continuing relevance of this rich method of Newton's to the later development and application of testing frameworks for relativistic theories of gravity. It also affords an example of Newton's method at work in response to an earlier challenge from a proposal to alter the inverse-square law to account for the excess precession of Mercury's orbit.

1. *The classical Mercury perihelion problem*

We have seen that Newton's precession theorem affords measurements of the inverse-square variation of solar gravity from absence of precession for orbits of planets. Precession that can be accounted for by perturbation due to forces toward other bodies can be ignored in using stable apsides to measure inverse-square variation of the centripetal force toward the sun maintaining planets in their orbits. For each such planet, the zero leftover precession counts as an agreeing measurement of the inverse-square variation of gravitation toward the sun.

[11] See Smith G.E., forthcoming.

[12] Here are some estimates of the mass of Venus in units equal to (1/the mass of the sun), the reciprocal of the mass of the sun:

Table 10.1

Laplace (1802)	383,130 (Bowditch III, 179)
Later corrected value	405,871 (Bowditch III, 180)
Newcomb (1895)	404,924 (1895, 117)
Roy Edward Laubscher (who did the last fully Newtonian calculation, from observing the perturbations of Venus on Mars) (1981)	408,735 ± 775 (Laubscher 1981, 392)
Current estimate (2009)	408,565 (NASA's "Venus Fact Sheet" and "Sun Fact Sheet" on NASA's website (standard errors not listed))

Laubscher's estimate is from a perturbation of Mars by Venus. This was calculated before any orbiters had been sent to Venus. We can see how it agrees with the last estimate, which has the advantage of direct measurement from an orbiting satellite. This illustrates that perturbations were able to provide quite good estimates.

[13] I have published earlier treatments of these issues in Harper 2007a and Harper 2009.

In 1859 Le Verrier found 38 arc seconds/century of Mercury's precession to be not accounted for by Newtonian perturbations.[14] In 1882 Newcomb gave a revised corrected value of 43 arc seconds/century.[15] This extra perihelion precession was a departure from the developing Newtonian model that resisted attempts to account for it by interactions with known bodies.

In 1894 Asaph Hall, appealing to a formula of Bertrand's which is equivalent to Newton's precession theorem,[16] proposed to account for the extra 43 arc seconds/century by revising the inverse-square to the -2.00000016 power that would be measured by it.[17]

THE

ASTRONOMICAL JOURNAL.

No. 319.

VOL. XIV. BOSTON, 1894 JUNE 2. NO. 7.

A SUGGESTION IN THE THEORY OF *MERCURY*,

By A. HALL.

Applying Bertrand's formula to the case of *Mercury* I find, taking Newcomb's value of the motion, or 43″, that the perihelion would move as the observations indicate by taking

$$n = -2.000\,000\,16$$

Figure 10.1 Hall's suggestion

In 1895 Newcomb, by then the major doyen of predictive astronomy, cited Hall's proposal as "provisionally not inadmissible" after rejecting all the other accounts he had considered.[18] He also introduced precession in accord with Hall's hypothesis into his predictive theory.[19]

[14] See Roseveare 1982, 20–3.

[15] Newcomb 1882, 473. One factor contributing to Newcomb's more accurate calculation is that he was able to take advantage of a more reliable estimate of the mass of Mars afforded by the discovery of its moons. (1882, 469)

[16] Valluri et al. 1997, 13–27.

[17] Hall 1894, 49–51. Here is the calculation from Newton's formula: According to *ESAA*, 704 the period of Mercury is 0.24084445 Julian years. This gives $(1/.2408445)(100) = 415.205748$ revolutions per Julian century and $43/415.205748 = 0.10356$ sec/rev or $0.10356/60^2 = 0.0000288$ degrees per revolution. On Newton's formula, the corresponding power law for the centripetal force is as the $(360/360.0000288)^2 - 3 = -2.00000016$ power of distance.

[18] Newcomb 1895, 110–21.

[19] Here is what he said:

> What I finally decided on doing was to increase the theoretical motion of each perihelion by the same fraction of the mean motion, a course which will represent the observations without committing us to any hypothesis as to the cause of the excess of motion, though it accords with the result of Hall's hypothesis of the law of gravitation; (Newcomb 1895, 174)

In 1903 Ernest Brown refined the Hill-Brown theory of the lunar orbit with sufficient precision to rule out Hall's hypothesis by empirically constraining the absolute values of departures δ from the inverse-square to less than 0.00000004.[20]

Mon. Not. r. astr. Soc. **62,**

On the Verification of the Newtonian Law.
By Ernest W. Brown, F.R.S.

If the new theoretical values of the motions of the Moon's perigee and node are correct, the greatest difference between theory and observation is only o″·3, making δ<·00000004. Such a value for δ is quite insufficient to explain the outstanding deviation in the motion of the perihelion of *Mercury*. It appears, then, that this assumption must be abandoned for the present, or replaced by some other law of variation which will not violate the conditions existing at the distance of the Moon.

Haverford College, Pa., U.S.A.:
1903 *May* 2.

Figure 10.2 Brown's new result

This is an example of all three of the features that make Newton's method richer than merely hypothetico-deductive method. It is a striking realization of Newton's richer ideal of empirical success, as accurate measurement of parameters by phenomena. The more precise limitations on the motion of the perigee and node of the earth's moon afford a new measurement precise enough to fix the inverse-square out to least seven decimal places. It successfully turns a theoretical question into one that can be answered empirically by measurement from phenomena, and is one where the outcome of that measurement is accepted as a guide to further research. It is the more precise measurement of the −2 power provided by Brown that rules out Hall's hypothesis as a solution to the Mercury perihelion problem.

2. Einstein's solution as support for General Relativity: an answer to Kuhn's challenge on criteria across revolutions

Thomas Kuhn counted the radical transformation from Newton's theory to Einstein's as an example of the sort of transition from one paradigm to another that he counted as a scientific revolution.[21] The following quotation challenges the very idea that there

[20] Brown 1903, 396–7.

[21] In the following quotation, Kuhn expanded upon the need to change familiar concepts of space, time, and mass in the transition from Newton's theory to Einstein's.

> This need to change the meaning of established and familiar concepts is central to the revolutionary impact of Einstein's theory. Though subtler than the changes from geocentrism to heliocentrism, from phlogiston to oxygen, or from corpuscles to waves, the resulting conceptual transformation is no less decisively destructive of a previously established paradigm. We may even come to see it as a prototype for revolutionary reorientations in the sciences. Just because it did not involve the introduction of additional objects or concepts, the transition from Newtonian to Einsteinian mechanics illustrates with particular clarity the scientific revolution as a displacement of the conceptual network through which scientists view the world. (Kuhn 1970, 102)

can be non-question-begging standards of theory assessment that apply across scientific revolutions.

> Like the choice between competing political institutions, that between competing paradigms proves to be a choice between incompatible modes of community life. Because it has that character, the choice is not and cannot be determined merely by the evaluative procedures characteristic of normal science, for these depend in part upon a particular paradigm, and that paradigm is at issue. When paradigms enter, as they must, into a debate about paradigm choice, their role is necessarily circular. Each group uses its own paradigm to argue in that paradigm's defense. (Kuhn 1970, 94)

Contrary to the skeptical implication strongly suggested by this quotation from Kuhn's famous book,[22] the revolutionary change from his theory to General Relativity is in accordance with the evaluative procedures of Newton's own methodology.

The successful account of this extra precession,[23] together with the Newtonian limit which allows it to recover the empirical successes of Newtonian perturbation theory, makes General Relativity do better than Newton's theory on Newton's own criterion of empirical success. Einstein's extreme excitement over this result,

> The first result was that his theory 'explains . . . quantitatively . . . the secular rotation of the orbit of Mercury, discovered by Le Verrier, . . . without the need of any special hypotheses.' This discovery was, I believe, by far the strongest emotional experience in Einstein's scientific life, perhaps in all his life. Nature had spoken to him. He had to be right. 'For a few days, I was beside myself with joyous excitement'. Later, he told Fokker that his discovery had given him palpitations of the heart. What he told de Haas is even more profoundly significant: when he saw that his calculations agreed with the unexplained astronomical observations, he had the feeling that something actually snapped in him . . . (Pais 1982, 253),

is appropriate to the fact that with this result he had grounds for confidence that his theory would beat Newton's theory on Newton's own standard of empirical success. He had good grounds for counting his theory as better than Newton's without any question begging appeal to some new standard that would favor it.

One essential part of this is that he knew the Newtonian limit of General Relativity could recover the 531 seconds per century of Mercury's precession that had been successfully accounted for by Newtonian perturbations. Without this recovery of the

[22] There is now a veritable industry of Kuhn interpretation. It may well be that one could offer interpretations of Kuhn's incommensurability thesis that would avoid the skepticism about non-question-begging criteria that apply across scientific revolutions. That Newton's own scientific method would endorse the transition from his theory to Einstein's shows that the radical skepticism about non-question-begging standards suggested by the quoted passage from Kuhn does not apply to that transition. It does not show that there are not important and illuminating contributions to understanding the practice of science in Kuhn's very influential work on scientific revolutions.

Indeed, we shall quote below a passage from Kuhn that illuminates a conception of progress through successively more accurate theories without commitment to Laplace's ideal limit of a final theory of everything. We shall argue that such a conception is endorsed by Newton's scientific method.

[23] Einstein ([1915] 1979). See Earman et al. 1993. See also Roseveare 1982.

Newtonian perturbations, his accounting for those 43 seconds would not have counted as a solution of the Mercury precession problem. The Newtonian limit recovers all the empirical successes of Newton's theory, including all the agreeing measurements of parameters such as the relative masses of the sun and planets.[24] In the years since 1915 many post-Newtonian corrections to solar system motion phenomena have afforded additional empirical successes that clearly favor Einstein's theory over Newton's without relying on any question begging appeal to new standards.[25]

3. *The Dicke-Goldenberg challenge to General Relativity and Shapiro's radar time delay measurement*

In 1966 Robert Dicke and H. Mark Goldenberg carried out measurements of solar oblateness which suggested a rapidly rotating inner core of the sun which would generate a quadrupole moment that would account for about 4 of the extra 43 seconds of Mercury's perihelion precession.[26] This would allow an alternative relativistic gravitation theory, the Brans-Dicke theory, to do better than General Relativity. A key difference between General Relativity and the alternative Brans-Dicke theory can be represented in the parametrized post-Newtonian (PPN) formalism for comparing alternative metrical gravity theories.[27] In the Brans-Dicke theory, the PPN parameter γ representing the amount of space curvature per unit rest mass is given by $\gamma = ((1+\omega)/(2+\omega))$, where ω is an extra parameter representing contributions of distant stars to

[24] Kuhn argues (1970, 101–3) that the Newtonian limit results in general relativity don't show that Newton's theory remains approximately true. He claims that in the Newtonian limit of general relativity, the terms for space, time, and mass which satisfy to good approximation the Newtonian equations are still representing parameters of general relativity, not those of Newton's theory.

In Harper (1997, 68–9), I pointed out that granting this point of Kuhn's would just make it even more obvious that the Newtonian limit results make Einstein's theory take over for its parameters the empirical successes of Newton's theory. On Kuhn's own assumption, the phenomena which accurately measured parameters of Newton's theory are now accurately measuring parameters of Einstein's theory; thereby undercutting his claim that it required new standards, different from Newton's, to count Einstein's theory as doing better.

Kuhn gives as his basis for claiming that corresponding symbols in the two theories cannot refer to the same parameter that the theories make incompatible claims about the parameter that would be referred to. This hardly appears to be sufficient basis on which to rule out co-reference of theoretical terms. There are, of course, the well-known accounts by Kripke (1972) and Putnam (1975) which do allow this sort of reference Kuhn finds problematic. It may also be of interest to note that Carnap, reacting to difficulties David Kaplan raised for his account of testability and meaning, proposed Hilbert's ε - calculus, a formalism which also allows incompatible claims to refer to the same theoretical object, as a formalism for expressing references of scientific terms. (See Kaplan 1971, xlv–xlvi).

[25] In addition to the famous three basic tests there are now a great many post-Newtonian corrections required by the more precise data made available by such new observations as radar ranging to planets and laser ranging to the moon. These provide not just predictions but also measurements of parameters, such as those of the PPN testing framework, which support General Relativity. See Will 1993, 332–43 and Will 2006.

[26] See Will (1993, 181–2); Dicke and Goldenberg (1967, 313; 1974, 131–82). See also Dicke (1974, 419–29).

[27] Work by Kenneth Nordtvedt and Clifford Will (Nordtvedt 1968a; Will 1971; Will and Nordtvedt 1972) led to the development of the parametrized post-Newtonian (PPN) formalism. The assumptions of the PPN formalism are approximations detailed enough to make solar system phenomena measure parameters on which alternative metrical theories of gravity can differ. See Will 1993, 86–99.

local space curvature.[28] In General Relativity the value of γ is fixed so that $\gamma=1$. This makes General Relativity unable to accommodate alternatives to the 43 seconds extra precession for Mercury. In contrast, setting its parameter ω at 5 would let the Brans-Dicke theory account for only 39 seconds of Mercury's perihelion precession.[29] Ironically, Mercury's perihelion, the first big success of General Relativity, now threatened to be its undoing.

In 1964 Irwin Shapiro proposed radar time delay as a test of General Relativity.[30] On a relativistic gravity theory there is a round-trip time delay for radar ranging to planets due to the gravitational potential of the sun along the path of the radiation, when that path passes close to the sun.[31] This round-trip time delay measures γ, the parameter representing space curvature at issue above.[32] By 1979 such measurements were precise enough to rule out any versions of the Brans-Dicke theory with ω much below 500.[33] It was some time later, perhaps into the late 1980s, before further investigation established that effects of solar rotation were not great enough to undercut General Relativity's account of Mercury's perihelion motion.[34]

As in the classical case, a more precise measurement of a parameter ruled out a proposed alternative to the accepted gravitational theory as a solution to the Mercury perihelion precession problem before an adequate solution was found. In the null experiment generated by Brown's lunar theory, the bound on lunar precession not accounted for by perturbation counts as a phenomenon measuring bounds of $\delta < 0.00000004$ for divergences from the -2 power for gravity toward the earth at the lunar distance. This counts as a measurement establishing the -2 power to that precision for gravity generally. It, thereby, rules out Hall's proposed alternative power law of -2.00000016 as a solution to the excess Mercury precession problem.

In the positive detection of Shapiro's time delay, the exhibited pattern counts as a phenomenon.

The quantity Δt is not an observable but is indicative of the magnitude and behavior of the measurable effect as predicted by general relativity. (Shapiro et al. 1971, 1132)

[28] The Brans–Dicke theory differs from General Relativity in postulating an extra scalar field that represents the contribution of distant masses to local curvature. The parameter ω represents this extra scalar field. See Brans and Dicke 1961 reprinted as appendix 7, 77–96 in Dicke 1965.

[29] The result for the precession of Mercury for the Brans–Dicke theory is

$$[(4+3\omega)/(6+3\omega)] \times (\text{value of GR})$$

See Dicke 1965, 88. For $\omega = 5$ this yields a precession of about 38.9 seconds per century. See also Dicke 1974.

[30] Shapiro 1964.

[31] See Shapiro op. cit. or Shapiro et al. 1971.

[32] Shapiro's round-trip time delay $\Delta\tau$ for radar ranging to planets measures γ:

$$\Delta\tau = (2r_0/c)((1+\gamma)/2)\ln\,((r_e + r_p + R)/(r_e + r_p - R))$$

where $r_0 = 2GM/c^2$, M is the solar mass, r_e and r_p are respectively the distances of the earth and the target planet from the sun, and R is the distance of the target planet from the earth. See Reasenberg et al. 1979.

[33] See Reasenberg et al. 1979, L221.

[34] See Will 1993, 334.

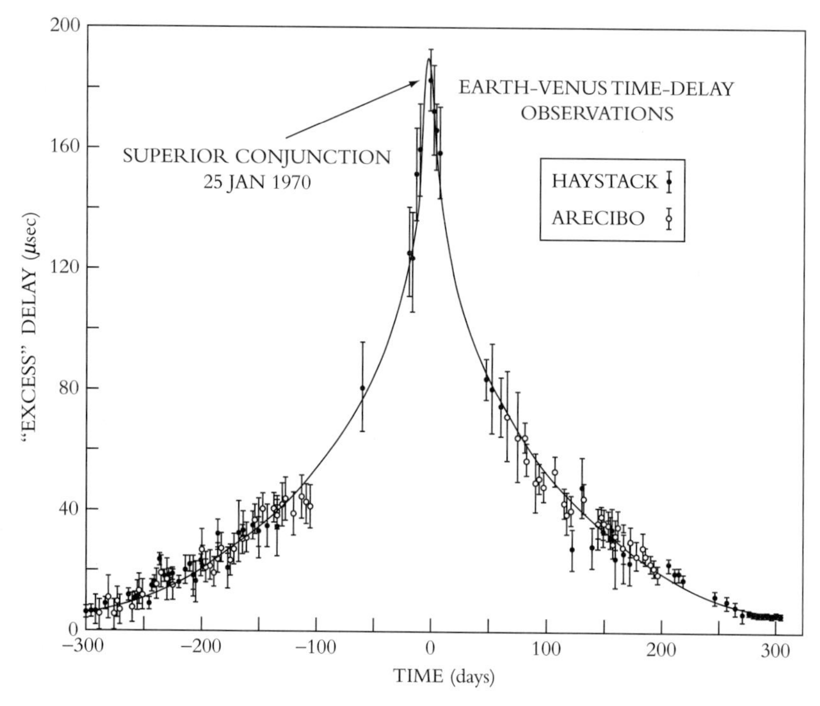

Figure 10.3 Radar time delay[35]

In the 1979 Viking experiment with Mars, such a time delay phenomenon was established with sufficient precision to measure $\gamma = 1 \pm 0.002$.[36] It was these measurements of γ that empirically ruled out any version of the Brans-Dicke theory that set its parameter ω much below 500.

The development and applications of testing frameworks for relativistic gravitation theories is very much an illustration of Newton's method. The many successful tests of General Relativity measure parameters that constrain alternative theories to approximate General Relativity for scales and field strengths similar to those explored by the phenomena exhibited in those tests.[37]

4. Some conclusions from the Mercury perihelion problem

Newton's scientific method was able to deal with the daunting complexity of solar system motions by successive corrections to the phenomena in which discrepancies from motion in accord with the model developed so far counted as *theory-mediated* phenomena carrying information to be exploited. Our review of the Mercury perihelion problem suggests that Newton's scientific method also informs the radical

[35] Reprinted from Shapiro et al. 1971, 1132, figure 124.1.

[36] See Reasenberg et al. 1979.

[37] See Will 1993; Will 2006.

theoretical transformation to Einstein's theory and continues to inform the development and application of solar-system testing frameworks for relativistic theories of gravitation today. All three of the features that make Newton's scientific method richer than the usual hypothetico-deductive model are evident in our story. The richer notion of empirical success as accurate measurement of parameters by phenomena, rather than just prediction alone, is achieved in Shapiro's measurement of the space-time curvature parameter γ, just as it is exhibited in Brown's measurement limiting deviations from the inverse-square. The exploitation of *theory-mediated* measurements from phenomena to give empirical answers to theoretical questions is the heart of the enterprise of developing and applying testing frameworks for relativistic gravitation theories. Finally, the provisional acceptance of theory as a guide to further research is evident both in the role of the background assumptions of the testing frameworks and in the application of outcomes of measurements as constraints that have to be recovered by any viable alternative theory.

In our postscript we shall see General Relativity play a role in guiding research in cosmology today, which is very much analogous to the role Newton's theory of gravity played in guiding research into the structure and details of solar system motions.

III Our Newton vs. Laplace's Newton

We have noted that Laplace followed up his successful treatment of the, long recalcitrant, great inequality in Jupiter-Saturn motions as a periodic perturbation with influential arguments for the stability of the solar system.[38] After the publication of his monumental treatises on celestial mechanics, there came to be very general acceptance of the solar system as what was taken to be a Newtonian metaphysics of a clockwork deterministic system of bodies interacting under forces according to laws. The successively more accurate approximations were seen as successively better approximations to an exact characterization of the forces and motions of the bodies in a stable solar system in which all perturbations were periodic. The following famous quotation is focused on the determinism that would characterize such a system.

> We ought then to regard the present state of the universe as the effect of its anterior state and as the cause of the one which is to follow. Given for one instant an intelligence which could comprehend all the forces by which nature is animated and the situation of the beings who compose it – an intelligence sufficiently vast to submit these data to analysis – it would embrace in the same formula the movements of the greatest bodies of the universe and those of the lightest atom; for it, nothing would be uncertain and the future, as the past, would be present to its eyes. The human mind offers, in the perfection which it has been able to give to astronomy, a feeble idea of this intelligence. Its discoveries in mechanics and geometry, added to that of universal gravity, have enabled it to comprehend in the same analytical expressions the past and future states of the system of the world. (Laplace [1814]; translation by Truscott and Emory [1902] 1995, 4)

[38] See note 100 from chapter 1 section VI.

Figure 10.4 Blake's depiction of Newton.
Here, Newton appears to convey a belief that his theory, represented by the diagram on which his attention is so sharply focused, gives him something approaching a God-like access to the whole world from the outside.

The depiction of Newton in William Blake's famous 1795 painting expresses what I think Blake regarded as the hubris of this Laplacian ideal about the prospects of science.

I want to suggest that Laplace's clockwork ideal of what is to be counted as a Newtonian system is one truer to himself than to Newton. Newton actually suggested that on his theory of gravitation one would expect non-periodic perturbations sufficient to threaten the stability of the solar system.[39] He also conjectured that this could provide an argument for the existence of God, who would be needed to intervene to maintain the solar system.[40]

We don't need Newton's appeal to stability threatening perturbations and God to have reason to doubt that progress ought to be construed as progress toward a Laplacian final theory of everything.[41] Robert Batterman has impressively argued that the classic

[39] This is suggested by the italicized lines in the following passage from query 31 of Newton's *Opticks*:

> For while Comets move in very eccentrick Orbs in all manner of Positions, blind Fate could never make all the Planets move one and the same way in Orbs concentrick, *some inconsiderable Irregularities excepted, which may have risen from the mutual Actions of Comets and Planets upon one another, and which will be apt to increase, till this System wants a Reformation.* Such a wonderful Uniformity in the Planetary System must be allowed the Effect of Choice. (Newton [1730] 1979, 402)

Italics added.

[40] Leibniz certainly took this interpretation in the first paper published in the Leibniz–Clark correspondence. See Alexander 1956, 11–12. Samuel Clark (1675–1729), Newton's defender who was in close touch with Newton, did not disavow Newton's advocacy of this argument. Instead, he defended the argument against Leibniz's objections to it. Ibid., 13–14.

[41] Indeed, Laplace's stability result and its impressive extension by Poisson showed that Newton's theory does not support short-term prospects for stability-threatening perturbations. See chapter 1 section VII.4 note 100.

case of theory reduction – the reduction of thermodynamics to statistical mechanics – requires appeal to singularities formulated in thermodynamics, the theory that was supposed to be reduced to the more fundamental theory.[42] Mark Wilson has provided a wide range of additional compelling examples where such reductions to fundamental theory are not informative and need not be possible.[43]

At a public debate on the interpretation of quantum mechanics Lucian Hardy used the foregoing Blake depiction of Newton as an expression of the confidence he had at one time maintained about the prospects for what science could, in principle, achieve.[44] He used a version of the following more modest depiction of Newton to express the more limited view about the prospects for science that he had been driven to by the problems raised by quantum mechanics.

Figure 10.5 Newton investigating light

[42] See, e.g., Batterman 2002 and 2005.

[43] Wilson, M. 2006.

[44] Lucian Hardy is a physicist at the Perimeter Institute for Theoretical Physics. This very well-attended debate was sponsored by the Perimeter Institute as its May Public Event on May 5, 2004 at the Waterloo Collegiate Institute in Waterloo, Ontario.

This second, more modest, depiction is of Newton investigating light and colors experimentally. I am suggesting that the correspondingly more modest view of the prospects for scientific progress is one that fits Newton's own view of his theory of gravity, as well as of his investigations of light and colors. The Newton that Blake reacted to is the one that many philosophers of science will recognize in their characterization of ideal Newtonian systems. This, however, is more a creature true to Laplace's aspirations than to the more modest views of the author of the *Principia* about the prospects for science.

In his very influential book, Kuhn argued that understanding scientific revolutions undercuts the excessively progressive view built into what has been the textbook tradition for teaching the history of science to science students. There has been an unfortunate tendency on the part of many historians and philosophers of science to regard Kuhn as having undercut the whole idea that science yields progressively better understanding of nature. As I see it, Newton's method of successively better approximations exemplifies the following characterization of scientific progress by Kuhn:

> The developmental process described in this essay has been a process of evolution *from* primitive beginnings – a process whose successive stages are characterized by an increasingly detailed and refined understanding of nature. But nothing that has been or will be said makes it a process of evolution *toward* anything (Kuhn 1970, 170–1)

This is an optimistic characterization of scientific progress. As Kuhn points out, one does not need to postulate a Laplacian final theory in order to be able to claim that science is delivering successively more accurate models that afford an increasingly detailed and refined understanding of nature.[45]

[45] I have been struck by the deep emotional satisfaction many people have taken toward the idea that Kuhn has shown that the pretensions of natural science as progress toward a Laplacian final theory are not warranted. It appears to me that Kuhn's discussion of scientific revolutions has tapped into an emotionally charged, deep-seated, resentment of having had to give the application of mathematics to the world in the exact sciences privileged status as the high road to Knowledge. On this resented scientific image, Eddington's table, which is said to be mostly empty space between fundamental particles, shows that the everyday conception of the table as a solid body is wrong. We have seen that Newton's more modest conception of scientific progress does without Laplace's conception of progress as progress toward a final theory of everything.

When I was visiting at Pittsburgh in 1974–5, I met with Wilfred Sellars every other week or so to discuss Kant on intuition and geometry. One thing we discussed was Kant's appeal to geometry to underwrite empirical knowledge of shaped bodies located with respect to human observers in three-dimensional public space. I later came to argue that this important part of Kant's appeal to Euclid is not undercut by the non-Euclidean space required by relativistic space-time, because all it required was that three-dimensional Euclidian geometry held at the scales and tolerances needed for our everyday interactions with the bodies we see, touch, and manipulate (Harper 1984; Harper 1996). I now want to suggest that our everyday conception of the table as solid makes no commitments about its features at the scales corresponding to the sizes of fundamental particles. The conception of improved knowledge of the world that goes with success according to Newton's scientific method does not undermine any legitimate applications of the conceptions with which we lead and find meaning in our everyday lives.

IV Approximations and Laudan's confutation of convergent realism

We have seen that central to Newton's method of piecemeal, successively more accurate, approximations is his exploitation of systematic dependencies that make phenomena measure parameters that explain them, independently of any deeper explanation accounting for these parameters and dependencies. These lower level dependency-based explanations are robust not only with respect to approximations in the phenomena. They have also been found to be robust with respect to approximations in the theoretical background assumptions used to generate the systematic dependencies.[46] To realize Newton's ideal, a successful alternative theory needs to recover correspondingly successful or improved convergent agreeing measurements of its parameters as well as improve on the accuracy and scope of the phenomena explained by its predecessor.[47] I want to suggest that the large-scale cumulative improvement in accuracy of phenomena and the systematic dependencies that explain them at this lower level is not undermined by Larry Laudan's confutation of convergent realism.

Laudan challenges realists to get clearer about the nature of the success they are trying to explain.

> The first and toughest nut to crack involves getting clearer about the nature of that 'success' which realists are concerned to explain. Although Putnam, Sellars and Boyd all take the success of certain sciences as a given, they say little about what this success amounts to. So far as I can see, they are working with a largely *pragmatic* notion to be cashed out in terms of a theory's workability or applicability. On this account, we would say that a theory is successful if it makes substantially correct predictions, if it leads to efficacious interventions in the natural order, if it passes a battery of standard tests. One would like to be able to be more specific about what success amounts to, but the lack of a coherent theory of confirmation makes further specificity very difficult. (Laudan 1981, 23)

[46] George Smith has argued in detail that the Newtonian limit in Einstein's theory preserves these dependencies through which Newtonian theory identified specific orbital configurational features, like those explaining the Jupiter–Saturn 900-year period mutual interaction. These features include the masses of the bodies involved. The subjuctive conditionals corresponding to the effects made by these features are preserved by General Relativity. (Smith, forthcoming)

[47] Perhaps understanding Newton's method can contribute insight to an improved understanding of how there can be progress through scientific revolutions. In his paper titled "Objectivity, Value Judgment and Theory Choice," Kuhn articulates five, relatively familiar, criteria for what are to be counted as good theories. The first of these is accuracy.

> First, a theory should be accurate: within its domain, that is, consequences deducible from a theory should be in demonstrated agreement with the results of existing experiments and observations. (Kuhn 1977, 320)

Our investigation of Newton's argument suggests that an ideal of accuracy should be extended to include not just accurate prediction of the phenomena to be explained but also accurate measurement by those phenomena of the parameters which explain them.

Newton's stronger conception of empirical success clearly adds significant content to what Laudan has described as the largely pragmatic characteristics he has cited. Except for his citing of efficacious intervention, his criteria correspond to the hypothetico-deductive model of success. Newton's method of provisional theory acceptance in an enterprise of seeking deviations to be exploited as carrying information to guide construction of a more accurate successor model is very much richer than the sort of hypothetico-deductive model of scientific inference Laudan considers.[48]

Laudan also criticizes appeal to approximations. He points out that realists have not provided any philosophically satisfactory analysis of approximate truth; but, his main criticism is based on a connection between approximate truth and reference that he claims realists must accept. On this basis, he then argues that explanatory success cannot be taken as a rational warrant for approximate truth.

> To see why, we need to explore briefly one of the connections between 'genuinely referring' and being 'approximately true'. However the latter is understood, I take it that *a realist would never want to say that a theory was approximately true if its central theoretical terms failed to refer.* If there were nothing like genes, then a genetic theory, no matter how well confirmed it was, would not be approximately true. If there were no entities similar to atoms, no atomic theory could be approximately true; if there were no sub atomic particles, then no quantum theory of chemistry could be approximately true. In short, a necessary condition – especially for a scientific realist – for a theory being close to the truth is that its central explanatory terms genuinely refer. (An *instrumentalist*, of course, could countenance the weaker claim that a theory was approximately true so *long* as its directly testable consequences were close to the observable values. But as I argued above, the realist must take claims about approximate truth to refer alike to the observable and the deep-structural dimensions of a theory.) (Laudan 1981, 33)

The remark in parentheses suggests a fairly sharp dichotomy between the theory and its observable consequences. Newton's richer notion of empirical success allows for, and invites, an account of approximate truth based on a combination of prediction together with measurement of parameters. As we have seen, Newton's ideal of empirical success as agreeing measurements from diverse phenomena requires more than accurate prediction of observable values of directly testable consequences. Its empirical support for inferences, like that in the moon-test to identify the centripetal force maintaining the moon in its orbit with terrestrial gravity, exhibits inferences to common causes that appear to go beyond the more limited commitments of constructive empiricism. On the other hand, Newton's method of piecemeal successively more accurate approximations exploits systematic dependencies that make phenomena measure parameters that explain them, independently of any deeper explanation accounting for these parameters and dependencies. This suggests that Newton's scientific method supports a conception of progress through theory change as a middle way between

[48] Laudan's remark about the lack of a coherent theory of confirmation suggests that he does not himself endorse the hypothetico-deductive model. Laudan is criticizing arguments to defend scientific realism by appeal to a second order hypothetico-deductive explanation of the success of science.

structural empiricism and the commitments Laudan refers to as "the deep-structural dimensions of a theory." We have seen that this Newtonian conception of scientific progress is independent of any commitment to count it as progress toward Laplace's ideal limit of a final theory of everything.

Consider Newton's transition from gravity as inverse-square acceleration fields toward planets to gravity as a universal force of pair-wise interaction between bodies. On Newton's later theory of universal gravity, the inverse-square acceleration fields are only approximations. They ignore the contribution of the mass of the attracted body. They are quite good approximations for bodies small enough to be treated as test bodies with respect to a given planet and close enough to it that attractions toward other solar system bodies can be ignored. Newton's theory of universal gravity explains his initial inverse-square acceleration fields as approximations generated by these spherically symmetrical mass distributions. The spherically symmetric mass distribution is the cause of the inverse-square acceleration components directed toward its center. Universal gravity shows that ratios of the strengths of such acceleration fields measure the ratios of the total masses of such central bodies. We have seen that the center of mass frame for solar system bodies will assign to any body acceleration components toward each of the other bodies that are directly proportional to their masses and inversely proportional to the squares of its distances from them.[49]

Now let us return to Laudan's main claim that a necessary condition for a theory being close to the truth is that its central explanatory terms genuinely refer. I doubt that one need be committed to Laudan's claim here. One might very well say that the gravities construed as inverse-square acceleration fields toward planets don't really exist. The appeal to Newton's richer ideal of empirical success to justify the transition to universal gravity need not require continued acceptance of the actual existence of these inverse-square acceleration fields of gravity toward planets. The sense in which they are treated as approximations could be as merely useful idealizations that are kicked away, like a ladder that is no longer needed, once one has climbed to the full theory of universal gravity.

We have noted that the practice of acceptance characterized by Newton's Rule 4 does not endorse or require any specification of necessary conditions for counting a proposition as gathered from phenomena by induction. The examples cited by Laudan appear to be compatible with counting the superseded theories as informative approximations.[50] I am quite confident that Galileo's resolution of motions into orthogonal components would allow his circular inertia to be recovered as an informative approximation by the more general theory of what we call Galilean transformations. The historian of chemistry Mel Usselman gave a presentation to our philosophy of science group at Western at which he compellingly argued that the phlogiston theory of chemistry contributed significant information that was taken up and added to by the

[49] See chapter 8 section I.3. [50] See Laudan (1981, 33).

oxygen theory. One would need to look at the details of the transitions from these cited superseded theories; but, I expect that each can be counted as an informative approximation by its successor theory.

Let us now consider the transition from universal gravity to General Relativity. Einstein's solution to the Mercury perihelion precession problem recovers the extra precession for geodesic motion of a test body at Mercury's distance from a spherically symmetric mass-energy distribution having the rest-mass of our sun. In the corresponding Schwarzschild solution of the Einstein field equation, motions in accord with the basic Kepler orbits of the other planets (with smaller but non-negligible precessions for Venus, the earth, and Mars) are recovered to accurate approximations by geodesic motions in the curved-space four-dimensional space-time manifold corresponding to that spherically symmetric mass distribution.[51] The inverse-square centripetal accelerations are recovered to good approximations for motion relative to the sun. In the model for the full solar system, however, geodesic motion for a body at an arbitrary location is represented by relativistic corrections added to the *n*-body point mass formula representing the Newtonian acceleration components toward the solar system bodies in a reference frame fixed by the center of mass of the solar system.[52] These calculations are not carried out in the metric corresponding to an *n*-body solution to Einstein's field equation. So far no one has been able to produce such an *n*-body solution for Einstein's field equation.

I remember a conversation in which David Malament expressed his disappointment at this failure of the transition to General Relativity to fully supersede appeal to Newtonian calculations in applications to motions of solar system bodies. Unlike the suggestion implied by Kuhn's remark that the recovery of calculations of Newton's theory as a special case shows why "Newton's Laws ever seemed to work,"[53] the appeal to Newtonian calculations in applications of General Relativity to solar system

[51] See, e.g., Ohanian and Ruffini (1994, 391–7; 401–8).

[52] See *ESAA*, 281.

[53] Here is the passage from Kuhn:

> Our argument has, of course, explained why Newton's Laws ever seemed to work. In doing so it has justified, say, an automobile driver in acting as though he lived in a Newtonian universe. An argument of the same type is used to justify teaching earth-centered astronomy to surveyors. But the argument has still not done what it purported to do. It has not, that is, shown Newton's Laws to be a limiting case of Einstein's. For in the passage to the limit it is not only the forms of the laws that have changed. Simultaneously we have had to alter the fundamental structural elements of which the universe to which they apply is composed. (Kuhn 1970, 102)

The fact that no one yet knows how to replace the Newtonian calculations in applications of General Relativity to solar system motions by calculations carried out in the framework of Einstein's theory makes it different from the case for teaching earth-centered astronomy to surveyors.

In discussing the application of the spherical mass solution to Einstein's field equation we pointed out that it was the rest-mass of the sun that curved the space around it to make an appropriately precessing orbit at Mercury's distance count as geodesic motion. The need to specify rest-mass is an example of the change in the concept of mass between Newton's theory and Einstein's that Kuhn is appealing to. On Putman's account, the reference of mass for Newton is the same parameter referred to in General Relativity. Newton's theory is wrong about some of its features.

motions cannot yet be replaced by calculations carried out in the framework of Einstein's theory. This makes it harder to regard Newton's theory as a ladder that can be thrown away once one has used it to arrive at Einstein's theory.

The important point is that Laudan's inductive skeptical worry is irrelevant to Newton's method of successive approximations. George Smith has suggested that the appropriate skeptical worry for this enterprise is what he has called the garden path problem.[54] A sequence of revisions using deviations from models to afford information to guide construction of more accurate successor models is a garden path if at some stage a deviation results in having to give up the conclusions reached by research predicated on the propositions that had been accepted at earlier stages. Smith has pointed out that Newton's investigation of resistance forces in book 2 failed to realize any very significant sequence of successively more accurate revisions.[55] We certainly have cases where accepting a theory that was later superseded by a more successful successor theory was seen to have prevented exploring what turned out to be more promising alternative investigations. The incorrect initial measurements of the age of the universe from Hubble's red shift observations conflicted with estimates of the age of the earth.[56] There may not be many clear cases of any very extended sequences of successively more accurate revisions that have turned out to be garden paths.

On the other hand, there appears to be no compelling general argument showing that our present research is not a garden path. An important part of Newton's method is that it does not rely on any such argument. The skeptical hypothesis that we are on a garden path carries no weight until it is backed up by an alternative that realizes Newton's ideal of empirical success sufficiently well to offset the successes that have been reached by that path, or delivers phenomena that show that the propositions that have been accepted are liable to exceptions sufficient to undermine that whole sequence of inferences from phenomena. I want to suggest that Newton's scientific method appears to be the way we would ever find out that we have been on a garden

Newton's scientific method does not require such commitment to continued existence of the same fundamental theoretical parameters to count Einstein's theory as more accurate than his own theory of universal gravity. As we have seen, the important connections have been the systematic dependencies affording measurements from orbital phenomena of the mass of the appropriately spherically distributed mass of our sun in Newton's theory and the appropriately corresponding systematic dependencies that make those phenomena measure what Einstein's theory counted as the rest-mass of the appropriately spherically symmetric mass-energy distribution of our sun. These agreeing measurements allowed the recovery of the extra precession of Mercury as geodesic motion at its distance in general relativity, together with the appropriate Newtonian limit, to make Einstein's theory more accurate than Newton's.

I want to remark that, though the account of progress through theory change exemplified by Newton's scientific method does not require this, it also does not rule out the sort of acceptance of acceleration fields and of Newtonian spherical mass distributions as approximations that Putnam's account of scientific reference would endorse. I expect that the more free-wheeling account of meaning argued for by Mark Wilson (2006) would be able to be applied to fruitfully explore and contrast contexts in which this would and would not be appropriate.

[54] Smith G.E. (2002a, 162–7).

[55] Smith G.E. (2002a, 163–4; 2001a).

[56] See, e.g., Rowan-Robinson (1993, 39–46).

path in gravitation and cosmology, and that his method would be the appropriate way to respond to such a situation if it did develop.

V Postscript: measurement and evidence – Newton's method at work in cosmology today

In 2004 a first paperback edition, and fourth printing, of Robert Kirshner's book *The Extravagant Universe: Exploding stars, dark energy and the accelerating cosmos* was published with a new epilogue by the author. The first edition had been published in 2002. The new epilogue contains a wonderful illustration of Newton's method at work in cosmology today. It begins with the following description of a colloquium presentation:

"We've done these calculations in a standard Λ-cold dark matter universe." The energetic young speaker at the front of the Philips Auditorium at the Center for Astrophysics, Kathryn Johnson, a professor from Wesleyan, was setting the stage for presenting her new results on galaxy cannibalism. There were 100 people in the room for the Thursday Astronomy colloquium, Kathryn had a lot of new results to share, and she wasn't wasting any of her time or theirs by justifying the cosmology she had assumed.

Nobody blinked. Nobody asked a question. But my mind, always unreliable after 4 p.m. in a darkened room, started immediately to drift into speculation. How could a "Λ" universe, two thirds dark energy and one third dark matter, be the "standard" picture in the autumn of 2003? Just 5 years earlier, cosmic acceleration had seemed unbelievable, and dark energy, in its guise as the cosmological constant, had been a notoriously bad idea, personally banished by Albert Einstein. What had changed? (Kirshner 2004, 262)

Kirshner went on to review and assess the contributions afforded from different agreeing measurements of parameters of the background framework assumed by this speaker.

Here is a brief rough characterization of these parameters.[57] They are set in a solution to Einstein's field equation that corresponds to having a universe that has the structure of flat space at scales for which clusters of galaxies are counted as though they were individual specks of dust. This dust solution determines the dynamics of space-time based on a parameter Ω, which is the sum of all the specific contributions to the total mass-energy from different sources. Each contribution to the mass-energy of the system influences the overall dynamics of the system in a different systematic way according to the metric for the FLRW space-time structure for such systems. The letters stand for Friedman, Lemaître, Robertson, and Walker, each of whom contributed to the development of this important representation of space-time structure for such systems. The dust-model can take differing values of Ω. Having this total $\Omega = 1$ is the special case of flat Euclidean space-time at these enormous scales. In the FLRW

[57] See, e.g., Perlmutter and Schmidt (2003).

metric, the cosmic microwave background measurements combine with other measurements to empirically support the assignment of **a total of** $\Omega = 1$, composed of Ω_m of about 0.3 and Ω_Λ of about 0.7. In this framework assumed by the speaker, Ω_m represents the total of the mass-energy sources of gravity **due to matter**. Having Ω_m be about 0.3 requires an enormous amount of dark matter, as the visible sources of ordinary mass-energy in stars would count for only about 0.03 of that total. Having Ω_Λ be 0.7 corresponds to having galaxies rushing away from one another at rates so coordinated with distances as to count as an **accelerated** expansion of space itself. The details of such a model allow calculation of the age of the universe from its origin in a big bang.

Kirshner's review begins with the supernova measurements supporting a dark energy value of Ω_Λ of 0.7.

> Part of the answer is that the supernova results had gelled to become more solid. (Kirshner 2004, 262)

He reviewed significant improvements, as well as removals of alternatives that would have counted as sources of error. He then told his readers

> Despite these improvements in the supernova data, I don't think that's the reason why our speaker spent no time in setting out the pros and cons of a Λ cosmology. The real reason was a sudden convergence of many independent lines of research on the very same values for the contents and age of the universe, weaving a web of evidence. (Kirshner 2004, 263–4)

He then went on to point out measurements from galaxy clustering

> ...the new data pointed to a universe with $\Omega_m \sim 0.3$. This was the same value we were getting from the supernova analysis. (Kirshner 2004, 264)

He also discussed the fitting together of estimates of the age of the universe with estimates of the age of the oldest globular clusters of galaxies. These developments, impressive as they are, do not get the main credit.

Kirshner goes on to give the main credit to the agreeing estimates afforded from the improved measurements of the cosmic microwave background.

> All this was very satisfying, but the most powerful new set of information came from better measurements of the cosmic microwave background (CMB). (Kirshner 2004, 264)

After reviewing promising preliminary results from as early as 1998, he focused on measurements of cosmic microwave background radiation using detectors carried on an orbiting satellite.

> But the best was yet to come. In 2001, the Wilkinson Microwave Anisotropy Probe (WMAP), a satellite to measure the CMB, was launched into a unique orbit at 4 times the distance to the moon. From this superb perch, it patiently mapped the whole sky for a year.
>
> The first data from WMAP were released in February 2003, and those results changed the tone of the discussion in observational cosmology from cautious and tentative to confident and

quantitative. Just as the earlier measurements had indicated less precisely, the WMAP results confirmed that the universe has the large-scale geometry of flat space. (Kirshner 2004, 264–5)

...

Even though the CMB measurements don't detect cosmic acceleration directly, as the supernova measurements do, taken together, they point with good precision to a universe with both dark matter and dark energy. Things were fitting together – and the better you measured them the better they fit. Quantitative agreement is the ring of truth. This is the reason why, by the autumn of 2003, our colloquium speaker didn't bother to make the case that a Λ-dominated universe was the right picture. (Kirshner 2004, 265)

Kirshner's remarks strongly suggest that Newton's methodology continues to guide cosmology today. It appears that it was the striking realizations of Newton's ideal of empirical success as convergent accurate measurements of parameters by diverse phenomena that turned dark energy from a wild hypothesis into an accepted background assumption that guides further empirical research into the large-scale structure and development of our universe.

These developments illustrate a feature of agreeing measurements from diverse phenomena that is especially important for turning data into evidence. To the extent that the sources of potential systematic error of the different measurements can be regarded as independent, the agreement of the measurements contributes additional support for counting them as accurate rather than as mere artifacts of systematic error.[58]

[58] The systematic errors of the supernova measurements and the systematic errors of the Cosmic Microwave Background measurements can be very much expected to be independent. This is compellingly argued in a dissertation which Dylan Gault defended in December 2009. See Gault (2009). Wayne Myrvold and I were co-supervisors.

References

Airy, G.B. ([1834], 1969). *Gravitation: An Elementary Explanation of the Principal Perturbations in the Solar System.* Ann Arbor: NEO Press.

Aiton, E.J. (1954). Galileo's theory of the tides. *Annals of Science*, 10: 44–57.

—— (1955a). The contributions of Newton, Bernoulli and Euler to the theory of the tides. *Annals of Science*, 11: 206–23.

—— (1955b). Descartes's theory of the tides. *Annals of Science*, 11: 337–48.

—— (1972). *The Vortex Theory of Planetary Motions.* New York: American Elsevier.

—— (1989a). Polygons and parabolas: Some problems concerning the dynamics of planetary orbits. *Centaurus*, 31: 207–21.

—— (1989b). The Cartesian vortex theory. *GHA 2A*, 207–21.

—— (1995). The vortex theory in competition with Newtonian celestial dynamics. *GHA 2B*, 3–21.

Alexander, H.G. (ed.) (1956). *The Leibniz-Clark Correspondence.* Manchester: Manchester University Press.

Aoki, S. (1992). The moon-test in Newton's *Principia*: Accuracy of inverse-square law of universal gravitation. *Archive for History of Exact Sciences*, 44: 147–90.

Batterman, R. (2002). *The Devil in the Details: Asymptotic Reasoning in Explanation, Reduction, and Emergence.* New York: Oxford University Press.

—— (2005). Critical phenomena and breaking drops: Infinite idealizations in physics. *Studies in History and Philosophy of Modern Physics*, 36(2): 225–44.

Bayes, T. (unpublished). Letter to John Canton. *Canton Papers, Correspondence.* vol. 2, folio 32. London: Royal Society Library.

Belkind, O. (2007). Newton's conceptual argument for absolute space. *International Studies in the Philosophy of Science*, 21(3): 271–93.

Berlinski, D. (2000). *Newton's Gift: How Sir Isaac Newton Unlocked the System of the World.* New York: Simon & Schuster.

Bertoloni Meli, D. (1991). Public claims, private worries: Newton's *Principia* and Leibniz's theory of planetary motion. *Studies in History and Philosophy of Science*, 22: 415–49.

—— (1993). *Equivalence and Priority: Newton versus Leibniz.* Oxford: Clarendon Press.

Biener, Z. and Smeenk, C. (forthcoming). Cotes' queries: Matter and method in Newton's *Principia.* In: Janiak, A. and Schliesser, E. (eds.), *Interpreting Newton: Critical Essays.* Cambridge: Cambridge University Press.

Blackwell, R.J. (trans.) (1977). Christiaan Huygens: The motion of colliding bodies. *Isis*, 68 (244): 574–97.

—— (trans.) (1986). *Christiaan Huygens's The Pendulum Clock or Geometrical Demonstration, Concerning the Motion of Pendula as Applied to Clocks.* Ames: The Iowa State University Press.

Bos, H.J.M. (1972). Huygens, Christiaan. In: Gillispie, C.C. (ed. in chief), *Dictionary of Scientific Biography: Volume VI.* New York: Charles Scribner's Sons, 597–613.

Boulos, P. (1999). From natural philosophy to natural science: Empirical success and Newton's legacy. Ph.D. dissertation, University of Western Ontario.

Bowditch, N. (trans.) (1829; 1832; 1834; 1839). *Laplace's Celestial Mechanics*, vols. 1–4. Boston: Hilliard et al. Publishers. Reprinted: Bronx, N.Y.: Chelsea Publishing Company.

Boyer, C.B. (1968). *A History of Mathematics*. New York: John Wiley & Sons Inc.

Brackenridge, J.B. (1995). *The Key to Newton's Dynamics: The Kepler Problem and the Principia*. Berkeley: University of California Press.

Brading, K. (2011). On composite systems: Descartes, Newton, and the law-constitutive approach. In: Jalobeanu, D. and Anstay, P. (eds.), *Vanishing Matter and the Laws of Motion from Descartes to Hume*. London: Routledge.

Brans, C. and Dicke, R.H. ([1961], 1965). Mach's principle and a relativistic theory of gravitation. *The Physical Review*, 124(3): 925–35. Reprinted: Dicke, R.H (1965). *Appendix* 7, 77–96.

Bricker, P. and Hughes, R.I.G. (eds.) (1990). *Philosophical Perspectives on Newtonian Science*. Cambridge, Mass.: The MIT Press.

Brown, E.W. (1903). On the verification of the Newtonian Law. *Monthly Notices of the Royal Astronomical Society*, May: 396–7.

Brown, J. R. and Mittelstrass, J. (eds.) (1989). *An Intimate Relation: Studies in the History and Philosophy of Science*. Dordrecht: Kluwer Academic Publishers.

Buchwald, J.Z. and Cohen, I.B. (eds.) (2001). *Isaac Newton's Natural Philosophy*. Cambridge, Mass.: The MIT Press.

Buck R.C. and Cohen, R.S. (eds.) (1971). *Boston Studies in the Philosophy of Science*, vol. VIII. Dordrecht: D. Reidel Publishing Company.

Butts, R. (ed.) (1988). *William Whewell, Theory of Scientific Method*. Indianapolis: Hackett Publishing Company.

Butts, R.E. and Davis, J.W. (eds.) (1970). *The Methodological Heritage of Newton*. Toronto: University of Toronto Press.

Cajori, F. (trans.) (1934). *Sir Isaac Newton's Mathematical Principles of Natural Philosophy and his System of the World: Motte's Translation Revised by Cajori*. Los Angeles: University of California Press.

Cartwright, N. (1983). *How the Laws of Physics Lie*. Oxford: Clarendon Press.

—— (1989). *Nature's Capacities and their Measurement*. Oxford: Clarendon Press.

—— (1999). *The Dappled World: A Study of the Boundaries of Science*. Cambridge: Cambridge University Press.

Caspar, M. ([1948], 1993). *Kepler*. Hellman, C.D. (trans.). New York: Dover.

Chandrasekhar, S. (1995). *Newton's Principia for the Common Reader*. Oxford: Clarendon Press.

Christensen, D. (1983). Glymour on evidential relevance. *Philosophy of Science*, 50: 471–81.

—— (1990). The irrelevance of bootstrapping. *Philosophy of Science*, 57: 644–62.

Churchland, P.M. and Hooker, C.A. (eds.) (1985). *Images of Science*. Chicago: University of Chicago Press.

Cohen, I.B. (1940). Roemer and the first determination of the velocity of light. *Isis*, 31: 325–79.

—— (ed.) (1958). *Isaac Newton's Papers and Letters On Natural Philosophy*. Cambridge, Mass.: Harvard University Press.

—— (1971). *Introduction to Newton's Principia*. Cambridge: Harvard University Press.

—— and Smith G.E. (eds.) (2002). *The Cambridge Companion to Newton*. Cambridge: Cambridge University Press.

—— and Whitman, A. (trans.) (1999). *Isaac Newton, The Principia, Mathematical Principles of Natural Philosophy: A New Translation*. Los Angeles: University of California Press. (C&W)

Dalitz, R.H. and Nauenberg, M. (eds.) (2000). *The Foundations of Newtonian Scholarship*. Singapore: World Scientific Publishing Co.

Damour, T. (1987). The problem of motion in Newtonian and Einsteinian gravity. In: Hawking, S. and Israel, W. (eds.), *300 Years of Gravitation*. Cambridge: Cambridge University Press, 128–98.

—— and Vokrouhlický, D. (1996). Equivalence principle and the moon. *Physical Review D*, 53: 4177–201.

Danby, J.M.A. (1988). *Fundamentals of Celestial Mechanics*. Richmond: Willmann-Bell, Inc. 2nd edn.

Débarbat, S. and Wilson, C. (1989). The Galilean satellites of Jupiter from Galileo to Cassini, Römer and Bradley. *GHA 2A*, 146–57.

Densmore, D. (1995). *Newton's Principia: The Central Argument*. Santa Fe: Green Lion Press.

—— (1999). Cause and hypothesis: Newton's speculation about the cause of universal gravitation, *The St. John's Review*, XLV(2): 94–111.

Descartes, R. (1954). *The Geometry of Rene Descartes*. Smith, D.E. and Latham, M.L. (trans.). New York: Dover Publications.

—— (1955). *Philosophical Works of Descartes: In Two Volumes*. Haldane, E.S. and Ross, M.A. (trans. and eds.). New York: Dover Publications.

—— (1983). *Descartes' Principles of Philosophy*. Miller, V.R. and Miller, R.P. (trans.). Dordrecht: D. Reidel Publishing Company. (*Principles*)

Dicke, R.H. (1965). *The Theoretical Significance of Experimental Relativity*. New York: Gordon and Breach Science Publishers.

—— (1974). The oblateness of the sun and relativity. *Science*, 184(4135): 419–29.

—— and Goldenberg, H.M. (1967). Solar oblateness and general relativity. *Physical Review Letters*, 18: 313.

—— and Goldenberg (1974). The oblateness of the sun. *Astrophysics Journal Supplement Series*, 27(341): 131–82.

Dickey, J.O., Bender, P.L., Faller, J.E., Newhall, X.X., Ricklefs, R.L., Ries, J.G., Shelus, P.J., Veillet, C., Whipple, A.L., Wiant, J.R., Williams, J.G., and Yoder, C.F. (1994). Lunar laser ranging: A continuing legacy of the Apollo Program. *Science*, 265: 482–90.

DiSalle, R. (2006). *Understanding Space-time*. Cambridge: Cambridge University Press.

—— Harper, W.L., Valluri, S.R. (1996). Empirical Success and General Relativity. In: Jantzen, R.T. and MacKeiser, G. (eds.), Ruffini, R. (series ed.), *Proceedings of the Seventh Marcel Grossmann Meeting on General Relativity*. Singapore: World Scientific, Part A, 470–1.

Dobbs, B.J.T. (1991). *The Janus faces of genius*. Cambridge: Cambridge University Press.

Dorling, J. (1979). Bayesian personalism, the methodology of research programs and Duhem's problem. *Studies in History and Philosophy of Science*, 10: 177–87.

Dreyer, J.L.E. (1953). *A History of Astronomy from Thales to Kepler*. New York: Dover Publications, Inc.

Duhem, P. (1962). *The Aim and Structure of Physical Theory*. Wiener, P. (trans.). New York: Atheneum.

Earman, J. (ed.) (1983). *Testing Scientific Theories*. Minneapolis: University of Minnesota Press.

—— (1989). *World Enough and Space-Time*. Cambridge, Mass.: The MIT Press.

—— and Glymour, C. (1988). What revisions does bootstrap testing need: A reply. *Philosophy of Science*, 55: 260–4.

—— Glymour, C., and Stachel, J. (eds.) (1977). *Foundations of Space-Time Theories*. Minneapolis: University of Minnesota Press.

Earman, J., Janssen, M., and Norton, J.D. (1993). Einstein's explanation of the motion of Mercury's perihelion. In: Earman, J., Janssen, M., and Norton, J.D. (eds.), *The Attraction of Gravitation: New Studies in the History of General Relativity*. Boston: Bärkhäuser, 129–72.

—— —— and —— (eds.) (1993). *The Attraction of Gravitation: New Studies in the History of General Relativity*. Boston: Bärkhäuser.

—— and Norton, J.D. (eds.) (1997). *The Cosmos of Science: Essays of Exploration*. Pittsburgh: University of Pittsburgh Press.

Edidin, A. (1988). From relative confirmation to real confirmation. *Philosophy of Science*, 55: 265–71.

Edleston, J. (ed.) ([1850], 1969). *Correspondence of Sir Isaac Newton and Professor Cotes*. London: John W. Parker. 2nd edn. Frank Cass and Company Limited.

Einstein, A. ([1915], 1979). Explanation of the perihelion motion of Mercury by means of the general theory of relativity. *Prussian Academy Proceedings*, 11: 831–9. Doyle, B. (trans.) In: Lang, K.R. and Gingerich, O. (eds.), *A Source Book in Astronomy and Astrophysics, 1900–1975*. Cambridge: Harvard University Press.

—— Lorentz, H.A., Weyl, H., and Minkowski, H. ([1923], 1952). *The Principle of Relativity*. New York: Dover.

Ekeland, I. (1990). *Mathematics and the Unexpected*. Chicago: University of Chicago Press.

Erlichson, H. (1992). Newton's polygon model and the second order fallacy. *Centarus*, 35: 243–58.

Feingold, M. (2004). *The Newtonian Moment: Isaac Newton and the Making of Modern Culture*. Oxford: Oxford University Press.

Feyerabend, P.K. (1970). Classic Empiricism. In: Butts, R.E. and Davis, J.W. (eds.), *The Methodological Heritage of Newton*. Toronto: University of Toronto Press, 150–70.

Forbes, E.G. (1971). *The Euler–Mayer Correspondence (1751–1755)*. London: Macmillan.

—— and Wilson, C.A. (1995). The solar tables of Lacaille and the lunar tables of Mayer. *GHA 2B*, 55–68.

—— Murdin, L., and Willmoth, F. (1997). *The Correspondence of John Flamsteed, the First Astronomer Royal*. Philadelphia: Institute of Physics Publishing.

Forster, M. (1988). Unification, explanation, and the composition of causes in Newtonian mechanics. *Studies in History and Philosophy of Science*, 19(1): 55–101.

—— and Sober, S. (1994). How to tell when simpler, more unified, or less *ad hoc* theories will provide more accurate predictions. *British Journal for the Philosophy of Science*, 45: 1–35.

Freedman, D., Pisani, R., and Purvis, R. (1978, [3rd edn 1998]). *Statistics*. New York: W.W. Norton & Company.

French, A.P. (1971). *Newtonian Mechanics: The MIT Introductory Physics Series*. New York: W.W. Norton & Company.

Friedman, M. (1983). *Foundations of Space-Time Theories: Relativistic Physics and Philosophy of Science*. Princeton: Princeton University Press.

—— (1992). *Kant and the Exact Sciences*. Cambridge, Mass.: Harvard University Press.

Galilei, G. ([1632], 1970). *Dialogue Concerning the Two Chief World Systems – Ptolemaic and Copernican*. Drake, S. (trans.). Los Angeles: University of California Press.

—— ([1638], 1914). *Two New Sciences*. Crew, H. (trans.). New York: Dover.

Garisto, R. (1991). An error in Newton's determination of planetary properties. *American Journal of Physics*, 59: 42–8.

Gault, D. (2009). Contemporary cosmology as a case study in scientific methodology. Ph.D. dissertation, University of Western Ontario.

Gauss, K.F. ([1809; 1857 (trans.)], 1963). *Theory of the Motion of the Heavenly Bodies Moving About the Sun in Conic Sections*. Davis, C.H. (trans.). New York: Dover.

Geire, R. (1999). *Science Without Laws*. Chicago: University of Chicago Press.

Gillispie, C.C. (1970–80) (ed. in chief). *Dictionary of Scientific Biography* (sixteen vols.). New York: Charles Scribner's Sons.

Gingerich, O. and MacLachlan, J. (2005). *Nicholas Copernicus: Making the Earth a Planet*. Oxford: Oxford University Press.

Glymour, C. (1980a). *Theory and Evidence*. Princeton: Princeton University Press.

—— (1980b). Bootstraps and probabilities. *Journal of Philosophy*, LXXVII(11): 691–9.

—— (1983). On testing and evidence. In: Earman, J. (ed.), *Testing Scientific Theories*. Minneapolis: University of Minnesota Press, 3–26.

Grice, P. (1989). *Studies in the Way of Words*. Cambridge, Mass.: Harvard University Press.

Gunderson, K. (ed.) (1975). *Language, Mind and Method*. Minneapolis: University of Minnesota Press.

Gurzadyan, V.G., Jantzen, R.T., and Ruffini, R. (eds.), Ruffini, R. (series ed.) (2002). *Proceedings of the Ninth Marcel Grossmann Meeting on General Relativity, Part C*. Singapore: World Scientific Publishing Co.

Hall, A. (1894). A suggestion in the theory of Mercury. *The Astronomical Journal*. xiv: 35–46.

Halley, E. ([1691], 1809). On the visible conjunctions of the inferior planets with the sun. *Philosophical Transactions* (abridged edn.), vol. 3: 448–56.

—— (1694). *Monsieur* Cassini *his new and exact tables for* eclipses *of the first* satellite *of* Jupiter, *reduced to the* Julian *stile, and Meridian of* London. *Philosophical Transactions*, vol. 18(214): 237–56.

Harper, W.L. (1977). Rational Conceptual Change. *PSA 1976 (Proceedings of the 1976 Biennial Meeting of the Philosophy of Science Association)*. Michigan: Philosophy of Science Association, vol. 2: 462–94.

—— (1979). Conceptual change, incommensurability and special relativity kinematics. *Acta Philosophica Fennica*. 30(4): 431–61.

—— (1984). Kant on space, empirical realism and the foundations of geometry. *Topoi*, 3(2): 143–61. Reprinted in Posy, C.J. (ed.) 1992, Kant's Philosophy of Mathematics. Dordrecht: Kluwer Academic Publications, 257–91.

—— (1989). Concilience and natural kind reasoning. In: Brown, J.R. and Mittelstrass, J. (eds.), *An Intimate Relation: Studies in the History and Philosophy of Science*. Dordrecht: Kluwer Academic Publications, 115–52.

—— (1993). Reasoning from phenomena: Newton's argument for universal gravitation and the practice of science. In: Theerman, P. and Seef, A.F. (eds.), *Action and Reaction: Proceedings of a Symposium to Commemorate the Tercentenary of Newton's Principia*. Newark: University of Delaware Press, 144–82.

—— (1995). Kant, Rieman and Reichenbach on Space and Geometry, Robinson, H. (ed.) *Proceedings of the Eighth International Kant Congress* vol. I. Milwaukee: Marquette University Press, 423–55.

—— (1997). Isaac Newton on empirical success and scientific method. In: Earman, J. and Norton, J.D. (eds.), *The Cosmos of Science: Essays of Exploration*. Pittsburgh: University of Pittsburgh Press, 55–86.

Harper, W.L. (1998). Measurement and approximation: Newton's inferences from phenomena versus Glymour's bootstrap confirmation. In: Weingartner, P., Schurz, G., and Dorn, G. (eds.), *The Role of Pragmatics in Contemporary Philosophy*. Vienna: Hölder-Pinchler-Tempsky, 265–87.

—— (1999). The first six propositions in Newton's argument for universal gravitation. *The St. John's Review*, XLV(2): 74–93.

—— (2002a). Howard Stein on Isaac Newton: Beyond hypotheses? In: Malament, D.B. (ed.), *Reading Natural Philosophy*. Chicago: Open Court, 71–112.

—— (2002b). Newton's argument for universal gravitation. In: Cohen, I.B. and Smith, G.E. (eds.), *The Cambridge Companion to Newton*. Cambridge: Cambridge University Press, 174–201.

—— (2007a). Newton's methodology and Mercury's perihelion before and after Einstein. *Philosophy of Science*, 74: 932–42.

—— (2007b). Acceptance and scientific method. In: Harper, W.L. and Wheeler, G. (eds.), *Probability and Inference: Essays in Honour of Henry E. Kyburg, Jr.* London: King's College Press, 33–52.

—— (2009). Newton's methodology. In: Myrvold, W.C. and Christian, J. (eds.), *Quantum Reality, Relativistic Causality, and Closing the Epistemic Circle: Essays in Honour of Abner Shimony*. New York: Springer Science and Business Media, 43–61.

—— (forthcoming). Measurement and method: Some remarks on Newton, Huygens and Euler on natural philosophy. In: Janiak, A. and Schliesser, E. (eds.), *Interpreting Newton: Critical Essays*. Cambridge: Cambridge University Press.

—— and DiSalle, R. (1996). Inferences from phenomena in gravitational physics. *Philosophy of Science*, 63: S46–S54.

—— and Smith, G.E. (1995). Newton's new way of inquiry. In: Leplin, J. (ed.), *The Creation of Ideas in Physics*. Dordrecht: Kluwer Academic Publishers, 113–66.

—— Valluri, S.R., and Mann, R.B. (2002). Jupiter's moons and the equivalence principle. In: Gurzadyan, V.G., Jantzen, R.T., and Ruffini, R. (eds.), Ruffini, R. (series ed.), *Proceedings of the Ninth Marcel Grossmann Meeting on General Relativity, Part C*. Singapore: World Scientific Publishing Co., 1803–13.

—— and Wheeler, G. (eds.) (2007). *Probability and Inference: Essays in Honour of Henry E. Kyburg, Jr.* London: King's College Press.

Heath, T.L. ([1908; 1925], 1956). *Euclid's Elements*, 2nd edn. New York: Dover Publications, Inc.

Herivel, J. (1965). *The Background to Newton's Principia: A Study of Newton's Dynamical Researches in the Years 1664–84*. Oxford: Clarendon Press.

Holton, G. (1990). Einstein's scientific program: Formative years. In: Woolf, H. (ed.), *Some Strangeness in the Proportion*. Reading Mass.: Addison-Wesley.

Huygens, C. (1669). A summary account of the Laws of Motion (1665–1678). *Philosophical Transactions*, vol. 4: 925–8.

—— (1690). *Discourse on the Cause of Gravity*. Bailey, K. (trans.), Bailey, K. and Smith, G.E. (annotations), manuscript.

Jalobeanu, D. and Anstay, P. (eds.) (2011). *Vanishing Matter and the Laws of Motion from Descartes to Hume*. London: Routledge.

Janiak, A. (ed.) (2004). *Newton: Philosophical Writings*. Cambridge: Cambridge University Press.

—— (2007). Newton and the reality of force. *Journal of the History of Philosophy*, 45: 127–47.

—— (2008). *Newton As Philosopher*. Cambridge: Cambridge University Press.

—— and Schliesser, E. (eds.) (forthcoming). *Interpreting Newton: Critical Essays*. Cambridge: Cambridge University Press.

Jardine, N. (1984). *The Birth of History and Philosophy of Science: Kepler's A Defence of Tycho against Ursus, with Essays on its Provenance and Significance*. Cambridge: Cambridge University Press.

Jungnickel, C. and McCormmach, R. (1986). *Intellectual Mastery of Nature: Theoretical Physics from Ohm to Einstein*. Chicago: University of Chicago Press.

Kant, I. ([1786], 2004). *Metaphysical Foundations of Science*. Friedman, M. (trans. and ed.). Cambridge: Cambridge University Press.

Kaplan, D. (1971). Homage to Rudolf Carnap. In: Buck R.C. and Cohen, R.S. (eds.), *Boston Studies in the Philosophy of Science*, vol. VIII. Dordrecht: D. Reidel Publishing Company, xlv–xlvii.

Kepler, J. (1937). *Gesammelte Werke*. Munich: C.H. Beck.

—— (1992). *New Astronomy*. Donahue, W.H. (trans.). Cambridge: Cambridge University Press.

—— (1995). *Epitome of Copernican Astronomy and Harmonies of the World*. Wallis, C.G. (trans. and ed.). Amherst, New York: Prometheus Books.

King, H. ([1955], 1979). *The History of the Telescope*. New York: Dover Publications, Inc.

Kirshner, R.P. (2004). *The Extravagant Universe: Exploding Stars, Dark Energy, and the Accelerating Cosmos, with a New Epilogue by the Author*. Princeton: Princeton University Press.

Knudsen, J.M. and Hjorth, P.G. (1996). *Elements of Newtonian Mechanics: Including Nonlinear Dynamics*, 2nd revised and enlarged edn. New York: Springer.

Kollerstrom, N. (2000). *Newton's Forgotten Lunar Theory*. Santa Fe: Green Lion Press.

Koslow, A. (ed.) (1967). *The Changeless Order: The Physics of Space, Time and Motion*. New York: George Braziller.

Koyré, A. (1965). *Newtonian Studies*. Chicago: University of Chicago Press.

—— (1968). *Metaphysics and Measurement: Essays in Scientific Revolution*. Cambridge: Harvard University Press.

—— and Cohen, I.B. (eds.) (1972). *Isaac Newton's Philosopiae Naturalis Principia Mathematica: The Third Edition (1726) with Variant Readings*. Cambridge, Mass.: Harvard University Press.

Kripke, S.A. (1972). *Naming and Necessity*. Cambridge, Mass.: Harvard University Press.

Kuhn, T.S. (1957). *The Copernican Revolution*. Cambridge, Mass.: Harvard University Press.

—— ([1962], 1970). *The Structure of Scientific Revolutions*. Chicago: University of Chicago Press, 2nd revised edn.

—— (1977). *The Essential Tension: Selected Studies in Scientific Tradition and Change*. Chicago: University of Chicago Press.

Kyburg, H.E. (1990). *Science and Reason*. Oxford: Oxford University Press.

Lakatos, I. (1978). *The Methodology of Scientific Research Programs: Philosophical Papers*, vol. 1. Worrall, J. and Currie, G. (eds.). Cambridge: Cambridge University Press.

La Lande, J.J.L. (1756). Première mèmoire sur la parallaxe de la lune et sur sa distance à la terre, *Mèmoires de l'Acadèmie des Sciences, annèe 1752* (Paris), 78–114.

Lang, K.R. and Gingerich, O. (1979). *A Source Book in Astronomy and Astrophysics, 1900–1975*. Cambridge, Mass.: Harvard University Press.

Laplace, P.S. ([vol. 1–4, 1789–1805; vol. 5, 1825; trans., 1829–1839], 1966). *Celestial Mechanics*. Bowditch, N. (trans.). New York: Chelsea Publishing Co.

—— ([1814; 1902 (trans.)], 1995). *A Philosophical Essay on Probabilities*. Truscott, F.W. and Emory, F.L. (trans.). Toronto: Dover Publications.

Laskar, J. (1990). Chaotic behavior of the solar system: A numerical estimate of the size of the chaotic zones. *Icarus*, 88: 266–91.

—— (1995). Appendix: The stability of the solar system from Laplace to the present. *GHA 2B*, 240–8.

Lattis, J.M. (1994). *Between Copernicus and Galileo: Christopher Clavius and the Collapse of Ptolemaic Cosmology*. Chicago: University of Chicago Press.

Laubscher, R.E. (1981). *The Motion of Mars 1751–1969*. Washington: United States Government Printing Office.

Laudan, L. (1981). A confutation of convergent realism. *Philosophy of Science*, 48: 19–38, 45–9.

Laymon, R. (1983). Newton's demonstration of universal gravitation and philosophical theories of confirmation. In: Earman, J. (ed.), *Testing Scientific Theories*. Minneapolis: University of Minnesota Press, 179–99.

Leplin, J. (ed.) (1995). *The Creation of Ideas in Physics*. Dordrecht: Kluwer Academic Publishers.

Machin, J. (1729). The laws of the moon's motion according to gravity. In: Motte, A. (trans.), ([1729], 1995), 373–6.

Maglo, K. (2003). The reception of Newton's gravitational theory by Huygens, Varignon, and Maupertuis: How normal science may be revolutionary. *Perspectives on Science*, vol. 11, no. 2, 135–69.

Malament, D.B. (ed.) (2002). *Reading Natural Philosophy: Essays in the History and Philosophy of Science and Mathematics*. Chicago: Open Court.

Matthews, M.R. (1989). *Scientific Background to Modern Philosophy*. Indianapolis: Hackett.

Mayo, D.G. and Spanos, A. (eds.) (2010). *Error and Inference: Recent Exchanges on Experimental Reasoning, Reliability, and the Objectivity and Rationality of Science*. Cambridge: Cambridge University Press.

McGuire, J.E. (1970). Atoms and the 'Analogy of Nature': Newton's third rule of philosophizing. *Studies in History and Philosophy of Science*, I: 3–58, 1970.

—— (1995). *Tradition and Innovation: Newton's Metaphysics of Nature*. Dordrecht: Kluwer Academic Publishers.

McKeon, R. (1941). *The Basic Works of Aristotle*. New York: Random House.

McMullin, E. (2001). The impact of Newton's *Principia* on the philosophy of science. *Philosophy of Science*, 68: 279–318.

Mermin, D.N. (2009). *It's About Time: Understanding Einstein's Relativity*. Princeton: Princeton University Press.

Milgrom, M. (1994). Dynamics with a non-standard inertia–acceleration relation: An alternative to dark matter in galactic systems. *Annals of Physics*, 229: 384–415.

Morando, B. (1995). Laplace. *GHA 2B*, 131–50.

Motte, A. (trans.) ([1729], 1995). *The Principia: Isaac Newton Translated by Andrew Motte*. Amherst, New York: Prometheus Books.

Moulton, F.R. (1970). *An Introduction to Celestial Mechanics* (1914), 2nd revised edn. New York: Dover.

Murray, G., Harper, W.L., and Wilson, C. (2011). Huygens, Wren, Wallis, and Newton on rules of impact and reflection. In: Jalobeanu, D. and Anstey, P. (eds.), *Vanishing Matter and the Laws of Motion from Descartes to Hume*. London: Routledge, 153–91.

Myrvold, W.C. (1999). *The Kid is Alright: Connecting the Dots in Newton's Proof of Proposition 1*. Manuscript of presentation.

—— (2003). A Bayesian account of the virtue of unification. *Philosophy of Science*, 70: 399–423.

—— and Christian, J. (eds.) (2009). *Quantum Reality, Relativistic Causality, and Closing the Epistemic Circle: Essays in Honour of Abner Shimony*. New York: Springer Science and Business Media.

—— and Harper, W.L. (2002). Model selection, simplicity, and scientific inference. *Philosophy of Science*, 69: S135–S149.

Nagel, E. (1961). *The Structure of Science: Problems in the Logic of Scientific Explanation*. New York: Harcourt, Brace & World, Inc.

Nauenberg, M. (2001). Newton's perturbation methods for the three-body problem and their application to lunar motion. In: Buchwald J.Z. and Cohen I.B. (eds.), *Isaac Newton's Natural Philosophy*. Cambridge: Cambridge University Press, 188–224.

Neugebauer, O. (1975). *A History of Ancient Mathematical Astronomy*. New York: Springer-Verlag.

Newcomb, S. (1882). Discussion and results of observations on transits of Mercury from 1677 to 1881. *Astronomical Papers Prepared for the Use of the American Ephemeris and Nautical Almanac* I. Washington: United States Government Printing Office, 367–487.

—— (1895). *The Elements of the Four Inner Planets and the Fundamental Constants of Astronomy*. Washington: United States Government Printing Office.

Newton, I. ([1687], 1965). *Philosopiae Naturalis Principia Mathematica*. Londini 1687, facsimile reproduction, Bruxelles: Impression Anastaltique Culture et Civilisation.

—— (1728). *A Treatise of the System of the World, Translated into English*. London: printed for F. Fayram.

—— ([1730], 1979). *Opticks: Or A Treatise of the Reflections, Refractions, Inflections and Colors of Light*, 4th edn. New York: Dover Publications, Inc.

—— (1959–77). *The Correspondence of Isaac Newton*. Turnbull, H.W. (ed., vols. I–III), Scott, J.F. (ed., vol. IV), Hall, A.R. and Tilling, L. (ed., vols. V–VII). Cambridge: Cambridge University Press. (*Corresp.*)

Nordtvedt, K. (1968a). Equivalence principle for massive bodies II. *Physical Review*, 169: 1017–25.

—— (1968b). Testing relativity with laser ranging to the moon. *Physical Review*, 170: 1186–7.

Ohanian, H.C. and Ruffini, R. (1994). *Gravitation and Space-time*. New York: W.W. Norton & Company.

Pais, A. (1982). *Subtle is the Lord: The Science and the Life of Albert Einstein*. New York: Oxford University Press.

Palter, R. (ed.) (1970). *The Annus Mirabilis of Sir Isaac Newton 1666–1966*. Cambridge, Mass.: The MIT Press.

Perlmutter, S. and Schmidt, B.P. (2003). Measuring cosmology with supernovae. In: Weiler, K. (ed.), *Supernovae and Gamma-Ray Bursters*. Berlin: Springer, 195–218.

Peterson, I. (1993). *Newton's Clock: Chaos in the Solar System*. New York: W.H. Freeman and Company.

Popper, K. (1963). *Conjectures and Refutations: The Growth of Scientific Knowledge*. New York: Harper & Row Publishers.

—— (1972). *Objective Knowledge: An Evolutionary Approach*. Oxford: Oxford University Press.

Pourciau, B. (2003). Newton's argument for proposition 1 of the *Principia*. *Archive for History of Exact Sciences*, 57: 267–311.

Pourciau, B. (2004). The importance of being equivalent: Newton's two models of one-body motion. *Archive for History of Exact Sciences*, 58: 283–321.

—— (2006). Newton's interpretation of Newton's Second Law. *Archive for History of Exact Sciences*, 60: 157–207.

Putnam, H. (1975). The meaning of 'meaning'. In: Gunderson, K. (ed.), *Language, Mind and Knowledge*. Minneapolis: University of Minnesota Press, 131–93.

Reasenberg, R.D., Shapiro, I.I., MacNeil, P.E., Goldstein, R.B., Breidenthal, J.C., Brenkle, J.P., Cain, D.L., Kaufman, T.M., Komarek, T.A., and Zygielbaum, A.I. (1979). Viking Relativity experiment: Verification of signal retardation by solar gravity. *The Astrophysical Journal*, 234: L219–221.

Rigaud, S.P. (ed.) (1972). *Miscellaneous Works and Correspondence of James Bradley*. New York: Johnson Reprint Corporation.

Roberts, J.T. (2005). Measurability and physical laws. *Synthese* 144: 433–47.

—— (2008). *The Law-Governed Universe*. Oxford: Oxford University Press.

Rosenkrantz, R. (1983). Why Glymour is a Bayesian. In: Earman, J. (ed.), *Testing Scientific Theories*. Minneapolis: University of Minnesota Press, 69–97.

Roseveare, N.T. (1982). *Mercury's perihelion from Le Verrier to Einstein*. Oxford: Oxford University Press.

Rowan-Robinson, M. (1993). *Ripples in the Cosmos: A View Behind the Scenes of the New Cosmology*. New York: W.H. Freeman and Company Limited.

Rynasiewicz, R. (1995a). By their properties, causes and effects: Newton's Scholium on time, space, place and motion – I. The text. *Studies in the History and Philosophy of Science*, 26(1): 133–53.

—— (1995b). By their properties, causes and effects: Newton's Scholium on time, space, place and motion – II. The context. *Studies in the History and Philosophy of Science*, 26(2): 295–321.

Schliesser, E. (2005). Wonder in the face of scientific revolutions: Adam Smith on Newton's 'proof' of Copernicanism. *British Journal for the History of Philosophy*, 13(4): 697–732.

—— (2011). Without God: Gravity as a relational quality of matter in Newton's *Treatise*. In: Jalobeanu, D. and Anstey, P.R. (eds.). *Vanishing Matter and the Laws of Motion form Descartes to Hume*. London: Routledge, 153–91.

—— and Smith, G.E. (forthcoming). Huygens, 1688 report to the Directors of the Dutch East Indian Company on the measurement of longitude at sea and the evidence it offered against universal gravity. In: *Archive for the History of the Exact Sciences*.

Schmeidler, F. (with additions by Sheynin, O.) (1995). Astronomy and the theory of errors: From the method of averaged to the method of least squares. *GHA 2B*, 198–207.

Seidelmann, P.K. (ed.) (1992). *Explanatory Supplement to the Astronomical Almanac*. Mill Valley: University Science Books. (*ESAA*)

Shapiro, A.E. (2002). Newton's optics and atomism. In: Cohen, I.B. and Smith, G.E. (eds.), *The Cambridge Companion to Newton*. Cambridge: Cambridge University Press, 227–55.

Shapiro, I.I. (1964). Fourth test of general relativity. *Physical Review Letters*, 13: 789–91.

—— (1989). Solar system tests of general relativity: Recent results and present plans. In: Ashby, N., Bartlett, D.F., and Wyss, W. (eds.), *General Relativity and Gravitation*. Cambridge: Cambridge University Press, 313–30.

—— Ash, M.E., Ingalls, R.P., Smith, W.B., Campbell, D.B., Dyce, R.B., Jurgens, R.F., and Pettengill, G.H. (1971). Fourth test of general relativity: New radar result. *Physical Review Letters*, 26(18): 1132–5.

Shapley, H. and Howarth, H.E. (eds.) (1929). *A Source Book in Astronomy*. New York: McGraw-Hill Book Company.

Shimony, A. (1993). *Search for a Naturalistic World View Volume I: Scientific Method and Epistemology*. Cambridge: Cambridge University Press.

Simpson, T. (1755). A letter to the Right Honourable George Earl of Macclesfield, President of the Royal Society, on the advantage of taking the mean of a number of observations, in practical astronomy. *Philosophical Transactions of the Royal Society*, 44: 82–93.

Singh, S. (2004). *Big Bang: The Origin of the Universe*. New York: Harper Perennial.

Smart, W.M. ([1931], 1977). *Textbook on Spherical Astronomy*, 6th edn revised by Green, R.M. Cambridge: Cambridge University Press.

Smith, G.E. (1996). Essay review: Chandrasekhar's *Principia*. *Journal of History of Astronomy*, xxvii: 353–61.

—— (1997). Huygens's empirical challenge to universal gravity. Manuscript notes and overheads for presentation.

—— (1999a). How did Newton discover universal gravity? *The St. John's Review*, XLV(2): 45–63.

—— (1999b). The motion of the lunar apsis. In: C&W, 257.

—— (2000). Fluid resistance: Why did Newton change his mind? In: Dalitz, R.H. and Nauenberg, M. (eds.), *The Foundations of Newtonian Scholarship*. Singapore: World Scientific Publishing Co., 105–42.

—— (2001a), The Newtonian style in Book II of the *Principia*. In: Buchwald, J.Z. and Cohen, I.B. (eds.), *Isaac Newton's Natural Philosophy*. Cambridge, Mass.: The MIT Press, 249–311.

—— (2001b). Comments on Ernan McMullin's 'The Impact of Newton's *Principia* on the Philosophy of Science'. *Philosophy of Science*, 68: 327–38.

—— (2002a). The methodology of the *Principia*. In: Cohen, I.B. and Smith, G.E. (eds.), *The Cambridge Companion to Newton*. Cambridge: Cambridge University Press, 138–73.

—— (2002b). From the phenomenon of the ellipse to an inverse-square force: Why not? In: Malament, D.B. (ed.), *Reading Natural Philosophy: Essays in the History and Philosophy of Science and Mathematics*. Chicago: Open Court, 31–70.

—— (forthcoming). Closing the loop: Testing Newtonian Gravity, Then and Now. Also at <http://www.stanford.edu/dept/cisst/visitors.html> accessed 26 April 2011.

Smith, S. (2002). Violated laws, *Ceteris Paribus* clauses, and capacities. *Synthese*, 130: 235–64.

Solmsen, F. (1960). *Aristotle's System of the Physical World: A Comparison with His Predecessors*. Ithaca New York: Cornell University Press.

Stanley, J. (2005). *Knowledge and Practical Interests*. Oxford: Oxford University Press.

Stein, H. ([1967], 1970a). Newtonian space-time. *The Texas Quarterly*, X(3): 174–200. Reprinted with revision in: Palter, R. (ed.), *The Annus Mirabilis of Sir Isaac Newton 1666–1966*. Cambridge, Mass.: The MIT Press, 258–84.

—— (1970b). On the notion of field in Newton, Maxwell, and beyond. In: Stuewer, R.H. (ed.), *Historical and Philosophical Perspectives of Science*. Minneapolis: University of Minnesota Press, 264–87.

—— (1977). Some philosophical prehistory of general relativity. In: Earman, J., Glymour, C., and Stachel, J. (eds.), *Foundations of Space-Time Theories*. Minneapolis: University of Minnesota Press, 3–49.

Stein, H. (1990). On Locke, the Great Huygenius, and the incomparable Mr. Newton. In: Bricker, P. and Hughes, R.I.G. (eds.), *Philosophical Perspectives on Newtonian Science*, 17–47.

—— (1991). 'From the Phenomena of Motions to the Forces of Nature': Hypothesis or Deduction? *PSA 1990* (*Proceedings of the 1990 Biennial Meeting of the Philosophy of Science Association*), Michigan: Philosophy of Science Association, vol. 2: 209–22.

—— (2002). Newton's metaphysics. In: Cohen, I.B. and Smith, G.E. (eds.), *The Cambridge Companion to Newton*. Cambridge: Cambridge University Press, 256–307.

Stuewer, R.H. (ed.) (1970). *Historical and Philosophical Perspectives of Science*. Minneapolis: University of Minnesota Press.

Sussman, G.J. and Wisdom, J. (1992). Chaotic evolution of the solar system. *Science*, 257: 56–62.

Swerdlow, N.M. and Neugebauer, O. (1984). *Mathematical Astronomy in Copernicus's De Revolutionibus*. New York: Springer-Verlag.

Taton, R. and Wilson, C. (1989). *The General History of Astronomy, vol. 2, Planetary Astronomy from the Renaissance to the Rise of Astrophysics, Part A: Tycho Brahe to Newton*. Cambridge: Cambridge University Press. (*GHA 2A*)

—— and —— (1995). *The General History of Astronomy, vol. 2, Planetary Astronomy from the Renaissance to the Rise of Astrophysics, Part B: The Eighteenth and Nineteenth Centuries*. Cambridge: Cambridge University Press. (*GHA 2B*)

Theerman, P. and Seef, A.F. (eds.) (1993). *Action and Reaction: Proceedings of a Symposium to Commemorate the Tercentenary of Newton's Principia*. Newark: University of Delaware Press.

Thoren, V.E. (1990). *The Lord of Uraniborg: A Biography of Tycho Brahe*. Cambridge: Cambridge University Press.

Todhunter, I. ([1873], 1962). *A History of the Mathematical Theories of Attraction and the Figure of the Earth*. New York: Dover Publications, Inc.

Truesdell, C. (1970). Reactions of late Baroque mechanics to success, conjecture, error and failure in Newton's *Principia*. In: Palter, R. (ed.), *The Annus Mirabilis of Sir Isaac Newton 1666–1966*. Cambridge, Mass.: The MIT Press, 192–232.

Valluri, S.R., Harper, W.L. and Biggs, R. (1999). Newton's precession theorem, eccentric orbits and Mercury's orbit. In: Piran, T. (ed.), Ruffini, R. (series ed.), *Proceedings of the Eighth Marcel Grossmann Meeting on General Relativity, Part A*. Singapore: World Scientific Publishing Co., 485–8.

—— Wilson, C., and Harper, W.L. (1997). Newton's apsidal precession theorem and eccentric orbits. *Journal for the History of Astronomy*, 28: 13–27.

van Fraassen, B.C. (1980). *The Scientific Image*. Oxford: Oxford University Press.

—— (1983). Glymour on evidence and explanation. In: Earman, J. (ed.), *Testing Scientific Theories*. Minneapolis: University of Minnesota Press, 165–76.

—— (1985). Empiricism in the philosophy of science. In: Churchland, P.M. and Hooker, C.A. (eds.), *Images of Science*. Chicago: University of Chicago Press, 245–308.

—— (1989). *Laws and Symmetry*. Oxford: Clarendon Press.

—— (2002). *The Empirical Stance*. New Haven: Yale University Press.

Van Helden, A. (1985). *Measuring The Universe: Cosmic Dimensions from Aristarchus to Halley*. Chicago: University of Chicago Press.

—— (trans.) (1989). *Sidereus Nuncius or The Siderial Messenger: Galileo Galilei*. Chicago: University of Chicago Press.

—— (1995). Measuring solar parallax: The Venus transits of 1761 and 1769 and their nineteenth-century sequels. *GHA 2B*, 153–68.

Waff, C.B. (1976). Universal gravitation and the motion of the moon's apogee: The establishment and reception of Newton's inverse-square law. Ph.D. dissertation Johns Hopkins University. Available from Ann Arbor: University Microfilms.

—— (1995a). Clairaut and the motion of the lunar apse: The inverse-square law undergoes a test. *GHA 2B*, 35–46.

—— (1995b). Predicting the mid-eighteenth-century return of Halley's Comet. *GHA 2B*, 69–82.

Weingartner, P., Schurz, G., and Dorn, G. (eds.) (1998). *The Rôle of Pragmatics in Contemporary Philosophy*. Vienna: Hölder-Pinchler-Tempsky, 265–87.

Westfall, R.S. (1973). Newton and the fudge factor. *Science*, 179(4057): 751–8.

—— (1980). *Never at Rest: A Biography of Isaac Newton*. Cambridge: Cambridge University Press.

Whewell, W. (1860). *On the Philosophy of Discovery*. Cambridge: Cambridge University Press.

Whiteside, D.T. (ed.) (1967–81). *The Mathematical Papers of Isaac Newton*, 8 vols. Cambridge: Cambridge University Press. (Math Papers)

—— (1976). Newton's lunar theory: From high hope to disenchantment. *Vistas in Astronomy*, 19: 317–28.

Will, C.M. (1971). Theoretical frameworks for testing relativistic gravity.II. Parametrized post-Newtonian hydrodynamics and the Nordtvedt effect. *Astrophysical Journal* 163: 611–28.

—— (1986). *Was Einstein Right? Putting General Relativity to the Test*. New York: Basic Books.

—— (1993). *Theory and Experiment in Gravitational Physics*, 2nd revised edn. Cambridge: Cambridge University Press.

—— (2006). The confrontation between general relativity and experiment. *Living Rev. Relativity*, 9, 3. Online Article: cited 2010.09.10, http://www.livingreviews.org/lrr-2006-3.

—— and Nordtvedt, K. (1972). Conservation laws and preferred frames in relativistic gravity I. Preferred-frame theories and an extended PPN formalism. *Astrophysical Journal* 177: 757–74.

Wilson, C.A. (1970). From Kepler's laws, so called, to universal gravitation: Empirical factors. *Archive for History of Exact Sciences*, 6(2): 115–16.

—— (1980). Perturbations and solar tables from Lacaille to Delambre: The rapprochement of observation and theory, part I. *Archive for History of Exact Sciences*, 22(3): 54–188.

—— (1985). The great inequality of Jupiter and Saturn from Kepler to Laplace. *Archive for History of Exact Sciences*, 33(1–3): 15–290.

—— (1987). D'Alembert versus Euler on the precession of the equinoxes and the mechanics of rigid bodies. *Archive for History of Exact Sciences*, 37(3): 233–73.

—— (1989a). *Astronomy from Kepler to Newton: Historical Studies*. London: Variorum Reprints.

—— (1989b). The Newtonian achievement in astronomy. *GHA 2A*, 233–74.

—— (1993). Clairaut's calculation of the eighteenth-century return of Halley's Comet. *Journal for the History of Astronomy*, 24: 1–15.

—— (1995). The precession of the equinoxes from Newton to d'Alembert and Euler. *GHA 2B*, 47–54.

—— (1999). Redoing Newton's experiment for establishing the proportionality of mass and weight. *The St. John's Review*, XLV(2): 64–73.

Wilson, C.A. (2000). From Kepler to Newton: Telling the tale. In: Dalitz, R.H. and Nauenberg, M. (eds.), *The Foundations of Newtonian Scholarship*. Singapore: World Scientific Publishing Co., 223–42.

—— (2001). Newton on the moon's variation and apsidal motion. In: Buchwald, J.Z. and Cohen, I.B. (eds.), *Isaac Newton's Natural Philosophy*. Cambridge, Mass.: The MIT Press, 138–88.

—— (2002). Newton and celestial mechanics. In: Cohen, I.B. and Smith, G.E. (eds.), *The Cambridge Companion to Newton*. Cambridge: Cambridge University Press, 202–26.

Wilson, M. (2006). *Wandering Significance: An Essay On Conceptual Behavior*. New York: Oxford University Press.

Wren, C. (1668). Dr. Christopher Wren's theory concerning the same subject 'The Law of Nature concerning Collision of Bodies'. *Philosophical Transactions* (1665–78), vol. 3: 867–8.

Yoder, J. (1988). *Unrolling Time: Christiaan Huygens and the Mathematization of Nature*. Cambridge: Cambridge University Press.

Zytkow, J. (1986). What revisions does bootstrap testing need? *Philosophy of Science*, 53: 101–9.

Acknowledgements for use of Images and Text

The author would like to thank the following for the creation of images and for permission, where applicable, to reproduce copyrighted material:

Figure 1.1 Retrograde motion of Mars; Photographer Tunç Tezel for the use of the image located at http://www.nasaimages.org/luna/servlet/detail/NVA2~4~4~5999~106525:Z-is-for-Mars.

Figure 1.2 Epicycle deferent, Figure 1.3 Equant: "Nicolaus Copernicus: Making the Earth a Planet" by Owen Gingerich and James H. MacLachlan (2005), pp. 29 and 41 by permission of Oxford University Press.

Figures 1.4 Ecliptic, 1.5 The twelve constellations, 1.6 The precession of the equinoxes, 1.7 Precession of equinoxes explained, 1.12 The Harmonic Rule as illustrated by satellites of the sun, 2.2 The Harmonic Rule as illustrated by Jupiter's moons, 2.3 The Harmonic Rule as illustrated by Saturn's moons, 2.6 The Harmonic Rule as illustrated by satellites of the sun using both Kepler's and Boulliau's data, 3.2 The Harmonic Rule as illustrated by satellites of the sun using both Kepler's and Boulliau's data, 4.1 Empirical support added to Paris pendulum estimates from the agreeing moon-test estimates, 5.2 Empirical support added to Paris estimate + estimates from other latitudes from moon-test estimates of edition 3, 6.1 Diurnal parallax, 6.2 Empirical support added to all ten cited pendulum estimates from the eleven cited syzygy corrected moon-test estimates, 7.1 Polarized orbit, 7.2 Polarized orbit, and the image in proof of lemma in Chpt. 8 appendix 3.2: Suomi Ghosh.

Figure 1.8 Retrograde motion of Mars: The anonymous author who made their work available on the Wikimedia commons under the following licenses: the GNU Free Documentation License, Version 1.2 or any later version published by the Free Software Foundation, with no Invariant Sections, no Front-Cover Texts, and no Back-Cover Texts; the Creative Commons Attribution-Share Alike 3.0 Unported license; and the Creative Commons Attribution-Share Alike 2.5 Generic, 2.0 Generic and 1.0 Generic license. The image has been altered as allowed by these licenses, without the knowledge of the original author of the work. The transformed images appearing in this book fall under the same licenses.

Figure 1.9 The Tychonic system and (the identical) Figure 2.5 The Tychonic system: Fastfission, who made their work available on the Wikimedia commons and donated their work to the Public Domain.

Figure 1.11 Ellipse with the Area Rule: Stw, who made their work available on the Wikimedia commons under the following licenses: the GNU Free Documentation License, Version 1.2 or any later version published by the Free Software Foundation, with no Invariant Sections, no Front-Cover Texts, and no Back-Cover Texts; the Creative Commons Attribution-Share Alike 3.0 Unported license; and the Creative Commons Attribution-Share Alike 2.5 Generic, 2.0 Generic and 1.0 Generic license. The image has been altered as allowed by these licenses, without the knowledge of the original author of the work. The transformed images appearing in this book fall under the same licenses.

Figure 1.14 Cartesian coordinates: Soumi Ghosh and Alkarex. Both works are available under the GNU Free Documentation License, Version 1.2 or any later version published by the Free Software Foundation, with no Invariant Sections, no Front-Cover Texts, and no Back-Cover Texts; and the Creative Commons Attribution-Share Alike 3.0 Unported license. The original image (of the Leaning Tower of Pisa) by Alkarex was altered by Soumi Ghosh without Alkarex's knowledge as allowed by these licenses. The transformed images appearing in this book fall under the same licenses.

Figure 1.17 Precession of planetary orbit by p degrees per revolution, Figure 3.3 Precession per revolution: James Overton (altered from original).

Figure 1.18 (other than the legend) The distances explored by each planet in our solar system, Figure 3.4 (other than the legend) The distance 7Au is not explored by any planet, Figure 6.2 Empirical support added to all ten cited pendulum estimates from the above cited eleven syzygy corrected moon-test estimates of d, and the Figure Plotting Inverse 1.5 Law, Inverse square Law, and Inverse 2.5 Law attraction: Wayne Myrvold.

Figures in proof of 1.1 of appendix 1 of Chpt. 3: Gemma Murray (altered from original).

The diagram in the proof of proposition 3 of Appendix 2 of Chpt. 3, and the diagram for Chandrasekhar's integral in appendix 2. of Chapter 8: *Newton's Principia for the Common Reader* by Chandrasekhar S. (1995), pp. 71 and 287, by permission of Oxford University Press.

Figure 3.6 N = 9/4, Figure 3.7 N = 7/4: Wayne Lau, Michael Hoskin (editor of the *Journal for the History of Astronomy*), and the *Journal for the History of Astronomy*, originally published in Valluri, S.R., Wilson, C., and Harper, W.L. (1997). Newton's apsidal precession theorem and eccentric orbits, *Journal for the History of Astronomy*, 28: 13–27.

Figure 5.1 Newton's channels and extensive quotations of text from the *Principia* are from: Cohen I.B. and Whitman A. trans. Isaac Newton *The Principia, Mathematical Principles of Natural Philosophy: A New Translation*, Los Angeles: University of California Press, 1999, with permission.

Figure 8.1 and 9.1 A two-body system of the sun (Helios) and Jupiter: From Smith G. "How did Newton discover universal gravity?" *The St. John's Review*, vol. XLV, number two, 1999, pp. 46–9, by permission of *The St. John's Review*.

Figure 10.3 Radar Time-Delay: Irwin Shapiro and the American Physical Society for the use of Figure 1 (124.1) from Shapiro, I.I., Ash, M.E., Ingalls, R.P., Smith, W.B., Campbell, D.B., Dyce, R.B., Jurgens, R.F., Pettengill, G.H., "Fourth Test of General Relativity: New Radar Result" Physical Review Letters, 26, no. 18, 1132–5, 1971. Copyright (1971) by the American Physical Society.

Figure 1.10 Kepler's diagram depicting the motion of Mars on a geocentric conception of the solar system, Figure 1.13 The phases of Venus, Figure 1.15 Huygens's telescope, Figure 1.16

From Area Rule to centripetal force, Figure 1.18 The distances explored by each planet in our solar system (just the legend), Figure 2.1 Telescope observations of Jupiter's moons, Figure 2.4 Phases of Venus as evidence against Ptolemy, Figure 2.7 Kepler's pretzel diagram, Figure 2.8 Kepler's data for Mars at aphelion and at perihelion, figure 2.9 The Area rule for an eccentric circle, Figure 2.10 the Area Rule in Kepler's elliptical orbit, Figure 3.1 Newton's illustration of gravity, the diagram in Newton's proof of Cor. 1 of The Laws of Motion, all three figures in the proof of 1.2 of appendix 1 Chpt. 3, the figures in the proofs of propositions 1 and 2 in appendix 2 of Chpt. 3, the figures in the proofs of Newton's proposition 4 and its corollaries in appendix 2 of Chpt. 3, the illustration of inverse-square force and spring force orbits in appendix 3 of Chpt. 3, the image for Huygens's rotating cylindrical vessel experiment in Chpt. 5, sec I.4, Figure 8.2 Huygens's representation of the relative sizes of bodies in the solar system, the diagram illustrating Newton's proof of proposition 69 in appendix 1 of Chpt. 8, the figure from Newton's proof of proposition 70 book 1 in appendix 2 and in appendix 3 of Chpt. 8, the figure from Newton's proof of proposition 73 of book 1 in appendix 2 of Chpt. 8, the figure from Newton's proof of proposition 71 book 1 and its repetition in appendix 3 of Chpt. 8, the figure from Newton's argument to construe terrestrial gravity as attraction in Chpt 9 section II.2.iii, Figure 10.1 Hall's suggestion Figure 10.2 Brown's new result, Figure 10.4 Blake's depiction of Newton, Figure 10.5 Newton investigating light, and the symbols (e.g., astronomical, Greek) used in images created by Soumi Ghosh from the Wikimedia commons are all in the Public Domain either due to copyright expiry or donation by creators.

Index

Figures, tables etc are given in italics